77 Birthday
From Karyn

5-11-05

P9-APS-668

DICTIONARY OF TOOLS

used in the woodworking and allied trades, *c*. 1700–1970

DICTIONARY OF TOOLS

used in the woodworking and allied trades, *c.* 1700–1970

R. A. Salaman

Foreword by Joseph Needham
F.R.S., F.B.A.
Master of Gonville and Caius College, Cambridge

CHARLES SCRIBNER'S SONS · NEW YORK

© George Allen & Unwin Ltd 1975

Copyright under the Berne Convention

All rights reserved. No part of this book
may be reproduced in any form without the
permission of Charles Scribner's Sons.

1 3 5 7 9 11 13 15 17 19 I/C 20 18 16 14 12 10 8 6 4 2

Printed in Great Britian

Library of Congress Catalog Card Number 75–35059
ISBN 0–684–14535–9

TO MIRIAM SALAMAN

CONTRIBUTORS

The following authors have contributed entries about particular trades and tools:

W. L. GOODMAN *Plane Making; Sash Making; Historical Notes on basic tools; many Joiner's tools.*

F. A. DIXON *Mining Carpentry*

HERBERT L. EDLIN *Forester's Tools*

CHARLES HAYWARD *Coffin Making; Veneering*

L. JOHN MAYES *Chairmaking*

PHILIP WALKER *The Plane*

ILLUSTRATIONS

The drawings of tools and equipment are by

CHARLES HAYWARD *and* JACK LAIRD

The drawings of men at work in various trades are by

JAMES ARNOLD, CHERYL WELLS, *and* SUSAN LEWINGTON

Most of the engravings are taken from *The Sheffield Illustrated List* of 1888 (published by Pawson & Brailsford Ltd.) and from trade catalogues issued by firms who, at the dates shown below, were trading under the following names:

William Gilpin, Senr. & Co. (Cannock 1868)
William Hunt & Sons, The Brades, Ltd. (Birmingham 1905)
Richard Melhuish Ltd. (London 1912)
Stanley Rule & Level Co. (USA 1902)
Ward & Payne Ltd. (Sheffield 1911)

Also, a small number of engravings from Buck & Ryan Ltd. (London 1925); Goldenberg & Cie (France 1875); William Marples & Sons Ltd. (Sheffield 1938); Alexander Mathieson & Sons Ltd. (Glasgow 1900); Isaac Nash & Sons (Stourbridge 1899); Charles Nurse & Co. Ltd. (London 1934); Robert Sorby & Sons Ltd. (Sheffield c. 1900); Stahlschmidt Tool Co. Ltd. (Germany 1911)

FOREWORD

by Joseph Needham, FRS, FBA
Master of Gonville and Caius College, Cambridge

It is a great honour for me to have been asked to write a foreword for Raphael Salaman's *Dictionary of Tools*; doubly an honour because I can thus pay homage to a devoted scholar concerned with the tools of an age-old craft, and therefore also indirectly to the practitioners of that art itself, among whom was once a Jewish carpenter, the founder of the most characteristic religion of the Western world. Personal friendship also I may claim, since the Salaman family of Barley was well known to me when I was an undergraduate. Raphael's father, the geneticist, and what would now be called an ethno-botanist, inspired me by his book on the history of the potato with the conviction that there need be no line of separation between the sciences and the humanities. As a food-plant the potato had wide-ranging influences on the peoples of the Old World as well as the New. So also did wood-working tools.

Occasional references to China and other far-away cultures will be found in this book, reminding me that Lu Gwei-Djen and I collaborated with Raphael Salaman in two papers on the art of the wheelwright in China during the centuries surrounding the beginning of the Christian era. In one of these we presented a translation of the relevant passages in the *Chou Li* (Institutions of the Chou Dynasty); and in the other we examined a number of workshop scenes which have come down to us in stone-carved reliefs of the Han period. In this work we were able to take back the origin of 'dishing', long considered a medieval European invention, to the fourth century B.C. in China. It was a particular pleasure in these papers to combine the academic and the workaday in the ascription: 'From the Engineering Department of Messrs. Marks & Spencers Ltd., and Gonville and Caius College, Cambridge.'

Raphael Salaman belongs to the lineage of those scholars and educated men who did not despise the manual crafts practised in their various cultures. What they did despise was the conventional values of class-stratified society. Ignoring all such barriers they sought devotedly to describe the kinds and names of the tools and machines used daily in the technical operations of the tradesmen.

When I cast about in my mind for predecessors I thought, of course, of Peter the Great of Russia, who did not disdain to work with his own hands when learning the shipwrights' techniques in Holland, upon the basis of which, and many other technical arts that he learnt, he laid the foundation of the greatness of modern Russia. In the same century there was the Encyclopaedia movement in France, when Denis Diderot edited highly detailed accounts of all aspects of trade and husbandry. Here there was a good parallel with Chinese culture, for rather earlier, about 1635, Sung Ying-Hsing had taken his brush in hand to describe all the arts and techniques used in the Chinese Empire, which he did with the help of beautiful woodcut illustrations. This was the *T'ien Kung K'ai Wu* (Exploitation of the Works of Nature), a book still invaluable at the present day. Pressing further backward into the Renaissance, there were men like Benvenuto Cellini and Leonardo da Vinci, equally artists and technicians; and that raised the whole question of the 'higher artisanate', men such as Nicoló Tartaglia, Robert Norman and Humphrey Cole. There can be no doubt that men such as these had a very important part to play in the origins of the Scientific Revolution. The interesting thing is that it was not the lack of them which prevented the rise of modern science in Chinese culture. All through the Middle Ages there were great patrons of artisans. Take, for example, the statesman-scientist Shen Kua in the eleventh century, who was patron and protector of that remarkable man Pi Shêng, the inventor, about 1060, of movable-type printing. Pi Shêng was also a metallurgical artisan of note and in his earlier years had been the assistant of Wang Chieh, that remarkable adept who made quantities of artificial gold for his emperor, yet never deceived him. Moreover, the biographies of eminent craftsmen are almost a genre in Chinese literature. For example, there is the *Mo Shih* (History of Ink-Makers) written by Lu Yu of the Yuan dynasty; and then in our own time the important *Chê Chiang Lu* (Records of Philosophical Artisans) which we owe to Chu Ch'i-Ch'ien and his collaborators.

I cannot end without a reference to the exquisite drawings which are provided for so many entries in the present dictionary. In the East we have a proverb *wan tzu pu ju i chien*, which is, being interpreted, 'a thousand words are not as good as one look' at the thing, and this is certainly what the diagrams give us. I should think that there will also be many people using this book who will be impelled to try out the tools themselves in their own workshop; and that reminds me of another proverb that we have: 'I hear, and I imagine; I see, and I understand; I do, and I remember for ever.' Thus, every good wish to our lexicographer of wood-working tools; and in the old phrase: 'to all joiners and wheelwrights and master-coopers, good luck!'

March 1974

CONTENTS

FOREWORD   *page*   9

INTRODUCTION   13
   Period covered and historical notes
   Trades
   Nomenclature and etymology

LIST OF TRADES INCLUDED   16

NOTES ON USING THE DICTIONARY   18
   To find the name of a tool
   Alphabetical arrangement
   Naming the component parts of a tool
   Illustrations
   Technical and trade terms
   References

METRIC EQUIVALENTS   19

ACKNOWLEDGMENTS   21

TEXT   23

REFERENCES AND BIBLIOGRAPHY   537

INTRODUCTION

I have been interested in tools and trades ever since, as a boy, I used to watch tradesmen at work in my home village; and since then, during spare hours in the busy life of an engineer, I have sought out over 400 town and country workshops (most of which have now disappeared), and spoken with their owners. Most of what I know about tools and trades I have learnt from these men. In a later paragraph I have named a few of them: some of their tools, which I collected at the time, are exhibited in the St. Albans City Museum and in the London Science Museum.

Anyone whose boyhood began before the First World War will remember that at that time, whether one lived in town or country, one could watch men working at a dozen different trades within a mile of one's door-step. Their disappearance is one of the harsher changes of recent years: it has removed a source of social intercourse without which life, particularly for young people, is less interesting and certainly more lonely.

Perhaps it will interest the reader – especially if he has not lived during the times I speak of – if I give a list of the occupations of those I happen to remember in the village of Barley, Hertfordshire, a parish of 500 souls, where I was brought up. This remarkable range of occupations illustrates the fact that villages, as recently as 50–60 years ago, were almost self-sufficient in the skills needed in their daily life:

Blacksmith and farrier; builder (who was also the undertaker and well-sinker); cabinet maker; carpenter and joiner; hurdle maker (formerly a brick maker); plumber and glazier; shoemaker (who was also the village barber); wheelwright, cartwright, and ploughwright (he made his own iron fittings but employed travelling sawyers to work in the saw pit). I have since been told that a tailor and coffin maker were also working in the village. Just outside the boundary of the parish there were a miller and a coach-builder.

In addition to these tradesmen, I can remember a baker, butcher, carrier, and grocer; and among the farm and other workers with special skills I remember a groom, forester, fencer, horsekeeper, stockman, two road menders, and a traction engine driver.

My aim when compiling material for this Dictionary has been to describe every tool used in the woodworking trades from about 1700 to the present time, and to explain its purpose; to give the reader some idea of the graceful shapes imparted to the tools by the men who designed and made them; and to record some of the methods and sayings of the tradesmen[1] who used them. I have also included tools belonging to trades allied to woodworking (e.g. tools of the sailmaker and rigger, because they are often found in a shipwright's kit); metal working tools used in the making of associated metal parts; and tools imported from the U.S.A. and elsewhere[2] if commonly found in British woodworking shops.

Period covered and historical notes

I have chosen 1700 as a starting point mainly because it was at about that time that hand tools became increasingly differentiated to meet the demand from the growing multiplicity of separate trades – a tendency which culminated in the flood of specialised tools which fill the pages of the toolmaker's catalogues of the nineteenth century. In dating particular tools I have used trade catalogues with

---

1. I have used the term 'tradesman' for the men engaged in these trades because this is what they call themselves. The term 'craftsman', though often used by writers, is seldom heard in the workshops.

2. When describing tools imported from the Continent, the catalogue of Goldenberg & Co. of Alsace is often referred to because this firm's tools are typical of those made by European tool-makers whose products were exported to many countries during the nineteenth and early twentieth centuries.

caution: when the demand for a tool became limited, firms often continued to make them long after they were dropped from the catalogues.

When describing the primary tools (e.g. Adze, Axe, Hammer, Chisel, Saw) their previous history has been briefly indicated. If my account of their more recent development appears a little sparse, it should be remembered that after 50,000 years of trial and improvement during the stone ages, the primary tools reached an advanced, almost a final, stage of evolution in the classical Mediterranean civilisations of Egypt, Greece, and Rome. During the period covered by this book, modification and improvement affected details rather than basic principles.

Trades

Except for trades employing a very large number of tools of general type, such as those of the shipwright, cooper, and wheelwright, the tools of each specialised trade are grouped together, with cross-references. This is done because one cannot easily understand the significance of a specialised tool without some acquaintance with the special trade or process for which it is intended. There is also included, when judged necessary, a diagram of the final product, giving the names of each part.

The lists of special tools used in each particular trade are as complete as I have been able to make them. These lists are intended for museum curators and teachers who may need information on a group of tools, in the context of a trade which is of special interest to the locality concerned; and for curators and collectors when searching for tools belonging to a particular trade and arranging them for exhibition. I have included only the briefest note on trade methods and processes. For a more comprehensive description of these the reader may consult works such as those listed in the Bibliography under the following authors: J. Arnold; H. L. Edlin; G. Ewart Evans; J. Geraint Jenkins; K. S. Woods; N. Wymer.

Nomenclature and etymology

In naming a tool I have endeavoured to combine tool-makers' and workshop usage with the demands of alphabetical arrangement. In general I have followed the nomenclature adopted by the toolmakers and merchants as exemplified by the Sheffield Illustrated Lists of 1862–1910. During the nineteenth and early twentieth centuries, many tools were made in different patterns according to (a) the trade for which they were intended, and (b) the part of the country in which they were offered for sale. Regional varieties were given geographical prefixes such as 'London', 'Glasgow', or 'Liverpool'. (The 1905 catalogue of William Hunt & Sons, The Brades, Ltd., illustrates forty-two regionally named varieties of the bill-hook.)

I have endeavoured to include the more important variants. Though distribution and local naming of tools still conform surprisingly closely to geographical boundaries, one naturally finds exceptions: for instance, one may find a so-called 'Glasgow pattern' tool in a London workshop because the user travelled southwards to find work, or because a London tradesman bought the tool from the widow of a man who came from the North.

It may not be out of place here to consider the reasons for this proliferation of types. The regional differences were seldom intended to meet any special local conditions; I believe they were more often caused by local toolsmiths making the tools to their own individual design. Local tradesmen became accustomed to a local pattern, and when the emerging factories of the eighteenth century determined to capture the business they had to follow the same designs.

In general, the factories preserved the graceful traditional lines of the earlier hand-made tools, although the men who made them often had to live and work in the gloom and ugliness of the industrial towns.

In the case of the trade variants, there are many real differences of function: e.g. a cooper needs a small, short-handled Adze and could not use the type made for a shipwright. But there are many other variations which do not seem to serve a useful purpose. For instance, the tool catalogues of the nineteenth century list several varieties of Chisel including a Coachmaker's, a Millwright's, a Car-

14

penter's, and a Ship's Block Maker's. There were differences in the length, thickness, and range of widths, but the tradesmen concerned could (and probably did) exchange chisels without suffering inconvenience. As in the case of the regional variants, I think the factories had to copy the designs invented by former tool makers in order to attract their customers.

When I began work on the text of the Dictionary I added a note on the etymology of each tool's name, if known. I obtained this information mainly from the usual sources, such as the *Oxford English Dictionary* and Wright's *English Dialect Dictionary*. However, it soon became apparent to me that without special knowledge in this field one might fall into all kinds of traps which lie in wait for the amateur. I therefore decided, reluctantly, to leave these researches to those more competent to undertake them. Instead I have recorded all the alternative names known to me, without giving their sources except to draw attention to well-known names of Scottish or American origin.

R. A. Salaman
1972

TRADES INCLUDED

Basket Maker:
  Osier
  Spale
  Trug
Bell Hanger
Block Maker: *see* Print and Block Cutter; Hat-Block
  Maker; Ship's Block Maker
Boat Builder: *see* Shipwright
Bodger: *see* Chairmaker
Bowl Turner
Box and Case Maker
Broom Maker
Brush Maker

Cabinet Maker: *see* Carpenter, Joiner, and Cabinet
  Maker
Carpenter, Joiner, and Cabinet Maker
Carpet Layer: *see* Upholsterer
Carver: *see* Woodcarver
Cellarman
Chairmaker
Clog–Sole Maker
Coachbuilder
Coffin Maker
Comb Maker
Cooper
Coppice Trades: *see* Woodland Trades

Engraver: *see* Wood Engraver

Fishing-Rod Maker
Forester: *see* Tree Feller

Glazier
Gunstocker

Handle Maker
Hat-Block Maker
Hoop Maker
Hurdle Maker

Intarsia and Intaglio: *see* Veneering and Marquetry

Joiner: *see* Carpenter, Joiner, and Cabinet Maker

Ladder Maker
Lath Maker

Marquetry: *see* Veneering and Marquetry
Mast and Spar Maker
Millstone Dresser

16

Millwright
Mining Carpenter
Musical Instrument Maker: *see* Organ Builder;
   Pianoforte Maker and Tuner; Violin Maker;
   Woodwind Instrument Maker

Organ Builder

Painter
Pattern Maker
Pianoforte Maker and Tuner
Pipe and Pump Maker
Plane Maker
Print and Block Cutter

Rake Maker
Reed Maker: *see* Woodwind Instrument Maker

Sailmaker
Sash Maker: *see* Window Making
Saw Doctor and Sharpener
Sawyer
Shingle Maker
Ship's Block Maker
Shipwright
Spade and Shovel Maker
Spoon Maker
Stamp Cutter: *see* Print and Block Cutter

Thatch-Spar Maker
Tree Feller
Trugger: *see* Basket Maker
Turner

Undertaker: *see* Coffin Maker
Upholsterer

Veneering and Marquetry
Violin Maker

Wheelwright and Wainwright
Window Making
Wood Carver
Wood Engraver
Woodland Trades
Woodwind Instrument Maker

NOTES ON USING THE DICTIONARY

To find the name of a tool

Tools and trades are arranged alphabetically according to their usual or family names, and after each entry any known alternative names have been added and cross-referenced. By 'family name' is meant the name of the type-group, e.g. Axe, Plane, Saw, etc. Thus the Chisel known as a *Ship's Slice* will be found under *Chisel, Shipwright's*. If the reader is not aware that the Slice is a Chisel, he will find it entered also under *Slice*, and also in the list of tools under *Shipwright*.

If the reader wishes to search for the name and purpose of some unknown tool he should begin by looking under one of the family groups and compare the unknown specimen with the illustrations in the text. If he is unsuccessful he should make a guess at the trade to which the tool belongs, examine the list of tools used in that trade, and then refer to the appropriate entry.

Alphabetical arrangement

The alphabetical sequence of entries may cause difficulty in finding the entry for a tool if part of its name is the same as the 'family' name of a series of tools. Thus the entry *Plane Maker* will not be found among the family of Planes which occupy several pages from *Plane, Astragal* to *Plane, Whip*. Instead, it will be found after *Plane, Whip* at the end of the series. Similarly, entries such as *Saw Bench* and *Sawyer* will not be found in the long series *Saw, Back* to *Saw, Woodcutter's*; instead, they will be found after the last of the Saw family, i.e. after *Saw, Woodcutter's*.

Naming the component parts of a tool

The names given to the parts of a tool (e.g. face, sole, tooth, cutting iron, etc.) will be found on the 'family portrait' of the tool under its appropriate entry. This applies particularly to primary tools such as Adze, Axe, Chisel, Hammer, Plane, Saw, etc.

Illustrations

The illustrations are not to a uniform scale, but approximate dimensions are usually included in the text. Metric equivalents will be found on a later page. (The specimens illustrated by line drawings are mostly from the author's collection which is now in the St. Albans City Museum.)

The following Figure Numbers have not been included: 52, 137, 138, 388, 414, 480, 646, 729.

Technical and trade terms

Those commonly used in the workshops are recorded and explained. Terms relating to metals will be found under *Metals used in toolmaking*.

References

References given in brackets refer to publications, trade catalogues, or personal communications, of which particulars will be found in the Bibliography and Reference pages.

METRIC EQUIVALENTS

Length			Weight	
Inches	*Approx.* *Equivalent*		*oz*	*gram*
$\frac{1}{16}$	1·6 mm		1	28·35
$\frac{1}{8}$	3·2		2	56·7
$\frac{3}{16}$	4·8		3	85·05
$\frac{1}{4}$	6·35		4	113·4
$\frac{5}{16}$	7·9		5	141·7
$\frac{3}{8}$	9·5		6	170·1
$\frac{7}{16}$	11·1		7	198·4
$\frac{1}{2}$	12·7		8	226·8
$\frac{9}{16}$	14·3		9	255·1
$\frac{5}{8}$	15·9		10	283·5
$\frac{11}{16}$	17·5		11	311·8
$\frac{3}{4}$	19·05		12	340·2
$\frac{13}{16}$	20·6		13	368·5
$\frac{7}{8}$	22·2		14	396·9
$\frac{15}{16}$	23·8		15	425·2
1	2·54 cm		16	453·6
2	5·08		*lb*	*kilogram*
3	7·62		1	0·45
4	10·2		2	0·91
5	12·7		3	1·36
6	15·2		4	1·81
7	17·8		5	2·27
8	20·3		6	2·72
9	22·9		7	3·17
10	25·4		8	3·63
11	27·9		9	4·1
12	30·5		10	4·5
			11	5·0
			12	5·4
			13	5·9
			14	6·35

ACKNOWLEDGMENTS

Most of what I have learnt about tools and trades comes from my visits to tradesmen in their workshops. I am very grateful to all of them. Some, from whom I have obtained specific information, are named in the Bibliography and Reference pages.

I wish to offer my warmest thanks to those who have given me written material and valuable advice. The most extensive contributor, especially on the subject of the joiner's work, is Mr. W. L. Goodman to whom I am greatly indebted. The historical notes on tools are largely based on material which he kindly put at my disposal. The section on the plane maker is taken directly from his original account of this trade.

Mr. Philip Walker contributed material on the Plane and other tools, and in addition made a critical inspection of almost every entry in the Dictionary – a formidable task for a busy professional man. For me this was an invaluable service.

Mr. Charles Hayward made many of the line drawings, and arranged over a thousand other drawings and engravings with all the expertise such work entails. He also contributed material for several entries, including veneering, carving, and coffin making.

I also want to thank the following for their help: the late Mr. Philip Clarke, of Manchester, for much useful material on coachbuilding; Mr. F. A. Dixon, of the National Coal Board, Gateshead, for a contribution on the mining carpenter; Mr. Herbert L. Edlin, of the Forestry Commission, for material on the tools used by the foresters, and I am also much indebted to his published works on the woodland trades; Mr. Kenneth Kilby of Luton, for extensive information on the cooper's trade, and for commenting on my entries relating to this subject; Mr. L. John Mayes, Director of High Wycombe Museum, for much original material on chairmaking, and for enlightening me on many aspects of this trade; Mr. Alan O'Connor-Fenton, Director of Wm. Marples & Sons Ltd., the Sheffield tool makers, for answering my questions over a period of many years, and for an account of steels used in toolmaking; Mr. F. Seward, for much valuable material on hand tools made in the U.S.A. during the last 100 years; the late Mr. George Weller, of Sompting in Sussex, for material on the work of the wheelwright, and for introducing me to tradesmen in other villages in the period 1947 to 1950 – just before their workshops disappeared for ever.

During the many years in which I have gathered material for this work I have spoken and corresponded with many other firms and experts some of whom I have named in the Bibliography and Reference pages in token of my gratitude.

In expressing my gratitude to these helpers I want to make it clear that the final presentation of both fact and opinion, including any possible mistakes, is my responsibility alone.

For the illustrations I am indebted not only to Mr. Charles Hayward, whom I have already mentioned, but also to Mr. Jack Laird for many of the drawings of tools from my collection, to Mr. James Arnold for the drawings of sawyer's, wheelwright's, and cooper's work, and to Miss Cheryl Wells and Mrs. Susan Lewington for sketches of tools used in other trades.

I am grateful to Messrs. Evans Brothers for permission to reproduce drawings by Mr. Charles Hayward which appeared originally in *The Woodworker* and in his *Tools for Woodwork*.

The engravings reproduced are mostly taken from nineteenth-century trade catalogues and from the Sheffield Illustrated List of 1888. For this generous help I wish to thank the following firms:

Brades Skelton & Tyzack Ltd., Oldbury
Buck & Ryan Ltd., London
Robert Dyas Ltd., Morden. (For illustrations from the catalogues of Richard Melhuish)

Wm. Gilpin, Senr. & Co. (Tools) Ltd., Cannock
Goldenberg & Cie., Zornhoff, Bas Rhin, France
William Marples & Sons Ltd., Sheffield
Alexander Mathieson & Sons Ltd., Glasgow
Pawson & Brailsford Ltd., Sheffield, Publishers of the Sheffield Illustrated List
Robert Sorby & Sons Ltd., Sheffield
Stanley Works (Great Britain) Ltd., Sheffield
Ward & Payne Ltd., Sheffield

I have quoted frequently from the works of those remarkable recorders of workshop practice –
Joseph Moxon (1677), Diderot and D'Alembert (1751) and Charles Holtzapffel (1847). I want also to
acknowledge my debt for extracts quoted from a number of recent works, including the following:

Ronald Blythe, *Akenfield* (Allen Lane, The Penguin Press)
H. L. Edlin, *Woodland crafts in Britain* (David & Charles Ltd.)
George Ewart Evans, *The Farm and the Village* (Faber & Faber Ltd.)
W. L. Goodman, *The History of Woodworking Tools* (G. Bell & Sons Ltd.)
Charles Hayward, *Tools for Woodwork*, and other works (Evans Brothers Ltd.)
Henry C. Mercer, *Ancient Carpenter's Tools* (The Bucks County Historical Society, U.S.A.)
G. P. B. Naish, 'Ships and Shipbuilding', *A History of Technology* © 1957 Oxford University Press;
    by permission of the Clarendon Press, Oxford.
Edward H. Pinto, *Treen* (G. Bell & Sons Ltd.)
Walter Rose, *The Village Carpenter* (Cambridge University Press)
Bernard Shaw, *The Doctor's Dilemma* (Society of Authors on behalf of the Shaw Estate)
George Sturt, *The Wheelwright's Shop* (Cambridge University Press)
Rex Wailes and John Russell, 'Windmills in Kent', *Transactions of the Newcomen Society, 1955*

Among the many museums and their staff from whom I have obtained help, I should like to mention:

Brixham Museum (J. E. Horsley)
Curtis Museum, Alton (the late W. H. Curtis)
English Rural Life Museum, Reading (C. A. Jewell; J. Anstee)
High Wycombe Art Gallery and Museum (L. John Mayes)
National Maritime Museum (Basil Greenhill; G. P. B. Naish)
Science Museum, London (K. R. Gilbert)
St. Albans City Museum (the late A. H. V. Poulton)
Shelburne Museum, Vermont, U.S.A. (Frank H. Wildung; H. R. Bradley Smith)
Welsh Folk Museum, St. Fagans Castle (J. Geraint Jenkins)

In preparing the text I am indebted for help given me by Mr. Noel Carrington, the late Mr. Cyril
King, Sir Bernard Miles, Professor Roy Pascal, Mrs. R. Phillips, and my brother Dr. M. H. Salaman.
I want also to acknowledge my debt to Dr. Gunther Nagelschmidt for undertaking the final critical
reading of the text; and to thank Mrs. P. I. Painter for the arduous labour of typing and retyping
which she has performed with such extraordinary patience over many years.

Finally, it is a pleasure to offer my gratitude to those who have contributed towards the cost of the
work: the Nuffield Foundation; the Trustees of the Leverhulme Research Fellowships; the Early
American Industries Association; Messrs. Marks and Spencer Ltd.; and my publishers, Messrs.
George Allen & Unwin Ltd. My thanks are also due to the staff of these organisations for their helpful
interest and encouragement.

R.A.S.
1972

Accessories and Appliances: see list under *Workshop Equipment.*

Action Regulator: see *Pianofòrte Maker.*

Addes: see *Adze.*

Addis Tool

A name sometimes given to Carving Chisels after the well-known maker Addis & Sons. See *Chisel, Carving.*

Adze (Earlier names include Addes, Jennet, and Thixel. A Scots form is Eatche. 'Let me hae a whample at him wi' mine eatche – that's a' – Sir Walter Scott, *The Bride of Lammermoor*, XXV Fig. 1.) The Adze differs from the Axe in having its blade set at right angles to the handle. Like the Axe, it began its long history in stone and reached a high degree of development when made in bronze by the Egyptians, and later in iron by the Romans.

The modern Adze is a steel forging with a tapered socket or eye, usually rectangular but occasionally round. The tapered eye allows the handle to be re-leased upwards when it is necessary to sharpen the blade. Early handles were straight, but in recent times they have been given a graceful double curve which makes the tool easier to control. According to its purpose, the blade may have a straight or hollow cutting edge.

Weights and widths of blade:

No.	Weight (*lb*)	Cutting Edge (*in*)
00	$3\frac{1}{4}$	3
0	$3\frac{1}{2}$	$3\frac{1}{4}$
1	$3\frac{3}{4}$	$3\frac{1}{2}$
2	4	$3\frac{3}{4}$
3	$4\frac{1}{4}$	4
4	$4\frac{1}{2}$	$4\frac{1}{4}$
5	$4\frac{3}{4}$	$4\frac{1}{2}$

Adzes are used for removing heavy waste, levelling, shaping, or trimming the surfaces of timber. Moxon (London, 1677) writes: 'Its general Use is to take thin Chips off Timber or Boards, and to take off those Irregularities that the Ax by reason of its form cannot well come at; and that a Plane (though rank set) will not make riddance enough with.' Judging by medieval and later pictures, the Axe was used more often than the Adze for hewing flat timber used in building. Machine planing has now restricted the use of the Adze in this country mainly to coopering, wheel-making, waterfront carpentry, and shipbuilding.

In operation, the Adze is swung in a circular path in a manner and direction which varies with the trade and work. The workman uses his chest or thigh as

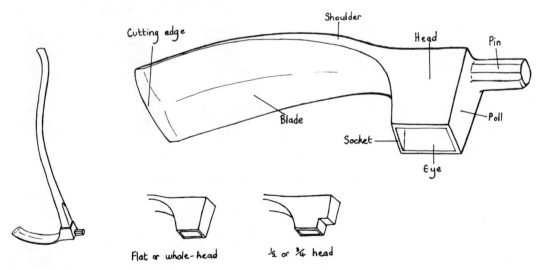

Fig. 1 Diagram of an Adze blade

a 'stop' to prevent the arm from swinging too far and causing an accident. Mr. Philip Clarke relates (Manchester, 1966) that his fellow coachbuilders used to say, 'The Adze is the only tool that the Devil is afraid to use'.

Adze, American *Fig. 2*
Variants include:

(*a*) '*Yankee' Ship Adze*
With a tapering spike instead of the square pin poll of the English Adze.

Fig. 2a

(*b*) *Lipped or Dubbing Adze*
The edges of the blade are turned upwards on each side to give a trenching cut, and the pin poll is spiked. It cuts a square rather than a rounded channel; this is said to prevent side-slipping when trimming. It was also used for the process known as 'dubbing' – the cutting of scores or notches across the grain at intervals along the beam down to a chalk line, after which the wood between is quickly hewn off with an Axe.

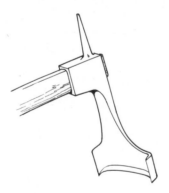

Fig. 2b

Adze, Blunt: see *Adze, Cooper's Nailing.*

Adze, Bowl: see *Adze, Hollowing.*

Adze, Brazil (Spanish Adze) *Fig. 3*
Round eye, square pin. A nail-pull slot is sometimes provided on the shoulder. Made for the South American market.

Fig. 3

Adze, Brazil, Hollowing (Brazil Gouge)
A name given to hollowing Adzes made for the South American market. They are like the Cooper's Howel or Chairmaker's Adze except for the round, shallow eye apparently favoured in this market.

Adze, Brazilian Slot: see *Adze, Slot.*

Adze, Canoe Howel *Fig. 4*
A gouge-shaped blade, square eye, and flat poll. Used for general hollowing work.

Fig. 4

Adze, Carpenter's (House Carpenter's Adze) *Fig. 5*
(*a*) Flat poll; (*b*) Half poll; (*c*) Pin poll.
Still used by carpenters on the waterfront, railways, and elsewhere for shaping and trimming heavy timbers.

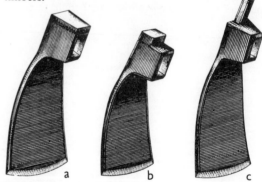

Fig. 5

Adze, Chairmaker's (Seat Adze)
A slightly gouge-shaped blade designed for hollowing chair seats. Often re-shaped by the blacksmith from an ordinary Carpenter's Adze.
L. J. Mayes (High Wycombe, 1965), an authority on chairmaking, writes: 'The operator straddled the bottom (i.e. seat) and gripped it with his feet. The adze was used across the grain with a quick chopping

stroke. In spite of the great skill of the "bottomers" as the adze men were called, accidents did happen, and one veteran bottomer was known as Billy Notoes. This same name was applied to the first successful adzing machine.'

Adze, Chairmaker's Howel *Fig 6*
A saucer-shaped blade, rather wider than the Cooper's Howel. It may have been used for hollowing chair seats.

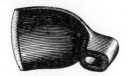

Fig. 6

Adze, Chequered: see *Adze, Cooper's Nailing.*

Adze, Claw: see *Adze, Cooper's Nailing.*

Adze, Cleaving
A name given to an adze-like tool with a short blade on an 18-in handle. The sides of the blade are sharpened as well as the front.

Used as an alternative to the Billhook or Froe for cleaving thin rods, for example, hazel rods for wattle hurdles and for hoop making. For illustration see under *Woodland Trades.*

Adze, Coachbuilder's: see *Adze, Wheelwright's.*

Adze, Cooper's *Fig. 7*
These tools are fitted with very short handles adapted for swinging with one hand within the radius of a

cask in order to cut the chime bevel or howel surface. (See diagram under *Cooper.*) This handle is often wedged into the head to lean sharply towards the blade so that the cooper's hand is almost imprisoned between the handle and the blade.

The Cooper's Adze has become quite distinct from the Carpenter's Adze: the blade is narrow, and the square poll, used for hammering, is hollowed underneath to reduce weight. The eye is usually rectangular and tapered.

Adze, Cooper's Bunging
Similar to the Cooper's Rounding Adze, but kept for the roughest work, including driving and cutting off bungs.

Adze, Cooper's Chimney-Pipe
Appears in a *List of Prices for Forging Edge Tools* (Sheffield, 1853) with 'round or square eye'. Appearance and purpose unknown.

Adze, Cooper's Howel (Butt Howel; Gouge Adze; Howle) *Fig. 8*
A short, slightly hollowed blade with a flat poll. Used at one time to pare down and smooth the inside surface of the cask at the stave ends – the surface known as the 'chiv' or 'howel' in which the croze groove is subsequently cut to take the head of the cask.

The use of this tool declined after the adoption of the Chiv Plane, except when the correct size of Chiv was not available. Today, in the absence of a Chiv or in repair work, the job would be done by a Jigger. (See Drawing Knife, *Cooper's Jigger.*) For a historical note and other information see *Cooper* and *Plane, Cooper's Chiv.*

Fig. 7

Fig. 8

Adze, Cooper's Nailing (Claw Adze; Scots – Blunt Adze or Chequered Adze) *Fig. 9*
Similar in appearance to the other Cooper's Adzes, but with the blade cloven like a Claw-Hammer. The poll has a chequered face for nail driving. In the Liverpool pattern (*a*), the socket forms a hexagon and the poll is strengthened by a central rib. The Scottish pattern (*b*) has a squared socket and the poll is deeply hollowed underneath for lightness. It is used for cutting and nailing wooden hoops.

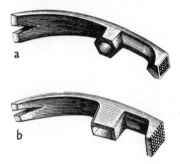

Fig. 9

Adze, Cooper's Notching *Fig. 10*
Similar to a Sharp Adze, but occasionally fitted with a claw on the end or side of the blade. It is used to cut V-shaped notches in both ends of a split hazel rod so that the notches interlock with one another to form a complete wooden hoop. A few nails or sprigs are then driven in to hold the wooden hoops in position. (See also *Hoop Notching Knife* under *Cooper (3) Hooping Tools*.)

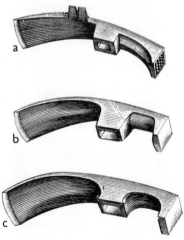

Fig. 10　(*a*) London Pattern with claw
　　　　(*b*) Scotch Pattern
　　　　(*c*) Irish Pattern

Adze, Cooper's Rounding *Fig. 11*
Heavier than most other Cooper's Adzes and often more sharply curved. Used for rough cutting the chime (the bevel at the stave ends) to 'knock the lumps off' before finishing with the Sharp Adze. The cutting is performed across the grain, after the staves are drawn up. Coopers often grind the blade at a 'skew' angle. In this way an inward and slicing direction is given to the blade which prevents the grain from breaking away on the inside edge of the staves.

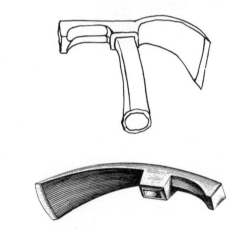

Fig. 11

Adze, Cooper's Sharp (Trimming Adze) *Fig. 12*
Like the Cooper's Rounding Adze in appearance, but smaller and lighter. After the chime has been roughed out with the Rounding Adze, this tool, which is kept very sharp, is employed for finishing.

Fig. 12

Adze, Cooper's Trussing (Scots – Stowing Adze) *Fig. 13*
Though appearing to be a heavier version of the Cooper's Adze, this tool is not an Adze at all and has a blunt thick pane instead of a cutting edge. Used for beating on or knocking off Truss Hoops (which are used for drawing the staves together) as an alternative to using the Cooper's Hammer. (See also *Cooper (3) Hooping Tools*.)

26

Fig. 13

Adze, Dished: see *Adze, Hollowing.*

Adze, Dubbing: see Lipped or Dubbing Adze under *Adze, American.*

Adze, Fencer's *Fig. 14*
Similar to the Wheelwright's Adze but with an axe-type eye.

Fig. 14

Adze, Fruit
Like an Adze in appearance but not used as one. See *Hammer, Fruiterer's.*

Adze, Gouge
A name sometimes given to Adzes designed for hollowing. See *Adze, Hollowing.*

Adze, Gum *Fig. 15*
With a flat poll and probably made for tapping gum-producing trees.

Fig. 15

Adze, Guttering: see *Adze, Spout.*

Adze, Hammer: see *Adze, Platelayer's.*

Adze, Hammer Head: see *Adze, Scotch.*

Adze, Herring Barrel
A term sometimes used for various Adzes including the *Adze, Cooper's Rounding.*

Adze, Hollowing (Dished, Gouge, or Saucer Adze)
A number of Adzes are made with a blade gouge-shaped in cross section and designed for hollowing. They include: *Adze, Brazil, Hollowing; Adze, Canoe Howel; Adze, Chairmaker's; Adze, Cooper's Howel; Adze, Spoon-Maker's* (Ladle Adze); *Adze, Spout* (Guttering Adze); *Adze, Stripping.*

Adze, House Carpenter's: see *Adze, Carpenter's.*

Adze, Howel: see *Adze, Cooper's Howel; Adze, Chairmaker's Howel.*

Adze, Joiner's: see *Adze, Carpenter's.*

Adze, Joiner's Norfolk *Fig. 16*
With a half poll.

Fig. 16

Adze, Ladle: see *Adze, Spoon-Maker's.*

Adze, Lipped Ladle: see *Adze, American.*

Adze, Loop or Stirrup: see *Adze, Slot.*

Adze, Marking: see *Hammer, Marking.*

Adze, Nail: see *Adze, Cooper's Nailing.*

Adze, Orange Chest *Fig. 17*

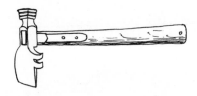

Fig. 17

A strapped Hammer with an adze-like pane provided with a side claw. See *Hammer, Grocer's and Warehouse.*

27

Adze, Oriental: see *Adze, Turkish.*

Adze, Platelayer's (Hammer Adze) *Fig. 18*
A heavier version of the Carpenter's Adze with whole, half, or hammer-type poll. Used by plate-layers for trimming sleepers etc. and for driving wedges. See *Adze, Trackmen's Spiking.*

Fig. 18

Adze, Saddle-Tree Maker's *Fig. 19*
A short head without a poll. The inner edges of the hollow blade are chamfered.

Fig. 19

Adze, St. Michael's (S. Miguel)
A Slot Adze made for the South American market with a thin, flat, round-shouldered blade from 4–5 in wide and 6 to 8 in long. It is not known for what work it was used, nor how it acquired its unusual name.

Adze, Saucer: see *Adze, Hollowing.*

Adze, Saw Mill (Timber Rafting Adze)
A name given to a heavy Adze with a deep, square socket and a stout spike on the poll. Used presumably for rafting and handling timber at saw mills.

Adze, Scotch (Hammer Head Scotch Adze) *Fig. 20*
With a round-faced hammer-head poll, instead of the usual pin.

Fig. 20

Adze, Seat: see *Adze, Chairmaker's.*

Adze, Sharp: see *Adze, Cooper's Sharp.*

Adze, Shipwright's (Ship Adze; Ship Carpenter's Adze) *Fig. 21*
The 9 in blade of the Shipwright's Adze is longer by an inch or more, and is rather flatter, than those of the Adzes made for other trades. It is usually provided with a peg poll. The handle is often given a double curve so that its lower end is brought forward to a point almost in line with the cutting edge of the blade. The purpose of this curve is not clear though shipwrights assert that it would be impossible to control an Adze with a straight handle. Nevertheless, Rålamb (Stockholm, 1691) illustrates an Adze described as 'English' with a blade almost identical with the modern pattern with a perfectly straight handle. The same straight handles appear in other early pictures of shipbuilding. It may be noted that similarly Axes did not acquire the curved handle until the development of the Wedge Axes in the early nineteenth century.

The normal length of handle is 2 ft 7 in, but in many early pictures, such as *The Building of the Ark* by M. de Vos *c.* 1580, Adzes with only very short handles are shown; indeed, they resemble more nearly the Cooper's Adzes. Shorter handles are also used today where a sharp inward curve in the timber makes this design necessary.

Shipwrights use the Adze for all kinds of shaping and finishing, including the trimming of both flat and curved framing and planking, and for the rough

Fig. 21

shaping of masts and spars. In one method of working they start a cut in one direction, and then begin another cut from the opposite direction so that the cuts meet. This method avoids too deep a cut if the grain runs inward. For general trimming work on upright surfaces the end of the handle is commonly held with one hand on the knee, while the Adze is hinged back and forth from that point; the curved handle is suitable for this way of holding. Holtzapffel (London, 1847, vol. 1, p. 473) writes of work on horizontal timbers as follows:

'In coarse preparatory works, the workman directs the adze through the space between his two feet, he thus surprises us by the quantity of wood removed; in fine works, he frequently places his toes over the spot to be wrought, and the adze penetrates two or three inches beneath the sole of the shoe, and he thus surprises us by the apparent danger yet perfect working of the instrument, which in the hands of the shipwright in particular, almost rivals the joiner's plane; it is with him the nearly universal paring instrument, and is used upon works in all positions.'

The pin poll is used as a punch to 'set' or drive spikes and nails below the surface. When using the Adze, the shipwright often encounters nails, broken off perhaps when temporary scaffolding was removed after framing. It is necessary to drive these metal obstructions below the surface to avoid striking them and so spoiling the edge of the Adze. An elderly shipwright, Mr. G. Worfolk (Kings Lynn, 1966), when describing this operation, declared: 'If you come across a spike, turn your adze and hit it; the old shipwrights used to say "Our only enemies are deal knots and rusty nails"'. (See also *Adze, American.*)

Adze, Slot (Spanish Slot Adze; Brazilian Slot Adze; Loop or Stirrup Adze) *Fig. 22*
A round-shouldered blade, without eye or poll, but instead a narrow wedge-shaped tongue for attaching to the handle. This is shown in (*a*) and (*c*).

There are several types of handle: one shown in (*b*) has a knee-shaped block and short haft; another is made like a saw handle with a shoulder for the strap. In both cases the blade and handle are held together by forcing the metal ring or strap over the wedge-shaped tongue.

These unusual Adzes were listed until recently by English as well as Continental tool makers. An ancestor from Graeco-Roman times (300 B.C.–A.D. 400) is shown in (*d*).

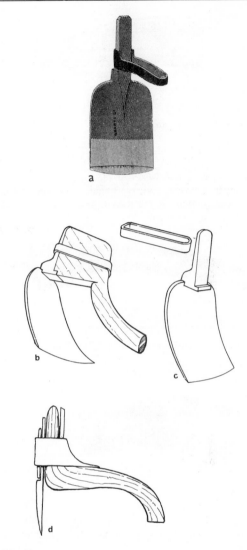

Fig. 22

Adze, Spanish: see *Adze, Brazil.*

Adze, Spoon-Maker's (Ladle Adze)
Small and short-handled, with a gouge-shaped blade. Used for rough hollowing the bowls of wooden spoons and ladles and for other small hollowing work. See *Spoon Maker.*

Adze, Spout (Guttering Adze) *Fig. 23 overleaf*
A deep gouge-shaped blade, with a square pin. For hollowing out the trough of wooden spouting or guttering.

29

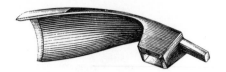

Fig. 23

Adze, Stirrup: see *Adze, Slot.*

Adze, Stowing: see *Adze, Cooper's Trussing.*

Adze, Stripping *Fig. 24*
An unusually long, concave blade, narrower at the cutting edge than at the throat, and with a square pin. Purpose unknown.

Fig. 24

Adze, Timber-Rafting: see *Adze, Saw Mill.*

Adze, Trackmen's Spiking *Fig. 25*

Fig. 25

A heavy Adze with a long pin. Used, presumably, for track laying in railway work. cf. *Adze, Plate-layer's.*

Adze, Trimming: see *Adze, Cooper's Sharp.*

Adze, Trussing: see *Adze, Cooper's Trussing.*

Adze, Turkish (Oriental Adze) *Fig. 26*
A light, straight blade with a shallow, round eye and a flat poll with chequered face. Usually provided with a key-hole shaped slot in the middle of the blade for extracting nails.

This tool is still made and used in Greece and the Eastern Mediterranean countries for general purposes for which an English carpenter or other woodworker would use a light hatchet.

Fig. 26

Adze, Wheelwright's (Wheeler's Adze) *Fig. 27*
A strong blade with a flat poll either 'whole-head' or 'half-head'. A variant of the whole-head pattern has an axe-eye, with an oval socket and pointed lugs.

This tool differs from the Carpenter's Adze in that the blade is usually more curved, and therefore suitable for hollowing the inside of felloes or the curved parts of the coach frame; and the head normally has a plain poll and so can be used for hammering. (For precautions against accidents in use, see *Adze.*)

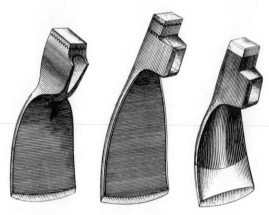

Fig. 27

Adze, Yankee: see *Adze, American.*

Allongee: see *Chisel, Carving.*

Alsene
Obsolete name for *Awl.*

American Tools
Tools made in the U.S.A. are referred to under many entries, including *Hammer, Adze Eye; Plane, Stanley-Bailey; Saw, Disston; 'Yankee' Tools.*

Angular Boring Machine: see *Wood Boring Machine.*

Angular Brace or **Bit Stock:** see *Brace, Corner.*

Annealing: see *Iron and Steel.*

Anvil: see Anvil (Bick Iron) under *Cooper: Hooping Tools;* Saw Maker's Anvil under *Saw Doctor and Sharpener.*

Appliances: A list is given under *Workshop Equipment.*

Apron Hook (Apron Pin) *Fig. 28*
A metal hook about 1½ in long, with a pointed hook at one end and a brass button in the form of a trefoil, heart, shield or flower at the other. Used in pairs for fastening an apron across the back of the wearer.

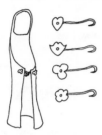

Fig. 28

Archimedean Tools: see *Drill, Archimedean; Screwdriver, Spiral Ratchet.*

Astragal: For Astragal Punch see *Chisel, Astragal.* For Astragal Moulding see *Plane, Moulding.*

Attice
A word quoted by Halliwell (London, 1847) meaning 'a carpenter's tool; an adze'.

Auger *Fig. 29 overleaf*

An iron shank 12–25 in (or more) long, with a T-shaped handle at one end and a boring device at the other end. Until recent years Augers were the only tools for boring large, deep holes in wood. In his *Shipbuilder's Assistant* of 1711 William Sutherland wrote: '. . . If ever any Person was Deify'd for Inventing, I should highly recommend the Author of an Augre deserving of that Glory.'

The Egyptians and Greeks do not appear to have used this tool: the earliest known Augers are of Roman date and made of iron. The body was usually shell or spoon shaped, with a square or flat tang to take the handle. The spoon-shaped Auger was also common in the Middle Ages; and it was about this time that the Nose Auger was introduced (*Fig. 29, Nos. 11–15*) which has a body shaped like a half cylinder with a small, flat, in-bent horizontal blade at the tip to clean out the hole.

Twist Augers (*Fig. 29, Nos. 3–8*) became common in the late eighteenth century and soon largely replaced the Shell patterns, though these were often preferred by tradesmen such as wheelwrights and shipwrights owing to their capacity for boring long holes without 'wandering'.

The chief types of Auger are described under the following entries: *Auger, Shell* (or *Nose); Auger, Taper; Auger, Twist.*

Auger handles Fig. 30 overleaf
There are two methods of fitting the handle: either with an eye, or with a flat tang which goes right through the handle and is clenched over. Typical examples illustrated include:

(*a*) and (*b*) *Winged handles* (drawn from eighteenth and nineteenth century examples)
This shape looks well and is easy to grasp.

(*c*) *Handle with centre boss*
Probably the commonest of the home-made handles.

(*d*) *Handle with centre swell*
Usually factory-made and commonly fitted to Taper Augers.

(*e*) *Straight, turned handle*

(*f*) *Eyed Handle*
This enabled one handle to serve any number of Augers. The eye may be either barrel-shaped ('barrel-eye') or with the top of the eye wider than the bottom ('bonnet-eye').

(*g*) *Interchangeable handle* (illustrated in *Fig. 29, No. 6*)
A square hole in the centre of the handle is fitted over the shank of the Auger and secured with a thumb screw.

(*h*) Some Augers are made with a plain shank left for the user to weld an extension. See *Fig. 29, No. 14.*

Auger, American: see *Auger, Cooper's Bung Borer.*

Auger, American Ship: see *Auger, Twist* (Single Twist).

Auger, Annular: see *Bit, Annular.*

Auger, Bell: see *Auger, Brick; Bellhanger's Jumper Drill;* Bell-Hanger's Gimlet under *Gimlet.*

Auger, Bit: see *Auger, Twist.*

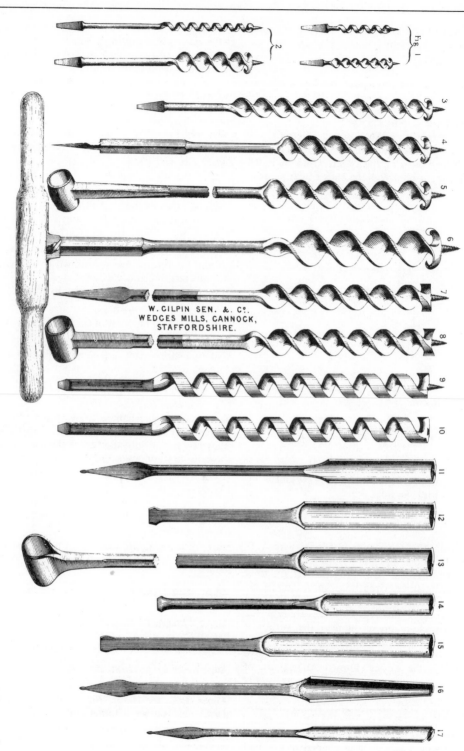

W. GILPIN SEN. &. C°.
WEDGES MILLS, CANNOCK,
STAFFORDSHIRE.

Fig. 29

Fig. 29   Some Auger blades etc. illustrated by William Gilpin (Cannock, 1868.)
1. Dowelling Bits of Gedge pattern
2. Brace Bits (Auger Bits) of Gedge pattern
3. 'Railway Carriage' Bit of Gedge pattern
4. Auger of Gedge pattern (tanged)
5. Auger of Gedge pattern (with barrel eye)
6. Auger of Gedge pattern (with interchangeable handle)
7. Auger of Scotch pattern (tanged)
8. Auger of Scotch pattern (with barrel eye)
9. Single Twist 'American Ship Auger' (L'Hommedieu's pattern)
10. Single Twist 'American Ship Auger' (L'Hommedieu's pattern) without screw lead known as 'Barefoot'
11. 'Carpenter's or House Auger', Shell pattern (tanged)
12. 'Ship Auger', Shell pattern*
13. 'Ship Auger' (with bonnet eye)
14. 'Dodds or Scotch Ship Auger' of Shell pattern*
15. 'Long-pod or Trenail Auger' of Shell pattern*
16. 'Wheeler's Taper or Dowelling Auger' (tanged)
17. Skewnose Auger 'for plate layers' (tanged)
   *With shank left for welding on an extension.

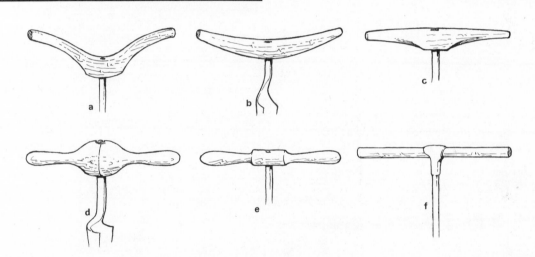

Fig. 30   Typical Auger handles

Auger, Blockmaker's (Blockmaker's Scillop) *Fig. 31*
Slightly tapered, half-funnel-shaped body with a flat circular disc at the nose end. It is listed by Mathieson (Glasgow, 1900) in sizes from $\frac{1}{2}$ to $1\frac{13}{16}$ in and called a 'Blockmaker's Tapering Shell Auger or Scillop'. It may have been used when making ship's blocks but we have been unable to find the purpose of the disc at the nose.

Fig. 31

Auger, Boring Machine: see *Wood Boring Machine*.

Auger, Breast *Fig. 32 overleaf*
A tool used in Northern Europe from the eleventh century onwards, mainly by shipbuilders. It consists of a wooden stock in which a Shell or Nose Bit is fitted. It is turned by a wooden bar passing through the stock.
   This tool is still used by rural craftsmen in Baltic countries, but has not been known in Britain since the sixteenth century.

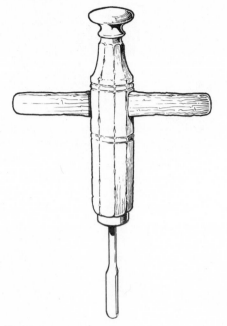

Fig. 32

Auger, Brick (Bell Auger) *Fig. 33*
A short, tapering, double-twist body and flat point. The square shank is sometimes pierced for turning with a tommy bar. Diameter at the widest part from $\frac{3}{8}$ to $1\frac{1}{2}$ in. Used for boring through bricks or stone for bell wires, plugs, etc.

Fig. 33 Brick Auger

Auger, Bull Nose: see *Auger, Twist*.

Auger, Bung: see *Auger, Cooper's Bung Borer; Bit, Cock Plug*.

Auger, Carpenter's
A term applied to Augers of standard sizes and of Shell or Twist type.

Auger, Centre Bit *Fig. 34*
18 in long overall, with a head in the form of a very large Centre Bit with a screw point, boring a hole $3\frac{3}{4}$ inch in diameter. Found in a Glasgow cooperage, it may have been used for boring bung holes in vats.

34

Fig. 34

Auger, Chairmaker's: see *Auger, Taper*.

Auger, Clogger's: see Sabot Tools under *Clog-Sole Maker*.

Auger, Coak Boring (Coak Engine)
A name given to various types of Auger, e.g. *Auger, Deck Counterboring*, used for boring holes to take the heavy wooden trenails known as coaks. These are used for joining heavy timber in ships, lock gates, etc.

Auger, Cooper's Bung Borer (Scots: Scillop) *Fig. 35*
A tapered hole is bored through the side of a cask at a point near the greatest diameter. The hole is stoppered with a tapered bung of wood or with a wooden disc known as a shive. Another tapered hole is bored through the head of the cask and then corked until the tap (fawcet) is fitted by the innkeeper or cellarman. Since the liquor will not flow from this tap without an air inlet, a small hole is bored through the shive and stoppered by a tapering peg known as a spile, which, when lifted out, allows the liquor to flow.

(a) Taper Auger
After boring a pilot hole, these Augers are used to enlarge both bung and tap holes. (See also *Auger, Taper*).

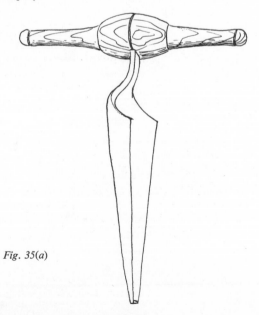

Fig. 35(a)

(*b*) *Bung Burner* (Burning Iron; Tap Burner)
An iron rod up to 3 ft long, with a ring at one end for hanging, and a solid iron cone at the other. Several different sizes were kept in the cooper's shop. After heating they were inserted into the bung or tap hole which had been previously bored. Some coopers declare that no tapered bung hole could be made perfectly true without the use of this tool; others say that burning out the hole was done to prevent rot.

Flush Boring Tool.

Six Jaw Square Shank Fixing Key.

Fig. 35(b)

Fluted Fixing Key

Fig. 35(c)

(*c*) *Bung Bushing Tools*
The bung holes of modern beer casks are lined with metal self-tapping bushes into which the shives are driven. There are a variety of tools for shaping the bung hole and inserting these bushes, including:

Taper Borer: a conical body fitted with a cutting knife, the upper end of which gives a small countersink to the tapered bung hole. Used for preparing the bung hole to take the simpler types of metal bush.

Flush Borer (Flush Boring Tool): similar to the Taper Borer, with the addition of a vertical spur cutter and a horizontal radial cutter mounted on the top of the body, which counterbore the tapered hole to take the lip of a screwed bush.

Fixing Key: a solid tapered body with sharp flutes or serrations instead of the single sloping cutter. The shank is square so that it can be turned with a wrench by hand. Used for gripping the self-tapping metal bush and screwing it into the bung hole.

(*d*) *Bung Auger with rasp cut*
A continental pattern with either a twisted or a plain tapered body. The outside surface is cut like a rasp, presumably to smooth off the bung hole to a true taper.

Taper Boring Tool.

Fig. 35(d)

(e) Cylindrical or American Bung Borer
Made in several patterns, this tool consists essentially of a hollow conical 'pod' or body with a slot ½ in wide down its length, one edge of which is sharpened to form a cutter. The nose of the tool is sometimes provided with a short length of Twist Auger, and the body is extended above to hold a wooden cross-handle. The pod is closed at the bottom to receive the chips, and this is one of its advantages, for the chips do not fall into the cask. Later examples have separate cutting irons, graduated cylinders, and a device to stop the Auger after it has penetrated a certain distance.

Fig. 35(e)

(f) Taphole Auger (U.S.A.: Ring Punch)
This tool resembles other Taper Augers, but is much smaller, and is adapted for working with one hand like a Gimlet when cutting the tap hole in the head of a cask. It is nearly always provided with a screw start. Made in sizes to cut holes up to about 2 in. in diameter.

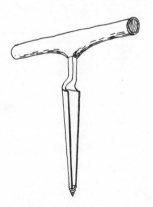

A Continental type has a half-funnel tapered body with a knife or spur hanging downwards from a circular disc or plug which closes the funnel at the wide end. This disc may be intended to act as a stopper. (See *Bit, Cock Plug*.)

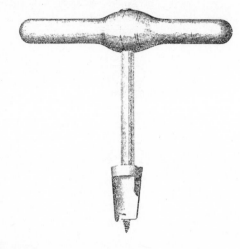

Fig. 35(f)

(g) Thief Auger
Not an Auger, but used when boring bung holes. The shank has the usual cross-handle and the bottom end (the 'tongue') is bent at right angles, and ground to a half-round shape with sharp edges. When a bung hole is bored, the jagged ends inside the cask (the 'spelch') can be pared off by this sharpened tongue. Possibly the name thief suggests the method of use: the tool must be introduced stealthily. Another type was provided with knives which, after insertion into the cask, could be opened like an umbrella, pulled backwards and turned.

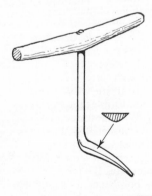

Fig. 35(g)

Auger, Crank: see *Auger, Raft; Brace, Platelayer's*

Auger, Crown Saw *Fig. 36*
This tool, used on the Continent, is similar to a
Crown Saw and is sometimes made by nailing a short
length of saw blade round the end of a cylindrical
piece of wood. It was used by coopers and others for
boring holes. A tool of this kind was also used for
making bungs or shives for casks. (See *Saw, Crown*.)

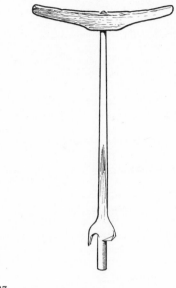

Fig. 36

Fig. 37

Auger, Dodd's Pattern *Fig. 29, No. 14* (page 32)
A Shipwright's Shell Auger of sturdy make, with a
short, shallow pod. The shank is left for welding an
extension to any length desired.

Auger, Deck Dowelling (Counterboring Auger) *Fig.
37*
This Shipwright's Auger is made in sizes up to *c.*
1½ in. in diameter. One type has the nose of a Centre
Bit, with a plain or screwed plug. The other type is a
Twist Auger, of the single twist pattern, with the
same plain or screwed plug. Its purpose is to countersink deck bolt holes to make room for the bolt head.
In use the plug is dropped (or screwed) into bolt
holes previously bored in the deck, after which the
nose cutter excavates a recess. After the bolt head
is sunk into this recess, a dowel is driven over it as a
cover. The dowels used for this purpose (called
pellets by joiners) are cross-grained so that deck
and dowel wear down at the same rate.

A home-made Auger for doing the same work consists of a stout iron stem (which acts as the plug to fit
into the bolt hole) with a horizontal slot about 2 in
from the foot in which a short cutting iron is secured.
This cutter excavates the recess in the deck or elsewhere for the bolt heads. For illustration see *Auger,
Hub Counter-Boring*, which it resembles.

Auger, Expanding *Fig. 38*
A strongly made boring tool about 24 in long with
cross-handle up to 18 in long. A pilot hole is bored
first, after which the router-like cutters, which are
adjustable laterally, will cut holes up to 5 in. in
diameter. Used by shipwrights and others for boring
large holes beyond the capacity of other Augers,
when counter-boring for very large deck bolts.

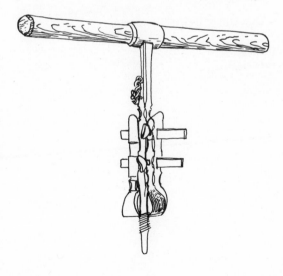

Fig. 38

37

Auger, Gedge's Pattern: see *Auger, Twist.*

Auger, Hollow: see *Bit, Hollow.*

Auger, Hommedieu: see Auger L'Hommedieu under *Auger, Twist.*

Auger, Hook (Hook-Nose Nave Auger) *Fig. 39*
A large Taper Auger provided with a stout hook at the nose.

A tool used by wheelwrights on the Continent for enlarging the hole in wheel hubs. It is believed that the hook is used to draw the Auger into the wood by means of a cord passed through a small pilot hole and pulled by hand if used horizontally, or by means of weights if used vertically. The effective cutting diameter of the Auger can be increased by packing it out with a special shell piece (or 'liner') attached to the trailing edge.

Hooked Augers are illustrated by Diderot (Paris, 1763) and have been found on Roman sites.

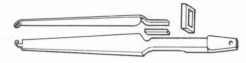

Fig. 39

Auger, Hub Counter-Boring *Fig. 40*
Handled like an Auger, this tool is designed for enlarging the bore of the hub to make room for the lynch-pin or axle-cap. Wooden plugs of differing size are fitted over the lower end of the shank to fit the bore of the hub. This holds the tool central, while the adjustable L-shaped cutter bar fitted just above it, when made to revolve, enlarges or counterbores the hub to the required depth. It was only used, presumably, when a Hub Boring Machine was not available.

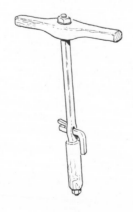

Fig. 40

Auger, Irwin Pattern: see *Auger, Twist.*

Auger, Jenning's Pattern: see *Auger, Twist.*

Auger, Ladder: see *Auger, Taper.*

Auger, L'Hommedieu: see *Auger, Twist.*

Auger, Long Pod (Trenail Auger) *Fig. 29 No. 15* under *Auger.*
A Shipwright's Shell Auger with an extra-long pod or body.

Auger Machine: see *Wood Boring Machine.*

Auger, Nose: see *Auger, Shell.*

Auger, Parrot Nose *Fig. 41*
A type of Spoon Auger with a shell body from $\frac{1}{2}$ to $1\frac{1}{2}$ in. in diameter, with half the nose ground to form a rounded point on the centre line, the other half ground away. Used for boring long holes in the end grain, e.g. for printing and mangle rollers, standard lamps, etc.

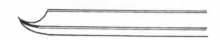

Fig. 41

Auger, Pipe and Pump *Fig. 42* See *Pipe and Pump Maker* for illustrations and methods.
Some examples of Pump and Pipe Augers are shown below.

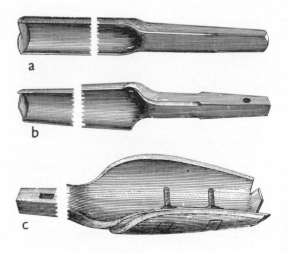

Fig. 42 Pipe and Pump Auger
    (*a*) First Pump Bit
    (*b*) Taper or Reaming Bit
    (*c*) Taper Bit with liner
    (*d*) Turning Handle

Auger, Pod
A common term among country tradesmen and others for Shell and Spoon Augers. Mercer (U.S.A., 1929) uses the term to denote the type of Auger described under *Auger, Snail*.

Auger, Raft (Crank Auger)
A Brace and Auger of the twist type made in one piece about 4 ft long with the upper arm of the crank extended sideways beyond the centre line. Described by Mercer (U.S.A., 1929), it was used by lumbermen for boring holes in the ends of floating logs to yoke them together into rafts. A similar tool is described under *Brace, Platelayer's*.

Auger, Railway
A long Twist Auger of the Scotch pattern used by Platelayers. See also *Bit, Wagon Builder's*.

Auger, Reamer: see *Auger, Taper; Auger, Cooper's Bung Borer*.

Auger, Saw: see *Bit, Annular; Auger, Crown Saw*.

Auger, Scillop
A Scots term for a half funnel-shaped Auger as described under *Auger, Taper* and *Auger, Cooper's Bung Borer*.

Auger, Scotch Pattern: see *Auger, Twist*.

Auger, Screw: see *Auger, Twist*.

Auger, Screw-Nose: see note under *Auger, Shell*.

Auger, Shell (Nose Auger; Pod Auger; Split-Nose Auger) *Fig. 29* under *Auger; Fig. 43; Fig. 44*.
A general term for Augers with the blade in the form of a half-cylinder with an in-bent horizontal cutter on the nose.

The carpenter's and shipwright's Augers of the Roman and Medieval periods appear to have had a plain shell body with the nose shaped like a gouge, but since the seventeenth century at least Shell Augers of most types have had a blunt mouth and the flat cutting lip. This makes the tool more efficient and less laborious in use, especially in the larger sizes. The survival of the Shell Bit with a plain gouge-shaped nose for use with the Brace may be accounted for by the possibility it offers of applying heavy and continuous pressure. With the Auger, every time the tool is stopped to change hands the pressure is released and has to be built up again for the next half turn. This is achieved more readily when the nose is already biting in the hole. Though all but the smaller sizes must be started by making a depression with a gouge, these shell-type Augers are still preferred by some tradesmen for boring long holes. (See *Auger, Shipwright's*.)

The confusion between the terms shell-type and nose-type boring tools is dealt with under *Bit, Shell*. For variants see *Auger, Carpenter's; Auger, Dodd's Pattern; Auger, Long Pod; Auger, Skew-nose*.

Note. A Shell Auger with a screw head mounted on the end of the inbent lip is illustrated by Richard Timmins (Birmingham, *c*. 1800). It is called a Screw-nose Shell Auger and made in sizes $\frac{1}{2}$–2 in. We have never seen an example, nor met one like it in any other maker's catalogue.

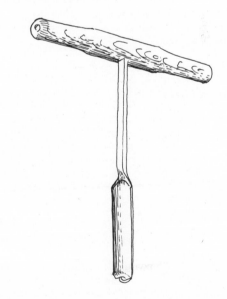

Fig. 43 A typical smith-made example (See Fig. 29, Nos. 11–15, for factory-made examples.)

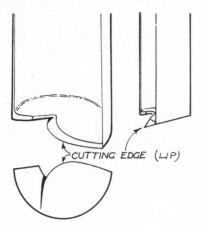

Fig. 44 Diagram of Shell-Auger Nose

Auger, Shipwright's (Ship Carpenter's Auger)
A term applied to several types of long, strong Augers of both the shell and twist types. In the shell types the shoulders of the pod are usually thicker to give added strength, and the shank is often left for welding to rods of any desired length according to the work. Many shipwrights, including those working in H.M. Dockyard, Portsmouth (1969), declare that they prefer Shell Augers to the twist varieties because they are less liable to 'wander' or follow the grain. This may not be important when boring holes for trenails, but it is essential when, for instance, bolt-holes are bored as long as 15 ft into the keel.

Very long Augers used for this purpose may be seen in the Museum at Brixham in Devon. They are of the shell type, $2\frac{1}{2}$ in wide with shanks 8 ft long.

At the time of Rålamb (Sweden, 1691) Shipwright's Augers were of the shell or pod type, some having a rudimentary nose or lip at the end. The same patterns were used by the ship's carpenters who perished on the ill-fated Novaya Zemblya Expedition of 1596.

In his *Ship builder's Assistant* (London, 1711) William Sutherland writes: 'The Augre is an Instrument of singular Use and Service in Ship-building and so very fitly adapted for composing the same, that if ever any Person was Deify'd for Inventing, I should highly recommend the Author of an Augre deserving of that Glory.'

Augers used by Shipwrights include: *Auger, Coak Boring; Auger, Deck Dowelling; Auger, Deck Counterboring; Auger, Dodd's Pattern; Auger, Expanding; Auger, Long Pod* (Trenail Auger); *Auger, Twist* (Single Twist). See also *Trenail Tools.*

Auger, Side-Cutting: see *Auger, Slotting.*

Auger, Single Twist or **Worm:** see *Auger, Twist.*

Auger, Skewnose *Fig. 29 No. 17* under *Auger.*
Similar to the Carpenter's Shell Auger, but made with a skewed nose, presumably so that it can start without a previously bored hole. According to Gilpin (Cannock, 1868) it is listed for use by plate-layers on the railways.

Auger, Slotting (Side-cutting Auger; Traversing Auger)
A Machine Auger (or Bit) with side-cutting flutes on the side of the body so that the wood may be cut laterally for making a mortice or slot. (See also *Bit, Mortice.*)

Auger, Snail (Pod Auger) *Fig. 45*
This Auger looks rather like an open seed-pod, with one side sharpened, and twisted towards the nose so as to grasp the wood and hold the shavings.

The Auger illustrated was probably imported from the Continent. It is listed by Goldenberg (Alsace, 1904) under various names including Tyrolean and Hungarian, and by Peter Fleus (Germany, 1910) who calls it a Swiss Snail Auger. The same type of Auger is illustrated for boring pumps (see *Pipe and Pump Maker*).

Fig. 45

Auger, Solid Centre, Nose or **Wing:** see under *Auger, Twist.*

Auger, Spiral: see *Auger, Twist.*

Auger, Split-Nose: see *Auger, Shell.*

Auger, Taper (Reamer or Rimer; Scots: Scillop; Taper Shell Auger) *Fig. 46*
A general term for Augers made in the form of a tapered, conical half-funnel, with one or both of the exposed edges side-sharpened to cut laterally.

The lower end or nose is either open, spoon shaped, provided with a side-cutting lip, or, exceptionally, with a point or screw lead. A 'dent' must be cut in the wood to start all but the last type.

The following variants are illustrated:

(*a*) The so-called 'tap' Auger with a screw lead.

Fig. 46(a)

(*b*) A reamer type with open nose, for enlarging a previously bored hole.

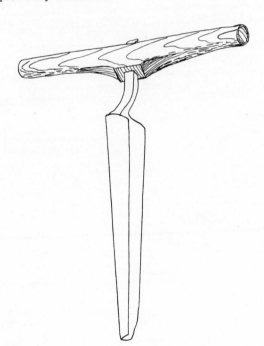

Fig. 46(b)

(*c*) A Wheelwright's Auger of the 'nose' type with inbent lip.

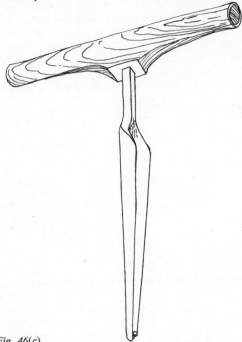

Fig. 46(c)

(*d*) A home-made, all-wooden version of the Taper Auger, which is found occasionally in country districts. It consists of a conical wooden stock fitted with a double-tanged blade of the kind used in a Spokeshave. The stock is bored out to receive the chips. This tool may perhaps be regarded as the direct ancestor of the American pattern bung-boring Auger (see *Auger, Cooper's Bung Borer*).

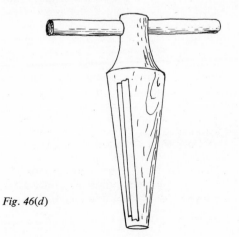

Fig. 46(d)

41

(*e*) The so-called Ladder Maker's Auger is illustrated under *Augers, Fig. 29 No. 16*. It enlarges holes up to about 1½ in. It is used by many different tradesmen, including ladder makers for boring tapered holes in ladder sides for the rungs; wheelwrights for enlarging the spoke-tongue holes in felloes; and windsor-chair makers (who often call the tool a 'Roomer') for opening up through-bored holes to make room for a slightly oversize leg.

Other kinds of Taper Augers are described under: *Pipe and Pump Maker; Auger, Hook; Auger, Cooper's Bung Borer*.

Auger, Tenon

A form of *Hollow Bit* described by Knight (U.S.A., 1875) for shaping the circular tenons on the ends of movable blind-slats, chair legs, etc.

Auger, Trenail: see *Auger, Long Pod*.

Auger, Twist (Screw Auger; Spiral Auger) *Figs. 47; 48; 49 overleaf*

Note: This entry refers also to Twist or Auger Bits which are illustrated in *Fig. 130* under *Bit, Twist*, and in *Fig. 29* under *Auger*.

Nomenclature. The Sheffield Illustrated Lists of 1862 and 1888, and Smith's *Key to the Various Manufactories of Sheffield* (1816) and all toolmakers' catalogues we know of from this period refer to these tools as Screw Augers. Among reference books, however, Holtzapffel (1847) uses the term 'screw', while Spon (1893), Coventon (1953), Sturt (1942), and Hayward (1946) use the term 'twist'.

In America these Augers are fairly widely described as 'spiral' – a term which we think is the best of the three, for it avoids confusion with the screw lead or nose or with the term 'single or double twist' for describing the form of the body of the tool. But the term 'spiral' is hardly ever encountered over here, so we have settled for 'twist', which incidentally matches the Twist Drill, the Auger's counterpart in the engineering world.

The terms 'screw', 'twist', or 'spiral' are applied to Augers and Auger Bits which have a body twisted into a spiral shape. There are two main types of body: the so-called 'double twist' made from a flat steel bar which is made hot and then twisted; and the 'single twist' made of half-round or triangular bar which is coiled in an open spiral.

Twist Augers are from 12 to 25 in (or more) in length and bore holes from ¼ to 2½ in. in diameter. A smith-made example is shown in *Fig. 47*.

Auger Bits resemble the corresponding pattern of Twist Auger, but are usually only 8–10 in long and are provided with a tapered square tang for use in a Brace. They bore holes from *c*. ¼ to 1½ in. in diameter.

One of the main advantages of the twisted form is that shavings ascend the spiral body and therefore frequent withdrawal of the tool to clear them is unnecessary. Another advantage is that the effect of the screw lead and the twisted body is that the tool pulls itself to some extent through the hole, and thus the amount of pressure required is reduced. The following advertisement from the 1909 catalogue of the Irwin Bit Company of America adds yet more virtues:

> 'No chocking, no breaking,
> No swearing, no tearing.
> A bit without bending.
> A joy without ending.'

The first Twist Auger of modern type known to us is illustrated in William Bailey's *The Advancement of Arts, Manufactures & Commerce* (London, 1772) and is described as 'Mr. Phineas Cooke's New constructed Spiral Auger'. It has a screw lead. A Twist Auger of the 'Scotch' Pattern appears in Smith's *Key to the Various Manufactories of Sheffield* (1816). William Bailey writes that these tools 'require no picking with a gouge' to start them; and 'they do not want to be drawn out of the wood to discharge the chip which is also the occasion of much labour and loss of time in boring with a common auger'.

Twist Auger and Auger Bit Patterns. Nomenclature – Diagram on page 44; Some examples of Auger Bits – *Fig. 29* under *Auger* and *Fig. 130* under *Bit, Twist*

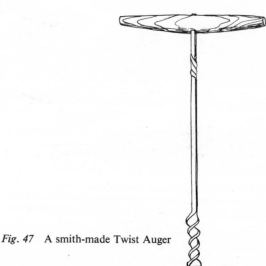

Fig. 47   A smith-made Twist Auger

Auger Bits (*Fig. 49 overleaf*)

Fig. No.	Pattern	Description	Special uses
49 (*a*) 130 (*f*) (Page 87)	*Jennings or Russell Jennings* (Patent by R. Jennings in 1855)	Body: double twist or solid centre. Nose: flat cutting edge with spurs; screw lead.*	Clean accurate holes, e.g. for cabinet work.
49 (*b*) 130 (*c*)	*Scotch*	Body: double twist. Nose: flat cutting edge with side wing (x) but no spurs; screw lead.*	Boring hard woods, and for all kinds of rough constructional work.
49 (*c*)	*Solid Centre* or *Irwin* (Patented by Charles Irwin in 1884)	Body: a stem with a fin winding round it. Nose: Scotch; Solid-Nose pattern for heavy work; Jennings pattern for lighter work.	General, heavy or fine work according to the nose design.
49 (*d*) 130 (*e*)	*Gedge*	Body: double twist. Nose: cutting edge curved upwards without spurs; screw lead.*	End-grain work; also for boring askew, for there are no spurs to foul the wood before the lead is engaged.
49 (*e*) 130 (*d*)	*Solid Nose* (Bull Nose; Solid Wing; Unbreakable)	Body: double twist. Nose: cutting edge continued round to meet the spiral, leaving two circular holes for chip release (these probably gave rise to its alternative name 'bull nose'); screw lead.*	Being without protruding cutters or spurs it is valued for tough jobs; also for its ability to enlarge a previously bored hole, or to bore at an angle.
49 (*f*) 29 (9) (10) Page 32	*Single Twist* or *L'Hommedieu* (American Ship Auger; Spiral Ribbon Auger; Single Worm Auger) Patented by Ezra L'Hommedieu 1809	Body: an open spiral. Nose: flat cutting edge without spurs; if without lead called 'Bull nose' in England, 'Shell mouth' in Scotland, and 'Barefoot' in the U.S.A.	Favoured by shipwrights who claim that it is less liable to get 'stuck in' than a double-twist Auger.
		Note: L'Hommedieu's original design had a screw lead. See Fig. 29 (9) under *Auger*.	

* A point is sometimes provided instead of a screw lead.

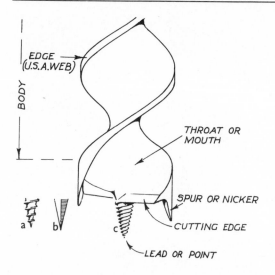

Fig. 48 Diagrams of a Twist Auger's nose with:
    (*a*) Single thread lead
    (*b*) Brad or diamond lead
    (*c*) Double thread lead

Auger, Unbreakable: see *Auger, Twist* (Solid Nose).

Auger, Wheelwright's and **Coachbuilder's**
These tradesmen use Augers of many different types including the large Taper Augers used for forming the tapered axle holes in wheel hubs. Other boring tools used include: *Auger, Hub Counter-Boring; Bit, Hollow; Bit, Spoke Trimmer; Burning Rods; Hub Boring Machine.*

Auger, Woodwind: see *Woodwind Instrument Maker.*

Awl (Obsolete form: Alsene or Elsin) *Fig. 50*
Awls are sharp metal spikes used for making holes in wood, leather, and other soft material. With the possible exception of the Square Birdcage Awl, they penetrate by squeezing the material apart – unlike the Gimlet which starts a hole in this way but later opens it up by side-cutting.

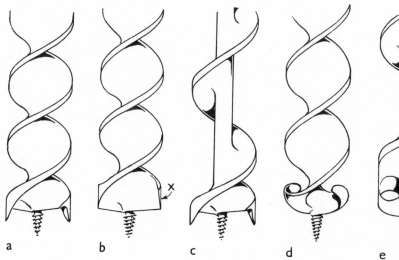
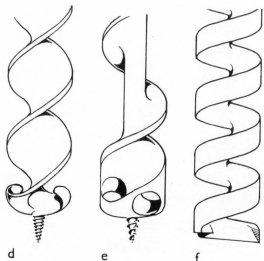

Fig. 49 Diagram of Auger and Auger Bit patterns
    (*a*) Jennings or Russell Jennings pattern
    (*b*) Scotch pattern
    (*c*) Solid Centre or Irwin pattern
    (*d*) Gedge pattern
    (*e*) Solid Nose or Bull Nose pattern
    (*f*) Single Twist or L'Hommedieu pattern

Awl, Birdcage: see *Awl, Square.*

Awl, Bird's Eye
A name sometimes given to a saddler's and harness-maker's awl (of oval cross-section) and not to be confused with the Birdcage Awl.

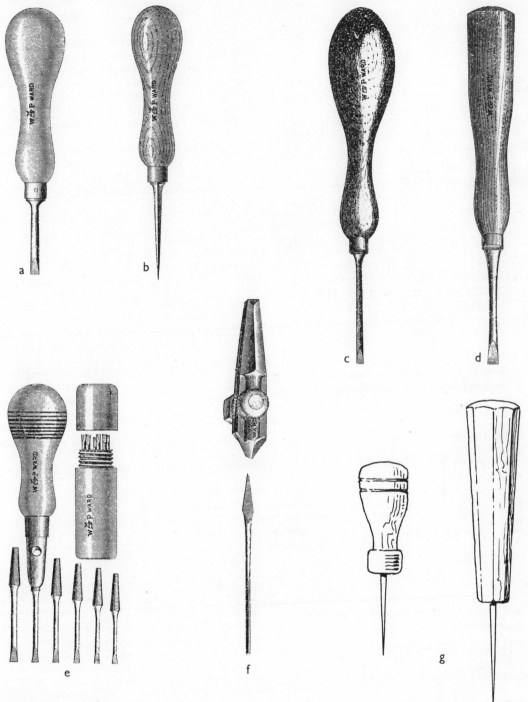

Fig. 50 Various Awls
(*a*) Bradawl   (*b*) Square Awl   (*c*), (*d*) Flooring Awls   (*e*) Awl Pad   (*f*) Boatbuilder's Piercing Awl   (*g*) Home-made Awls

45

Awl, Boat-Builder's Piercing *Fig. 50 (f)*
Usually a shell body with a flat, spear-shaped tang which can be held in a wooden handle, but is more often held in the interchangeable wooden pad of the Shipwright's Brace. Alternatively, it can be used in a modern screw-chuck brace if first secured in a Boat Bit Holder; this consists of a short steel pad made in two halves, held together by a bolt and nut. The forward face of the pad is sharpened for countersinking. These awls are used for boring the planks of small boats to take the nails.

Awl, Brad – **The Bradawl** (Scots: Brog or Stob; Sprig Bit) *Fig. 50 (a)*
A round blade 1–3 in long with a chisel point. The bolstered tang is usually fitted in a turned beech handle with a metal ferrule, which in the best qualities is 'through-pinned' to prevent twisting. Used for boring pilot holes for nails or screws. The tool is started with the chisel point across the grain of the wood; then, by twisting back and forth through the wood, the grain is squeezed aside without producing any shavings.

Awl, Entering: see *Upholsterer.*

Awl, Flooring *Fig. 50 (c, d)*
A large Bradawl with a 2¼–3 in blade. It is used for making pilot holes for flooring nails etc.

Awl, Garnish: see *Upholsterer.*

Awl, Marking: see *Marking Awl and Knife.*

Awl Pad (Pricker Pad; Tool Pad) *Fig. 50 (e)*
A wooden handle with a solid metal pad to take small tapered-shank Bits, with a hole drilled through the pad to enable the bits to be ejected. The interchangeable Bits, up to 12 in number, sometimes include a Gimlet and Screwdriver as well as a set of Awls of different sizes. In later patterns the Bits are carried inside the hollow handle.

Awl, Pegging
The name of a small round or square Awl with a blade ½–1¼ in long, used by shoemakers for piercing holes in the sole to take wooden pegs.

Awl, Pritch: see *Upholsterer.*

Awl, Sail: see Pricker under *Sailmaker.*

Awl, Sprig
A name given to a Bradawl of small size. A sprig is a small nail with a very small head.

Awl, Square (Birdcage Awl) *Fig. 50 (b)*
A tapered, pointed blade of square cross-section with shouldered tang. Used for boring hard wood, especially of thin or delicate section, such as veneers. It can be used close to the edge without splitting the wood, for which purpose it is preferred to the Bradawl. This was described by Moxon (London, 1677) as follows: 'The square Angle in turning it about breaks the Grain, and so the Wood is in less danger of splitting.'
    According to Messrs. Wm. Temporal, Toolmakers (Sheffield, 1969), these awls were used to bore holes in the wooden frame of a birdcage to take the steel wires – hence their alternative name.

Awl, Upholsterer's, Stabbing, Stiletto, Straining, Trimmer's: see *Upholsterer.*

Axe (An earlier term for Axe is Belt: 'Wherefore, seyd the belte, With grete strokes I schalle hym pelte', from *Debate of the Carpenter's Tools*, a fifteenth-century document in the Bodleian Library.) *Fig. 51. Opposite.*
This most fundamental of all woodworking tools began its long history as a rough stone that was chipped on one side to form an edge and grasped with the bare hands. Much later – about 6000 B.C. – flint and stone axes were fitted to wooden shafts, perforated to take the butt, and bound with thongs.
    The earliest metal Axes were of copper and later of bronze, cast to a similar shape as the flint and with the butt held between the prongs of a knee-shaped handle. To give a more rigid fixing, flanges were cast down the side, and finally the head was cast with a hollow socket to take the end of the bent shaft. This sequence is shown in the Bronze Age of the West, up to about 1000 B.C., but in the Middle East the smiths had been casting Axes and other tools with holes for the shafts for some centuries before that.
    The Iron Age Axes of about 900 B.C. were forged with sockets, copying the shape of the bronze tools they superseded, but by the Roman period the axehead had taken its modern form, with an eye for the shaft and a flat poll.
    During the later Middle Ages specialised axes were developed, including the long-headed Tree Felling Axes, and the broad Side Axes with a sharpening bevel on one side only, used for hewing flat sides of the timber. Those developed for particular trades include the Cooper's and Shipwright's Axes, which have remained almost unchanged in design until today. But the Axe was still a general-purpose tool, as Thomas Tusser makes clear in his *Five*

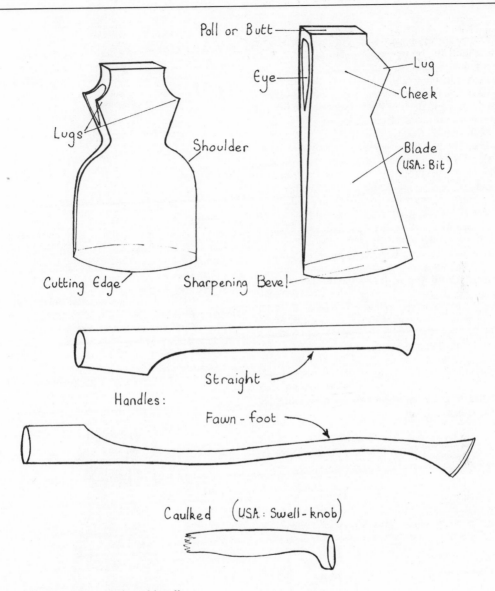

Fig. 51 Diagram of Axe blade and handles

hundreth pointes of good husbandrie (London, 1573): 'An ax and a nads, to make troffe for the hogs'.

About 1800, or possibly earlier, an Axe was developed in England with a short, heavy, wedge-shaped blade, which in Smith's *Key* (Sheffield, 1816) is called a Wedge Axe. It became popular in America, and under the common title Yankee, or American Axe, it spread to most parts of the world.

Apart from those Axes made for particular trades, most of them are grouped under the following entries:

(*a*) Felling Axes (including the Wedge Axe)
(*b*) Mortice Axes
(*c*) Side Axes
(*d*) Axes for Export
(*e*) The lighter forms of Axe, weighing up to about 2 lb, which are described under *Hatchet*.

47

Axe Handles

In Roman and Medieval times, and indeed right up to the middle of the nineteenth century, large Axes were fitted with plain, straight handles, round or oval in section and up to three feet long. Smaller Axes sometimes had gently curved handles, with a swelling under the eye, and the Side Axes, used for side paring, had their handles off-set (i.e. bent to one side) to protect the hands from injury through contact with the work. The handles of some of the smaller Axes, as well as some Hatchets and Billhooks, have a thickening at the foot; they are then described as 'caulked' (U.S.A.: 'swellknob'). This arrangement prevents the tool from flying out of the hand.

Modern English Axes have a smoothly curved handle, oval in cross-section and thickened slightly at the foot. The double-curved 'fawn-foot' pattern handle appears to have been designed originally for American Felling Axes in the mid-nineteenth century. It permits a fixed grip with one hand near the foot and a sliding grip with the other; with a heavy head, one hand must be raised well up the handle, in order to lift it, and then brought down smoothly for the working stroke. Furthermore, the fawn-foot itself, like the caulked handle, tends to prevent the Axe from slipping out of the hand. The resulting curves, although purely functional, have considerable appeal to the eye, and have led some writers to assume that this shape of handle has had a much longer history than the evidence warrants.

Axe, African: see *Axe, Export.*

Axe, American
A name given to a number of Axes of American pattern, see *Axe, Broad* and *Axe, Felling, Wedge Type.* (For South American Axes, see *Axe, Export.*)

Axe, Amoor Fantail: see *Axe, Export.*

Axe, Army
Illustrated in Smith's *Key,* (Sheffield, 1816) with a wide flared blade, round lugs, and square poll.

Axe, Australian: see *Axe, Export.*

Axe, Banbury *Fig. 53*
A Kent type of Axe but with a greater slope in the shoulder of the blade.

Fig. 53

Axe, Barge
A term used in the sixteenth and seventeenth centuries, (Goodman 1972). Probably a cooper's T-shaped Axe similar to the type still in use.

Axe, Bearded
A term applied to Axes when the lower part of the blade overhangs the socket. Commoner on the Continent of Europe, but an English example is the Coachmaker's Axe (*Fig. 61*).

Axe Beating *Fig. 54*
A heavy blade with a single lug below the eye. Purpose unknown.

Fig. 54

Axe, Bench
A term used to describe a small, general purpose Axe or Hatchet used for rough-surfacing, chopping, and sometimes nailing.

Axe, Bill Hook: see *Bill Hook.*

Axe, Blocking (Squaring Axe) *Fig. 55*
An asymmetrical, parallel-sided blade, sometimes as long as 22 in when new. The poll is light and flat, the lugs pointed below and flat on top, and the eye is sometimes set at an angle so that the blade leans towards the handle. Ground either as a Side Axe or with a double bevel.

Used for squaring heavy timbers, for example in shipbuilding and dock works. The extra length of blade is needed for reaching the bottom of wide pieces. The word blocking probably refers to the 'block' – a word used for the 'heeling' of the main mast, and also to other solid pieces of wood on a ship and elsewhere.

and smash through a bulkhead. We have one in the Museum which comes from H.M.S. *Captain*. The ship capsized in the Bay of Biscay in 1870 and was an iron clad . . . It was a weapon of destruction, but, of course, still basically an axe with a spike at the back. It was also useful in emergencies for cutting away ropes or spars which threatened to endanger the ship.'

Fig. 55

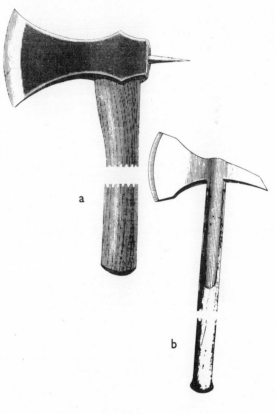

Fig. 56

(*a*) From Sheffield List 1888
(*b*) From William Gilpin 1868

Axe, Boarding (Boarding Hatchet) *Fig. 56*
Illustrated in the Sheffield List of 1888 as a light Axe (1¾ lb), with a flared blade and a square poll on which is mounted a sharp spike. Mr. G. P. B. Naish (National Maritime Museum, 1970) has informed us: 'The axe was carried by boarders and used to damage the enemy's ship, cut up his rigging, hack through a spar

Axe, Broad *Figs. 57; 58*
A name sometimes given to short-handled Axes with one characteristic in common – a rather broad blade, i.e. wide in relation to its length. They are mainly of the Side Axe group, such as the Wheelwright's and Cooper's Axes, and are used for squaring logs or for making railway sleepers, and for squaring

structural timbers of all kinds. Distinctive types of foreign design include:

(*a*) *Pennsylvania Squaring Axe* (*Fig. 57*)
A broad blade and made in sizes from 3 to 7 lb.

(*b*) *A Continental European pattern* (*Fig. 58*)
A heavy Side Axe of handsome appearance, with the top of the blade pointed or flared upwards (rather like the Goose Wing Axe) and the lower part 'bearded', that is, overhanging the socket. In this it resembles the English Coachmaker's Axe except that there is sometimes insufficient space between the 'beard' and the socket for the user's fingers. The cutting edge, from 10 to 15 in long, is bevelled on one side only. The long socket has a ridge on one side and is flat on the other, and is sometimes off-set to protect the knuckles. It is still in fairly common use on the Continent of Europe.

Fig. 57

Fig. 58

50

Axe, Butcher's Pole (Butcher's Felling Axe) *Fig. 59*
A light flared blade, usually with round lugs and a cylindrical pin, with a cupped end projecting from the square poll. Used for slaughtering cattle, but now superseded by more humane methods. The pin was driven into the front of the skull to stun the animal. The blade was used subsequently for removing hooves.

Fig. 59

Axe, Camp
A term found mainly in Continental tool catalogues for Axes and/or Hatchets which are mostly small, with flared blades and square poll. One called a Cavalry Hatchet has a poll in the form of a round hammer-head.

Axe, Canadian (Canada Axe): see *Axe, Felling, Wedge Type.*

Axe, Ceylon: see *Axe, Export.*

Axe, Chairmaker's (Chair Bodger's Axe) *Fig. 60 and Fig. 190* under *Chairmaker*
Two special Axes are used by chairmakers and chair bodgers:

(*a*) A tool known as a *Splitting-out Hatchet*, which is really a wedge with an eye to take a wooden handle. It is never used as an Axe, but struck on the poll with a mallet or club. It was used to split out sections of a tree trunk into billets for the production of turned chair parts: legs, forefeet, fronts and stretchers. The timber to be split was supported on a splitting-out block. Skill was required in the use of this tool, since too large a billet meant waste of wood and of energy (all surplus wood had to be removed in subsequent operations) and too small a billet was an obvious loss. Illustrated under *Chairmaker (2) The Bodger.*

(*b*) A tool sometimes known as a *Chair Axe*, which is a Side Axe with a flared blade and a short handle, and is used to shape chair leg billets etc. before shaving and turning.

Fig. 60

Axe, Chinese: see *Axe, Export.*

Axe, Chisel: see *Axe, Mortising.*

Axe, Coachmaker's (Coach Side Axe) *Fig. 61.*
A wide, side-cutting blade, with a flat poll. The cutting edge is carried well below the socket, and the bottom edge overhangs the socket (known as 'bearded') to meet the pointed lug below the eye. It weighs 3–7 lb. It is used in general for the same purposes as the broad Wheeler's Axe (*Axe, Wheelwright's*), but it is specially favoured for delicate work, such as paring down an oak wedge. For this purpose it is held in the 'throttle' position, i.e. the handle is grasped at the top, where it enters the eye of the Axe, with the forefinger placed on the side of the blade to keep it upright. The late Mr. G. W. Weller (1953) describes this operation as follows: 'Many of the wedges for wheels were only 1 in wide, 3 in long, and $\frac{1}{2}$ in thick. These were held upright on the chopping block with the heel of the thumb, and with the rest of the thumb and fingers held well clear of the wedge, the knee supporting the wrist, the wedge was chopped to the appropriate shape, evenly from both sides. A tricky, dangerous job, especially if one's attention was distracted. Even so, it did not do to have too small an axe, as it was the weight of the tool which did the work. No force or exertion was required beyond just lifting, dropping, and guiding the axe.'

Axe, Cooper's *Fig. 62*
(a) A thin flat T-shaped blade, 10–12 in measured along the cutting edge, without a poll, and ground on one side only. The tapered socket springs from the middle of the back of the blade; in some patterns it is an extension of the blade, bent over to form a triangular socket. The handle is offset (to prevent the user from grazing his knuckles) by forming a bend in the handle itself or by a twist made in the socket. And the blade is often bent downwards so that the cutting edge forms an angle of about 20° with the axis of the handle, instead of being parallel with it.

This tool is used for listing staves, i.e. chopping off the 'waney' irregular edges and trimming them; this is done with the stave held almost horizontal on the block. It is also used for rough-shaping the circular outline of the heads and for chopping away the bevel on the back of the heads before using the Heading Knife (*Drawing Knife, Cooper's Heading*).

The rather strange shape of this Axe seems to be a survival of the T-shaped Axes of the Middle Ages (see historical note under *Cooper*). It lends itself to a lighter method of construction than the Side-Axes used by wheelwrights and coachbuilders, in which the eye is forged into the blade itself. But heavy Axes of the latter type are still used by coopers on the Continent; they are illustrated by Messrs. William Gilpin Senr. & Co. of Cannock in their 1868 Catalogue as follows:

(b) *Irish Cooper's Bench Axe*
A fan-shaped blade with a circular cutting edge and flat poll and probably ground on one side only.

(c) *London Cooper's Bench Axe*
Shaped like a Coachmaker's Axe, with a bearded blade, but whether side or double ground is not clear.

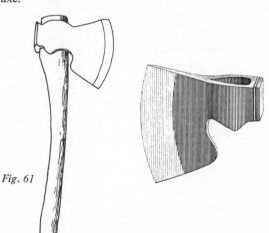

Fig. 61

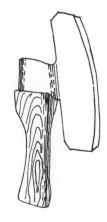

a

Fig. 62(a)

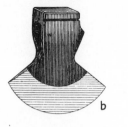 

Fig. 62(b) and (c)

Axe, Cratemaker's *Fig. 63*
A broad blade with a flat poll, pointed lugs on both sides of the eye, and a curved cutting edge. Weight 2–7 lb. Purpose unknown.

Fig. 63

Axe, Cricket Bat Maker's
A Side Axe with a wide, asymmetrical blade similar to the Coachmaker's Axe, with rounded lugs below the eye and a flat poll. Used for rough shaping willow blanks for cricket bats.

Axe, Cross: see *Axe, Grubbing* and *Axe, Twybill.*

Axe, Cumberland: see *Axe, Felling, Long.*

Axe, Deputy: see *Axe, Miner's Deputy.*

Axe, Derbyshire: see *Axe, Felling, Long.*

Axe, Dill: see *Froe.*

Axe, Double-Bitted *Fig. 64*
A wedge-type head with two cutting edges. Weight 3–5 lb.
    Used widely in the U.S.A. and Canada for felling etc. The advantages of the double blade are: one edge can be kept for lopping and topping, the other

for 'laying in' or felling proper; both edges can be ground and sharpened and used alternately to last all day; if the tool needs sharpening on the job, it can be stuck upright in a stump and honed easily.

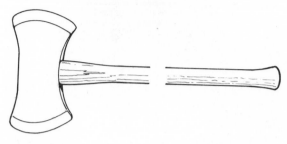

Fig. 64

Axe, Dutch: see *Axe, Export.*

Axe, Export *Fig. 65*
Many English toolmakers make Axes for export, and some of these are occasionally found in English workshops. We have included some examples that are significantly different in appearance from those made for the home or North American markets. Judging by catalogue entries, South America must have been an important market for British toolmakers during the period 1890–1910. Some of the woods of those countries are extremely hard and would have provided a severe test of quality and workmanship.

(a) Australian
Heavy versions of Kent type.

Fig. 65(a)

(*b*) *African*
Long cheek and poll.

b

(*c*) *Black Congo*
Absence of eye suggests hafting by natives as for previous stone implements.

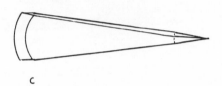

c

(*d*) *Ceylon* (Round Eye Axe)
Circular eye and no poll.

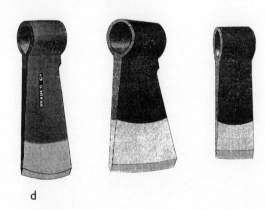

d

(*e*) *Chinese*
Wedge shaped with very flat cheeks.

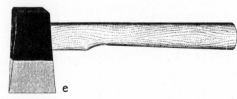

e

Fig. 65(b)–(h)

(*f*) *Dutch Side*
A flared blade with a pointed lug on lower edge, of unknown purpose.

f

(*g*) *Russian* or *Amoor Fantail*
Similar to the Dutch Side Axe with the pointed lug. (Amoor is possibly a mis-reading of the Amur region where the Siberian Pine comes from.)

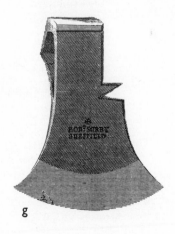

g

(*h*) *Siam*
Narrow parallel blade.

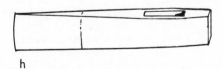

h

(*i*) South American

Most have straight-sided blades with no lugs and no thickening at the poll. The shape of blade and eye derives mainly from the Spanish and Portuguese tradition; some are listed as 'Basque' or 'Biscayan' patterns. They include:

(1) *De Tumba* – A Wedge type Axe used for felling.
(2) *Media Labor* ⎫ Side axes for hewing.
(3) *Labor Entera* ⎭
(4) *Guarabu Axe* made for Brazil.

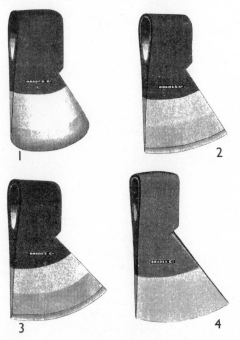

Fig. 65(i)

Axe, Felling (often pronounced and spelt 'falling') Felling Axes are distinguished by the greater length of handle (about 3 ft), which enables both hands to be used, and by the weight of the heads, which is greater than that of many other axes.

The Felling Axe is regarded as the tree feller's master tool, but at the present day it is very rarely used in Britain for complete felling. Instead its use is confined to the 'laying-in' and 'rounding', i.e. removing the spurs and forming a 'mouth' or 'sink' and a ledge on which the Felling Saw will rest when starting the cut. (See *Tree Feller*.)

In use, one hand grasps the base of the handle and the other is taken up the handle close to the head at the start of the stroke. The tool is then poised, aimed, and swung at the trunk, and as the head flies through the air, describing an arc with the base of the handle as the centre, the 'head' hand is allowed to slip down the handle towards the base. In this way the axe-man can use nearly every muscle of his arms, body, and legs, and put great force behind the blow.

The main types are described below.

Axe, Felling, English (Kent Felling Axe) *Fig. 66*
This group are mostly of the Kent type and are now subject to a British Standard Specification (2945:1958). The blade is shorter than the obsolete Long Felling Axe and differs from the popular Wedge Axe by having a flared blade and pointed lugs below the eye.

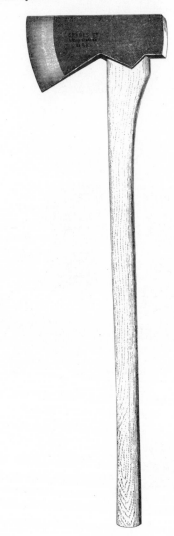

Fig. 66

Axe, Felling, Long (Pitching Axe; Merrick Axe; also called after certain counties, e.g. Northumberland, Cumberland, Westmorland or Yorkshire Axe) *Fig. 67*

This type has a narrow blade about 15 in long, sometimes curved longitudinally. The poll is usually square and the lugs pointed beneath the eye. Weight of head 3½–7 lb.

Though commonly used for felling before the middle of the nineteenth century, it is now practically obsolete. The long blade gave an advantage when undercutting the 'sink' (which was cut into the side of the tree to direct the fall), for it penetrated further; and it tended to produce a 'cupped out' stump which would hold rainwater and consequently rot quicker.

Variants include:

(*a*) Long Axe

(*b*) Merrick type

(*c*) Westmorland type (a so-called Scotch type is similar)

(*d*) Round lugged type

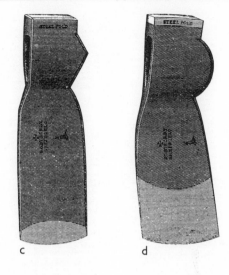

Fig. 67

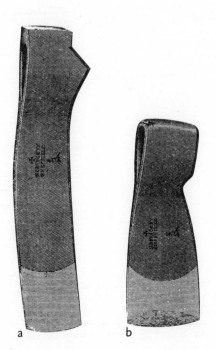

Axe, Felling, Wedge Type (American Axe; American Felling Axe; Square Axe; Yankee Axe; but see also list of variants below) *Fig. 68*

A plain, stocky, wedge-shaped blade weighing from 2 to 7 lb, with somewhat swollen sides and a heavy flat poll.

An outstanding characteristic of the Wedge Axe is the size of the poll; it is said that the heavy poll leads to a steadier and more reliable blow. The swollen sides make it easier to extricate the head when wedged tight in the cut.

Though used for many purposes, as a Felling Axe it has largely superseded all other types in England (as well as America) and its dimensions are the subject of a British Standard Specification (2945: 1958). 'It is curious to reflect', writes W. L. Goodman (London, 1964), 'that this, the final product of 6000 years of incessant trial and experiment, is perhaps the simplest and most functional of all.' One might add that axes made in stone by neolithic man bear a striking resemblance in outward appearance to the modern wedge axe; both are smooth wedges with swollen sides.

Note: Mercer (U.S.A., 1929) calls the wedge-type of Axe an American Axe, and states that it 'differed from the ancient axes not only conspicuously in appearance, but radically in construction; since, with the European ancestral instruments, the bit always outweighed the poll; in the new axe, the poll outweighed the bit'. But Mercer was presumably not aware that a Wedge Axe was illustrated in Smith's *Key to the Various Manufactories of Sheffield* in 1816.

55

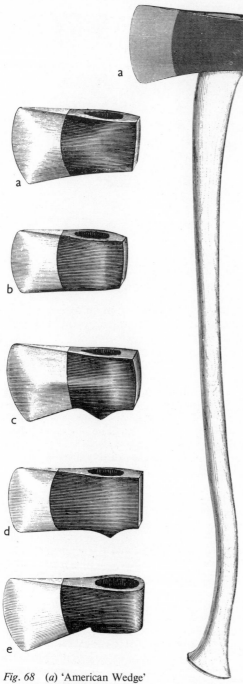

It is numbered No. 265 and has pointed lugs below the eye like the Kentucky Wedge Axe (*Fig. 68 (c)*). In the later copy of Smith's *Key* marked 'James Cam and Marches & Shepherd' (1787–1835), this same axe is listed as a 'Wedge Axe', and Plate 1061, described as an Improved Felling Axe, is practically identical with the modern Wedge Axe.

It would appear that fellers and other woodworkers in the U.S.A. and Canada must have taken a fancy to the Wedge Axe as shown in Smith's *Key* of 1816, probably because of its heavy poll, and ordered it from Sheffield or copied it themselves.

Variants are illustrated.

Axe, Fireman's *Fig. 69*

A flared blade, with a strong pick-poll, and with the head securely fitted to the handle with solid languets or straps. In another pattern the head is made solid with the handle, which has a scale grip. An early nineteenth century pattern was provided with a short saw which folded into the handle.

Fig. 69

Fig. 68 (*a*) 'American Wedge'
     (*b*) 'Canadian Wedge'
     (*c*) 'Kentucky Wedge'
     (*d*) 'Turpentine Wedge'
     (*e*) 'Spanish Wedge'

Axe, Gardener's (A combination tool): see *Hammer, Gardener's.*

Axe, Gent's

A name given to a number of small Axes intended for amateurs. (See, for example *Axe, Turner's.*)

56

Axe, Goose Wing *Fig. 70*

A name given to a Side Axe with a long blade shaped like a bird's wing, with the eye or socket near the lower, wider end. There is virtually no poll, and the cutting edge is usually bevelled on one side only. The handle is often off-set to prevent damage to the knuckles when the tool is used as a Side Axe for edge-trimming boards and planks. Common in Continental Europe from the Middle Ages onwards, and carried to the U.S.A. by immigrants.

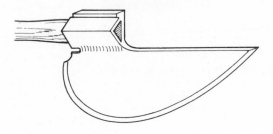

Fig. 70

Axe, Grafting: see *Grafter's Froe.*

Axe, Grubbing (Cross Axe; Grubbing Mattock; Two Bill) *Fig. 71*

A mattock type of tool with two blades set at right angles to one another – used for grubbing out tree roots and included here because it resembles a wood-worker's tool.

Fig. 71

Axe, Guildford *Fig. 72*

A Kent type of Axe but apparently heavier and wider.

Fig. 72

Axe, Hammer: see *Hammer-Pincers.*

Axe, Hand

A term often applied to lighter Axes and Hatchets used about the homestead, garden, farm, and work-shop.

Axe Handles: see under *Axe.*

Axe, Hedging *Fig. 73*

Hedgers use an Axe for heavy cutting when laying a hedge. The cutting strokes are directed upwards at first to avoid splitting the stool. The flat poll is used for driving in the stakes. Variants include:

(a) Common Hedging Axe

A head of the Kent type, but the blade is longer and narrower. There are pointed lugs above and below the eye. In the modern version, the blade is sometimes flared ('fantail').

(b) Yorkshire Hedging Axe

A head similar to the above, but there are lugs only below the eye.

a   b

Fig. 73

57

Axe, Hewing: see *Axe, Side.*

Axe Holders for grinding: see *Grinding Appliances.*

Axe, Hunter's: see *Hatchet, Hunter's.*

Axe, Hurdlemaker's: see *Hurdlemaker.*

Axe, Irish *Fig. 74*
A flared blade with straight sides and flat poll, and with pointed lugs above and below the eye. Weight 2–8 lb.

Fig. 74

Axe, Kent *Fig. 75*
Up to the end of the nineteenth century there were many regional variations in the shape of the axe head (the catalogue of Messrs. William Gilpin of 1868 lists no fewer than 47). These were often named by counties, such as the Suffolk, Norfolk, Yorkshire, and Lincolnshire, as well as Scotch, Irish, etc.; of these the Kent is almost the sole modern survivor. However, the name Kent has come to include a number of variations, though with the exception of the felling pattern all have a symmetrical, round-shouldered blade, a flat poll, and pointed lugs both below and above the eye. The weight varies from $1\frac{1}{2}$ to 8 lb.

Axes of the Kent type are illustrated under *Axe, Banbury; Axe, Felling, English; Axe, Guildford; Axe, Kent, Side; Axe, Mahogany, Squaring; Axe, Manchester; Axe, Norfolk; Axe, Plymouth; Axe, Suffolk; Hatchet, Kent.*

Fig. 75

Axe, Kent, Side *Fig. 76*
A Kent Axe ground with a single bevel for use as a Side Axe.

Fig. 76

Axe, Kentucky: see *Axe, Felling, Wedge type.*

Axe, Lathing: see *Hatchet, Lathing;* also *Froe.*

Axe, Lincolnshire (Lincoln Side Axe) *Fig. 77*
A long asymmetrical blade with a flat poll and pointed lugs below the eye. Weight 2–6 lb.

Fig. 77

Axe, Lopping: see *Axe, Topping.*

Axe, Mahogany, Squaring *Fig. 78*
A large wide-bladed Axe of Kent type, with the blade thinning gradually to the cutting edge. Used for squaring mahogany logs to save weight and space in shipping.

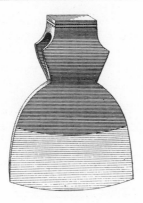

Fig. 78

Axe, Manchester *Fig. 79*
A Kent type of Axe but with the edges of the blade parallel.

Fig. 79

Axe, Marking: see *Hammer, Marking.*

Axe, Mast Maker's (Mast Axe; Mast and Spar Axe) *Fig. 80*
A long blade (up to 12 in) with a flared-shaped cheek and no poll. It is usually ground as a Side Axe. It is used for paring and trimming masts and spars. It is sometimes found with a shorter blade probably as the result of constant re-sharpening, which gradually reduces its length. When describing the long hours of labour making masts and spars, a King's Lynn shipwright (Worfolk Bros, 1966) told us that his axe 'shone like silver'.

The mast makers at H.M. Dockyard at Portsmouth (1969) do not use this type of Axe. Their Axes resemble the Ship Axe, with a blade measuring about 11 × 4 in, ground both sides, and with a square poll. The handles are 1 ft 11 in long, and they are held near the head. They were used to chop off the corners of square timbers and were followed by an Adze. (See *Mast and Spar Maker.*)

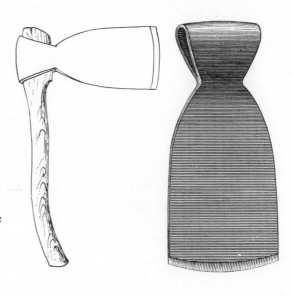

Fig. 80

Axe, Merrick: see *Axe, Felling, Long.*

Axe, Miner's Deputy (Deputies Axe; Scots: Doggie's Hawk) *Fig. 81*
A long blade, flat on the upper edge and rounded on the lower, which carries a notch for nail pulling. The poll is a hammer with a round face. Weight $3\frac{1}{2}$–7 lb. The flat upper edge enables the axe to be used close to the roof of the roadway.

According to Mr. F. A. Dixon of the National Coal Board (Gateshead, 1968) this axe was carried by the miners' deputy in order to 'cut and notch timber as required, and to erect or withdraw the timber props using the hammer end of the tool for this purpose. The deputy was responsible for the installation and maintenance of the light railway track in his section of the mine, and his axe and handle was used to measure the rail gauge of two feet, and the axe's hammer-poll to fix the steel rails by nails or spikes to the wooden sleepers. The small notch in the cutting blade of the axe was used to withdraw the nails or spikes when it became necessary to relay or repair the light railway track.' The top of the axe was also used to test the roof (see *Mining Carpenter*).

59

Fig. 81

Axe, Miner's Welsh *Fig. 82*
A long blade with square poll, flat top, and pointed lugs below the eye. Weight 2–8 lb.

Fig. 82

Axe, Mortising (Post Axe) *Fig. 83*
These Axes have long narrow blades suitable for chopping inside a mortice hole. The following are variants, but see also *Axe, Twybill* and Mortising Knife under *Hurdlemaker*.

(*a*) A parallel-sided blade 1½–2 in wide and 12 in long overall, with pointed lugs on both sides of the eye and a square flat poll. Weight 2–7 lb. One made with lugs below the eye only is known as a Single-Cheek Mortising Axe. Used for cutting large, rough mortices in posts, farm buildings, and the like. One method was to make two auger holes and then chop out the wood between them.

(*b*) A continental pattern with an 18-in blade. It is of lighter construction than (*a*) and is intended for paring the sides of a mortice rather than for chopping out the waste. It is still made in Germany. The nomenclature is uncertain: Mercer (U.S.A., 1929) calls it a Mortice Chisel Axe; Goldenberg (Alsace, 1875) calls it a 'Twybill, Alsation Pattern'.

(*c*) Another continental pattern with a 'pick' type head 18–30 in long. Schmidt (Germany, *c.* 1890) calls this axe a Cross Axe and Mercer (U.S.A., 1929) a Twybill.

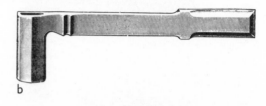

Fig. 83

Axe, Newcastle *Fig. 84*
The cutting edge of the blade is often out of square, and the poll is sometimes at an angle to the centre line of the head. Weight 1½–7½ lb.

Fig. 84

Axe, Newcastle Ship: see *Axe, Shipwright's.*

60

Axe, Norfolk *Fig. 85*
A Kent type of Axe but with a narrower blade with sloping shoulders.

Fig. 85

Axe, Northumberland: see *Axe, Felling, Long.*

Axe, Pennsylvanian Broad or Squaring
A name given to an American Side Axe with a very wide blade. See *Axe, Broad.*

Axe, Piano: see *Axe, Turner's.*

Axe, Pitching: see *Axe, Felling, Long.*

Axe, Plymouth *Fig. 86*
A Kent type of Axe but with a rather narrower blade.

Fig. 86

Axe, Pole: see *Froe.*

Axe, Poll: see *Axe, Butcher's.*

Axe, Post: see *Axe, Mortising.*

Axe, Quarter Moon *Fig. 87*
A name given to an Axe with a blade shaped like a quarter-ellipse. Weight 3–7 lb.

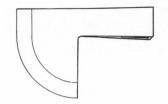

Fig. 87

Axe, Rafting *Fig. 88*
A short flared blade with a long square poll chamfered to form a hammer head, and with pointed lugs below the eye. Presumably made for export and used when building timber rafts for floating to the saw mill.

Fig. 88

Axe, Rending: see *Froe.*

Axe, Riving: see *Froe; Lath-making Tools.*

Axe, Round Eye
A name sometimes given to Axes with a circular eye and no poll. (See, for example, Ceylon or South American Axes under *Axe, Export.*)

Axe, Rounding
A term sometimes applied to Felling Axes used for 'rounding' or facing the foot of a tree, i.e. removing the spurs and forming a 'mouth' and a ledge on which the Felling Saw will rest when starting the cut.

Axe, Russian: see *Axe, Export.*

Axe, Scaffolder's: see *Hammer, Scaffolder's.*

Axe, Scotch *Fig. 89*
The ordinary pattern has a straight-sided blade with rounded lugs above and below the eye. Weight 2–8

lb. The Felling Pattern has pointed lugs below the eye and is similar in appearance to the Yorkshire Axe.

Fig. 89

Axe, Shipwright's (Boatbuilder's Axe; Ship Axe; Ship Carpenter's Axe) *Fig. 90*
These have long, heavy blades weighing up to 8 lb and are used for paring and trimming ship's timbers. They are to some extent interchangeable with the other special Axes used in the shipyards, i.e. the *Mastmaker's* and *Blocking Axes*. Variants include:

(*a*) *Ship Axe*
A round-shouldered blade *c.* 11–12 in long. It is usually made with double pointed lugs and a square poll, and is sometimes ground as a Side Axe. Weight 2–7 lb.

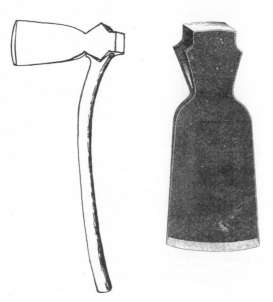

Fig. 90(a)

(*b*) *Newcastle Ship Axe* (Single-Cheek Ship Axe; Wear Ship Axe)
Similar to the Ship Axe and used for the same purpose. It has a rather longer blade with pointed lugs below the eye. Weight 2–7 lb. The longer blade may be an attempt to combine the Ship and Blocking Axe in one tool. When ground on one side only it is sometimes known as a Ship Side-Axe.

Fig. 90(b)

(*c*) *American pattern Ship Axe*
The blade is a little shorter than the English Ship Axe, with double-lugs, rather angular shoulders, and sometimes a flared poll.

Fig. 90(c)

Axe, Siam: see *Axe, Export.*

Axe, Side (Hewing Axe; Broad Axe) *Fig. 91*
A name given to Axes whose cutting edge is bevelled only on one side. This is normally the right-hand side away from the work (as marked by 'X'), but Side Axes ground and bevelled on the reverse side are made for left-handed users, or kept for those occasions when it is necessary to work on the other side of the timber. In order to avoid damaging the knuckles, the handles are often fitted at an angle to the cutting edge or face, either by off-setting the eye, or curving the handle sideways (as illustrated), or both. Such handles are known as 'bent' or 'swayed'.

The Side Axe was one of the main tools of the medieval builder and is still in occasional use, e.g. on the water front, or for squaring-up round timbers in the woods for the saw pit or saw mill. One advantage of the Side Axe over the Saw is its ability to follow the angles or taper of the timber, thus avoiding waste and retaining full strength.

For examples see the following: *Axe, Blocking; – Coachmaker's; – Cooper's; – Goose Wing; – Kent Side; – Mastmaker's; – Squaring; – Wheelwright's.*

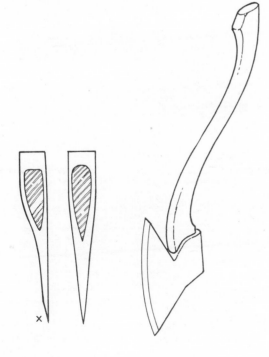

Fig. 91

Axe, Snedding: see *Axe, Topping* and *Axe, Trimming.*

Axe, Socket *Fig. 92*
A term used to describe an Axe, usually imported from the Continent, in which the handle is fitted into a tube-like socket forged on the back of the blade, instead of being secured in the eye of the blade itself. Illustrated by Goldenberg (Alsace, 1875) examples are described as a 'Paring Axe of Dutch pattern' and a 'Socket Paring Axe'.

Fig. 92

Axe, South American: see *Axe, Export.*

Axe, Spanish: see *Axe, Felling, Wedge Type,* also *Axe, Export.*

Axe, Split
An alternative name for a *Froe,* also a term used by Mercer (U.S.A., 1929) for a 'heavy, iron wedge-shaped, dull-bladed, sledge axe'. He also gives this tool the German name Holzaxt. It is used for very rough work such as opening a split when the wedges have become fast in the log, or for driving iron wedges as a substitute for the Beetle.

Axe, Square: see *Axe, Felling, Wedge Type.*

Axe, Squaring: see *Axe, Blocking* and *Axe, Side.*

Axe, Suffolk *Fig. 93*
A Kent type of Axe but with a narrower blade.

Fig. 93

Axe, Sugar: see *Hatchet, Sugar.*

Axe, Topping (Lopping Axe; Snedding Axe) *Fig. 94*
Various Axes, including Felling Axes, can be used for lopping and cutting away branchwood after felling (a process known as 'snedding'), but the following Axes are made specially for this purpose. ('Lop and Top' consists of branchwood and tree tops, usually tied up into cords, or burnt.)

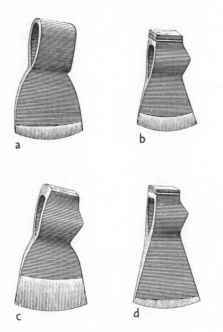

Fig. 94   (*a*) Chart's Lopping or Topping Axe
       (*b*) Wantage Lopping Axe
       (*c*) Chichester Lopping Axe
       (*d*) Marychurch's Lopping Axe

Axe, Trade
A term used by Mercer (U.S.A., 1929) to describe Felling Axes of the seventeenth century first sold to Indians by European traders. The blade is 8–9 in long, flat topped, but flaring out below, with no poll. Several thousands of these Axes have been found on Iroquois dwelling sites in New York State and Canada.

Axe, Trimming (Hand Axe)
A term used to describe a short-handled Axe weighing not more than 4 lb, and often less, and with heads of varying design including the Kent or wedge types. They are used by foresters to cut away small branches from the stem, and for felling small trees and underwood, and by other tradesmen for pointing stakes, cleaving, chopping fire-wood, and for many other purposes about the homestead, farm, and workshop.

Axe, Turner's (Gent's Side Axe; it is listed by Ward & Payne (Sheffield, 1911) as a 'Gent's Side or Piano Axe'.) *Fig. 95*
Small T-shaped Side Axe without poll and the handle 'caulked' at the foot. Presumably used by amateur woodworkers ('gents') and by turners for trimming. Its connection with piano-making remains a mystery.

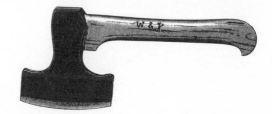

Fig. 95

Axe, Turpentine: see *Axe, Felling, Wedge Type.*

Axe, Two Bitted or Two Bill: see *Axe, Double-Bitted* and *Axe, Mortising.*

Axe, Twybill (Cross Axe; Two-Bill; Twibill; Twilbill; Twyvete) *Fig. 96*
Although fairly common on the Continent up to recent times, it is doubtful whether the Twybill has been used in Britain for woodworking for close on 300 years. Its reference as a carpenter's tool is given in the *O.E.D.* for 1686. The name Two-Bill apparently survived to modern times to denote a Hurdle-maker's Mortising Knife (see *Hurdlemaker*) and a semi-agricultural tool, the Grubbing Axe.

Fig. 96

Axe, Wear: see *Axe, Shipwright's*.

Axe, Wheelwright's (Wheeler's Side, Bench or Broad Axe) *Fig. 97*
A wide symmetrical Side Axe with lugs above and below the eye and a flat poll. Weight 3–7 lb. Variants are illustrated.

Wheelwrights used these Axes for trimming spokes especially the earlier 'plug' spoke with a tapered foot, and for paring down or shaping the sides of beams and boards. (Chopping is quicker than shaving.) The handle is often offset (bent to one side) so that the fingers are kept clear of the work. Wheelwrights also used the Coach Axe, especially for making wedges. (See *Axe, Coachmaker's*.)

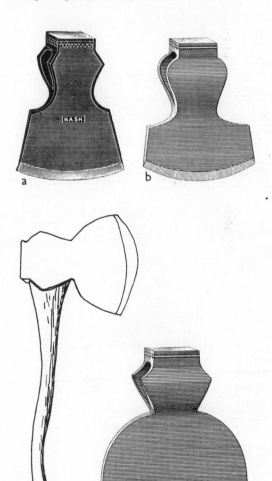

It is illustrated in use with a long handle for cutting mortices in one of Jost Ammann's woodcuts (Schopper, 1568), and is shown without a handle in the frontispiece to Diderot's 'Charpente' (1763).

The more recent examples consist of a long, double-bladed head with a tapering socket for a handle. No handle is inserted in the socket, for the tool is grasped by the socket itself and the long blade steadied with the other hand. The head measures 4 ft–4½ ft long overall; one blade is shaped and ground like a Mortising Chisel with the cutting edge at right angles to the socket, while the other has the form of a bevelled-edge Paring Chisel in the same plane as the socket. See also *Axe, Mortising*.

Fig. 97   (*a*) Manchester pattern
         (*b*) Newtown and Welsh Pattern
         (*c*) London pattern
         (*d*) 'Wheeler's' pattern

Axe, Wry-Necked
A term sometimes used for an Axe with an off-set handle. See *Axe, Side*.

Axe, Yankee: see *Axe, Felling, Wedge Type*.

Axe, Yorkshire *Fig. 98*
A round-shouldered blade with parallel sides, a flat poll, and pointed lugs below the eye. Weight 2–8 lb.

Fig. 98

Axle Key: see *Wrench, Coach Builder's*.

B

Backaroni Tool: see *Chisel, Carving (Fig. 204)*.

Back Check: see *Plane, Fillister*.

Back Iron (Cap Iron): see *Plane*.

Back Stop: see *Bench Stop*.

Badgered: see *Skewed*.

Bag Hooks *Fig. 99*
Hand hooks for lifting sacks and bales are often found in workshops.

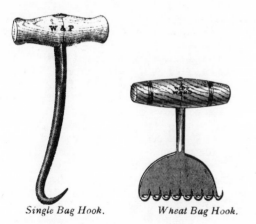

Single Bag Hook.   Wheat Bag Hook.

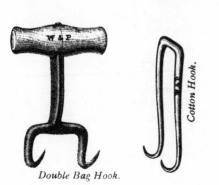

Double Bag Hook.   Cotton Hook.

Fig. 99

Balk (Baulk)
A squared log or hewn timber over 6 in square.

Bar Bender: see *Wrench, Bar Bending*.

Bar, Lengthening (Eke; Extension Piece): see *Cramp, Joiner's*.

Barefoot
U.S.A. term for the nose of an Auger or Bit with a cutting edge but with no protruding lead or spur.

Barge: see *Axe, Barge*.

Barking Iron (Wrong Iron; Bark Spud; Peeling Iron; Rinding Iron) *Fig. 100*
A blunt blade, often socketed, mounted on a wooden handle, or made in iron throughout. The blade is some 2 in across, and is either round, semi-circular, spade shaped, or shield shaped. It is wedge shaped in section, and usually rounded on the upper side for prising off the bark. The length varies from 30 in down to a smaller pattern about 10 in long which is sometimes called a Wrong Iron, since it is used to remove the bark from the 'wrongs', i.e. smaller branches or crooks. The handle of these smaller socketed irons is often a piece of naturally forked branch which gives a good grip. The longer tools have a plain knob, T-handle, or plain iron ring. Barking Irons were used for removing bark which was sold for tanning leather. They were also used for stripping bark from logs to be used as posts etc.

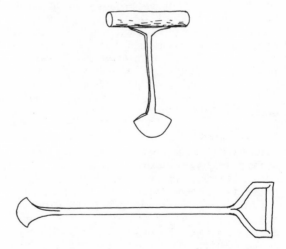

Fig. 100

67

Barrel Jarvis: see Downright under *Shave, Cooper's.*

Basil
A sloping edge; e.g. the grinding bevel of an edge tool.

Basket Maker (Osier) *Fig. 101*
The weaving of basket work has been carried on from prehistoric times to the present day, and the materials and methods used have varied little. Wicker-work chairs of circular shape with high backs were popular in Roman times and are still made today; hurdles made of interwoven rods were used before the Christian era for hut-making.

In addition to the almost infinite variety of baskets for household and garden, perhaps the greater part of the basket-maker's production goes to agriculture and the retail trade – fishermen and fishmongers, butchers and bakers, vegetable and fruit growers. There are many specialised uses such as basket-work furniture; travelling hampers for the theatrical companies; the great circular baskets for kettle drums; the passenger-carrying balloon basket. Even horse-drawn carriages were sometimes made in basket work.

The trade has so far escaped all efforts at mechanisation, but the demand for housewives' baskets has been greatly reduced since plastic baskets and bags came on the market.

Materials
Most English baskets are made from different varieties of willow ('withy') grown at one time in osier beds in many parts of this country; now, much of the material is imported.

The withies used for weaving baskets are called rods. The stouter sticks used as a framework are called stakes. The ribbon-like material known as skein (made by splitting a rod into three parts) is used for light baskets and for binding the handles. Cane from abroad is also used for this purpose, owing to its hard-wearing properties.

The workshop
Many basket makers work in a shed or back room of their own house, usually unfurnished except for a bench along one side. The basket maker often sits on the floor on a few sacks, with his back to the wall. His bench is the sloping Lap Board placed across his knees, or set between his legs and supported on a block. Bundles of willow (bolts) and most of his tools lie ready to hand on the floor; but the shaves used for trimming the skeins may be found on the bench, for they are held in a vice when the skeins are pulled through them.

Why the work is done sitting on the ground is not altogether clear, for it would seem possible to do it in a more comfortable position. However, basket makers assert that to sit at a bench would be 'impossible', for it would impede free movement and one could not reach all round the basket nor reach high enough to finish the top of a basket. Like the tailors and other tradesmen (such as slate splitters) who work sitting cross-legged, the posture may be traditional as well as functional.

The process
After harvesting, the removal of the bark is done on a Peeling Brake; but if green-coloured rods are wanted, the withies are boiled in water before peeling so that the colour from the bark enters the wood. Back in the workshop the rods are made pliable before use by being laid in the Soaking Trough.

The baskets are made by weaving fine rods or skeins on a frame of stouter rods. The first process – indeed the beginning of every basket – is to make the base. In the case of round and oval baskets this consists of a cross-work of rods (the 'slath') which are bound together and then spread out like the spokes of a wheel. Weaving begins around the slath; more rods are inserted as the work proceeds, and these are bent upwards to form the side frame on which the finer rods continue to be woven. The upright stakes can be held to the desired size by means of a loose withy hoop which can be moved up or down as the work proceeds. When the required height is reached, the upright stakes are turned over and interlaced to form a strong rim.

In the case of rectangular baskets, the base rods, known as bottom sticks, are held in a Basket Maker's Block made of two wooden bars bolted together at their ends. Thinner rods are then woven between the upright sticks. When a basket is finished all exposed rod ends are cut off with the Picking Knife.

Tools used include:

(a) Cleaver (Basket maker's Cleave)
A small egg-shaped piece of boxwood or horn about 4 in long, cut away at the narrower end to form three or four straight vanes or wings with sharp edges, sometimes tipped with brass inserts. Used for splitting down willow rods to make the skeins which are used for light baskets and for binding handles etc.

Note: A larger size of Cleave called a Bond Splitter is used for making the bands for tying birch brooms etc. See *Woodland Trades.*

(b) Commander
A solid iron bar about 18 in long overall, with a

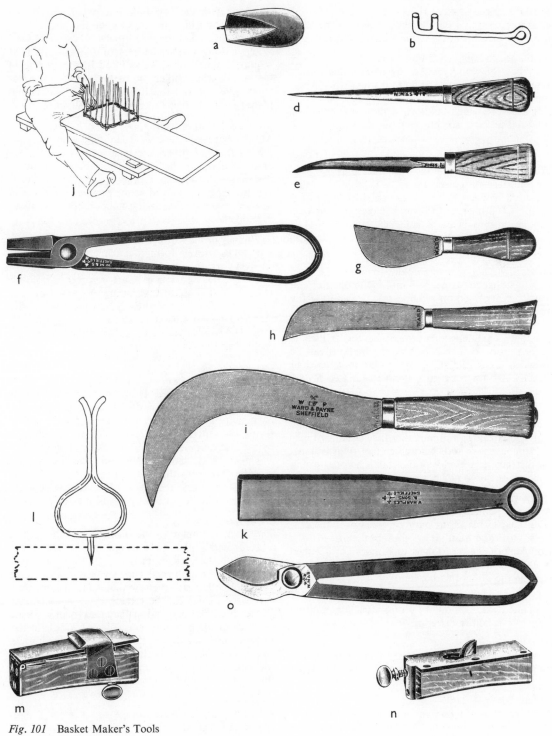

Fig. 101   Basket Maker's Tools

U-shaped piece fitted at right angles on one or both ends like a smith's Bar Bending Wrench. The short arms are used for straightening stakes and rods.

(c) Block or Cramp [Not illustrated]

Two bars of wood, bolted together at their ends. Used for holding erect the bottom sticks when weaving the base of a square or rectangular basket. A single bar, in which holes are drilled at intervals to hold the sticks, is sometimes used instead.

(d) Solid Bodkin

A round, tapering spike, 4–10 in long, of polished iron with a wooden handle. Its chief purpose is to open the weave to insert the rods. It has many other uses: for instance, it can be driven sideways with a small mallet in order to bring a stake into line or a handle-rod into place. A somewhat thicker tool of about the same length, known as a Staking Bodkin, is used for inserting the upright rods or stakes.

(e) Hollow Bodkin (Bent Bodkin; Shell Bodkin)

A gouge-shaped blade, 4–8 in long, pointed, and usually bent. Used on repair work to facilitate the insertion and weaving-in of new rods. The bent blade is passed under alternate stakes, and this provides a smooth channel along which the rod (which is pushed in towards the point of the bodkin) can be threaded under the stake. A similar tool, used in binding besoms, is called a Bond Poker (see *Broom Maker*).

(f) Kinking Tongs (Cane Squeezer)

Iron tongs about 13 in long overall, one jaw provided with a sharp ridge and the other flat. Used for kinking stout canes or rods before bending. This is done, for instance, to the ends of the stakes before turning them off at the top or bottom of the basket.

(g) Picking Knife

A stubby, pointed blade, about 3 in long. The 'goose wing' shaped blade has hardly altered over the centuries; with the Maul, it appears on the arms of the Company of Basket Makers of London, established in 1569. It is used for trimming off the ends of the rods on the unfinished basket.

(h) Shop or Hand Knife

A pointed blade, about 7 in long, used for pointing willow rods before insertion.

(i) Willow Knife

A curved, sickle-like blade $9\frac{1}{2}$ to 10 in long, with the tang of the blade riveted through the handle. Used for harvesting withies in the coppice.

(j) Lap Board

A board measuring about 4×2 ft and used as a bench by basket makers. The basket maker sits on the floor with the Lap Board across his knees and sloping away from him. Some keep the board propped up on a block between their outstretched legs. When a long run of baskets of the same type is required, a nail or spike is fitted in the middle of the board, which holds the basket but leaves it free to revolve. An old flat iron or stone is sometimes placed in the basket to keep it from slipping about.

(k) Maul (Beating Iron)

An iron bat-shaped tool, 7–10 in long, with one edge thicker than the other, and tapering towards the eye or knob serving as the handle. Used for beating down the weave during the course of making a basket. When a ring is provided at the handle end, this can be used as a Commander (see above) for straightening stakes.

(l) Peeling Brake (Stripping Brake; Stripper)

This takes various forms but consists essentially of an iron fork mounted on a beam. The prongs are close together and spring apart if a rod is inserted at the top. It is used for stripping the bark from the rods. The bark of the withy is first loosened at the butt end by pushing it between the prongs of the fork. The rod is then pulled through the brake, thus peeling off the rest of the bark.

(m) Basket Shave

A wooden stock $4\frac{1}{2}$ in long and $1\frac{1}{2}$ in square, with a loose steel sole hinged at one end. A small blade fixed in a slot in the middle of the sole or passing right across, bent round and screwed to the side, forms the cutter. The thickness of cut is adjusted by a thumbscrew regulating the slope of the sole.

The stock is held in a Vice and the skeins (previously cleft from the willow rods with a Cleaver) are pulled through the Shave. This reduces them to flat strips suitable for light baskets and for binding.

(n) Upright Shave

The stock is similar to the Basket Shave but it has a fixed metal sole. Two upright blades, set at an angle, with sharp edges, project from the sole. These blades are the upturned ends of a U-shaped steel bow hidden beneath the sole plate. The distance between the blades can be reduced by a thumbscrew (projecting from one end of the stock) which forces the ends of the bow together. There is a hole in the stock behind the blades to release shavings. It is held in the Vice like the Basket Shave and used for trimming the skeins to the same width throughout their length.

(o) Shears

A heavy pair of shears from 8 to 14 in long, with curved jaws. Used for cutting rods to length.

(p) Soaking Trough [Not illustrated]
A strong wooden trough, measuring about 12 ft ×
14 in and 2 ft deep, used for soaking the willow rods
to make them pliable and easier to handle.

Basket Maker (Spale) (Spelk Basket Maker; Slop or Skip Maker)

Like osier basket-making, the weaving of baskets
out of strips of wood, known as spale, has been
carried on from prehistoric times. The material still
used in this country is coppice-grown oak, about
6 in. in diameter, and hazel, oak, birch, or ash rods
for the frame or 'bool'. The oak poles are sawn up
into suitable lengths and then boiled and cleft into
quarters. They are split ('riven') with the Froe into
strips from 1 to 3 in wide and about 1/16 in thick,
and finally trimmed with a Drawing Knife on a
Shaving Horse or Brake.

A special plane was made for making wooden
strips, but it was probably used only for softer woods
– see *Plane, Scaleboard*. For other tools used in this
trade, see *Woodland Trades*.

Basket Maker (Trug)

The modern trug consists of two frames, one for the
rim and the other for the handle, of cleft ash or
chestnut, with the bark left on the outside and the
inner surface shaved with the Drawing Knife and
Spokeshave. After steaming, the frames are shaped
in a Setting Frame and the ends nailed together.

The bottom of the trug is made of white pollard
willow, cleft into strips and smoothed and shaped
in a Shaving Horse. After steaming and bending,
the strips are nailed inside the frame to form a
boat-shaped basket, each board slightly overlapping
the other one. Cross pieces are nailed to the bottom
to give the basket stability.

Bass: see *Tool Basket*.

Bastard

A term applied to files – those with serrations of
medium coarseness.

Batter: see *Mallet, Woodland Worker's*.

Bead and Beader

A small rounded moulding. See also *Plane, Moulding*;
Router, Coachbuilder's; *Router, Metal Types*.

Beam Drill: see *Drill, Press*.

Bearing

Of tools and machines – a support provided to hold
a revolving shaft.

Beater: see *Maul*.

Beck Iron (Bick Iron): see Anvil under *Cooper (3)
Hooping Tools*.

Bed Key (Bed Wrench) *Fig. 102*
Not a woodworking tool, but often found in older
workshops. Made in many different patterns, it has
two or more arms radiating in the form of a cross
or star. Some of these arms contain sockets for turning nuts, and another arm may end in a turnscrew,
plain or forked. In another pattern there is an arm
containing a tapered slot for gripping nuts of various
sizes. Richard Timmins (Birmingham, *c.* 1800) illustrates a Bed Key with a long-headed Claw Hammer
at one end.

They were used for turning the bed-bolts which
secured the head and foot-board of a bed to the
side rails, and enabled the bed to be taken to pieces.
(See also *Bit, Bedstead*.)

Fig. 102

Beetle (Beddle; Bittel; Commander; Stake Driver;
Wedge Bittle) *Fig. 103*
A large, cylindrical, wooden head about 1 ft long
and 6 in. in diameter, with metal rings round each
end to prevent splitting. The handle is about 3 ft
long. The head is normally of elm, but sometimes
of apple or beech. Used by foresters, woodland
workers, and other tradesmen for driving fence posts,
for driving wedges when cleaving or tree felling, for
framing up timbers, and for other similar purposes.

Pavior's Beetles (used for levelling flagstones), also
of wood, are much bigger, with heads 8–10 in. in
diameter and without rings.

Thomas Tusser (London, 1573) writes of a 'Dovercourt Beetle'. This is taken to refer to the celebrated

elm trees of Dovercourt in Essex, which were considered exceedingly durable and suitable for making Beetle heads.

Mercer (U.S.A., 1929) gives the name Commander to a very large, heavy, wooden Beetle with its head made from a natural log. It was used for framing up timber structures. The same word is used by Moxon (London, 1677) for a 'very great wooden *Malet*' used to 'knock on the Corners of Framed Work, to set them into their position'. The term is also used by basket makers for a wrench-like tool used for straightening the rods.

(*Note:* Mallets and Mauls of smaller size are described under *Mallet, Woodland Worker's.*)

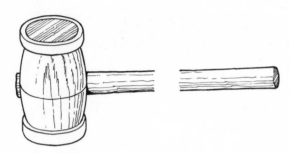

Fig. 103

Bell Hanger

Before electric bells were introduced, most ordinary house bells were operated by wires which ran from a pull in each room (or at the front door) to a group of bells situated in the servants' quarters. Each room was represented by its own bell, which after being 'pulled' continued to rock to and fro on its C-spring, thus indicating the origin of the call. The wires were guided through holes in the walls by various fittings including a triangular bell-crank. The bell hanger installed this apparatus; his special tools included the following: *Auger, Brick*; *Bell Hanger's Jumper Drill*; Bell Pliers (under *Pliers*); *Gimlet, Bell-Hanger's.*

Bell Hanger's Jumper Drill *Fig. 104*

An iron bar, about 2 ft long, drawn down at one end to form a flat pointed cutter. A short peg, forged at right angles near the other end, serves as a handle. Used by bell hangers for boring holes in plaster, brick, etc. to take bell-pull wires. The head of the tool is struck with a Hammer while it is rotated from side to side by means of the projecting handle.

Fig. 104

Bellied

A term used by tool makers for a convex edge or surface, e.g. of a Cross-Cut Saw, or the swollen sides of a handle.

Bellows, Cooper's

A pair of 'fire-side' type bellows, about 12 in. in diameter, but circular in shape and with a tapered spout placed centrally in one of the circular side-boards. According to information from the Cooper's Shop at the Mystic Seaport Museum (U.S.A.), they were used 'to test the barrel for being watertight'. The inside of the barrel was 'sloshed with water', the head put on, and air pumped through the bung hole with the Bellows. If water bubbled along the seams, it was not tight. We have not heard of this process in Britain.

Bellows, Piano: see *Pianoforte Maker.*

Bench: see *Bench, Woodworker's*; *Chair Maker* (3) *Workshop Equipment* (Framer's Block); *Clave*; *Cooper* (2) *Furniture*; *Sailmaker*; *Wheelwright's Equipment* (1) *Workshop Furniture.*

Bench, Woodworker's *Fig. 105* (See also *Vice, Woodworker's.*)

An early form of Woodworker's Bench used by the Romans consisted of a stout plank on four splayed legs. The work was held on the bench-top by means of L-shaped Holdfasts, or by pins driven into the bench top. There is no evidence that the Romans used their knowledge of the screw to make a Bench Vice. This simple type of Bench continued in use throughout the medieval period, but by the seventeenth century it was replaced by a massive framework of square, vertical legs, sometimes strengthened by cross-rails. Up to this period the Vice was not used. But by the seventeenth century one method of supporting a board laid on its edge for trueing was an L-shaped block of wood (a 'cleat') nailed to the left end of the bench top, with a wedge to secure the workpiece and a peg in the bench leg to support it from below. There was a row of holes down the front legs of the bench to take the peg according to the width of the board. Some time later, a screw passing through the cleat replaced the wedge for holding the workpiece tight.

The next stage was to dispense with the L-shaped cleat and to use instead moveable vice-cheeks (also called 'chops') drawn in towards the Bench by wooden screws which engaged in wooden nuts under the top of the Bench. The horizontal vice-cheek was either operated by two screws or was kept parallel by runners on each side of a single screw.

Since the eighteenth century, a Vice with a long, vertical, wooden cheek has been common in this country, either attached to the front of the Bench or standing separately – the Post Vice.

On the Continent of Europe an additional L-shaped Back or Tail Vice was fitted to the right-hand end of the bench top. (Benches so fitted are usually called German Benches.) By means of a stop fitted into one of a row of holes on the Bench and a similar stop on the Back Vice, work of any length could be held flat on the bench top for planing or other work. It may be wondered why the great convenience of this device for holding the workpiece never seems to have been fully accepted in this country.

Vice screws are now usually of steel, and various types of quick-action Vices with wood-faced steel jaws have replaced the upright wooden vice-cheek.

Devices for holding work flat on the Bench for planing are described under *Bench Stop* and *Hold-fast*.

Bench tops are normally made of timber at least 2 in thick, so as to be firm under hammering and also to allow for periodical planing off as they become uneven through wear. They often have a 'well' running parallel with the back edge, in which tools can lie below the working surface.

Bench Chops

A Bench Vice consisting of two cheeks connected by two wooden screws with wooden tommy-bars. It is used loose, or it can be fixed to the bench with thumb screws.

Bench Hook (Sawing Board; Side Hook) *Fig. 106*

A board of hardwood about 10 in by 6 in with blocks fitted at top and bottom at opposite ends, secured by wooden pins or dowels. The Bench Hook is held against the side of the bench or in the vice by means of the lower strip; the workpiece to be sawn is held firmly against the upper strip, which is frequently made shorter than the width of the board to prevent damage to the bench.

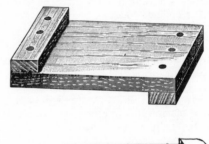

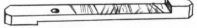

Fig. 106

Fig. 105

Bench Knife: see *Bench Stop* and *Knife, Bench.*

Bench Screw: see *Vice, Woodworker's.*

Bench Stop (Bench Peg) *Fig. 107*
A wood or metal pillar which projects above the surface of the Bench against which the workpiece is held when being planed. There are two types in common use:

(*a*) A small block or peg of wood, sometimes fitted with a leaf-spring on one side to help hold it in position. Adjustable by striking the head or foot of the peg.

(*b*) A wood peg or a metal device with a projection (or jaw) whose height above the bench surface is adjustable by means of a screw. The jaw has a toothed or serrated edge.

(*c*) When the bench is not provided with a Tail Vice (see *Bench, Woodworker's*), various devices are used to hold the workpiece against the forward Bench Stop. These include an old broken table knife which is stuck into the surface of the bench, and an attachment called a 'Bench Knife Stop', which consists of a plate with projecting pins fitting into a row of holes along the bench top, with a cam-operated arm designed to press the work up against the stop.

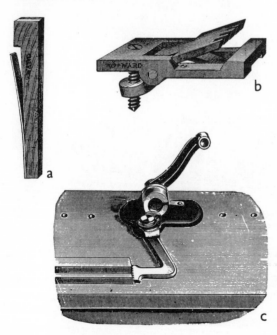

Fig. 107

74

Bending Tools: see Bending Iron under *Violin Maker*; Bending Mould under *Chair Maker* (*Equipment*); Setting Pin under *Handle Maker*; Wheelwright's *Equipment* (3) Tyring Tools; Wrench, Bar Bending.

Besom Grip: see *Broom Maker.*

Bettye: see *Saw, Bettye.*

Bevel: see *Square and Bevel.*

Bevel
An edge which is not at right angles; a sloping or canted surface (see also *Chamfer*); of an edge tool, e.g. Chisel or Axe, the grinding or sharpening bevel along the cutting edge.

Bevel Block (V-Block)
A holding device consisting of a block of wood with a V-shaped piece cut out of the upper surface. When laid on the bench it is used for holding a piece of timber while it has its corners planed off, e.g. when making whipple-trees, legs for milking stools, etc. (See also, *Rounding Cradle*.)

Bevel, Boxing: see Rim Gauge under *Gauge, Wheelwright's.*

Bevelling Tool: see *Cornering Tool.*

Bick Iron: see Anvil under *Cooper* (3) *Hooping Tools.*

Bilboquet: see *Saw, Armchair Maker's.*

Bilfie
Scots term for a heavy hammer used in shipyards. (See *Hammer, Ship.*)

Billet
A short piece of wood or metal roughly dressed, ready for finishing.

Bill Hook (Hand Bill; Bill) *Fig. 108*
A short-handled Axe with a long cutting face, sometimes curved at the end, and made in many different patterns. The handle is often caulked (swollen on the lower side at the foot) to prevent the tool from flying out of the hand. Though used mainly for cutting and laying hedges and for faggoting, it is included here because of its frequent use in other trades for harvesting the thinner material, for splitting poles when making hurdles etc., as a substitute for the Cleaving Adze or Froe, and for pointing stakes. (See *Woodland Trades.*)

John Evelyn in his *Sylva, or a Discourse of Forest Trees* (1664) writes, 'Unskilled woodmen and mischevious bordurers who go always armed with short hand-bills hacking and chopping all that comes their way'.

Tool makers of the nineteenth and early twentieth centuries, such as William Gilpin and Isaac Nash, listed up to a hundred different types of Bill Hooks, and even today the makers can supply a wide range. Whether differences in shape in general reflect any significant difference of function is doubtful, though in certain cases they do, e.g. a concave blade is obviously more suitable for cutting thin branches. It seems more probable that the remarkable variety in type was designed to meet a persistent local demand. Before the factories took over, farm workers and householders bought from a local tool-smith whose business was later captured by the factory only if the customer could still be supplied with the particular shape to which he and his forebears had become accustomed.

The illustrations represent some of the main varieties. The names are common, but by no means universal; and the Bill Hooks concerned may be found in areas well outside those to which the names apply.

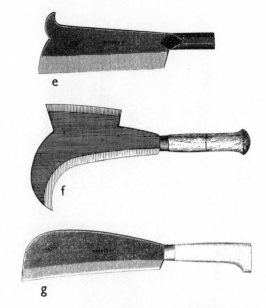

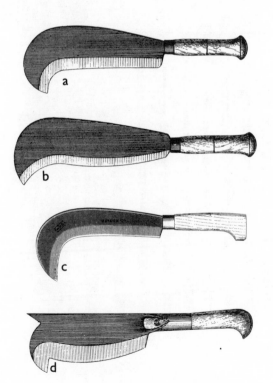

Fig. 108 (*a*) Dorset, Epsom, or West of England
(*b*) Bath, Ledbury, or Pontypool
(*c*) Spar or Hurdling. (The deeply curved blade is used for splitting poles and thatching spars.)
(*d*) Suffolk or Norfolk
(*e*) Hertfordshire or Block. (The backward pointing spike can be used for gathering the next piece of firewood for chopping.)
(*f*) Broom or Nottingham
(*g*) Rodding, Knighton, or Hereford

Billiard Table Bit: see *Bit, Screwdriver.*

Billot: see *Woodwind Instrument Maker.*

Billy: see *Jumper (Cooper's).*

Bit (Brace Bit) *Fig. 109 overleaf*
Note. The term 'bit' is sometimes applied to quite different tools, or parts of tools, such as the head of an Axe, or the head of a Soldering Iron or the cutting iron of a Plane or Router.

A Bit or Brace Bit is an interchangeable tool for boring, reaming, and other purposes, designed to be fitted into a Brace. A characteristic feature is the short, square-tapered tang; but some Bits have flat tangs for insertion into a wooden Pad or into the foot of a wooden Brace.

Bits for use in the spring-latch Brace chucks of the early nineteenth century had a small nick filed on one face of the tapered tang. As the size and position

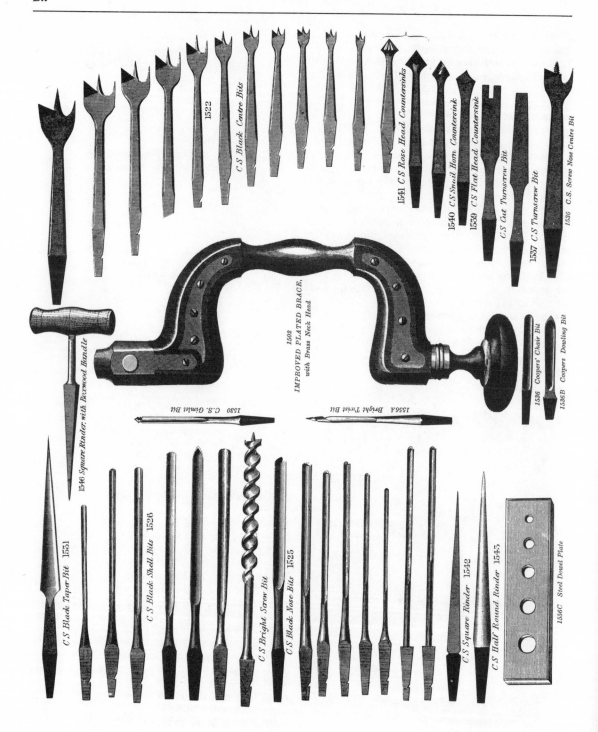

Fig. 109 Brace Bits (Sheffield Illustrated List, 1888)

of the latch in the chuck differed for each particular Brace, a set of Bits could only be used for one tool. This disadvantage was eliminated by the introduction of the Spofford and Barber Braces about 1868, but Braces with latch chucks were still being used by tradesmen up to the beginning of this century. (See *Brace*.)

Until recently, the trade catalogues offered Brace Bits as either Black, Bright, or Straw coloured. The Sheffield *List* of 1888 prices twelve assorted cast steel Bits as follows: Black 4/8ᵈ, Bright 5/3ᵈ, Straw coloured 5/11ᵈ. The term 'straw colour' may relate to the colour produced when tempering the steel to the proper temperature, and such Bits may have been preferred because the straw colour indicated that the tempering had been properly done. We have not found a better explanation, nor do we know the reason for the increased price – unless it was owing to the need to polish the Bits after hardening and before tempering.

There are a surprising number of different Bits but the chief types are described under the following entries: *Auger, Twist* (for *Auger Bits*); *Bit, Centre*; *Bit, Countersink*; *Bit, Gimlet*; *Bit, Nose*; *Bit, Shell*; *Drill, Twist*.

Bit, Adjustable: see *Bit, Expansive*.

Bit, Annular (Crown Saw Bit; Plug Bit; Tank Bit) *Fig. 110*
The following Bits are very small versions of the Crown Saw:

(*a*) *Annular Bit* (Pivot Bit)
A small Crown Saw, with a pointed, spring-loaded plug inside which prevents the Bit from wandering. Used for making cylindrical pegs from a board. A similar tool of larger size, without the plug, is made in the form of an Auger.

(*b*) *Broken Screw Borer* (Pivot Bit; Hollow Drill)
A small Crown Saw, often made by the worker himself out of a piece of iron tube with teeth cut on one end. Used by cabinet makers and others to remove the wood round an old nail or screw that has rusted or broken in. After this is done, the head of the nail or screw can be grasped for removal.

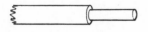

Fig. 110

Bit, Archimedean Drill *Fig. 111*
So called because of their use in Archimedean Drills, these small Bits have round shanks and the cutting nose is flat and ground on either side to a V-shaped point. The two facets of this point are sometimes bevelled on both sides thus enabling the drill to cut (or rather scrape) with equal facility in both directions. They are thus suited to drill-stocks which operate with a back-and-forth action, including the Archimedean, Centrifugal, and Bow Drills.

Fig. 111

Bit, Auger
These include the following types of Twist Bits: Jennings, Scotch, Irwin or Solid Centre, Gedge, Solid Nose, Single Twist or L'Hommedieu. For description and illustration see *Auger, Twist* and *Bit, Twist*. Other Twist-type Bits are listed under *Bit, Twist*.

Bit, Bedstead
A strong Nose Bit, about 8 in long and ½ in. in diameter, with the flute carried down only half way, of the type shown in *Fig. 109, No. 1525*. These Bits were listed in Smith's *Key* (Sheffield, 1816) and by Wynn Timmins (Birmingham, 1892), but we have not found an example in any list of later date.

They were probably used for boring holes through the bed post into the ends of the side rails. This was done to take the bed-bolts which held the head-and-foot boards to the rails and enabled the bed to be taken apart. (The Wrench used for turning these bolts is described under *Bed Key*.)

Bit, Bobbin (Scots: Bobbin Swarf or Scillop) *Fig. 112*
Nose or Spoon Bits used for boring holes through wooden Bobbins and made in diameters from $\frac{9}{16}$ to $1\frac{1}{2}$ in. Some are slightly tapered; some have a Parrot-type nose (as illustrated); many are made with round shanks for use in a machine.

Fig. 112

Bit, Brace
A term sometimes used for any kind of Bit used in a Brace.

Bit, Brace Drill: see *Bit, Smith's Drill*.

Bit, Brick *Fig. 113*
A Twist Bit with a plain V-shaped nose. Used for boring holes in brick or soft stone (see also *Auger, Brick*; *Bit, Stone*).

Fig. 113

Bit, Broken Screw or **Nail Remover:** see *Bit, Annular*.

Bit, Broughton Pattern: see *Bit, Countersink*.

Bit, Brushmaker's *Fig. 114*
A short round-nosed Spoon Bit with a flat tang similar to a Chair Bit, made in sizes up to ½ in. in diameter. For method of use and illustration see *Brush Maker*.
　Note. Smith's *Key* (Sheffield, 1816) and Richard Timmins (Birmingham, *c.* 1850) illustrate two distinct types. One is the Bit described above; the other resembles a rather narrow Taper Bit.

Fig. 114

Bit, Bullnose: see Bit, Solid Nose under *Auger, Twist*.

Bit, Button
Described by Mercer (U.S.A., 1929) as having two outer spurs and a central pointed lead like a Centre Bit. The cutters are flat and bevel-sharpened to smooth the surface of the button; the spurs make a circular incision and release the button-discs from thin plates of bone or wood.

Bit, Caning
Listed in the Sheffield *List* of 1888 as Nose, Shell, and Spoon Bits. Probably used for boring frames of chairs for cane seats.

Bit, Car
U.S.A. term for a large sized Auger Bit of the Jennings Pattern, with a 12 in twist; diameter ¼–2 in and overall lengths up to 20 in. See also *Bit, Wagon Builder's*.

Bit, Carriage Maker's (Bit, Coach Maker's) *Fig. 130* (*a*) under *Bit, Twist*
A Twist Bit of Gedge pattern with a 12 in twist and up to 1 in diameter. Used by coach builders and other tradesmen when a long, strong bit is required.

Bit, Centre *Fig. 109 No. 1522* on page 76; *Fig. 115*
A flat nose, with one half bent upwards to form a cutter almost at right angles to the axis of the bit; the other half is prolonged downwards on the outside edge to form a sharp spur-cutter or nicker. There is a spike or screw lead. In use the nicker describes a circle round the lead, and the cutter pares away the surface inside the circle. Used for boring clean, comparatively wide and shallow holes from ¼ to 3½ in. in diameter.
　Until the more general use of Twist Bits, the Centre Bit was the commonest woodworking Bit, probably because it was fairly simple for local smiths to make. In France, Centre Bits are sometimes known as 'English three-point bits', which suggests an English origin. There are several variants:

(*a*) For sizes over 2 in a second nicker inside the first is sometimes provided to break up the shaving and make the work easier.

(*b*) The so-called Screw Pin or Screwpoint centre Bit.

(*c*) A modern version has a half turn of helical nose like a Jennings Twist Auger but with a single spur. It is made in sizes ¼–2 in and is used for rapid working in soft wood.

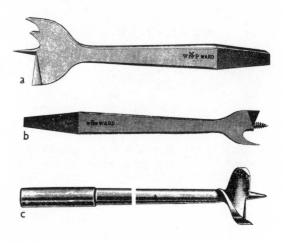

Fig. 115

Bit, Centre Nose Dowling: see *Bit, Dowel*.

Bit, Centre Plug: see *Bit, Deck Dowelling* (for Shipwright's tool) or *Bit, Cock Plug* (for Cellarman's tool).

Bit, Chairmaker's (Duck Bill Bit) *Fig. 116*
A short Spoon Bit, made up to about 1 in. in diameter, with a rounded nose. The tang is flat for wedging into a separate wooden pad, or directly into the foot of a wooden Chairmaker's Brace, one pad or Brace being kept for each size of Bit. They were used for boring chair parts to take legs, stretchers, sticks, or dowels.

Mr. L. J. Mayes (High Wycombe, 1960) wrote to the author as follows concerning the Chair Bit: 'Very nearly a perfect tool. No modern boring device known to the writer has all its advantages. These are: (1) It will bore to within a fraction of an inch of the full thickness of a piece of wood without breaking through. (2) A simple line can be filed across the back of the bit to form a depth gauge. (3) It does not split wood; there is no screw point to impose an arbitrary rate of feed. (4) It will start at virtually any angle – most important when boring a bow for the reception of the sticks in a Windsor chair, for example. (5) It will bore a true hole with a to-and-fro motion of the brace – again very useful when boring in confined situations . . . . Many men today cherish sets of spoon bits and will refuse to part with them.'

The smooth holes it produces makes the Spoon Bit especially suited for boring the holes for cane-seat chairs, and when this tedious task was mechanised the Spoon Bit was retained.

Bit Sharpener. A piece of triangular file sharpened to a point, and with the teeth ground off, was used by chairmakers to scrape the inner surfaces of the Bits to sharpen them. Continual sharpening steadily reduced the effective cutting diameter of the Bit, which explains why Spoon Bits were classified by use – Legging Bit, Stick Bit, Stump Bit, etc. - instead of by arbitrary sizes which would have been true only when the Bits were new. The change of size did not matter since the man who bored the holes also tenoned the material to fit them.

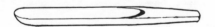

Fig. 116

Bit, Chamfer: see *Bit, Countersink*.

Bit, Clarke's Pattern: see *Bit, Expansive*.

Bit, Coachmaker's: see *Bit, Carriage Maker's*.

Bit, Cobra: see *Bit, Gimlet*.

Bit, Cock Plug (Plug Centre Bit; Tapping Bit) *Fig. 117*
A centre Bit $\frac{5}{8}$–1 in. in diameter, with a tapering skirt which acts as a plug to prevent the escape of liquid when boring into a full cask before fitting a tap. As soon as the Bit penetrates, the tapered skirt fills the hole until the stem of the tap is ready to be smartly screwed or driven into its place. Used by coopers and cellarmen.

Smith's *Key* (Sheffield, 1816) illustrates a Cock Bit in which the tapered body ends with a hollow tapered shell with a cutting nose.

Fig. 117

Bit, Cooper's (Cooper's Dowel or Dowling Bit) *Fig. 118*
A short Bit, usually of the spoon type with a pointed or blunt-pointed nose in sizes up to $\frac{1}{2}$ in. in diameter. The flat tang fits directly into the Cooper's Brace, one being kept for each size of Bit unless provided with interchangeable pads. It is also made with square taper tangs for use in modern braces. Those with more sharply pointed noses are sometimes ground flat on the face, leaving a sharp edge.

Cooper's Bits are not always of the spoon type: Smith's *Key* (Sheffield, 1816) lists them as Spoon or Nose Bits. James Cam (Sheffield, *c.* 1787–1835) illustrates an 'improved Cooper's Bit' which is a short Twist Bit with a Scotch type nose and point lead. These Bits are used for drilling the head sections of a cask to take the oak dowels which hold them together.

Fig. 118

Bit, Counterbore: see *Auger, Deck Dowelling*.

Bit, Countersink (Chamfer Bits) *Fig. 119*
These Bits are designed to produce a conical depression at the top of a previously bored hole in which the

head of a screw will lie flush with the surface of the wood. Variants include:

(a) Flathead Fig. 109 No. 1539
A flattened V-shaped head, with two cutting edges at an angle of 45° to the axis of the bit. These cutting edges are slightly hollow ground, which gives the head a twisted appearance. Intended for counter-sinking metal fixtures such as hinges.

(b) Rosehead (Cone Countersink) *Fig. 109 No. 1541; Fig. 119 (b)*
A conical head, with radial serrations varying in number according to the size and diameter of the head, ranging from $\frac{1}{4}$ to 1 in.

An earlier pattern has serrations on the head which are rather similar to the 'quarter dress' system of grooves in a millstone. Looked at in plan, there are four radial grooves which the intervening grooves ('harps') meet at an angle like the strings of a harp.

(c) Snailhorn Fig. 109 No. 1540
A conical head with only one or two cutting edges; these are exposed by grinding on the undercut which gives the head the appearance of a snail-shell.

(d) Broughton Pattern (Shell Countersink) *Fig. 119 (d)*
A countersinking attachment for fitting to a Bit, consisting of a short, hollow cylinder of steel with the ends ground to form a conical cutter. This is fixed by means of a screw engaging with the side of the Bit. An improved pattern has an adjustable ring which acts as a fence for regulating the depth of the countersinking.

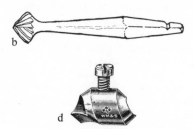

Fig. 119

Bit, Crown Saw: see *Bit, Annular.*

Bit, Cutler's: see *Drill, Passer.*

Bit, Deck Dowelling (Plug Bit; Centre Plug Bit) *Fig. 120*
A Centre Bit or a Twist Bit of the Single Twist type 1–1$\frac{1}{4}$ in. in diameter, with a plain, turned, or screwed centre plug in place of the usual centre spike or screw point. For counter-boring bolt holes in the deck for the insertion of cover-dowels. For details see *Auger, Deck Dowelling.*

Fig. 120

Bit, Dee
A name sometimes given to a shell-type Auger owing to its D-shaped cross-section.

Bit, Diamond: see *Bit, Gimlet.*

Bit, Dowel (Dowelling, Dowl, or Dowling Bit)
These Bits were shorter than normal, about 4–5$\frac{1}{2}$ in long and from $\frac{1}{4}$ to $\frac{3}{4}$ in. in diameter. Illustrations from catalogues of the early nineteenth century show that these Bits were of the nose, spoon, or shell-gimlet type. James Cam (Sheffield, *c.* 1787–1835) shows one with a twist body and a Scotch pattern nose; Gilpin (Cannock, 1868) illustrates a 'Twist Dowelling Bit' with a Gedge pattern nose (see *Fig. 29* under *Auger*); Ward & Payne (Sheffield, *c.* 1900) illustrate Twist Dowel Bits with either Gedge or Jennings pattern nose (see *Fig. 130 (g) and (h)* under *Bit, Twist*). Another type of Dowel Bit is illustrated under *Bit, Chairmaker's.*

They were used for boring accurate holes with a Brace to take the dowels (pegs) used in joining together the pieces of a cask head, the felloes of a wheel, certain parts of chairs, and other similar work. The shorter length helped to maintain accuracy. See also *Bit, Cooper's* and *Bit, Spoon.*

Bit, Dowel Trimmer (Dowel Shaver; Dowel Sharpener; Pointing Tool) *Fig. 121*

(a) A Bit with a hollow conical nose, slotted to form a cutter. It is used in a similar manner to the Spoke Rounding Bit for chamfering the end of dowel pegs about $\frac{1}{2}$–$\frac{3}{4}$ in. in diameter to facilitate their insertion into holes.

(b) Another pattern with a hollow cup-shaped nose for rounding the ends of the dowel is called a Dowel Rounder.

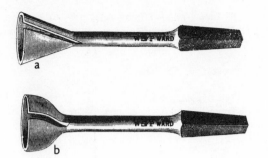

Fig. 121

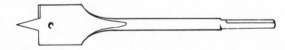

Fig. 122

Bit Drill: see *Drills* and *Bit, Archimedean Drill.*

Bit, Duck Nose: see *Bit, Chairmaker's.*

Bit, Expanding (Expansive Bit) *Fig. 122*
Since about 1890 Expansive Bits have been developed that are essentially Centre Bits with expanding cutting edges.
Variants include:

(*a*) 'Clark's Pattern': a flat blade, with a short screw point and a fixed spur. There are interchangeable cutters which can be moved across the nose of the Bit to produce the diameter of the hole required, and secured with a screw. (Named after Wm. A. Clark, New Haven, U.S.A., *c.* 1890).

(*b*) 'Steer's Pattern': this operates on the same principle as Clark's but the lower edge of the cutter has a rack which engages with a worm screw for adjustment. [Not illustrated.]

(*c*) 'Anderson Pattern': this is a Centre Bit with an additional cutter screwed to its back.

A simple form of Expansive Bit is illustrated by Mathieson (Glasgow, *c.* 1900), in the form of a *Washer Cutter*, which bores holes up to 6 in. in diameter.

a

Bit, Extension Holder (Long Arm)
A metal rod with a brace-type chuck at one end and a tang at the other. When fitted as an extension to a Brace, boring can be done in places where the crank would otherwise be obstructed. See *Brace, Corner.*

Bit, Fast Sheeting: see *Bit, Wagon Builder's.*

Bit, Fencing (Twist Auger Fencing Bit; Scotch Fencing Bit) *Fig. 130* (*p*) (page 87)
A relatively long Twist Bit of the Scotch pattern. Diameter $\frac{3}{16}$–$1\frac{1}{2}$ in and length up to about 12 in overall. Used on fencing and similar rough work.

Bit, Fipple
A Scots name for *Nose Bit.*

Bit, Flatbit *Fig. 123*
The cutting head resembles that of a Centre Bit, but the cutting edges are flat and each is bevel-sharpened on opposite sides. This is a recent development for boring wood at high speed with an Electric Drill. Made in sizes up to $1\frac{1}{2}$ in.

Fig. 123

Bit, Flat Head: see *Bit, Countersink.*

Bit, Forstner Pattern *Fig. 124*
A cylindrical nose with flat radial cutters, designed to be guided by its rim rather than by its centre. Sizes from $\frac{3}{8}$–2 in. in diameter. Since it cuts a very clean, accurate hole with a flat base, it is used for

boring coin cases, circular recesses for veneer and similar work, and in pattern making.

Fig. 124

Bit, French Screw (French Twist Bit)
A short Twist Bit, ground at the nose to a V-shape of about 50 degrees, like an engineer's twist drill. For boring hardwoods.

Bit, Gedge's: see *Auger, Twist.*

Bit, Gimlet *Fig. 125*
A twisted body like a Gimlet, used for boring pilot holes for screws and nails and for similar purposes when the cleanness of the hole itself is of minor importance.

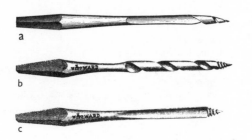

Fig. 125

(*a*) *Bit, Swiss Gimlet* (Twist Bit; Half Twist Bit; Twisted or Wilk Shell Bit; Persian Bit; Prussian Bit; Cobra Bit; Diamond Bit; Snail Bit; Swiss Pattern Twist Bit).
A pod-like shell body for about two-thirds of its length, after which the pod is given a half-twist, tapering off into a sharp screw lead. Sizes from $\frac{1}{16}$ to $\frac{5}{8}$ in.
   (*Note:* There is an Auger which is a larger edition of the Swiss Bit – see *Auger, Snail.*)
(*b*) *Bit, Twist Gimlet*
A cylindrical body in which a spiral groove has been cut, terminating with a screw lead, like an ordinary modern Gimlet. Sizes up to $\frac{3}{8}$ in. in diameter.

(*c*) *Bit, Shell Gimlet* (Gimlet Bit)
A shell body terminating in a screw lead. Sizes up to $\frac{3}{8}$ in. in diameter.

Bit, Gouge: see *Shell Bit.*

Bit, Half Moon *Fig. 126*
A flat semi-circular nose, horizontal cutting edge, and a spur on one side only. Probably imported.

Fig. 126

Bit, Half Twist: see *Bit, Gimlet.*

Bit Holder: see *Awl, Boatbuilder's Piercing*; Extension Bit Holder under *Brace, Corner.*

Bit, Hollow (Hollow Auger: this name is sometimes also given to a Hollow Mortising Chisel.) *Fig. 127*
Made in the form of a Brace Bit, it consists of a cylindrical body, enlarging to a cylindrical nose of larger diameter, with two radial cutters set on either side of a central hole sized from $\frac{3}{8}$ to $1\frac{1}{2}$ in. Though capable of making all kinds of round tenons, it is used mainly by wheelwrights for rounding and shouldering spoke tongues. This is done after tapering the end of the spoke with a Spoke Trimmer Bit.

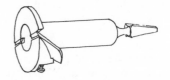

Fig. 127

Bit, Irwin: see *Auger, Twist.*

Bit, Jennings: see *Auger, Twist.*

Bit, Key: see *Pianoforte Maker.*

Bit, Ladder
A larger version of the Taper Bit, about $1\frac{1}{4}$ in wide at the broad end, tapering to about $\frac{1}{2}$ in at the nose. Used for reaming out tapering holes for ladder rungs. See also *Auger, Taper.*

Bit, Leadbeater's Pattern: see *Bit, Wagon Builder's.*

Bit, L'Hommedieu: see *Auger, Twist*.

Bit, Morse: see *Drill, Twist*.

Bit, Mortice (Machine Mortice Bit) *Fig. 130 (m) (n) (o)* on page 87
Twist-type Bits with round shanks for fitting into a Mortising Machine. Used for boring mortice holes. See also *Auger, Slotting*; *Mortising Machine*.

Bit, Nail or **Screw Removing:** see *Bit, Annular*.

Bit, Nose (Scots: Fipple Bit; Slit-nose Bit) *Fig. 109 No. 1525 under Bit*
A Shell Bit with slightly more than half the mouth in-bent at an angle with the centre line of the tool, forming a cutting lip which scoops out the bottom of the hole and breaks up the core left by the shell. This makes for quicker boring with less effort, which accounts for the popularity of the type. The diameter ranges from $\frac{1}{16}$ to $\frac{5}{8}$ in. (For an account of the properties and nomenclature of Nose-type boring tools, see *Auger, Shell* and *Bit, Shell*.)

Bit, Nut Wrench (U.S.A.: Brace Wrench) *Fig. 128*
A Box Spanner in the form of a Brace Bit, made in sets for square or hexagonal nuts or bolts from $\frac{1}{4}$ to $\frac{5}{8}$ in.

Used by wheelwrights and others for running nuts on bolts in places where an ordinary Spanner cannot be used, or when a more rapid method of nut-turning is required.

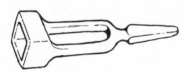

Fig. 128

Bit, Parrot-Nose
For description and illustration see *Auger, Parrot Nose*.

Bit, Passer
A bifurcated Bit. See *Drill, Passer*.

Bit, Persian: see *Bit, Gimlet*.

Bit, Pianoforte: see *Pianoforte Maker*.

Bit, Pin (Shutter Bit)
A name sometimes given to Shell and Spoon Bits. In the 1892 catalogue of Wynn Timmins it is illustrated as a normal Shell Bit with rounded end; in

Smith's *Key* of 1816 it appears as a short Bit of the shell type, but is cut off almost square at the nose end and with apparently little means of cutting downwards. This type of square-ended Shell Bit is also illustrated by Goldenberg (Alsace, 1875) where it is called a Shutter Bit.

Bit, Pivot: see *Bit, Annular*.

Bit, Plug (Plug Centre Bit)
The following are sometimes given this name: *Bit, Deck Dowelling*; *Bit, Cock Plug*; *Bit, Annular*.

Bit, Pod
A name given by Mercer (U.S.A., 1929) to a Bit with a short, pod-shaped, twisted body with a screw lead.

Bit, Prussian: see *Bit, Gimlet*.

Bit, Pump: see *Pipe and Pump Maker*.

Bit, Quill: see *Bit, Shell*.

Bit, Railway Carriage *Fig. 29 (3) under Auger*
A name given by Gilpin (Cannock, 1868) to a relatively long Twist Bit of the Gedge pattern from $\frac{5}{16}$ to $1\frac{1}{2}$ in. in diameter. See also *Bit, Wagon Builder's* and *Bit, Carriage Maker's*.

Bit, Rimer (Reamer, Rymer, Rhymer, Rinder, or Roomer Bit) *Fig. 109 Nos. 1542 and 1543*
A tapering blade which is either square, half-round, half-round and hollow, or five to eight-sided. The end is pointed, and it is used for enlarging existing holes. The spelling 'reamer' seems to be preferred by engineers, while woodworkers more often use the other alternative names. (See also *Rimer or Rinder, Hand*.)

Bit, Rinder: see *Bit, Rimer*.

Bit Roll *Fig. 129*
Canvas, felt, or leather holder with up to 36 compartments to take sets of Brace Bits.

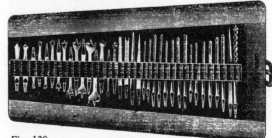

Fig. 129

Bit, Rosehead: see *Bit, Countersink.*

Bit, Russell Jennings: see *Auger, Twist.*

Bit, Sash *Fig. 130 (b)* page 87; *Fig. 727* under *Window Making.*
A Bit up to 18 in long and from $\frac{3}{8}$ to $\frac{1}{2}$ in. in diameter, made either in the form of a Nose Bit (*Fig. 109 No. 1525*), or in the form of a Twist Bit of the Gedge or Jennings pattern (*Fig. 130 (b)*). Used for boring holes down the stiles of a sliding sash for holding the cords.

Sash Bit Collars. Until recently these were listed for use with Sash Bits, and were made of brass rod, square in section, with a hole bored down the centre. Their purpose was to guide the bit. See *Window Making.*

Bit, Saw Handle Maker's
A type of Centre Bit consisting of a prominent central lead with flat cutting wings. The Bit is designed to cut a central hole surrounded by a recess (counterbore) which is needed to house the circular head and nut of the screws used to secure a saw handle to the blade.

Bit, Scillop
A Scots term for certain Bits – see *Bit, Taper*; *Bit, Bobbin.*

Bit, Scotch: see *Auger, Twist.*

Bit, Scouring
Listed but not illustrated by Howarth (Sheffield, 1884). It is listed after 'Pump Bit' as 'Scouring Bits to follow', and is probably a tapering Pipe Auger. (See *Auger, Pipe and Pump* and *Pipe and Pump Maker.*)

Bit, Screw: see *Bit, Twist, Fig. 130* on page 87.

Bit, Screwdriver (Turnscrew Bits) *Fig. 131*
A general term for Brace Bits with screwdriver tips. They are made with flat or round stems, and finished black or bright.

The advantage of tightening up or removing wood-screws with a Brace (preferably a Ratchet Brace) is not only that of speed: greater pressure can be brought to bear on the screw to prevent the blade from slipping out of the slot in the screw head, and greater turning-power can be exerted owing to the leverage given by the sweep of the Brace.

(a) Plain Turnscrew Bit
Used for turning ordinary wood screws.

(b) Fork End Turnscrew Bit
Used for tightening up or removing the type of screws used in saw handles.

(c) Billiard Table Bit
A flared blade made in widths from $\frac{5}{8}$ to 1 in. Used for driving the large number of bolts required in the construction of a billiard table. These bolts are usually too tight to turn by hand and their great number would make the work laborious. It is said that the exceptionally wide blade is less likely to slip off the bolt head, and also wears better.

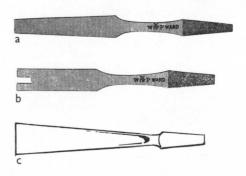

Fig. 131

Bit, Screw Point (or Screw Nose) Centre: see *Bit, Centre.*

Bit Sharpener: see *Bit, Chairmaker's.*

Bit, Shell (Gouge Bit; Quill Bit) *Fig. 132* and also see *Fig. 109 No. 1526* under *Bit.*
A shallow half-cylindrical body with the nose sharpened on the outside like a Gouge, but sometimes asymmetrically. Sizes $\frac{1}{16}$–$\frac{1}{2}$ in. in diameter.

Used for boring pilot holes for screws etc. and, in the larger sizes, by carpenters and joiners for the dowels or pins that secure frame and sash joints.

There is some confusion in nomenclature as between the Shell Bit and the Shell Auger. The term Shell is often applied to any Auger or Bit with a body of half-cylindrical form; but whereas a Shell Bit has a gouge-shaped nose as well as body, the Shell Auger has the same type of in-bent cutter at the nose as the Nose Bit. The Spoon Bit is also a form of Shell Bit but it has a spoon-shaped mouth.

Fig. 132

Bit, Shell Countersink: see *Bit, Countersink*.

Bit, Shipwright's: see *Bit, Deck Dowelling*.

Bit, Shutter: see *Bit, Pin*.

Bit, Single Twist: see *Auger, Twist*.

Bit, Slit-Nose: see *Bit, Nose*.

Bit, Slotting: see *Auger, Slotting*; *Bit, Mortice*.

Bit, Smith's Drill (Stock Drill Bit; Brace Drill Bit) *Fig. 133*
A flattened nose brought to a broad V-shaped point with its edges bevelled and sharpened. Commonly smith-made, these bits were used by many tradesmen for drilling metal. They are designed for use in a Brace, or in a Press or Bench Drill.

Fig. 133

Bit, Snail: see *Bit, Gimlet*; *Auger, Snail*.

Bit, Snailhorn: see *Bit, Countersink*.

Bit, Solid Centre: see *Auger, Twist*.

Bit, Solid Nose: see *Auger, Twist*.

Bit, Spiral: see *Auger, Twist*.

Bit, Spoke Trimmer *Fig. 134*
A funnel-shaped body with a cutting iron set in the side like a pencil sharpener. Used for tapering the tongue of a spoke before applying the Hollow Auger to round it, or for trimming the tongues before driving on the felloes. The Dowel Trimmer Bit is a similar tool designed for much smaller diameters.

Fig. 134

Bit, Spoon *Fig. 135*
A gouge-shaped (shell) body with a spoon-like nose with sharp edges. The nose is central and is in the same plane as the edges of the body; this tends to prevent wandering when starting the hole. The spoon-shaped nose usually tapers to a point. Those with a more rounded nose, such as the Chairmaker's Bit, are sometimes called 'Duck Bill'. Spoon Bits are much favoured for boring clean holes for dowels, pins, chair legs, and the like. That illustrated has a pointed nose. Other examples, usually with rounded noses, are described under *Bit, Brushmaker's*; *Bit, Chairmaker's* (under which the special virtues of the Spoon Bit are explained); and *Bit, Cooper's*.

Fig. 135

Bit, Sporle
A long Nose Bit up to $\frac{3}{8}$ in. in diameter, illustrated in Smith's *Key* (Sheffield, 1816). Possibly a misprint for 'Spool' Bit, in which case it may have been used for boring through bobbins. (See *Bit, Bobbin*.)

Bit, Sprig: see *Awl, Brad*.

Bit, Stone (Bath Stone Bit; Freestone Bit) *Fig. 130 (k)* page 87.
Made like a Twist Bit, but the nose is ground to a plain V-shaped point. Used for boring into soft stone or bricks.

Bit, Straw: see *Bit*.

Bit, Swiss Pattern Twist: see *Bit, Gimlet*.

Bit, Table
A spoon Bit with a pointed nose. Diameters vary from $\frac{1}{8}$ to $\frac{1}{4}$ in, and they were probably made up to

about 8 in long. In some lists it is shown as a Nose Bit up to $\frac{3}{8}$ in. in diameter and is grouped with the Sash Bits. Their use is uncertain, but they may have served for boring at an angle through the rails (or adjacent blocks) to take the so-called 'pocket screws' which secure the table top. A Bit longer than usual would be needed to keep the Brace clear of the work. They may also have been used for drilling the wooden knuckle-joint hinge of the brackets which support the drop leaf of a Pembroke or similar table.

Bit, Taper (Scots: Scillop) *Fig. 109 No. 1531* on page 76.
A Bit with a tapering half-funnel blade from $\frac{3}{8}$ to $1\frac{1}{2}$ in wide at the broad end, and tapering almost to a point. Used for boring tapering holes or for enlarging existing holes. See also *Bit, Rimer*; *Bit, Ladder*; *Auger, Taper*; *Pipe and Pump Maker*.

Bit, Trimming: see *Bit, Spoke Trimming*.

Bit, Tube
Term for a very long shell or nose type Bit for boring into end grain, e.g. wooden tubes for lamp standards.

Bit, Turnscrew: see *Bit, Screwdriver*.

Bit, Twist (Auger Bits; Screw Bits) *Fig. 130*
The main group under this heading, known as Auger Bits, are described under *Auger, Twist* with diagrams of the principal types.

Other Bits of the screw or twist type include: *Bit, Deck Dowelling*; *Bit, Dowel*; *Bit, Fencing*; *Bit, Gimlet*; *Bit, Sash*; *Bit, Wagon Builder's* (Fast-Sheeting Bit).

Bit, Unbreakable: see *Auger, Twist*.

Bit, Wagon Builder's *Fig. 130; Fig. 136*
Strong Bits designed for boring the main framework and chassis of railway and farm wagons. Variants are described below, but see also *Bit, Carriage Maker's* and *Bit, Railway Carriage*.

(*a*) *Wagon Builder's Plate Bit* (Wagon Builder's Twisted Centre Bit) *Fig. 130 (j)*
The nose of this unusual Bit consists of a single helical twist with a Scotch pattern cutting nose and a screw lead. Diameter $1–2\frac{1}{2}$ in. The term 'plate' was given to the main framework of the wagon's chassis.

(*b*) *Wagon Centre Bit Fig. 136*
An ordinary Centre Bit from 1 to $2\frac{1}{2}$ in. in diameter, but with a screw lead which is longer than normal.

(*c*) *Wagon Builder's Auger Bit* (Leadbeater's Pattern Wagon Bit) *Fig. 130 (q)*
See also *Bit, Car* and *Bit, Railway Carriage*.

An exceptionally long and strong Auger Bit of the Scotch pattern. Diameter $\frac{3}{8}–1\frac{3}{4}$ in, and up to about 17 in long with a 12 in twist. (According to Pigot's *Directory* of 1824, Thos. Leadbeater was a Brace and Bit maker of 40 Barford Street, Birmingham.)

(*d*) *Fast Sheeting Bit Fig. 130 (i)*
An Auger Bit of the Scotch type, $\frac{3}{8}–1\frac{1}{2}$ in. in diameter and up to about 12 in long overall with a 6–8 in twist. It has a special 'fast' lead which is explained as follows by Mr. R. H. McKears of Messrs William Ridgeway & Sons, Ltd., who make these Bits (Sheffield, 1968):

'The "Sheeting Bit" takes its name from the fact that it was designed for use in boring the wooden sheets on railway wagons – "sheets" being the wooden sides of the wagons. The word "fast" refers to the fact that the Bit is produced with a fast double thread on the screw point. A single thread is one which has a one-start lead – from one side of the cutting nose only. A double thread is a thread with two starts, i.e. one from each side of the cutting nose and a "fast double thread" indicates a pitch of thread which makes for fast boring whilst having the smoother boring qualities of the double thread. [See diagram of nose under *Auger, Twist*.]

With the advent of steel wagons the demand for Fast Sheeting Bits is now very very limited. In fact one of the main reasons it is kept in our catalogue is sentiment – Wagon Builders' Tools being those lines on which this Company's business was originally founded.'

Fig. 136

Bit, Washer Cutter: see *Washer Cutter*.

Bit, Wheeler's
A term applied to a relatively long Bit, usually of the shell-pattern gimlet type, with a screw lead.

Bit, Wilk Shell
An alternative name given to a Swiss Bit by Mathieson (Glasgow, 1900). See *Bit, Gimlet*.

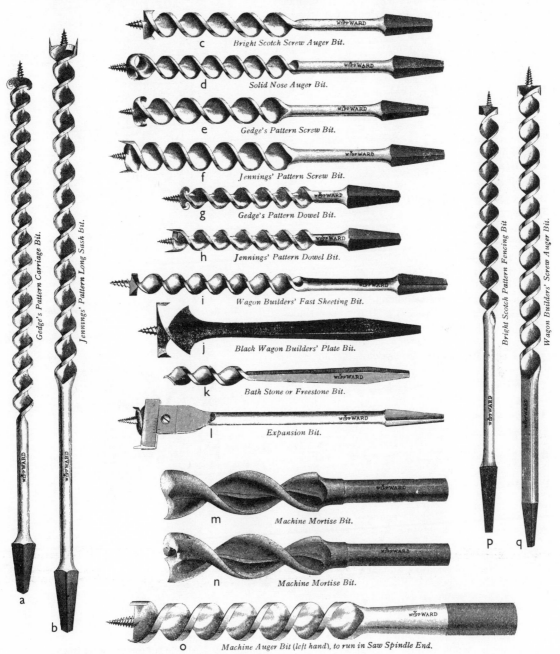

WARD & PAYNE, SHEFFIELD,

CORPORATE MARK.

Fig. 130 Twist Bits (Ward & Payne, Sheffield 1911)

a — Gedge's Pattern Carriage Bit.

b — Jennings' Pattern Long Sash Bit.

c — Bright Scotch Screw Auger Bit.

d — Solid Nose Auger Bit.

e — Gedge's Pattern Screw Bit.

f — Jennings' Pattern Screw Bit.

g — Gedge's Pattern Dowel Bit.

h — Jennings' Pattern Dowel Bit.

i — Wagon Builders' Fast Sheeting Bit.

j — Black Wagon Builders' Plate Bit.

k — Bath Stone or Freestone Bit.

l — Expansion Bit.

m — Machine Mortise Bit.

n — Machine Mortise Bit.

o — Machine Auger Bit (left hand), to run in Saw Spindle End.

p — Bright Scotch Pattern Fencing Bit

q — Wagon Builders' Screw Auger Bit.

Bit, Wimble
A name given to a Nose Bit by Wynn (Birmingham, 1810). These were short Nose Bits with flat tangs. They were probably fitted into separate pads, one for each size of Bit, as described by Moxon (London, 1678), a custom retained to modern times in the chairmaking trade.

Bit, Woodwind: see *Woodwind Instrument Maker.*

Bittel: see *Beetle*; *Mallet, Coppice Worker's.*

Black Tools
Tools 'left in the black', as opposed to 'bright', are those which are not polished to a bright finish after manufacture.

Block: see *Basket Maker (Osier)*; *Chairmaker (3) Workshop Equipment*; *Cooper (2) Furniture*; *Wheelwright's Equipment (1)*; *Workshop Furniture.*

Block, Chopping
A short piece of tree trunk about 18 in thick and 2 ft high, used mainly for trimming wood to shape with an Axe, e.g. in the woodland trades for paring and pointing pegs and stakes, and by wheelwrights for trimming spokes and chopping out wedges.

Block Hook: see *Cooper (2) Furniture.*

Blocking Knife: see *Knife, Bench.*

Blockmaker: see *Hat Block Maker*; *Print and Block Cutter*; *Ship's Blockmaker*; *Wood Engraver.*

Board Stick: see Log Stick under *Rule.*

Boat Builder: see *Shipwright.*

Boat Grip (Hutchit; Tongs) *Fig. 139*
Wooden Cramps used for holding together the planks of lighter-built boats while they are being riveted through. Usually home made, they consist of a pair of strong wooden legs about 2 ft long, joined at the top like a clothes peg. Another type has the legs connected by a leather hinge and drawn up by a central bolt and nut. The Boat Grip illustrated has side arms joined at their centre by a cross bar on which each arm is free to pivot. A wedge driven between the arms above causes the legs to close over the planks.

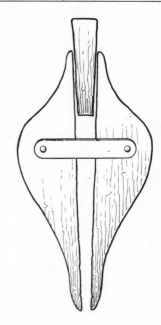

Fig. 139

Boat Hooks *Fig. 140*
Often found in boat-building yards. Mounted on a pole and used for handling small boats at the waterside.

Fig. 140

Boat Sway: see *Brace, Shipwright's.*

Bobbin Swarf: see *Bit, Bobbin.*

Bodger
A name given at one time to an itinerant tradesman or pedlar, but more recently the term was applied

to the men who cleave and turn chair legs and rails in the open woods. See *Chairmaker* (2) *The Bodger.*

Bodkin: see *Basket Maker.*

Bolection
A moulding which is rebated at the back, with its face standing above the face of the work it decorates. (Illustrated under *Plane, Moulding.*)

Bolster
A name for various tools, including a Cooper's Chincing Iron (see *Flagging Iron*), and for an all-steel Chisel used by masons and bricklayers for cutting stone or bricks. The bolster is also a term for part of a chisel (see diagram under *Chisel*) and for a part of a wagon bed.

Bolt Breaker *Fig. 141*
An iron bar about 10 in long with a hole of $\frac{1}{4}$–$\frac{3}{8}$ in. in diameter, drilled through a thickened boss at one or both ends. Used for breaking off the tail of an iron bolt. After fitting a bolt in some permanent position (e.g. in a fence or barn), the unwanted tail is nicked with a file. The head of the bolt breaker is then fitted over it and bent back and forth until the tail breaks off.

Fig. 141

Bolt Clam: see *Wrench.*

Bolting Iron: see *Chisel, Drawer Lock.*

Bond Poker: see *Broom Maker.*

Bond Splitter: see *Broom Maker;* Splitting Tools under *Woodland Trades.*

Boning Strips (Winding Strips; Parallel Strips; Trying Sticks)
A pair of narrow strips of well-seasoned hardwood, such as teak or mahogany, usually triangular in section with one face vertical, and up to 2 ft long, usually home made. Used by joiners to check that pieces are 'out of winding (or wind)', i.e. not twisted. The strips are set at right angles to the length and

sighted along the tops. Sighting was made easier by lines of black or white inlay on the upper edges. (Boning Rods are T-shaped stakes used when sighting levels and falls in excavations.)

Book: see *Millwright.*

Bookbinder's Plough
Not a wood-working tool but has the appearance of one. Two upright cheeks of wood are connected together by a wooden screw and by two stems in the manner of a Plough Plane. A knife is fixed in one of the cheeks, and when the tool is moved backwards and forwards in a guide over the book to be ploughed, the edges of the pages are trimmed.

Boring Machines (Hand): see *Drill, Bench; Drill, Cramp; Drill, Press; Hub Boring Machine; Wood Boring Machine.*

Boring Tools are included under the following entries: *Auger; Awl; Bit; Boring Machines; Brace; Drill; Gimlet.*

Borrier (Borryer; Boorrier)
Mentioned in five entries of the Apprentice Indenture Records, Bristol, 1535–1646, for carpenters and shipwrights. Probably a carpenter's or shipwright's Auger.

Bosh
A trough of water near the blacksmith's hearth, or in a wheelwright's yard, in which hot iron is quenched or tempered.

Boss
A protuberance or swelling, e.g. on an iron rod or bar, to take a bolt hole, bearing, etc.

Bottoming Iron: see *Shave, Chairmaker's.*

Boulle: see *Veneering and Marquetry.*

Bowl Turner
A country turner who made bowls for domestic use, mainly in sycamore or elm, using a lathe (usually a Pole Lathe) with the ordinary range of Turning Gouges and Chisels.
One speciality was the making of a set of nested bowls, each smaller than the other, from a single block. This was done with a special curved Chisel, which under-cut a narrow groove forming the inside of the larger and the outside of the smaller. (See

Hook Chisel under *Chisel, Turning*.) The bowls were then separated and finished on the lathe.

Box and Case Maker

Boxes and crates, usually in rough, unplaned wood, are made for holding all kinds of goods from the smallest components to complete machines or vehicles. Finely made cases for optical instruments, sporting guns, etc. belong to the cabinet maker's trade. But much ingenuity is also needed to construct the packing cases or tea chests that will stand the rough-and-tumble of transport by land and sea. They are expensive to produce and must earn their keep on the return journey and for many years to come before being broken up. Many boxes are strengthened and secured with metal hoops and straps. Among the special tools used are: *Hammer, Box Maker's; Hammer, Magnetic; Hoop tightener; Plane, Box Maker's; Plane, Box Scraper.*

Box and Case Opening Tools *Fig. 142*

Tools for opening boxes and cases include:

(*a*) *Box Openers*

These are usually lighter than the Case Openers which follow. They include the so-called 'Gents' Opener, a strip of steel about 8 in long and 1 in wide, with a claw at one end and a turned handle at the other.

(*b*) *Case Openers* (Goat's Foot; Wrecking Bar)

There are many types, the essential feature being a stout iron bar from 12 to 18 in long, with one end drawn down to form a chisel end, the other end furnished with a claw. Variants include those illustrated.

(*c*) *Combination Case Opener*

An all-steel hatchet with a scale handle, and with a head combining hammer, hatchet, and nail claw.

See also *Chisel, Case Opening; Cigar Box Opener; Hammer, Grocer's & Warehouse; Nail Extracting Tools.*

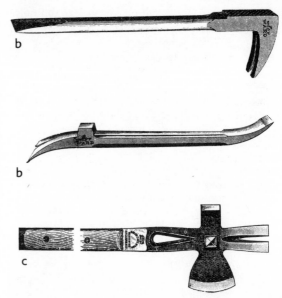

Fig. 142

Boxed

Of pincers – when interwoven at the joint.

Boxing

Of Planes – the provision of a boxwood insert in the sole of a Plane to resist wear (see *Plane; Plane Maker*). Of wheels – a hollow metal bearing inserted into the hub of a cart wheel to hold the axle (see *Wheelwright*). Of Rebates, see *Rebate*.

Boxing Machine (Boxing Engine): see *Hub Boring Engine.*

Boxing Router: see *Router, Coachbuilder's.*

Boxing Tools (Wheelwright's): see *Auger, Wheelwright's; Chisel, Bent; Chisel, Bruzz; Gouge, Boxing; Hub Boring Engine;* Rim gauge under *Gauge, Wheelwright's.*

Boxing Try: see *Rim Gauge* and *Gauge, Wheelwright's.*

Brace (Bit Brace; Bit Stock; Breast Stock or Wimble; Stock; Sway; Sweep; Wimble) *Fig. 143; Fig. 144*

The Brace is a tool for boring, consisting of a chuck or pad for holding the Bit at the foot, a head at the top for a hand-hold, and between the two a crank for rotating.

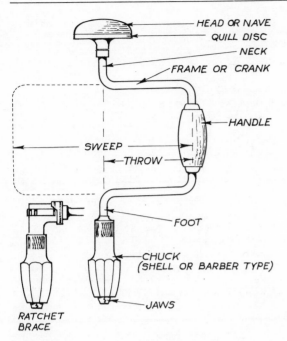

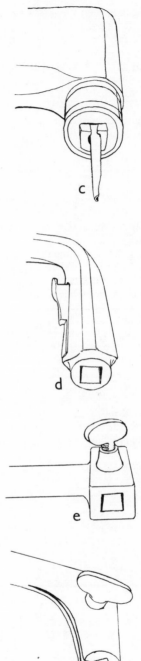

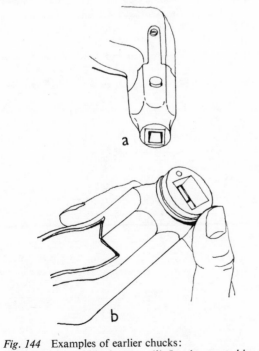

Fig. 143 Diagram of a modern Brace with a 'Barber' shell chuck

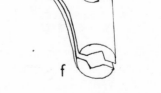

Fig. 144 Examples of earlier chucks:
(*a*) Latch operated by button   (*b*) Latch operated by push-ring   (*c*) The Bit is wedged   (*d*) Latch operated by lever
(*e*) Socket and thumbscrew   (*f*) 'Spofford' or split-chuck

Though crank motion was known in China in the first century A.D., the Brace did not appear in Europe until the fifteenth century. It is usually shown in illustrations of that period with a small Bit and accompanied by larger Augers, which suggest that it was used mainly to bore pilot holes for the Augers. Later, as the tool developed, it was used for boring larger holes, and gradually displaced the smaller Augers altogether. In most wooden Braces of the seventeenth century, each Bit had its own 'pad' of wood which fitted into a tapering hole in the stock and was secured by wedges or screws. A survival of this method may be seen in the home-made Braces used by chairmakers and shipwrights. The spring chuck, with a latch actuated by a button or lever, was introduced towards the end of the eighteenth century, using interchangeable notched Bits. The 'Barber' Brace, patented in 1864, had a screwed shell chuck which avoided the necessity for notching the shank of the bits. (See diagram *Fig. 143*.)

Brace, Adjustable: see *Brace, Chairmaker's*.

Brace, American
A name often given in nineteenth century tool catalogues to the metal Braces of a modern type. See *Brace, Joiner's*.

Brace, Angle: see *Brace, Corner; Drill, Hand*.

Brace, Archimedean
A term used in the Sheffield Standard List of 1862. See *Drill, Archimedean*.

Brace, Armourer's *Fig. 145*
An all-iron Brace, with a socket-chuck offset from the foot of the frame, and apparently without a

thumb screw. Intended presumably for use in a regimental or ship-armourer's workshop.

Brace, Barber: see *Brace, Joiner's*.

Brace, Belly
A shipwright's term for the Boat Brace or Sway. See *Brace, Shipwright's*.

Brace, Bit
The catalogues of the nineteenth and twentieth centuries often use the term Bit Brace for Brace; and also the term Brace Bit for Bit.

Brace, Boatbuilder's: see *Brace, Shipwright's*.

Brace, Chairmaker's (Sweep Stock) *Fig. 146*
A solid wooden Brace with very wide, stout arms to the sweep, a comparatively small head, and a socket chuck. The small revolving head fits into the hollow of the Breast Bib (see below) on which the full pressure of the body could be applied.

The Bits used were of the spoon type, and these were usually fixed permanently in the Brace. The general rule – one Brace, one Bit – ensured rigidity, an important advantage when boring holes at awkward angles by eye. Consequently, some twenty or thirty Brace-Bit units would be the usual equipment of a framer or chairmaker. An alternative to this arrangement was the so-called Adjustable Chairmaker's Brace in which each Bit was fitted in a wooden pad with a long, square, tapered shank. This had a shallow recess cut on one side to take a spring latch, which held the pad in place in the corresponding tapered hole in the foot of the stock. This had the advantage that only one stock was required. On

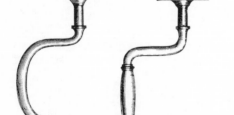

Fig. 145

Fig. 146

the other hand, the pads and Bits were far from rigid, however well fitted in the first place.

The Breast Bib
A piece of hardwood shaped to the chest and secured by a light leather harness. It has a recess on the front surface in which the head of the Brace is held. Its purpose was to distribute the pressure exerted by the head of the Brace over the area of the chest; and by locating the head of the Brace, it helped to steer it accurately at the right angle. Sometimes Breast Bibs are found with a groove across the recess on the front. This is where the Chairmaker used his chest, protected by the Bib, as a cramp to force parts of a chair together.

Fig. 146

Brace Chucks: see *Brace*.

Brace, Cooper's (Dowelling Stock) *Fig. 147*
Though Braces of many kinds were used in the cooperages for general work, the all-wood Brace is still preferred for boring the edges of the cask head-pieces for the dowels which hold them together. Usually made of beech throughout, it differs from the ordinary wooden Braces in having, like the Chairmaker's Brace, very stout (though deeply chamfered) arms to the sweep. This is done to add strength, for these braces are rarely, if ever, strengthened with metal plates. And it has a much larger head so that it can be safely lodged between work and chest while holding the work in one hand and turning the Brace with the other. The flat tang of the Spoon Bit is often permanently wedged in a simple hole in the foot of the Brace. Consequently, a separate Brace is kept for each size of Bit. (See *Bit, Cooper's*.) But Braces with facilities for changing Bits were available: Richard Timmins (Birmingham, *c.* 1850) illustrates three types – a replaceable pad, a spring jaw, and a plain square metal socket and thumbscrew. A feature of the early pads is that they were removed

by tapping the tail of the pad which protruded into the sweep space, i.e. inside the crank.
The following examples are illustrated:

(*a*) Factory-made in wood with a special spring-jaw chuck.

Fig. 147(a)

(*b*) An all-iron example of the so-called Scotch Brace, often listed as a Cooper's Brace.

Fig. 147(b)

(*c*) A home-made Cooper's Brace in wood with wedge-type chuck.

Fig. 147(c)

Brace, Corner *Fig. 148*
A general term for certain Braces and other tools, mostly developed during the present century, intended to enable holes to be bored in confined spaces and awkward corners where the crank of an ordinary Brace could not be turned. These tools include:

(*a*) *Drill Brace*
A Joiner's Brace, usually of the ratchet type, with a hand-drill attachment which, when put into gear, enables the Brace to be used as a Breast Drill.

(*b*) *Gear Frame Brace*
The Brace is held at an angle of 45° to the vertical by means of a metal frame, and drives the chuck through a bevel gear.

(*c*) *Angular Brace Extension*
The shank, passing through an iron hand-grip, is connected by universal joint to a chuck. The angle (up to 60°) between the extension rod and the chuck is adjustable.

(*d*) *Hand Ratchet Brace* (Corner Ratchet Brace)
A brace with a revolving wooden head with a chuck beneath it. It is turned by means of a bar operating through a ratchet to the stock.

(*e*) *Engineer's Ratchet Brace*
A metal-working boring tool often found in woodworking shops. A short stock has a chuck at the foot and a blunt point or cone at the top, which can be raised by screwing until it impinges on some fixed object in order to exert downward pressure on the drill. The stock is driven through a ratchet assembly by the back-and-forth motion of a bar as in (*d*) above.

(*f*) *Extension Bit Holder*
A metal rod with a Bit holder at one end and a tang at the other.

(*g*) See also *Brace, Ratchet; Drill, Hand.*

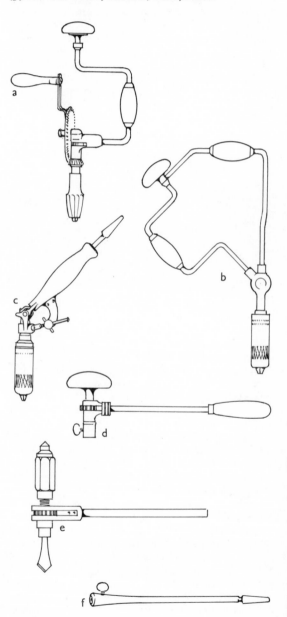

Fig. 148

Brace, Crank

An all-iron crank used mainly for drilling metal and held inside a drilling appliance. (See, for examples, *Drill, Cramp; Drill, Press*.) It is a Brace-like tool with a simple chuck to hold the bit at the foot; but instead of a head, the frame ends with a blunt point which engages with the pressing-down device of the drilling appliance.

Brace, Double Crank: see *Brace, Wimble*.

Brace, Folding: see *Brace, Undertaker's*.

Brace, Framed (Presentation Brace; 'Ultimatum' Brass-framed Brace; Metallic Frame Brace; Patent Framed Brace) *Fig. 149; Fig. 159* under *Brace, Wooden*

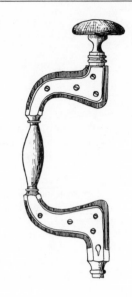

Fig. 149

This heavy, decorative Brace was patented by Wm. Marples of Sheffield about 1850–60 and was listed until about 1900. It consisted of two heavy brass castings which form the two angles of the sweep, the lower one containing the chuck. They are joined by a steel rod on which a wooden handle revolves, and the upper casting is surmounted by the revolving wooden head. Both castings are hollowed to take inserts of ebony or other hardwood. The Bits used in the Brace have notches in the tang and are held by an ingenious spring latch released by sliding the necked ferrule sideways, instead of the push-button or lever normally used in the plated Brace. (See *Fig. 144 (b)* under *Brace*.)

Mr. W. L. Goodman (London, 1964) writes as follows about this remarkable tool: 'In the Marples tool catalogue of 1864 this ebony brace, as the master-piece of its inventor, is given a page to itself, with a display of the complete set of bits to go with it. There seems to have been some doubt as to what to call it, as it is referred to in various places in the catalogue as the "Patent Metallic-Framed Ebony Brace", the "Patent Brass-Framed Brace", and the "Ultimatum Framed Brace". It was certainly the ultimate, in every sense, and in ebony, boxwood, or rosewood it was listed at 23s., while an identical model in beechwood was 20s. This was a very expensive tool compared to the plain wooden one at 9s., or the simple iron brace at about 1s. retail, especially when we remember that the wages of the average carpenter of the period were 37s. 8d. for a 56½ hour week.'

Brace, Gas Fitter's *Fig. 152 (g)*

A solid iron Brace of 8 in sweep, with a wood or iron head and plain sweep of round iron without handle. The Bits are held by a thumb-screw in a socket-chuck. The foot is extended upwards and terminates in a short point projecting inside the sweep. (See also *Brace, Iron*.)

Brace, 'German' (Common Ball or Thumbscrew Brace) *Fig. 150; Fig. 152 (b) and (c)*

A light Brace of varied design but mostly with a round iron sweep, a bulbous iron or wooden handle, and a comparatively large wooden head. The Bits are held in a square socket-chuck by means of a thumbscrew. Owing to their cheapness, they were sometimes referred to as 'Sixpenny Braces' – at which price they were sold in the early 1900's.

Fig. 150

Brace, Gearing: see *Drill, Hand.* (For Gear Frame Brace see *Brace, Corner.*)

Brace, Hurdle Maker's: see *Hurdle Maker.*

Brace, Iron *Fig. 151; Fig. 152*
When Braces first emerged in the fifteenth century, they were probably made only in wood. Among early metal Braces are the beautifully made Surgeon's Braces of the sixteenth century, known as 'trepans'; mostly made on the Continent, they were used for cutting out a small disc of bone from the skull. By the late seventeenth century, the village and town blacksmiths were making efficient all-iron Braces not only for their own use, but for other tradesmen who needed something stronger than the wooden tool. Thus began the evolution of the modern steel Brace which, unlike most of the early iron Braces, is provided with a wooden head and handle-grip.

A singular feature of the factory-made iron Braces is that the foot was sometimes extended upwards and terminated in a short point standing about $\frac{1}{2}$ in inside the sweep. (See *Fig. 152 (g)*.) Its exact purpose is unknown although it does enable the Brace to be hung securely on a nail or on the edge of a shelf. It may however be simply a vestigial spigot belonging to the replaceable pads fitted to certain wooden Braces.

Note. For all-iron Braces used in various drilling appliances see *Drill Crank.*

Fig. 151   Typical smith-made iron Braces
      (*a*) With plain head
      (*b*) With the so-called 'lantern' or 'squirrel cage' head

Brace, Joiner's (American Pattern Brace; Barber Brace) *Fig. 153*
This modern form of Brace, often referred to as the American Pattern Brace, consists of a steel frame of rectangular form, rotating hardwood head and handle, and a chuck in the form of a hollow, screwed shell, tapered inside, which forces together the two spring jaws holding the Bit. This type of chuck, which soon superseded all other forms, was patented by Barber in the U.S.A. in 1864. Later improvements included interlocking chuck jaws, ball-bearings in the head, and a ratchet drive.

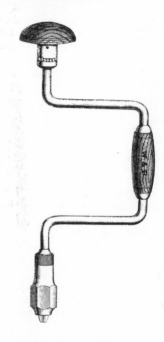

Fig. 153

Brace, Nut (Spanner Brace)
A tool of brace pattern in iron throughout with a socket hollowed out to take a hexagon nut. Used for detaching and fitting nuts, when assembling rail wagons etc. (A similar tool is provided for changing car wheels.)

Brace, Plated *Fig. 152 (i); Fig. 159*
A factory-made Brace with a solid wooden frame usually of beech, but strengthened by thin brass plates. The plates were either inlaid or screwed to the surface of the wood. The notched bits were usually held in the solid brass socket-chuck by a spring latch.

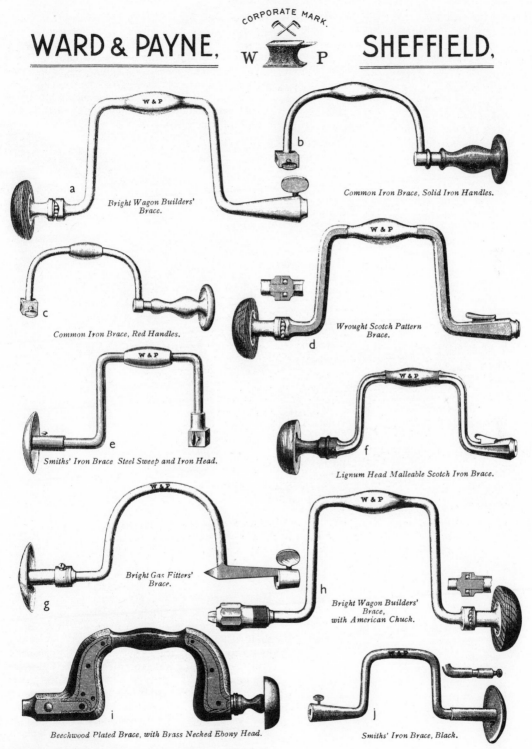

CORPORATE MARK.

WARD & PAYNE, SHEFFIELD,

Bright Wagon Builders' Brace.

Common Iron Brace, Solid Iron Handles.

Common Iron Brace, Red Handles.

Wrought Scotch Pattern Brace.

Smiths' Iron Brace Steel Sweep and Iron Head.

Lignum Head Malleable Scotch Iron Brace.

Bright Gas Fitters' Brace.

Bright Wagon Builders' Brace, with American Chuck.

Beechwood Plated Brace, with Brass Necked Ebony Head.

Smiths' Iron Brace, Black.

Fig. 152 Typical factory-made iron Braces (Ward & Payne Ltd., Sheffield, 1911)

97

This was the ordinary Brace used by joiners, carpenters and other tradesmen from about 1800 to 1910, when they were gradually replaced by the metal Joiner's ('American pattern') Brace, with a screwed shell-type chuck which could grip any Bit without the notch in its tang.

Brace, Platelayer's *Fig. 154*
A large iron crank with an Auger (usually of the screw type) welded to the end of a long extended foot. Used by platelayers for boring holes in railway sleepers etc.

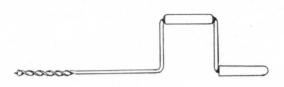

Fig. 154

Brace, Push: see *Drill, Push*.

Brace, Ratchet *Fig. 155*
A name given to a Brace provided with a ratchet device. This usually consists of a circular rack and pawl in the chuck, which enables rotary motion to be maintained while the handle is moved back and forth through part of its sweep only. Thus it can be operated in a corner or close to a projection. When the work is heavy, e.g. because a large Bit is being turned, the ratchet also enables the most advantageous leverage to be used. The ratchet was applied to the Barber chuck shortly after its introduction, about 1865.

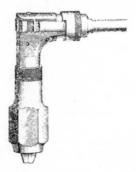

Fig. 155

Brace, Scotch *Fig. 156; Fig. 152 (d)* and *(f)*
An iron Brace with a wooden head and swollen iron handle. The frame is gracefully curved, chamfered, and hexagonal in section. The bits are held in a tapered socket by means of a thumb-operated lever which withdraws the latch from the notched tang of the bit. *Note:* A plain all-wood Brace illustrated by Richard Timmins (Birmingham, *c.* 1850) is also called a 'Scotch Brace'.

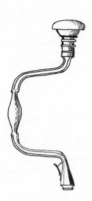

Fig. 156

Brace, Shipwright's *Fig. 157*
For light work shipwrights sometimes use a home-made wooden Brace often known as a Belly Brace or Boat Sway. These have a rather shallow sweep, and a square, tapered socket to take wooden pads. A stock of these pads is kept, each holding a different size of Awl or Bit, ready to be placed in the socket as required. They were used for boring the planks of smaller craft before nailing. For heavier work, shipwrights sometimes used a Brace of the type described under *Brace, Platelayer's* and *Brace, Wimble*.

Fig. 157

Brace, Smith's *Fig. 152 (e)* and *(j)*
A name given to various iron Braces with plain iron frame and large saucer-shaped iron head, and a tapered socket-chuck with thumbscrew.

Brace, Split Chuck: see *Brace, Spofford.*

Brace, Spofford (Split-Socket Pattern Brace; Split-Jaw or Split-Chuck Brace) *Fig. 144 (f)*
An iron Brace with a wooden head and loose hand grip. The foot of the frame was split, forming a square, tapered socket in which the Bit was secured by drawing the two halves together with a thumbscrew.

The Spofford Brace was patented in the U.S.A. in 1859. It had the advantage of accepting any Bit with a square tang, with or without notches, and was a forerunner of the Barber screwed chuck, patented in 1864.

Brace, 'Ultimatum': see *Brace, Framed.*

Brace, Undertaker's *Fig. 226* under *Coffin Maker*
A small iron Brace, with the head attached to the top of the sweep with a thumbscrew, so that it can be folded back into the sweep itself and carried in a small valise or in the pocket. An imported tool probably intended for use with a Screwdriver Bit for opening and closing a coffin lid – particularly in countries where it is customary for the coffin to be opened in church during the funeral.

Brace, Wagon Builder's *Fig. 152 (a)* and *(h)*
A strong, heavy iron Brace, with the crank swollen for a hand-hold and the lower end shaped to form a socket-chuck in which the Bit is held by a thumbscrew, or later by a screwed chuck. The sweep is larger than usual (16–17 in) to give the maximum leverage. Used for heavy work by wheelwrights, wainwrights, makers of railway wagons, and the like.

Brace, Washer Cutter: see *Washer Cutter.*

Brace, Wheel
A name given to an early version of the Hand Drill. (See *Drill, Hand.*)

Brace, Wimble (Double Crank Brace) *Fig. 158*
An iron Brace with that part of the frame which carries the head extended laterally to a distance about equal to the throw, so that leverage can be exerted on both handles. Apparently made and listed only in the U.S.A., the Stanley Works catalogue of 1941 describes it as designed specially for millwrights, ship carpenters, and farmers. It is shown with a Spofford (split) chuck.

Experiments suggest that the tool is very difficult to use without 'wobble', but an American shipwright (R. G. Paterson, 1968) informed us that they are not as awkward to use as they look, and that they were often made by the shipyard blacksmiths.

Fig. 158

Brace, Wooden Home-made examples: *Fig. 146* (Chairmaker's); *Fig. 147* (Cooper's); *Fig. 157* (Shipbuilder's). Factory-made examples: *Figs. 159* and *152 (i)*

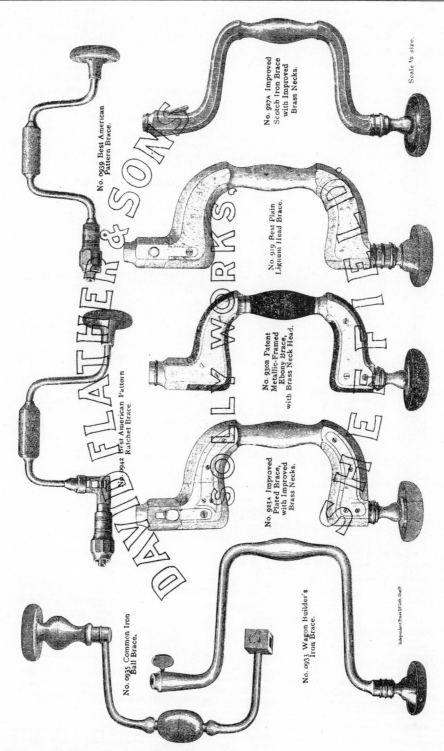

No. 0939 Best American
Pattern Brace.

No. 927A Improved
Scotch Iron Brace
with Improved
Brass Necks.

Scale ⅓ size.

No. 919 Best Plain
Lignum Head Brace.

No. 0942 Best American Pattern
Ratchet Brace.

No. 930B Patent
Metallic-Framed
Ebony Brace,
with Brass Neck Head.

No. 923A Improved
Plated Brace,
with Improved
Brass Necks.

No. 0935 Common Iron
Ball Brace.

No. 0933 Wagon Builder's
Iron Brace.

Fig. 159 Typical factory-made wooden and other Braces
(David Flather & Sons Ltd., Sheffield, c. 1900)

Up to the end of the nineteenth century many Braces were made of wood throughout, except for a metal ferrule or socket at the foot which held the Bit, and sometimes a metal neck or bearing for the head. These Braces were mostly home-made, some crude but others of great beauty. Factory-made wooden braces, mostly of beech, were also of pleasing design, with a handsome chamfer on the sweep, and a cast brass chuck with a latch to hold the notched tang of the Bit.

Brace, Wrench: see *Brace, Nut; Bit, Nut Wrench.*

Brace, 'Yankee'
A name given by Stanley Works (U.S.A., 1965) to a Joiner's Brace with a concealed ratchet mechanism and parallel jaws.

Bracket, Workshop *Fig. 160*
A right-angled natural bough was often flattened on one side and fixed to the wall for hanging patterns, clothes, or tools.

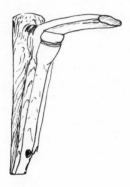

Fig. 160

Bradawl: see *Awl, Brad.*

Brake: see *Woodland Trades* (2) *Brakes; Handle Maker; Basket Maker.*

Brands: see *Burn Brands.*

Brass
The most widely used brass contains two parts of copper and one part of zinc. This has reasonable strength with malleability. Since it does not rust, it is much used for parts of hand tools, small machines, and fittings of all kinds.

Brazil Gouge: see *Adze, Brazil.*

Breast Bib or Plate: see *Brace, Chairmaker's; Drill, Passer.*

Breast Wimble: see *Brace.*

Brewer's Fret: see *Gimlet, Wine Fret.*

Bricklayer's Tools: see *Builder's Tools; Level; Scutch; Square-Try Square.*

Bridle
When applied to a tool, this term implies a yoke or link joining two parts of a tool or workpiece. For example, see Spoke Bridle under *Wheelwright's Equipment* (2) *Spoke Tools;* Bridle Plough under *Plane, Plough;* Horsing Iron under *Caulking Tools.*

Bright
Applied to the finish of a metal tool when ground and polished, as against 'black' when left unpolished.

Broach
A tapering steel rod of polygonal cross-section made from very small sizes up to $\frac{1}{2}$ in. in diameter or more. Used for opening up a previously bored hole in metal or wood. (See *Reamer.*)

Brog
Scottish joiners' term for *Bradawl.* (See *Awl, Brad.*)

Broken Screw or Nail Remover: see *Bit, Annular.*

Broom Maker (Besom maker) *Fig. 161* (See also *Brushmaker.*)

The tools described below belong to the besom maker or 'broom squire', a country tradesman who makes brooms from bundles of birch twigs (or, in some parts, from heather) which are bound on to a handle with skeins of cleft willow, oak, or other woods. Nowadays the brooms are bound with wire. Birch brooms are still widely used for sweeping leaves from grass. A typical Besom is illustrated.

The tools used include some of those listed under *Woodland Trades,* and also the following:

(a) Besom Grip or Broom Horse (Broom Maker's Vice; Broom Maker's Pincers; Knee Vice.)
Several different holding devices are used for compressing the bundle of birch twigs or other material

ready for binding. That illustrated consists of circular metal jaws mounted on a post. The top jaw is hinged and ends with an 18 in handle. The circular mouth of the grip is about 4 in. in diameter when closed over a bundle of birch twigs. Another type is made like a pair of huge pincers with circular jaws. One leg of the pincers is firmly fixed on a bench or block, the other is bent outwards for operation by knee or foot, thus leaving the hands free. Other broom makers assemble the bundle of twigs on a Shaving Horse, using the foot-operated vice to hold them.

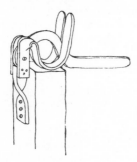

Fig. 161 (a)

(b) Bond Poker
A curved blade, about 7 in long, gouge-shaped in cross section like a basket maker's Hollow Bodkin, and fitted with a wooden handle. Used for threading the loose end of the skein or wire bond which holds a broom head together. The Bond Poker is pushed under the bond, and the end is passed down the hollow of the blade and then tied.

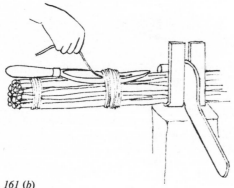

Fig. 161 (b)

(c) Bond Splitter
An egg-shaped piece of hardwood with the end formed into three cutting edges. It is used for splitting the withies for binding the brooms. Illustrated under *Woodland Trades* and under *Basket Maker* (Cleaver).

Brush Maker *Fig. 162*
The tools of the country 'Broom Squire', who makes brooms from birch or heather, are described under *Broom Maker*. The brush maker whose workshops are to be found largely in the towns makes brushes from clusters of animal bristles or vegetable fibres, usually set in a wooden stock. An enormous variety of brushes and brooms are made for the household as well as for industrial and commercial purposes.

Many woods are used for brush stocks and broom heads, including beech, birch, oak, elm, alder, and sycamore, The filaments used include costly hair, such as sable, used for the best paint brushes, the various kinds of hair, bristle, or nylon used for the ordinary run of household brushes, and the stiff cane (or even whale-bone) used for the yard or road broom.

The brush stocks were at one time trimmed with a Bench Knife, but later turned on a lathe and then halved. The face of the stock is bored with holes to hold the bristles. This is done with a Spoon Bit on the lathe-like tool illustrated.

Brush fibres are disentangled and sorted by drawing through a Comb or Hackle. The sorting and 'turning' of animal hairs (roots at one end, tapering flag at the other) is highly skilled work and must be done mainly by hand.

The setting of the bristles in the stocks is done by dipping a knot of fibres into hot pitch, tying with hemp, and then inserting into each hole in the stock. More valuable brushes (e.g. hair brushes) have their stocks threaded with wire to secure the fibres. Finally, the bristles are trimmed with very large scissor-like Shears.

The tools used include:

(a) (b) and (c) Brush Bits (Brush Drills)
These are usually Spoon Bits, or occasionally Taper Bits. They are used for drilling the holes in the wooden stocks (*m*) that hold the bristles. Spoon Bits have the advantage of drilling almost the full thickness of the wood without breaking through.

The drilling is often done on a treadle lathe (*n*). The Bit is set in a box-wood pad, the tail of which is gripped in the lathe chuck. The head of the pad is bored out, the Bit centred by small wedges, and the cavity filled with molten lead. Any minor adjustment can then be made by a tap from a Hammer and Punch.

(d) Scale
A jig which guides the Bit when boring the stock. A large number of these pattern-boards are kept hanging in the brush maker's workshop.

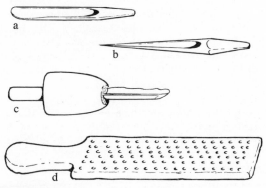

Fig. 162 (a)–(d)

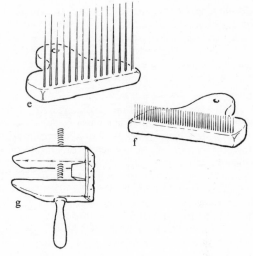

Fig. 162 (e)–(g)

(e) Hackle

The terms Hackle and Comb both relate to the comb-like tools used for untangling, sorting, and dressing the filaments used for brushes. The Hackle is the name usually given to those of larger size set with steel spikes up to 18 in long. The spikes are sometimes set in more than one row, and staggered; they are often square in section, set diagonally, and are removable so that those in the centre, which wear first, can be replaced. The Hackle is used for 'dragging', i.e. untangling, the longer filaments of coarser fibres such as bass.

(f) Comb (Brushmaker's Engine)

This is similar to the Hackle, but smaller and used for the finer filaments. The typical examples illustrated consist of a block of wood about 8 in long with a row of needles set upright along the front edge. Six or eight of these combs are required, with the needles uniformly spaced at varying distances apart. Starting with the widest, the bundles of bristles and other filaments are passed through the combs, thus sorting out the filaments into progressively smaller sizes, an operation known as 'dressing'.

A finer variety of the Comb, called an Engine, was adapted for dressing the bristle used in making paint brushes. It was also used for extracting 'turned hairs'. It is important that bristle used in making paint brushes should all lie one way, as the tapering tip or 'flag' would not be locked into the adhesive as securely as would the knob of the 'root'; and the harsh root, if it remained unturned in the brush, would cause a streak in the painted surface. The flag of the bristle passes through the finest Engine, but any turned hairs are caught between the pins.

(g) Glueing Screw

A small wooden Hand Screw with its jaws often hinged together at one end with a strip of leather. Used for cramping veneer on the back of a brush after glueing.

(h) Hand Shears

Heavy shears with wide jaws and straight cutting edges 8–10 in long. Used for cutting and trimming brush bristles.

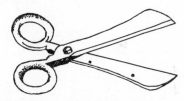

Fig. 162 (h)

(j) Bench Shears

Large shears consisting of a fixed blade 9–15 in long, mounted on an upright leg with a screw at the foot which passes through the bench or is fitted to a socket screwed to the bench top. The other blade of the shears is pivoted in the usual way and is provided

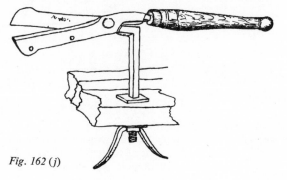

Fig. 162 (j)

with a long handle. It is used for trimming the bristles of brushes. The two holes in the lower blade are for securing a wooden fence at varying distances so that each row of bristles is trimmed to the same height. The fence makes contact with the face of the brush stock.

(k) Pitch Pan
A shallow pan about 15 in. in diameter, provided with four or more compartments to hold molten pitch. It was usually heated on a tall cylindrical stove, sometimes let into the centre of a table round which 4 to 8 workers sat. Each worker dips a 'knot' (i.e. bundle) of fibres into the hot pitch, ties it with 'thrum' (a hempen string), dips again, and then inserts the knob into a hole in the stock with a twisting action to ensure the spreading of the bristles. As the pitch hardens, the knot is securely held. The compartments in the pan enable each worker to dip without getting in the way of his neighbour.

(l) (m) Knot of bristle and brush stock

(n) Brushmaker's Boring Lathe
A small treadle lathe with a chuck and bit for boring the holes in brush stocks or boards.

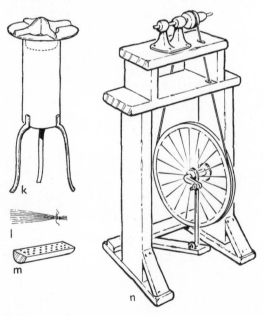

Fig. 162 (k)–(n)

Trepanning Needle
A long needle used in securing bristles into comparatively thin stocks made of valuable material such as ivory or ebony. Trepanning was the process of boring shallow holes for the bristles and connecting their bases by a long hole through which wire is threaded with a trepanning needle. The wire is picked out at each hole, and the doubled bristles are looped through the wire which is then drawn tight.

Hand Leather
A loop of leather with a hole cut out for the thumb. Used to protect the hand when making wire-drawn brushes. Several turns of wire are taken round the leather-protected hand and pulled tight.

Brushmaker's Donkey
A block of wood with a concave rest at one end on which to place a broom stock when making a pan-set broom, i.e. when inserting bundles of bristles which have been previously dipped in pitch.

Measuring stick
Brushmakers have a measuring board mounted on the bench and marked with graduations to show the height from the bench. Used for measuring the length of bundles of bristles.

Brush, Workshop *Fig. 163*
Brushes commonly found in woodworking shops include:

(a) Bench Brush (Dusting Brush; Jamb Brush)
A wide, soft brush set in a semi-circular stock. Used as an alternative to the ordinary household Bannister Brush for cleaning shavings, sawdust, etc. from the bench top or workpiece; and by painters and others for cleaning off dust.

(b) Glue Brush. See *Glue Pot*.

(c) Painting Brushes. See under *Painting Equipment*.

Fig. 163 (a)

Bruzz: see *Chisel, Bruzz*.

Bucker: see Spoke Bridle under *Wheelwright's Equipment* (2) *Spoke Tools*.

Buhl (or Boulle) Work: see *Veneering and Marquetry.*

Builder's Tools *Fig. 164*
Woodworking tools used by building workers will
be found under their appropriate entries. Building
tools commonly found in woodworking shops are
illustrated. (See also *Plumber's Tools.*)

   (*a*) *Brick Hammer.* For rough cutting bricks.

   (*b*) *Hawk.* On which plaster, cement, or putty is
'served', i.e. ready to be taken off and placed in
position.

   (*c*) *Lathing Hammer or Hatchet.* For cutting and
nailing plaster laths. (See separate entry under
Hatchet, Lathing.)

   (*d*) *Line Pin.* When stuck into the joint between
bricks it holds the line which guides the bricklayer.

   (*e*) *Lewis.* For raising heavy blocks of stone. The
three 'tails' form a dovetail which is fitted in a
dovetail-shaped mortice cut in the top of the stone.

   (*f*) *Mason's Mallet.* For driving chisels when cutting soft stone.

   (*g*) *Plasterer's Trowel.*⎫ For laying and levelling
   (*h*) *Plasterer's Float.* ⎭   plaster.

   (*i*) *Slater's Ripper.* For cutting off nails when removing broken slates.

   (*j*) *Slater's Saxe* (Home-made examples). Used
for trimming slates to size and pecking holes for the
nails.

   (*k*) *Trowel.* Used by bricklayers for both laying
the mortar and, when necessary, for cutting bricks
to size.

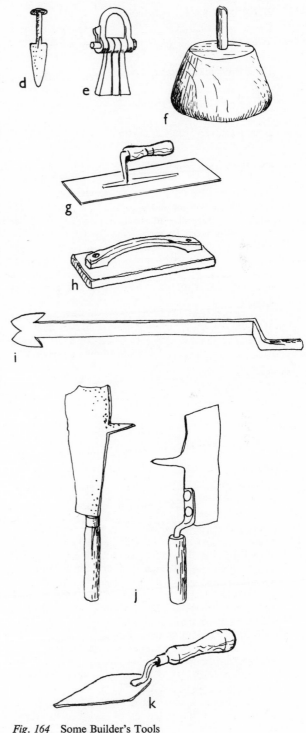

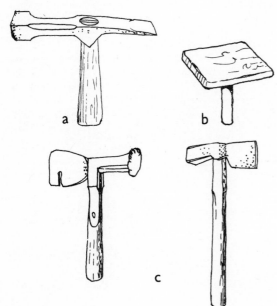

Fig. 164 Some Builder's Tools

Bullsticker: see *Wood Engraver*.

Bung Borer: see *Auger, Cooper's Bung Borer*.

Bung Burner: see *Auger, Cooper's Bung Borer*.

Bung Chain: see *Cellarman*.

Bung Removers: see *Cellarman*.

Burin: see *Wood Engraver*.

Burn Brand (Branding Iron; Marking Iron) *Fig. 165*
An iron rod about 2 ft 6 in long with a ring for
hanging at one end, and an oblong head at the other
end, on the face of which some device, number, or
letter is set in relief. The head must be fairly substantial in order to hold the heat. A later pattern
provides for loose or interchangeable letters or
figures of cast iron, fitted in a holder with a wooden
handle.

Used after heating to mark tools, timber, the heads
of casks, etc. by branding with indelible evidence of
ownership. (Those with very small heads are likely
to have been used for branding hoofs of horses.)

Fig. 165

Burning Irons, Cooper's: see *Auger, Cooper's Bung
Borer*.

Burning Rods, Wheelwright's
These are long iron spikes which are sometimes employed by wheelwrights for enlarging previously
bored holes, e.g. to enlarge a bolt hole, and for
burning away any roughness. According to Mr. C.
Spary (1955), some customers specified that all holes
were to be finished in this way since this sealed the
grain of the wood and would stop the rusting of the
iron bolts. Coachbuilders also used Burning Rods
to char the drain holes in the coach floor to prevent
rotting from frequent washing out.

Mercer (U.S.A., 1929) describes a Wheelwright's
Burning Iron. This is a tapering square spike with
a cross-handle; it was heated to burn out a square
hole to receive the upper square shank of a screw
bolt.

Burr
A hooked edge applied purposely to the edge of a
Scraper to act as a cutting edge (see *Scraper*).

Bush
A ring or collar for lining a hole, e.g. to serve as a
bearing.

Bushing Engine: see *Hub Boring Engine*.

Bushing Tools: see *Auger, Cooper's Bung Borer*.

Butcher's Felling Axe: see *Axe, Butcher's Pole*.

Butt
The thicker end of a tree or tool.

Butted (Butt Joint)
Timbers placed end to end with no special jointing
beyond squaring the ends.

Butter: see *Shingle Maker*.

Butter and Cheese Sampler (Butter or Cheese Hammer, Taster, Borer, or Tryer) *Fig. 166*
A gouge-shaped blade with an iron head shaped like
a Hammer, often with a claw on one end. In some
the head consists of an iron ring, with a claw on the
side. Though sometimes mistaken for a woodworking tool, it is used by grocers and others for
taking a sample of butter or cheese. The hammer
head and claw can be used for fixing labels or replacing hoops on casks. Those used for butter vary
in length from 14 to 28 in and those for cheese from
$3\frac{1}{2}$ to 5 in.

The blade, called by Plumrose, Ltd. (Denmark,
1969) a 'hollow sword', is pressed into the butter or
cheese, turned round and the core pulled out for
tasting or examination, after which the remainder
is replaced with the same tool. The largest Butter
Hammers are long enough to reach to the bottom
of a cask.

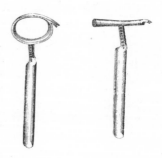

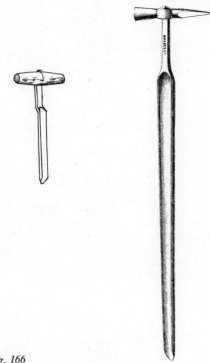

Fig. 166

Butt Howel: see *Adze, Cooper's Howel.*

Buzz: see *Scraper, Cooper's Buzz; Chisel, Bruzz.*

C

Cabinet Maker: see *Carpenter, Joiner, and Cabinet Maker.*

Calico Web: see *Web, Calico Printer's.*

Calliper Rule: see *Rule.*

Calliper

These tools, almost unchanged in appearance since Roman times, are used by many tradesmen for measuring workpieces or parts not directly accessible to the ordinary rule or scale, such as cylinders and circular holes. Sizes vary from about 4 to 30 in overall. The large iron Callipers used by wheelwrights when turning hubs were usually made by the local smith; similar Callipers in wood were often made by the tradesman himself.

The stiffness of the joint is often relied on to keep the legs apart, but in some cases a flat curved wing is riveted to one leg and secured at the desired position on the other with a screw. In another pattern the hinge end of the legs is made straight for a short distance, the upper part of one leg interleaving with the other. In others the joint is replaced by a spring, the distance between the points being regulated by a screwed rod and butterfly nut.

The shape of the legs depends on the kind of measurement to be made; but it is not uncommon to find the legs of some home-made Callipers made in the form of human legs, with the toes turned outwards.

Calliper, Bow: see *Calliper, Outside.*

Calliper, Coach-Hood *Fig. 181 (e)* and *(f)* under *Calliper, Wheelwright's*
A term sometimes applied to Callipers or Compasses made from old coach-hood arms. This was often done because they contained a ready-made hinged joint, and the arms only needed to be drawn down to a point and bent to a suitable curve.

Calliper, Double *Fig. 167*
Two calliper arms mounted on a central stem. Used by turners and others for testing two diameters, and by smiths when making nuts and bolts.

108

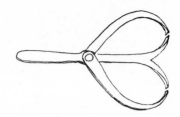

Fig. 167

Calliper, Double End (Double Sided Calliper) *Fig. 168*
Two Outside Callipers made like a figure-of-eight, and sharing a common riveted joint at the centre. Used for measuring when, owing to external projections, it is not possible to remove the Callipers from the work without opening them; the free end shows the measure taken by the other.

Fig. 168

Calliper, Double 'S' *Fig. 169*
A name given to Callipers with both legs shaped like a letter 'S', one reversed and sliding over the other. The object of this arrangement is probably to enable the user to open or close the legs with one hand.

Fig. 169

Calliper, Egg: see *Calliper, Outside.*

Calliper Gauge *Fig. 170*
A bar with a sliding jaw like an adjustable Spanner instead of the folding legs of an ordinary Calliper.

This is a smith's tool but is sometimes found in woodworking shops. The example illustrated is home made and is without calibrations. A Calliper Gauge combined with a rule is described under *Rule*.

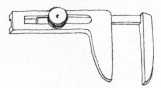

Fig. 170

Calliper, Hermaphrodite: see *Calliper, Jenny*.

Calliper, Hole and Socket: see *Calliper, Inside* and *Outside*.

Calliper, Index *Fig. 171*
Outside Callipers with a graduated scale attached to one leg and a pointer to the other, giving an automatic reading of the distance between the points. The legs are from 3 to 6 in long.

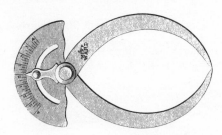

Fig. 171

Calliper, Inside (Straight Calliper) *Fig. 172*
A name given to Callipers with a riveted joint and straight tapered legs having a slight inward curve only at their ends. When the legs are crossed, the toes are turned outwards and consequently can be used for measuring the internal diameter of holes etc., and since the legs are straight, the tool can be used for measuring outside diameters within a confined space.

Fig. 172

Calliper, Inside and Outside (Hole and Socket Callipers) *Fig. 173*
The arms on one side of the joint circle round to form an 'outside' Calliper, while the legs extending on the other side, with feet turned sharply outwards, form an 'inside' Calliper. These vary in length from 3 to 10 in, but the home-made example (*b*) has legs almost 20 in long.

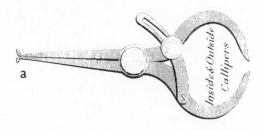

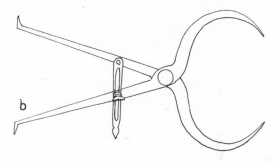

Fig. 173

Calliper, Jenny Leg (Odd Leg Callipers; Hermaphrodite Callipers) *Fig. 174*
Two straight legs, one tapering to a point, the other with a slight inward curve at the toe. Used for scribing lines parallel to the edge of a workpiece, along which the curved toe is made to bear. Normally used on metal, but useful to woodworkers when space does not permit the use of a Marking Gauge.

Fig. 174

Calliper, Lancashire *Fig. 175*

Factory-made Callipers are often known as 'Lancashire Pattern' (see also *Lancashire Tools*). In one type the legs are of flat steel with a riveted joint, the stiffness of which keeps the legs apart. The more characteristic Lancashire type is a well-shaped forging with tapered legs of a thicker, rectangular cross-section. There are two main varieties, Wing and Spring: the upper, straight part of the leg is shaped gracefully to provide for the wing in one case, or the screwed rod and spring in the other, with neatly chamfered corners giving the tool a very attractive appearance.

Callipers of similar design were imported from Germany and elsewhere during the nineteenth century and probably earlier. The continental pattern can be recognised in the case of some of the spring patterns by a low ridge raised on the outside of the spring. Some examples of Lancashire Callipers are shown below:

(*a*) Wing type
(*b*) Detail showing chamfers
(*c*) Small Spring Callipers.

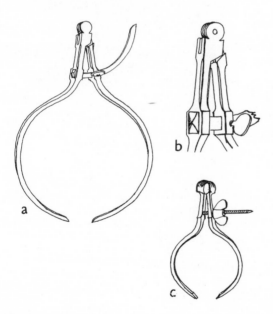

Fig. 175

Calliper, Mast *Fig. 176*

Large Callipers with round, tapered legs, up to 24 in long, are listed under this name for measuring the diameters of masts, but one would imagine that Callipers of other types would serve just as well.

Bright Callipers

Fig. 176

Calliper, Odd Leg: see *Callipers, Jenny Leg.*

Calliper, Outside (Bow Callipers; Egg Callipers) *Fig. 177*

A general term for Callipers with bowed legs, with the toes turned inwards. Used for measuring outside diameters.

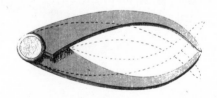

Fig. 177

Calliper, Rack Wing *Fig. 178*

Wing Callipers of the Lancashire type. Teeth are cut on the outer circumference of the wing to form a rack which appears to engage with an additional thumbscrew in the leg. It is not clear whether this thumbscrew provides a rack-and-pinion action or merely acts as a fixing screw. Long-established makers of Callipers and Compasses, such as Messrs. Peter Stubs of Warrington, have been unable to enlighten us on this point.

*Rack Wing
Callipers*

Fig. 178

(*b*) In this modern version, sometimes known as 'American Pattern', the legs are separate but connected at the top by a flat spring bent to a circular shape, fitting tightly into notches at the top of each leg, just above the centre of the loose stud which acts as a fulcrum.

Fig. 179 (b)

(*c*) A smith-made example, forged in one piece.

Calliper, Spring *Fig. 179*
Instead of a movable joint, the legs are connected at the top by a steel spring, called the bow. The distance between the legs is regulated by a wing nut fitted to the end of a threaded rod which is pivoted to the centre of one leg and passes through a hole in the other.

(*a*) The so-called Lancashire Pattern noted for its graceful design and fine forging. A similar pattern was imported from Germany during the nineteenth century. It can sometimes be recognised by a central ridge on the outside of the spring.

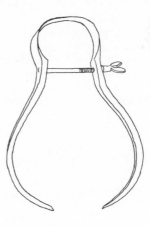

Fig. 179 (c)

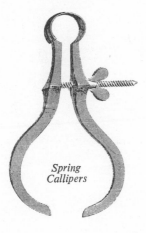

*Spring
Callipers*

Fig. 179 (a)

Calliper, Straight (Straight Leg)
A term sometimes applied to Callipers with straight legs, e.g. Inside or Jenny Leg Callipers.

111

Calliper, Timber (Log or Lumberman's Calliper) *Fig. 180*

A giant-sized Calliper Gauge made in metal or wood, 30 in or more long. One jaw is fixed, the other slides along the bar which is graduated and indicates the distance between the small projections facing each other at the end of the jaws. Used for measuring the size of timber, both in the log or sawn.

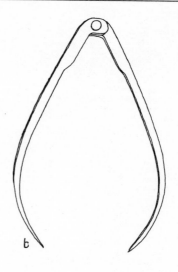

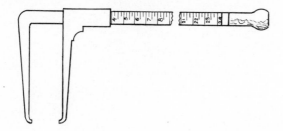

Fig. 180

Calliper, Wheelwright's *Fig. 181*

Wheelwrights use all kinds of Callipers, and those found in their workshops are frequently smith made. The larger ones, with legs up to 30 in overall, were used for measuring the diameter of hubs when turning them on the Lathe. The following are typical examples:

(*a*) Smith-made, with interleaved shoulders
(*b*) Smith-made, with flat legs
(*c*) and (*d*) Factory-made, Lancashire type
(*e*) Smith-made from old hood irons, with adjustable strut
(*f*) A small pair made from old hood irons.

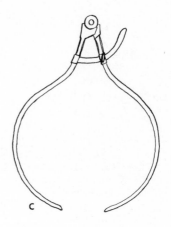

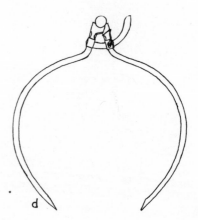

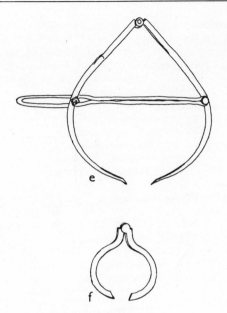

Fig. 181   Wheelwright's Callipers

Calliper, Wing *Fig. 182*

A term applied to Callipers in which a curved wing is let into one leg and passes through a hole in the other, and is used for fixing the legs in the required position with a small thumbscrew. Often of the Lancashire pattern, many are well forged and nicely finished, with the end of the wing cut to an ogee contour.

Black Wing Callipers

Fig. 182

Cam *Fig. 183*

A wheel, or projecting part of one, mounted eccentrically in order to move some other object, e.g. for lifting a bench stop, as illustrated; or for operating the lever-cap of a metal Plane; or (in engineering) for lifting the valve of an engine.

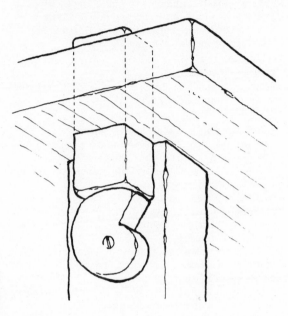

Fig. 183

Candle Holder: see *Lighting, Workshop.*

Cane Cutting Tools: see *Woodwind Instrument Maker.*

Cane Squeezer: see Kinking Tongs under *Basket Maker.*

Cant

A name for the outer pieces of a cask head; but also used to describe a bevelled or sloping surface, e.g. a Cant Chisel.

Cant Board, Coachbuilder's

A templet made in the form of a full-size plan of the carriage to indicate the shape and position of the various parts. The sizes of the parts were written on the board, including a cross-section of all the pillars.

Cant Dog or **Hook:** see *Timber-handling Tools.*

113

Cap Iron
The back plate of a Plane cutting iron. Described under *Plane*.

Capstan
An upright post fitted with handles radiating from it, used as a means of hauling or tightening by rope; or a turret with tools projecting radially, as on a Capstan Lathe; also, a name given to a Hooping Windlass. (See *Cooper* (*3*) *Hooping Tools*.)

Carcase
Main part of a structure, e.g. of a house; or the frame of a piece of cabinet work.

Carpenter, Joiner, and Cabinet Maker
The word carpenter is believed to have come originally via the Latin from the Celtic word for carriage maker, a trade at which the Celts were in some ways more advanced than their Roman conquerors. Up to the end of the sixteenth century, when most secular building in northern Europe was almost entirely in wood, the master carpenter was the leading tradesman in the industry.

Woodworkers in the building and allied trades today may be divided as follows: carpenters who work on the main structure; joiners who make the stairs, doors, and internal fittings; and cabinet makers, who specialise in the movable furniture.

(a) The Carpenter
Since carpentry is usually carried out on the site and is concerned with comparatively large timbers nailed or pinned together or framed up with simple joints, the carpenter's tools are confined in the main to the larger and basic types of Adze, Axe or Hatchet, Saws, boring tools, chisels and Hammers; and measuring and testing tools such as the Plumb Rule and Level, Chalk Line, and Rule. Ancillary equipment includes Sawing Horses, Steps, Trestles, Ladders, Hoisting Gear, and a simple form of Bench.

(b) The Joiner
In the late fourteenth and early fifteenth centuries, higher standards of domestic architecture were called for; and when improved tools and methods made possible the making of furniture and house fittings from smaller sections of wood, cunningly jointed together into frames, instead of the previous nailed-plank method, the art of the joiner became separated from that of the carpenter. The earliest references to 'joiner' in this sense in the *O.E.D.* are dated 1386 and 1412.

The joiner made sash windows, and the change of emphasis in social life from the large hall with an ornamental roof and no ceiling to smaller rooms on separate floors, often connected by elaborate staircases and decorated with panelling on walls and ceilings, helped to enhance the importance of the joiner. Other changes of fashion, such as the increased importance of mouldings, and the consequent refinement of methods of jointing, also served to widen the gap between the carpenter and the joiner and the number and range of the tools used.

The joiner's kit had to include not only the everyday tools of the carpenter, but also many special Planes, including the range of 60 or more Moulding Planes, including hollows and rounds, still to be found in the older workshops. These became a distinctive feature of the joiner's kit. In addition, the kit included fine-toothed Back Saws for cutting joints, Bow Saws for curved work, and a wide range of testing and marking tools.

Today only the basic tools are required; all sash and door stock is prepared by machine, and a wide range of work is done with portable electric tools based on the hand drill.

(c) The Cabinet Maker
According to the *Oxford English Dictionary*, the term cabinet maker was first used in 1681. The reason for this may be that it was not until the late seventeenth century that it was found necessary to have a special word in English to denote the joiner, who specialised in making movable furniture as opposed to the fixed, internal fittings of houses. It was also about this time that oak, and other home-grown woods which had been used up to then, were largely superseded by walnut and later in the eighteenth century by mahogany, satinwood, and other exotic woods from overseas. These woods were expensive, and consequently increasing use was made of them in the form of veneers, an art which was known to the Egyptians but had fallen into disuse. Consequently, although the cabinet maker used much the same tools and methods as a joiner, he required in addition a special range of equipment for making furniture, as well as for veneering, marquetry, and similar decorative work. (See also *Chairmaker* and *Veneering and Marquetry*.)

Carpenter's Furniture and **Equipment**: see *Workshop Equipment*.

Carpet Layer: see *Upholsterer*.

Carriage Builder: see *Coachbuilder*.

Carrying Stick (Carrying Cane; Frail Stick) *Fig. 184*
A crooked or S-shaped stick, often cut from the

hedge. One end of the crook lies on the shoulder, the other forms a handle. It is used for carrying a bag of tools, and a wooden pin is set near the end to prevent the bag from slipping off.

(*a*) A home-made carrying stick used by a Hertfordshire farrier when visiting farms and neighbouring forges.

(*b*) A factory-made version with an iron swivel-hook at the end of the stick from which to suspend a tool bag.

Fig. 184

Carvel

A method of boat construction in which the side planks are fitted edge to edge.

Carver: see *Wood Carver.*

Carver's Bench Screw *Fig. 185*

An iron screw with a square thread, a screwed point at one end, and a short, square section at the other. The nut has two square wrench holes in the wings. Made in sizes from 6 to 12 in long. It is used for fixing planks or blocks to the top of the Bench for carving. A small hole is bored in the underside of the wood and the screw passed through a hole in the Bench from below and screwed in firmly, the wing of the nut being used as a spanner. The nut is then screwed up tight, thus holding the wood firmly without anything being in the carver's way, as it would be if a Holdfast or G-Cramp were used.

Fig. 185

Carver's Clip *Fig. 186*

A turn-button with one end serrated and the other both clawed and serrated. Two or more Clips are screwed to the Bench to hold a workpiece while being carved.

Fig. 186

Case Hardening: see *Iron and Steel.*

Case Maker: see *Box and Case Maker.*

Case Opening Tools: see *Box and Case Opening Tools.*

Cask Pulley: see *Cellarman.*

Caul

A piece of flat or curved wood used for holding down a veneer or other material while the glue sets. (See *Veneering and Marquetry.*)

Caulk

Filling a joint to make it watertight. (See *Caulking Tools.*)

Caulked Handle *Fig. 51* under *Axe*

A handle with a swelling at the foot, on the same side as the cutting edge of the tool, as provided on some Billhooks and Axes. The calk is intended to prevent the tool from flying from the hand when making a long hard stroke. (The 'calkin' on a horseshoe is a thickening at the heel to prevent slipping.)

Caulking Iron, Cooper's: see *Flagging Iron.*

Caulking Tools *Fig. 187/1–3*

Ship's Caulking Tools are used to force stranded oakum into the seams between planks on the deck

115

and ship sides to make the ship watertight. For this purpose the edges of the planks are very slightly bevelled to a distance of one third of their thickness, thus presenting an open seam into which the oakum is forced. The outside planks are caulked in this way only if carvel-built (butted planks), and not when clinker-built (overlapping planks) (see *Shipwright*). The tools and equipment used include the following:

CAULKING MALLET (Ship-Carpenter's Mallet) *Fig. 187/1*.
A long-headed wooden Mallet, used for driving Caulking Irons. The head is made of beech, lignum vitae, or 'live oak' (*Quercus virens*, a very hard oak from the U.S.A.). The head measures about 13 in long and $1\frac{3}{4}$ in across the faces, which are circled with thick iron rings. These rings are usually made to taper in thickness from the face backwards. The long head allows for wear; the rings can be moved back when necessary. The central portion of the Mallet is enlarged and oval-shaped in section. It is bored centrally for a round handle, one end of which, either straight or tapered, is often left protruding above the head.

Fig. 187/1

A unique feature of these mallets is the longitudinal slots in the head. Holes are bored through the head at points about 2 in on each side of the eye. From these holes, a saw cut is made to a point about $1\frac{1}{2}$ in from each face. To prevent the head from splitting in half (as a consequence of being partly severed by these slots), the central boss of the Mallet is held together by two horizontal rivets, one on each side of the eye.

The Caulking Mallet illustrated in Smith's *Key* (Sheffield, 1816) is apparently not provided with these slots, but that illustrated by Rålamb (Stockholm, 1691) is so provided, and it also has a central boss with two transverse rivets, just as provided today.

Many shipwrights take a particular interest and pride in their Caulking Mallet. They will tell you that a good Mallet, properly looked after, made the work easier and, when correctly slotted, was a pleasure to listen to. The late Mr. C. Bunday, a shipwright of Burlesdon, Hampshire, told us that a shipwright always tried to get a better ring from his Mallet than his workmate did. This was done by adjusting the slots and ferrules. 'A good caulking mallet is worth looking after like a gold watch.'

Other reasons for slotting the head are given by two shipwrights as follows: Mr. Dornum of Salcombe, Devon: 'The old shipwrights liked to hear the mallet sing; the slots in the heads cause the singing. All these things were done for a good reason – if several men are caulking the deck together and they used ordinary mallets, they'd deafen each other.' Mr. G. Worfolk of King's Lynn said: 'The slots make the mallets whistle, like birds. You can stand a quarter of a mile away and if you hear them whistle, you know the caulking's been done right.' (This mention of birds may explain the strange name 'Chirping Caulking Mallets' used in the 1872 Tool Catalogue of James Howarth.)

This quality of the sound may be demonstrated by striking the head of a Caulking Iron first with an ordinary Mallet, and then with the Shipwright's Caulking Mallet. The former gives a wooden, jarring noise, the latter an almost musical note. That these sounds were familiar is evident from a sentence in Flaubert's *Madame Bovary*, Part 3, Chapter 3, (1856): 'It was the hour of day when you hear the caulkers' mallets ringing against the ships' hulls along the dockside.'

Note: An American Shipwright, Mr. R. Paterson, has informed us (1968) that a partridge wood Mallet head has more 'life' in it than lignum vitae and consequently is preferred by American caulkers.

CAULKING IRONS *Fig. 187/2*
These all-steel, chisel-like tools, usually about 6–7 in long, are mushroom-headed and their blades are mostly flared – a shape known as 'fantail'. Their edges are either sharp, blunt, or provided with grooves known as creases. They are driven by means of the Caulking Mallet.

The seam is first opened (when necessary) with a Reaming Iron, and the threads of oakum are then

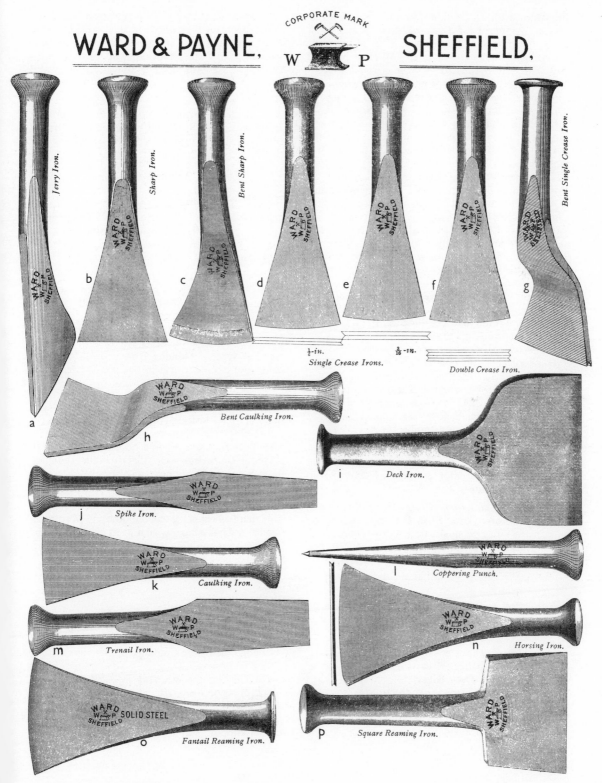

WARD & PAYNE,

CORPORATE MARK

W P

SHEFFIELD,

Jerry Iron.

Sharp Iron.

Bent Sharp Iron.

Bent Single Crease Iron.

½-in. ⅜-in.

Single Crease Irons.

Double Crease Iron.

a b c d e f g

Bent Caulking Iron.

h

Deck Iron.

i

Spike Iron.

j

Caulking Iron.

k

Coppering Punch.

l

Trenail Iron.

m

Horsing Iron.

n

Fantail Reaming Iron.

o

Square Reaming Iron.

p

Fig. 187/2 Caulking Irons (Ward & Payne Ltd., Sheffield, 1911).

driven in with a Caulking Iron. The oakum is further compressed ('hardened down') and sunk below the surface with a Making Iron. The seam is then filled ('payed') with pitch. Much experience and skill were needed to judge how much oakum should be forced into the seam; too little would not keep the water out, but too much could spring the planks apart and even shear off a bolt or trenail.

Caulking Irons have been made in their present form at least since medieval times. A thirteenth-century representation of caulking may be seen in a most remarkable and beautiful high-relief sculpture on the outside of the Cathedral of St. Mark in Venice. Shipbuilders are shown working on a hull with Caulking Iron and Mallet. In eastern Asia and China, Caulking Irons are sometimes made in the form of stout Chisels with flared blades and wooden handles, and are driven by a Hammer.

Caulking Irons are made in various shapes including the following:

(*a*) *Jerry Iron* (Hook Rave; U.S.A. – Reaping or Cleaning Iron; Meaking Iron. The *Oxford English Dictionary* states that according to Mr. G. Crocker of H.M. Dockyard, Devonport, the term Meaking Iron is now often misapplied to the Making Iron.) This tool has a diagonal edge, tapering in thickness from front to back in order to clear itself when being driven along the seam. About 12 in long and $1\frac{7}{8}$ in wide. Used for running old oakum out of the seams before re-caulking. See HOE below which performs a similar function.

(*b*) (*c*) *Sharp Iron*
Similar to the Set Iron but the edge is sharpened like a Chisel. Used for cutting out defective or unwanted threads of oakum.

(*d*) (*e*) (*f*) *Making Irons* (Crease Iron; Blunt Iron) With a flat or grooved edge. The grooves (known as creases) may be single, double, or treble. Used after caulking to dress down and compress the oakum, and thus leave sufficient room for the subsequent insertion of pitch.

(*g*) (*h*) *Bent Iron*
With offset shank. Used for caulking in places which cannot be reached with a normal Iron, e.g. around deck combings.

(*i*) (*o*) (*p*) *Reaming or Deck Iron* (Deck Iron; U.S.A. – Dumb Iron; Ream Iron; J. Howarth, Sheffield, 1884 – Deck River)
A wedge-shaped blade, either fantail or square-shouldered in outline, about 9 in long and 3 in wide. Used for opening a tight seam before caulking.

(*j*) *Spike Iron* (U.S.A. – Sharp or Butt Iron)
A narrow blade tapering down to about $\frac{3}{4}$ in width. Used for caulking in narrow spaces, e.g. the ends of deck planks where they taper off, or the corners of hatchways and around the shoes of guard stanchions.

(*k*) *Set Iron*
A name given to the most ordinary of Caulking Irons with straight fantail blade, either sharp or blunt, of $1\frac{3}{4}$–$2\frac{1}{2}$ in width. Used for driving the caulk.

(*l*) This is not a caulking tool. See *Punch, Coppering*.

(*m*) *Trenail Iron* (Trunnel Iron)
Like the Spike Iron but usually with a blunt edge, about 1 in wide. Used for splitting and spreading the head of a trenail before inserting a wedge or caulking material.

(*n*) *Horsing Iron* (Hausing Iron)
Like the Making Iron, but larger, and used for the same purpose when dealing with thicker planks. Operated by two men, one holding the neck of the tool by means of an iron rod (known as a 'Bridle') while the other strikes with a Beetle. Another type of Horsing Iron is made with the blade and handle forged in one piece, resembling a Butcher's Cleaver.

(*o*), (*p*). See (*i*) above.

(*q*) *Boot Iron.* Illustrated by C. Drew & Co., (U.S.A. *c.* 1920).
This has a foot splayed in a boot-shaped form. It is used to caulk deck seams that run under the sides of deck houses and under the cat-heads.

(*r*) *Caulking Wheel.* Listed by C. Drew & Co., (U.S.A. *c.* 1920).
This small wheel is mounted on the end of a short handle and has a narrow edge at its circumference. It is used for caulking small boats where a single strand of 'wicking' is being layed.

CAULKING BOX
A wooden box with a sliding lid, used by shipwrights for holding Caulking Irons. There is usually a ring on one end of the box for carrying over the shoulder with the handle of the Caulking Mallet.

OIL BOX (Oil Chock) *Fig. 187/3*
A block of wood measuring about $7 \times 3 \times 3$ in, with a short handle fixed at one end. The centre of the block is hollowed out and filled with linseed oil. Used for dipping Caulking Irons to prevent them

from becoming sticky from contact with oakum and so sticking in the seams.

Fig. 187/3

PITCH-MOP and LADLE *Fig. 187/4*
A small tufted brush at the end of a 5 ft handle. The Mop is dipped into hot pitch and then 'brushed' along the outer planks of a carvel-built ship to fill the seams after caulking. The unwanted pitch is scraped off afterwards. For the deck a Ladle about 2 ft 6 in long with a narrow spout is used. Hot pitch (sometimes called 'glue') is also poured into the seams between the deck planks after caulking, and the unwanted pitch scraped off.

Fig. 187/4

HOE (Rake; Rave Hook; Rove Hook)
A steel hook usually forged from an old file, and flat in section. Used for raking out 'old glue' (i.e. old pitch) and the rotted oakum below it. (See also Jerry Iron under *Caulking Irons* above.)

NOTE ON THE CAULKING OF IRON SHIPS
This is done with Cold Chisels whose ends are ground at a blunt angle for flush plates, and at a sharper angle for lapped plates. An indentation is made along the edges of the plates and around rivet ends, and this upsets and spreads the metal in order forcibly to close up any crevices.

Cavetto: see *Plane, Moulding.*

Cellarman *Fig. 188*
The cellarman is in charge of the cellars and warehouses where wines and liquors are stored. Their special tools are found among the equipment of the bottler, in the cellars of public houses and hotels, or in the pantry of a wine waiter. But they are also often found among the tools of the cooper – hence their inclusion here. Some of the tools used in this trade are described below:

(*a*) BUNG CHAIN [not illustrated]
A bunch of chain made up of three or more lengths, with small pointed plates woven into each link, used for cleaning out casks. After inserting the Chain through the bung-hole, the cask is revolved by rolling or by spinning on a machine made for the purpose. The Chain rubs off the unwanted deposits from the inside surfaces of the cask. A plug on the end of the Chain, which remains outside, enables the chain to be pulled out afterwards.

(*b*) BUNG REMOVERS *Fig. 188 (b)*
The bung-hole in the side of a cask was at one time stoppered with a tapered wooden plug; today the hole is usually lined with a metal bush and stoppered with a wooden disc known as a shive. The smaller hole in the head of the cask is corked until fitted with a tap by the innkeeper or cellarman. Since the liquor will not flow from the tap without an air inlet, a small hole is bored through the shive with a Brewer's Gimlet and is stoppered with a small tapering peg called a spile which, when lifted out, allows air to be drawn in and the liquor to flow. The following tools are used for removing bungs and shives from casks:

1. *Bung Chisel and Bung Pick*
A solid iron Chisel 5–7 in long with the sharp end bent at an angle. The chisel is driven into the heart of the bung or shive which is then levered out. This method of removal is used if the bung will not respond to the Tickler or Flogger, but an even rougher method is described by Brombacher (U.S.A., 1922). This is a 'Bung Pick' – a Hammer with a chisel-ended pick on the head.

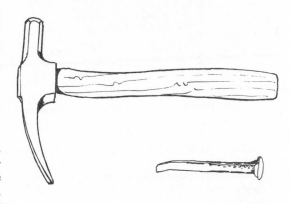

2. *Bung Tickler*

A curved blade 5–7 in long, flattened and pointed, with a wooden cross-handle or made in iron throughout. Used for easing out a bung by inserting the point and levering upwards.

3. *Flogger* (Bung Flogger; Bung Start; Starter)

A narrow headed Mallet mounted on a flexible handle about 2 ft long which is sometimes made of cane. Used for 'starting' the shive or bung by striking the bung-stave close to the bung. It is also used for 'sounding out casks'. If on striking the cask the sound is a dull thud, then the cask is full; but should the cask ring, then the cooper or cellarman knows that some of the contents are missing. Mr. Bob Gilding, a retired cooper, who had

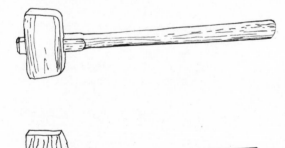

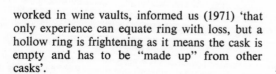

worked in wine vaults, informed us (1971) 'that only experience can equate ring with loss, but a hollow ring is frightening as it means the cask is empty and has to be "made up" from other casks'.

4. *Bung and Shive Extractors* (Shive Drawer; Shive Vice; Messrs. Wm. Gilpin (Cannock, 1868) calls this tool a Cooper's Travice.)

Earlier types consist of a U-shaped forging fitted with a central screw which looks and works like a skirted Cork Screw. A modern tool, of French design, is in the form of a Hammer with a hollow 14 in handle in which a captive rod, with a screw on the end, slides up and down. After screwing the rod into the shive, the hammer-head is grasped and pulled smartly upwards in a series of jerks, and this removes the shive.

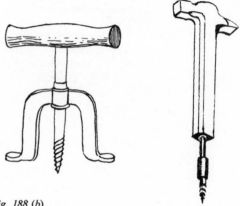

Fig. 188 (b)

(*c*) CHIME HOOK (Cleek; Cleck) [Not illustrated]
Like the Bag Hook, but with a flat, serrated claw instead of a hook. Used for lifting the barrel by catching the claw under the chime.

(*d*) COCK PLUG BIT (Plug Bit; Tapping Bit) *Fig. 188 (d)*
A Brace Bit $\frac{3}{4}$–1 in. in diameter, with a tapering skirt which acts as a plug to prevent the escape of liquor when boring a full cask before fitting a tap. (See *Bit, Cock Plug.*)

Fig. 188 (d)

(*e*) CORKING TOOLS *Fig. 188* (*e*)
Those used by bottlers include:

1. *Cork Driver*

A flat piece of hardwood, pared down at one end to form a grip, and used for driving corks.

2. *Cork Squeezer* (Cork Gripes; Cork Press)

A lever press, 8–12 in long, hinged at one end, with serrated openings to fit different sizes of cork. Used to compress and soften a cork before fitting it into the neck of a bottle.

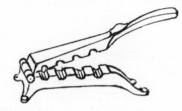

Fig 188 (*e*)

3. *Corking Tube*

A wooden tube about 6 in long, with a metal piston. Used for forcing a cork into the neck of a bottle.

(*f*) CORK REMOVERS *Fig. 188* (*f*)
Various instruments for removing corks from bottles (or from the heads of casks) include:

1. *Bottle Opener*

A spiked blade fitted to a wooden handle 6–7 in long overall. The blade is sometimes saw edged. It is used for breaking wire and sealing wax, and could also be used for easing out a cork. A short brush is sometimes fitted into the handle for removing dust from the bottle neck.

2. *Cork Lifters*

These instruments are used for removing corks accidentally pushed down into the bottle – a function which they perform with surprising efficiency. They are provided with barbs or hooks which hold the fallen cork while being pulled out.

There are three kinds all about 6–8 in long. One is the so-called 'Shot Rake' which consists of a flat iron bar with barbs at its lower end. This is an interesting example of a tool-name outliving its original meaning: before being adopted by the wine trade, the Shot Rake was employed for removing the wad from the barrel of a muzzle-loading gun after a mis-fire. The other type is known as a Cork Drawer. This consists of two springy steel legs with inward-pointing hooked toes enclosed by a loose ring and held in a short cross handle. After being lowered into the bottle, the ring is pressed downwards thus gripping the cork, after which it can be drawn out through the neck of the bottle. A third type, known as a Dumb Waiter, is similar to the Cork Drawer, but there are no barbs on the legs which are made of thin, flat spring steel. In order to inspect the contents of the bottle without damaging the cork, the legs are forced down on either side of the cork which is then eased upwards.

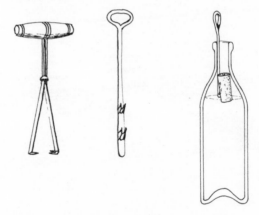

3. *Cork Screw* (Worm; Bottle Screw)

These common implements consist essentially of a spike formed into an open spiral which can be screwed into a cork. Of the innumerable varieties listed by the makers of 'Steel Toys' in Birmingham and elsewhere, some are illustrated here. The 'Cork Extractor' operates like a Cork Lifter.

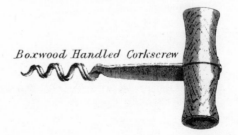

Boxwood Handled Corkscrew

121

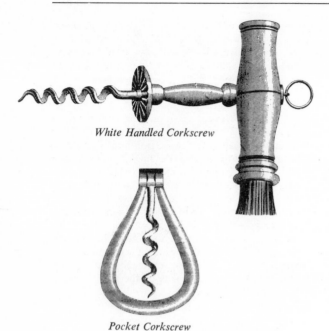

White Handled Corkscrew

Pocket Corkscrew

German Silver Case Corkscrew

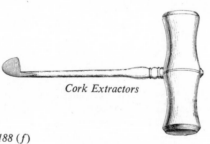

Cork Extractors

Fig. 188 (f)

1. *Brewer's Gimlet* (Spile Gimlet)
For boring a tapered vent hole in the shive which is subsequently stoppered with a spile.

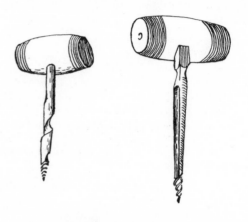

2. *Wine Fret* (Cooper's or Brewer's Fret)
A Gimlet used for boring holes in a cask in order to take a sample (see SAMPLING TUBES below).
This is done when the casks are piled one on top of another, so making it impossible to remove the bung. The hole is repaired afterwards by driving in a spile.

Fig. 188 (h)

(*g*) ERASING IRON [Not illustrated]
Tools for removing brands and other marks from casks and boxes are described under *Scrapers; Plane, Box Scraper; Drawing Knife, Cooper's Round Shave.*

(*h*) GIMLET *Fig. 188 (h)*
The following Gimlets are used by cellarmen (see also separate entries under *Gimlet*):

(*i*) PLIERS, PINCERS, etc. *Fig. 188 (i)*
Those used by cellarmen include:

    1. *Champagne Pliers*
Side cutting, sometimes with a brush on one of the handles. For cutting the wire which holds the cork.

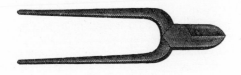

2. *Neck Tongs*
Made in several patterns, but essentially an instrument with long handles and rounded jaws to grip the neck of a bottle. It is used when a cork is stuck fast and cannot be drawn out. The Tongs are heated and applied to the neck, after which the neck can be broken off clean.

3. *Wax Tongs*
With round jaws, serrated on their inner surface. Used for removing wax from bottle necks.

4. *Spile Nippers*
With the jaws flattened so that they can grip a spile at the point where it emerges from the shive and so pull or lever it out.

5. *Cooper's Pincers*
Similar to the Carpenter's Pincers except that the foot of one handle is turned outwards at right angles to form a spike. The purpose of the spike is unknown, but it may be for easing out a shive from the bung hole.

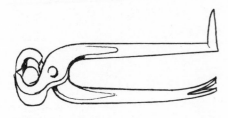

Fig. 188 (i)

(*j*) PUTTER [Not illustrated]
A pole with one end shod like a Boat Hook. Used for moving casks about.

(*k*) SAMPLING AND TASTING TUBES (Velincher etc.)
Fig. 188 (k)
These appliances are used for extracting a sample of liquor from a cask. This is done by introducing a tube called a Velincher through the bung hole; but if a cask is stacked in such a way that the bung is covered, a Tasting Tube is used instead. For this purpose a small hole is bored with a Wine Fret (*h*) above, usually through the head, and then plugged afterwards with a spile. Knight (U.S.A., 1877) says that the Velincher is sometimes called a thief-tube. 'The sucking-tube or monkey-pump, as sailors call it, is a straw or quill introduced through a gimlet-hole into a barrel, to draw the liquid therefrom. . . . It has also the merit – though it needs no extraneous recommendation – of being at least as old as Xenophon, who described this mode of pilfering from the wine jars of Armenia.' The author witnessed an instance of this when being shown over a cooperage and cask storage in Glasgow (Lowrie & Co., 1947). A small man suddenly darted from behind a cask. The Foreman caught him and snatched from him a brass tube which he carried. He had been sucking out the dregs from an old cask; a common practice, apparently, but dangerous, since the drinker is apt to smoke and set fire to the spirit-soaked timber. Many people have written about this practice of 'sucking the monkey', including the author's father who related how, when he was a young doctor at the London Hospital, men were brought in occasionally after being found dead-drunk in a wine warehouse. Lying prostrate on the floor, they had developed pneumonia from which, in most cases, they died.

1. *Velincher* (Valincher; Velinche; Sampling Thief)
A tube or pipette made in many different patterns but essentially a tube open at each end, the lower orifice being reduced in diameter so that when dipped into the liquor, a finger can be closed upon the upper end and the liquid will not flow out until the finger is released.

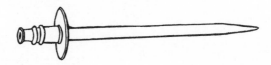

2. *Tasting Tube* (Sampling Tube)

A small horn-shaped tube, sometimes made of silver, about 7 in long. After boring a small hole with a Wine Fret, the thin end of the tube is inserted with the wide end upwards. When the horn is full, it is turned downwards, the sample collected in a glass, and the hole plugged.

Fig. 188 (*k*)

(*l*) SKID (Cask Pulley) *Fig. 188* (*l*)

Two strong posts, from 6 to 12 ft long, held apart at intervals by iron spacing rods. These are laid at an angle from a pavement hatch to the cellar floor beneath, and used for lowering casks.

(*m*) TILTING HAUNCH *Fig. 188* (*m*)

An iron tube, about 2–3 ft long, made like a Lifting Jack and extending to double this length by means of a screwed internal rod operated by a large fly-nut. A two-pronged fork is provided at top and bottom. Used in the cellar for tilting a cask so that the last drop can be drained from it. One of the forked ends is set against the wall, while the other catches under the chime hoop. As the haunch is lengthened, the cask tilts.

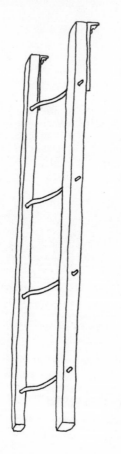

Fig. 188 (*l*)

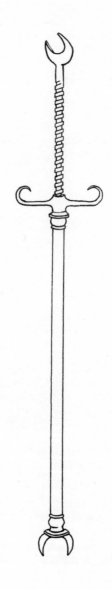

Fig. 188 (*m*)

(*n*) SCOTCH [Not illustrated]
A wooden or metal wedge placed at the side of a cask or wheel to prevent it from moving.

Centrifugal Drill Brace: see *Drill Archimedean.*

Chaif
Term used by Scots Coopers for Chiv. See *Plane, Cooper's Chiv.*

Chair Maker (1) MAKER OF WINDSOR CHAIRS
Fig. 189
This account is based on an article kindly contributed by Mr. L. John Mayes, formerly Director of the High Wycombe Library and Museum. It deals with the chairmaking trade in High Wycombe which started in the period 1780–1790. Long after most industries had become mechanised the chair trade stuck to hand work. One of the main reasons for this was the bottleneck in the supply of raw material; not the trees themselves, but in the first process – that of converting round timber into plank. At first the only method of doing this was by sawyers working over the traditional saw pit. (See *Sawyer.*) Chair manufacture became mechanised only in the second half of the nineteenth century after the development of the Band Saw.

Fig. 189   Windsor Arm-chairs

Materials
The following woods are most generally used in making Windsor chairs: turned parts (legs, stretchers, sticks) – beech; solid seats – elm; bent parts (bows for back and arms) – yew or ash; ornamental work – fruit woods from local orchards.

The Workshop
In the yard would be found the 'dipping tank', where the chairs were stained, and, most important, the 'stove'. This was a brick-built room provided with racks all around the walls to hold the chair parts and having a fire box with a long internal metal flue to distribute the heat. It was the forerunner of the modern drying kiln.

The inside of the factory would appear very cluttered and probably very gloomy to modern eyes – cluttered because every inch of wall space would be covered with full-sized patterns of chair pieces, or with tools, or both; and gloomy because most chair masters provided no glazed windows, just oiled calico. If the men wanted glass windows they paid for them, but many men did not. After dark the scene was equally gloomy since the chair master provided no artificial light and each man had his own oil lamp to light his own bench. Even when gas lighting became general, the men often paid a regular weekly sum for the gas consumed, and so there was no temptation to over-light the factory.

Making up the chair

Elm plank, two inch stuff, was the favourite material for chair seats and there were men known as 'bottomers' who specialised in providing sawn seats to the trade, 'bottomed' with the chairmaker's Hollow Adze. Bows for Windsor chairs were often bought in from men who specialised in bending them from sawn ash squares for common chairs, and from yew for better chairs.

All the parts, whether made in the factory or bought in, came at last to the chair framer. He bored the seats for the legs (see *Bodger* below) and the legs for the stretchers, He cut the mortices and tenons for the other assembly work and made all fast with wedges and glue and then finished the whole chair with Shave and Scraper, ready for staining and polishing. Cane or rush seating was often done by outworkers, mostly women; and children could be seen in the streets carrying bundles of chairs home or to the factory, or loaded with bundles of cane or rushes – a sight which gave considerable pleasure to many people who maintained that the only necessity for a contented working class was unlimited work from the cradle to the grave.

The more specialised tools of the chair maker are described under the following entries: *Adze, Chairmaker's; Bit, Chairmaker's; Brace, Chairmaker's; Chairmaker* (2) *Bodger,* and (3) *Equipment; Chisel, Chairmaker's; Hammer, Chairmaker's; Knife, Hooked* (*Hook Shave*); *Saw, Armchair-Maker's; Saw, Bettye; Scraper, Chairmaker's; Shave, Chairmaker's* (including straight; curved; Bottoming; Smoker Back; Travisher).

Chairmaker (2) THE BODGER *Fig. 190*

Many of the small factories bought in almost all the chair parts ready made, including the legs, stretchers and other turned parts which could be produced well and cheaply in the extensive local woods using beech, the 'Buckinghamshire Weed'. Most obligingly, this wood can be worked and turned while quite green, and can be subsequently dried without warping – it merely shrinks, and produces the characteristic oval form in the turned parts of chairs.

A pair of men bought a stand of trees, and after felling, moved their primitive hut into the clearing so formed. The logs were supported on Sawing Dogs, sawn into lengths, and then split into billets using a Splitting-out Hatchet. The billets were chopped to a rough leg shape with the Chairmaker's Axe and then transferred to the Shaving Horse, in which the billet was held while the operator shaped it with a Drawing-Knife. The pieces then went to the Pole Lathe where the final finish was given and the distinctive patterns cut. A pair of 'Bodgers' (the Wycombe name for the woodland turners) could produce from twelve to fifteen gross of legs and stretchers from tree trunk to finished article in a week – for which, in the 1914–18 period, a man could earn 18s. to 20s. including the money obtained from selling the waste wood for firing. This was only a very little above the labourer's rate.

The tools used by the chair bodger, who turned the legs and stretchers, are illustrated below and are described under their appropriate entries:

(*a*) *Measuring Stick* (*Dotter*)
(*b*) *Hand Saw*
(*c*) *Cross-Cut Saw*
(*d*) *Splitting out Hatchet*⎤
(*e*) *Mallet*   ⎬ See *Axe, Chairmaker's*
(*f*) *Side Axe*   ⎦
(*g*) *Drawing Knife*
(*h*) *Shaving Horse*
(*i*) *Froe*
(*j*) *Pole Lathe* (see *Lathe, Pole*)
(*k*) *Turning Chisels and Gouge*
See also *Shaving Horse; Woodland Trades*.

a

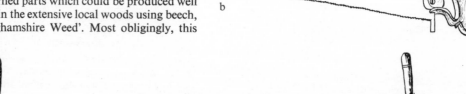

b

c

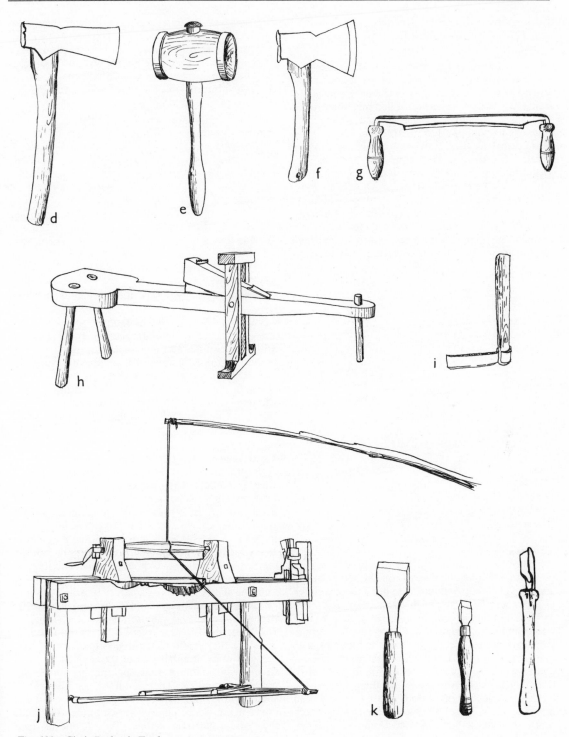

Fig. 190 Chair Bodger's Tools

Chair Maker (3) WORKSHOP EQUIPMENT
Fig. 191

(a) Bending Mould
Mr. L. J. Mayes (High Wycombe, 1955) relates:

'When bending bows for Windsor chairs the sawn square of timber, or the grown stakes in the case of best work, were steamed or boiled and then forced into shape around a wooden mould which much resembled a common chair seat. Every size and pattern needed its own individual mould. One of the greatest of the windsor chairmakers, the late Jack Goodchild of Naphill, was always experimenting with bow shapes, and to avoid having to make and scrap unlimited numbers of moulds, he obtained a piece of heavy steel plate punched all over like a piece of modern peg board on a larger scale. Into the holes he fitted wooden pegs to outline the shape he fancied and bent the stakes around the pegs.'

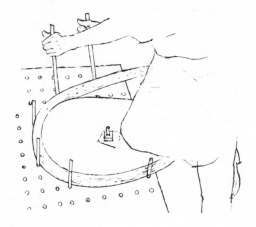

Fig. 191 (a)

(b) Framer's Block
A slab of hard wood, about 5 ft by 2 ft by 4 in, supported on stout legs so that its top surface would be about 18 in high. In its centre are fitted three wooden pegs (the 'cogs'), about one in. in diameter, and these, with the addition of a wooden wedge, were used as a vice to hold, for instance, a pair of legs for boring. Pegs were driven into the ends of the block to hold tools, and a piece of wood with a vee-notch cut in it was fastened to one end to hold round stuff when trimming the ends with a saw. In the centre of the block was a depression to hold grease for the Spoon Bits used for boring, and close to the sawing notch would be a grease box hollowed from a solid block of wood and having a saw kerf in it through which

the saw was passed to pick up a little of the grease within.

The Framer's Block was used to support Windsor chairs while they were being 'framed', or assembled and finished. Its lowness was an advantage, especially when boring with the Brace and Spoon Bit.

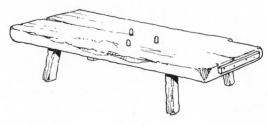

Fig. 191 (b)

(c) Donkey
A holding device. See separate entry under *Donkey*.

(d) Seat Holder [Not illustrated]
This consists of a seat blank (i.e. one not adzed-out) with a block screwed to its under surface by which it could be gripped in the Vice. Its upper surface was furnished with four sharp protruding metal points. When finishing off an adzed seat before 'legging-up', the seat would be set down on the four protruding points which would hold it laterally steady while work with Shave and Scraper proceeded.

Chalk Line and Reel (Snap Line) *Fig. 192*
A length of fine twine, usually kept on a turned wooden reel and used for setting out long straight lines on timber. The line is rubbed over a lump of chalk; or, when a black mark is required, over a burnt stick, known as a Smut Stick. The line is held in position at both ends and then lifted away from the wood and released sharply, leaving a straight mark between the two points. This operation is known as 'snapping the line'.

A line and ochre box was part of a carpenter's tool kit in ancient Greece, and it is also frequently depicted in medieval manuscripts. Moxon (London, 1677) describes its use as follows: '. . . at this Distance therefore they make with the points of their Compasses a prick at either end of the Stuff; Then with Chalk they whiten a Line, by rubbing the Chalk pretty hard upon it; Then one holds the Line at one end upon the prick made there, and the other strains the Line pretty stiff upon the prick at the other end; then whilst the Line is thus strained, one of them between his Finger and Thumb draws the middle of the Line directly upright, to a convenient height (that it may spring hard enough down) and then lets it go again, so that it swiftly applies to its first Position, and strikes so strongly against the Stuff, that the Dust, or Atoms of the Chalk that were rubbed into the Line, shake out of it, and remain upon the Stuff. And thus also they mark the under side of their Stuff: This is called *Lining of the Stuff*.'

(*a*) Chalk Line and Smut Stick.

(*b*) Factory-made Line Reels.

(*c*) Line Pin. For attaching the line to a point away from the timber itself, e.g. on the ground. Better known as a bricklayer's or mason's tool, it is about 6 in long with mushroom head and splayed foot.

(*d*) Chinese Line, used with ink (*c*. 1900).

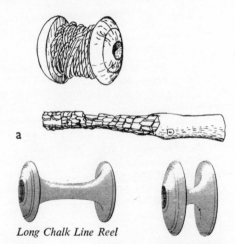

a

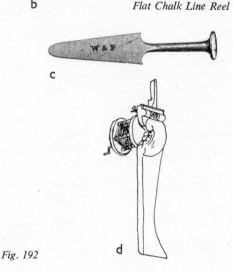

Long Chalk Line Reel

b   *Flat Chalk Line Reel*

c

Fig. 192   d

Chamfer (Bevel) *Fig. 193*

A chamfer is a flat surface, formed by the removal of a sharp edge, usually at 45°. A stop chamfer is one which stops some distance short of the end of a beam or at an intervening joint. For tools used see: *Cornering Tool; Drawing Knife, Cooper's Chamfering; Drawing Knife, Stop Chamfer; Drawing Knife, Wheelwright's; Plane Chamfer; Router, Coachbuilder's (Corner); Shave, Spokeshave (Metal).*

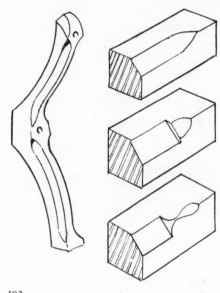

Fig. 193

Chase Wedge: see *Plumber's Tools.*

Check: see *Rebate.*

Chequered

Of a Hammer, Axe or Adze head, roughened by criss-cross lines incised on the face or poll. This is intended to avoid slipping off the heads of nails etc. when driving them.

Chequering Tool

A wood-engraving tool made in various forms, including that of an ordinary Graver, but with saw-edge. Used by gunstockers for the cross-hatching (known as chequering) on the butt-end and elsewhere on the stock. See *Gunstocker.*

Cheve: see *Plane, Cooper's Chiv.*

129

Chime Hook: see *Cellarman.*

Chime Maul (Beater): see under *Cooper* (3) *Hooping Tools.*

Chincing Iron: see *Flagging Iron and Chincing Iron.*

Chinese Tools: see *Axe, Chinese; Chalk Line and Reel; Chisel, Chinese; Drill, Bow; Plane, Chinese; Saw, Chinese.*

Chintz: see Chincing Iron under *Flagging Iron and Chincing Iron.*

Chip Carving Tool: see *Knife, Chip Carving.*

Chisel *Figs. 194; 195; 196; 197; 198.*
A tool with a steel blade mostly of rectangular section, with the end ground to a sharp edge, and usually fitted with a wooden handle. Some heavy-duty Chisels used by wheelwrights and others are solid steel throughout.

Chisels for woodworking have been used since neolithic times, and in the succeeding Bronze Age both tanged and socketed types were cast in stone

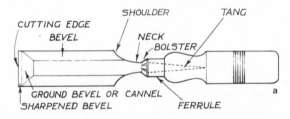

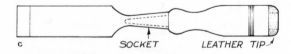

Fig. 194   Diagram of typical chisels
(*a*) Firmer Chisel
(*b*) A Chisel with 'Registered' handle
(*c*) A Chisel with socketed handle

130

moulds. The Romans further differentiated between Firmer Chisels and Mortice Chisels, both the tanged and socketed forms of the latter being practically identical with those in use to-day. Tanged and Bevelled-edge Chisels appear in Germany in the sixteenth century, the handles being usually square or hexagonal in section with no ferrule. Firmer Chisels of the seventeenth century often have splayed blades and are sometimes illustrated with very short handles. (A Chisel of this kind, found in a Sussex workshop, is illustrated.) Turned handles with ferrules appear to have been introduced in the early eighteenth century.

Fig. 195   Firmer chisel of seventeenth-century type

(*a*) *Types of Chisel*
Chisels are sub-divided according to the shape and thickness of the blade itself and the type of work for which it is normally used, as follows: (but see also separate entries).

Firmer Chisel. A general-purpose Chisel with a flat blade with parallel sides, strong enough to be struck with a Mallet, and used for general work.

Paring Chisel. A lighter blade, long and thin, frequently bevel-edged. Used without a Mallet by joiners, patternmakers, cabinet makers, and others for fine paring and trimming. (In current usage, the lighter Chisels of this sort are often called Bevel-Edged Chisels.)

Fig. 196 Opposite.   Typical Chisels and Gauges (reading from left to right)

2423–2407 Firmer
2428 and 2410 With 'Registered' handles
2471 Millwright's Chisel
2459 Butt Chisel
2405 } Paring Chisels
2420 }
2404 Turning Chisel
2412 Mortice Chisel – Sash
2411 Mortice Chisel – Joiner's
2417 }
2429 } Firmer Chisels with socketed handles
2415 }
2452 }
2474 } Firmer Gouges
2449 }
2448 Paring Gouge
2453 Heavy Gouge with socketed handle

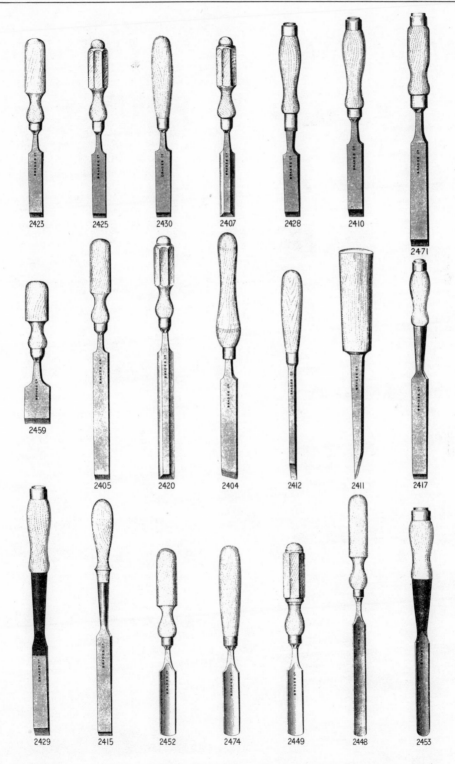

Fig. 196   Typical Chisels and Gouges (William Hunt & Sons, The Brades Ltd., Birmingham, 1905)

WARD & PAYNE,

CORPORATE MARK.

W P

SHEFFIELD,

Best Round Beech Chisel Handle.

Best Taper Octagon Box Chisel Handle.

Common Round Beech Chisel Handle.

Beech Socket Chisel Handle, not hooped.

Best Improved Round Beech Chisel Handle,
London pattern.

Bright Iron Hooped Ash Socket Chisel Handle.

Best Taper Round Beech Chisel Handle.

Iron Hooped Ash Socket Ship Chisel Handle.

Best Carving Pattern Beech Chisel Handle.

Double-hooped Ash Ship Chisel Handle.

Best Carving Pattern Box Chisel Handle.

Bright Double Iron Hooped Ash Registered Chisel Handle.

Plain Beech Octagon Chisel Handle.

Brass Hooped Plain Octagon Beech
Chisel Handle.

Beech Turning Chisel Handle.

Best London Octagon Box Chisel Handle.

Common Octagon Box Chisel Handle.

Oval Beech Mortice Chisel Handle.

Fig. 197 Chisel Handles (Ward & Payne Ltd., Sheffield, 1911). *See entry overleaf*

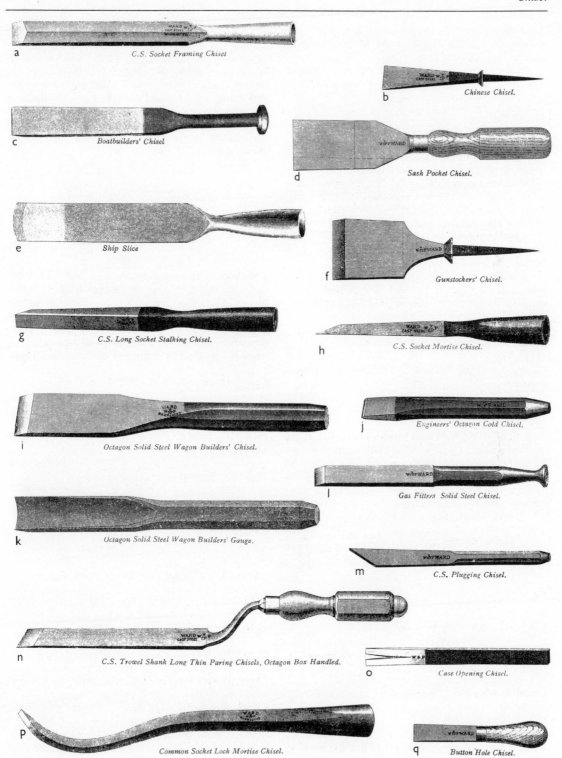

a   *C.S. Socket Framing Chisel*

b   *Chinese Chisel.*

c   *Boatbuilders' Chisel*

d   *Sash Pocket Chisel.*

e   *Ship Slice*

f   *Gunstockers' Chisel.*

g   *C.S. Long Socket Stalking Chisel.*

h   *C.S. Socket Mortise Chisel.*

i   *Octagon Solid Steel Wagon Builders' Chisel.*

j   *Engineers' Octagon Cold Chisel.*

l   *Gas Fitters Solid Steel Chisel.*

k   *Octagon Solid Steel Wagon Builders' Gouge.*

m   *C.S. Plugging Chisel.*

n   *C.S. Trowel Shank Long Thin Paring Chisels, Octagon Box Handled.*

o   *Case Opening Chisel.*

p   *Common Socket Lock Mortise Chisel.*

q   *Button Hole Chisel.*

Fig. 198   Chisels for special purposes (Ward & Payne Ltd., Sheffield, 1911) See appropriate entries

Mortice Chisel. A thick, stout blade and heavy handle to take the blows of the Mallet. Used for cutting mortices and similar work.

Special-purpose Chisels. Used by carvers, turners, wheelers, millwrights, and other tradesmen for particular jobs.

(b) Handles Fig. 197
The usual home-made handle was octagonal in section and tapered gracefully to fit the hand. This form, which lacks a ferrule, carried on a tradition of the seventeenth and eighteenth centuries.

Ordinary carpenter's Chisel handles are usually of beech or ash. Ash was also used for heavy-duty socketed wheeler's and millwright's tools, but joiner's and cabinet maker's Chisels often have boxwood handles. Carving and Turning Chisels were often fitted with the more exotic woods such as ebony, rosewood or hickory.

Country wheelwrights and others often made the long handles for their Turning Chisels without the usual bulb near the ferrule, resulting in a characteristically long, tapered plain handle.

Chisel, Allongee: see *Chisel, Carving*.

Chisel, American Framing or **Firmer:** see *Chisel, Framing*.

Chisel, Astragal (Astragal Punch; Sash Punch) *Fig. 199*
A small all-iron Chisel with two forked cutting edges bevelled inside to meet a semi-circular hole. Though listed in certain trade catalogues (e.g. Howarth, 1884 and Mathieson, *c.* 1900), its use is unknown. But from its appearance it may have been a form of the continental Chisel – *Chisel, Hinge*.

Fig. 199

Chisel Axe: see *Axe, Mortice*.

Chisel, Bargebuilder's: see *Chisel, Shipwright's; Chisel, Registered*.

Chisel, Bedding: see *Plane Maker*.

Chisel, Bent *Fig. 200*
A heavy wheelwright's Chisel, often home-forged by bending forward the lower part of the blade of an old Chisel. It is usually socketed for a wooden handle, but is sometimes all-iron. The cutting edge varies from about $\frac{1}{2}$ to 1 in wide. Used when excavating the wheel hub (boxing) to take the iron bearing known as the box. Its special purpose is to undercut and scoop out the wood to make room for the axlenut, and also to cut off the ends of the wedges which are driven into the hub around the box to centre and secure it within the hub. See also *Chisel, Wedging*.

Fig. 200

Chisel, Bevelled Edge *Fig. 196 Nos. 2407 and 2420*
A name given to Firmer or Paring Chisels with their blades bevelled on the long edges. Often fitted with octagonal handles, they are intended for hand use and are not driven by a Mallet. They give a better clearance when trimming dovetails, grooves, etc. Most lighter Chisels in current use are bevel-edged.

Chisel, Blockmaker's *Fig. 201*
A long, strong, socketed Firmer Chisel, about 16 in long overall and from $\frac{1}{2}$ to 2 in wide. Used by ship's block makers for cutting out the slots for the pulleys. (See also *Gouge, Blockmaker's; Print and Roller Cutter; Ship's Blockmaker*.)

Fig. 201

Chisel, Board Lifting: see *Chisel, Floor Cutting*.

Chisel, Boat: see *Chisel, Shipwright's*.

Chisel, Bolt: see *Chisel, Drawer Lock*.

Chisel, Boxing: see *Gouge, Wheelwright's*.

Chisel, Broad: see *Chisel, Slick; Chisel, Ship Slice*.

Chisel, Bruzz *Fig. 202* (Buzz; Bruzz Iron. Other variations include: Gilpin (1868) – Wheeler's Buzzer; Wm. Hunt (1875) – Wheeler's Buzzets; U.S.A. – Bur Chisel, Corner Chisel; David Steel

(1797) – Burr; Goldenberg (1875) – Brazze; Suffolk Millwright 1968 – Burze. The name Bruzz is said to be the same word as Birse, which according to Wright (1898) is a northern dialect name for a 'triangular chisel used to square out mortice holes', and is connected with a number of similar sounding words meaning bruise or crush.)

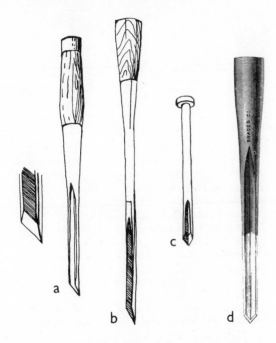

Fig. 202   (*a*)⎫ Socket Bruzzes with home-made handles
            (*b*)⎭
            (*c*) All-iron Bruzz
            (*d*) Socket Bruzz

A strong Chisel with a V-shaped blade, 10–26 in long overall. It is either socketed for a wooden handle, or made in steel throughout. The size of the V varies from $\frac{1}{2}$ to 1 in, and the angle from 50° to 90°. These Chisels have many uses, including chopping out the waste from deep mortices and, in particular, cleaning out the corners of spoke mortices. Bruzzes with the smaller angles are useful when cutting the dovetail-shaped mortice holes to be found in some wheel hubs. They are also extensively used for enlarging the axle-holes in hubs. Mr. G. W. Weller (1953) writes: 'For body building as well, the buzz or bruzz was equally indispensable. Nearly every tenon was tapered, rendering it almost impos-

sible to make good clean cuts at the corners of the mortices without the aid of this tool.'

Although the Bruzz has come to be regarded as almost symbolic of the wheeler's shop, it was also used by carpenters and millwrights for the deep mortices in wooden beams. It was still so used until recent times in the U.S.A., where it is called a Corner Chisel. V-shaped Chisels are used also by wood turners and wood carvers, and a chisel of similar design is used in bone surgery.

The Bruzz is depicted by Diderot under *Charon* (Paris, 1763). Made in solid iron and called a 'gouge quarrée', it is described as being used to 'hollow out the mortices of the hubs and felloes'. It is listed by Goldenberg (Alsace, 1875), but we have not seen it listed in other continental catalogues of the period 1900–30, possibly because the 'Square Gouge' was serving a similar purpose (see *Chisel, Hinge*).

Chisel, Bung: see Bung Removers under *Cellarman*.

Chisel, Bur or Buzz: see *Chisel, Bruzz*.

Chisel, Butt (Pocket Chisel) *Fig. 196 No. 2459*
A Firmer Chisel with blade 3 in long and from $\frac{1}{4}$ to $1\frac{1}{2}$ in wide, with tang or socketed handle and sometimes made with bevelled edges. Used for fine joinery or cabinet work, including the housings for butt hinges.

Chisel, Buttonhole *Fig. 198(q)*
A small Chisel with 3 in blade $\frac{3}{8}$–$\frac{1}{2}$ in wide, and ground on both sides or tapering gradually to the edge. According to Mr. A. Collier (1954), they were used for cutting buttonholes in linen shirts.

Chisel, Cabinet
A name given to a light-weight Firmer Chisel, $\frac{1}{4}$–2 in wide, with a socketed handle, which is sometimes leather-tipped. Used by cabinet makers and joiners as a Paring Chisel for fine work.

Chisel, Cant: see *Chisel, Framing*.

Chisel, Carving *Figs. 203; 204; 205.* See also *Woodcarver* and *Knife, Chip Carving*.
Carving Chisels and Gouges are made in a very wide range of patterns and sizes (there are in fact over 1000 of them), but the average trade carver would find about 60 to 70 tools sufficient for most purposes.

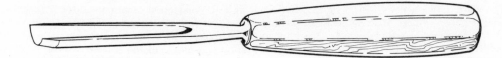

Fig. 203 Typical Carving Gouge

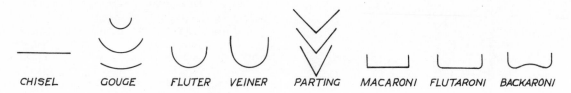

CHISEL GOUGE FLUTER VEINER PARTING MACARONI FLUTARONI BACKARONI

Fig. 204 Typical cross-sections of Carving Tools

It became the practice in the nineteenth century to give a number to all carving tools according to their longitudinal shape and cross-section. These numbers are the last two digits of the 'Article number' given by the Sheffield Illustrated List from about 1880 onwards. By this system all tools with the same shape and the same cross-section have the same number irrespective of width.

In general, Gouges are sharpened both outside and inside for hardwoods, with the larger bevel on the outside; but on the outside only for softwoods.

The handles of carving tools are sometimes made in octagonal shapes with a graceful swelling toward the centre (*Fig. 203*). Others are plain turned in beech, rosewood, or boxwood.

One of the best known makers of Carving Tools was S. J. (and later J. B.) Addis & Sons. His mark was acquired by Ward & Payne, Sheffield, in 1870.

Carving Tool types Figs. 204; 205
The chief variants in shape of blade and cutting edge are illustrated.

(*a*) Straight Chisel
(*b*) Skew Chisel (Corner Chisel)
(*c*) Straight Gouge (see below (*z*))
(*d*) Curved Gouge (Double Bent Gouge)
(*e*) Bent Chisel ⎰ (Spoon Bit; Entering Chisel)
(*f*) Bent Chisel ⎱
(*g*) Bent Chisel, Left Corner
(*h*) Front Bent Gouge (Front Bent Spoon Bit Gouge) ⎱ For acute
(*i*) Back Bent Gouge (Back Bent Spoon Bit Gouge) ⎰ curves and for deeply recessed detail.

(*j*) Straight Parting Tool ⎱ V-shaped.
(*k*) Curved Parting Tool ⎰
(*l*) Front Bent Parting Tool
(*m*) Unshouldered Spade Chisel. For lighter finishing operations. See also *Chisel, Print* and *Block Cutter*.
(*n*) Straight Macaroni Tool. For finishing sides of recesses.
(*o*) Curved Macaroni Tool. For finishing sides of concave recesses.
(*p*) Straight Flutaroni Tool. For finishing recesses with rounded sides. Also for modelling acanthus detail.
(*q*) Bent Spoon Flutaroni Tool. Uses as for (*p*).
(*r*) Fishtail Spade Chisel. For light finishing work and lettering.
(*s*) Fish Tail Spade Gouge. For light finishing work and lettering.
(*t*) Long Pod Spade Chisel. For finishing work.
(*u*) Long Pod Spade Gouge. For finishing work.
(*v*) Dog Leg Chisel. For finishing recessed work.
(*w*) Allongee Chisel. Used largely by sculptors.
(*x*) Allongee Gouge. Used largely by sculptors.
(*y*) Side Chisel. For deeply recessed cuts.
(*z*) Fluter and Veiner [not illustrated]. These are Gouges but with a somewhat deeper channel, like the letter U. The Veiner has longer sides than the Fluter.

Chisel, Case Opening *Fig. 198(o)*
A flat bar of steel, 12 in long with a bevel-ground claw at one end. Used for opening cases.

Chisel, Caulking: see *Caulking Tools.*

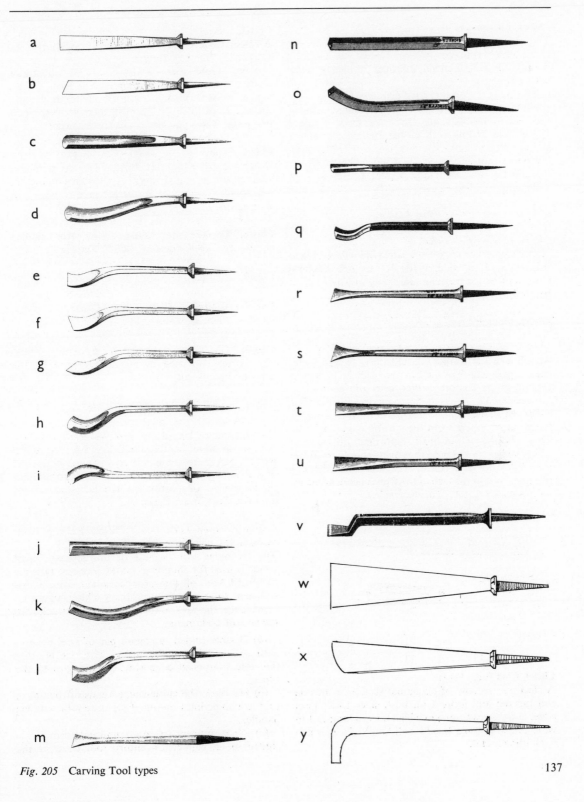

Fig. 205 Carving Tool types

Chisel, Chairmaker's

The chairmaker used ordinary Mortice Chisels for chopping out mortices; a short Gouge was used for cutting the round-ended mortices in heavy chair seats etc.

The bodgers (turners) used ordinary turning tools beginning with a Turning Gouge for 'hogging-down' the legs to rough shape, followed by a Turning Chisel, and ending with a Vee Parting Chisel for sharply defined cuts. The *Sheffield List* (1888) lists, but does not illustrate, a Chairmaker's Chisel described as an extra long Firmer Chisel $\frac{1}{4}$–$1\frac{1}{2}$ in wide, with bevelled edges.

Chisel, Cheeking: see *Plane Maker.*

Chisel, Chinese *Fig. 198(b)*

A small tanged or socketed Chisel with a flared blade from 1 to 3 in wide, made for export. Chinese Chisels of native manufacture come mostly in sizes below 1 in. See also *Gouge, Chinese.*

Chisel, Chip: see *Knife, Chip Carving.*

Chisel, Coachmaker's (Coach Firmer) *Fig. 206*

An extra strong Firmer Chisel from $\frac{1}{4}$ to 3 in wide, the standard length of the 1 in size blade being 8 in to the bolster. It was often fitted with a large double-hooped handle and sometimes made with bevelled edges. In some continental patterns the blade is flared, i.e. it tapers from the cutting edge upwards. This was one of the most important tools of the coachbuilder; it was used for paring shoulders at the joint to ensure a perfect fit. For this reason he was particular about the face of the Chisel being perfectly flat. (See also *Chisel, Wheelwright's.*)

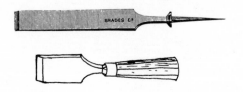

Fig. 206

Chisel, Cold *Fig. 198(j)*

A steel bar, usually of octagonal section, with one end tapered and ground on both sides to a chisel edge. Usual sizes from $\frac{1}{4}$ in wide and 9 in long to 1 in wide and 30 in long. Used by all kinds of tradesmen for cutting metal.

Chisel, Cooper's

All-steel Cold Chisels for cutting iron hoops. See *Cooper (3) Hooping Tools.*

Note. Goldenberg (Alsace, 1875) illustrates Cooper's Chisels for cutting wood. Their blades appear to be very slightly curved like a Gouge, and they taper in thickness from the ferrule to the cutting edge. Their purpose is unknown.

Chisel, Cope

Described by Knight (U.S.A., 1877) as a 'chisel adapted for cutting grooves'. Appearance unknown, but possibly a Sash or Scribing type of Gouge (see *Gouge, Sash*).

Chisel, Corner: see *Chisel, Bruzz* (for similar chisels see *Chisel, Carving; Chisel, Turning*).

Chisel, Cranked: see *Chisel, Bent; Chisel, Paring.*

Chisel, Diamond Point: see *Chisel, Turning,* (Scraping Tools).

Chisel, Dog-Leg: see *Chisel, Carving.* (These Chisels are also used by gunstockers and by the print and block cutter.)

Chisel, Double: see *Mortising Machine.*

Chisel, Double-Hooped *Fig. 196 Nos. 2410 and 2471*

A term for chisel handles fitted with a ferrule at the bottom end, close to the bolster, and another ferrule, or a metal ring, at the top end. Their purpose is to reduce the risk of splitting the handle by blows from a mallet. See *Chisel, Registered.*

Chisel, Drawer Lock (Bolt Chisel; Bolting Iron: Lock Bolt Chisel) *Fig. 207*

The following are variants of this special Chisel which is used for chopping out the recess to take the lock and bolt of drawer-locks when there is insufficient room to use an ordinary Chisel. In a confined space, the back of the Chisel can be struck with the side of a Hammer.

(*a*) Double Ended: a square bar of steel $6 \times \frac{3}{8}$ in with chisel ends bent 'at mock' with one another (i.e. one sharpened edge is at right angles to the other).

(*b*) Handled: the chisel-shaped ends of the above tool are sometimes mounted separately in wooden handles.

(*c*) What appears to be an unusual form is included in an eighteenth-century tool chest at the

Rochester Museum and listed in the inventory as a 'Bolting Iron'. It measures about 9 in overall. There are two chisel-like cutting edges: one in the plane of the shank and the other at right angles.

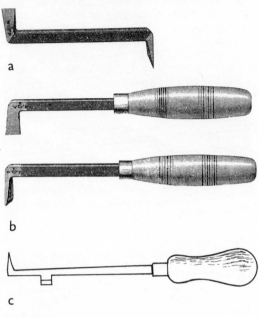

a

b

c

Fig. 207

Chisel, Electrician's: see *Chisel, Floor-cutting.*

Chisel, Engraving: see *Wood Engraver.*

Chisel, Eyeing: see *Plane Maker.*

Chisel, Firmer *Fig. 196 Nos. 2425, 2430, 2407*
A general term for the ordinary Chisel of fairly sturdy build. It has a parallel blade with a square cutting edge, from 1/16 to 3 in wide. It is normally made with a tang fitted to a wooden handle with a ferrule. It is used for the general run of carpentry and joiner's work and is strong enough to be driven by a Mallet.

As may be seen from the illustrations, these chisels are also made with bevel edges for better clearance when working in corners; and for the same reason are occasionally made with a rounded back to form two thin edges along the sides.

Moxon (London, 1677) uses the term 'Skew Former' for a Firmer Chisel with the blade ground on the skew. They are rarely used to-day except occasionally by woodcarvers and cabinet makers to clean out corners of angular mortices etc.

Chisel, Fish Tail: see *Chisel, Carving.*

Chisel, Flared (Splayed Chisel) *Fig. 195*
Chisels made with the blade flared, i.e. tapering outwards from the part nearest the handle up to the cutting edge. One found in a wheelwright's shop is illustrated under *Chisel.*

Many Chisels of the seventeenth and eighteenth centuries were flared in this way. Modern examples include Chinese Chisels and certain spade-type Carver's Chisels. A flared Firmer type of Chisel with bevelled edges is called a Brabant Chisel in some continental catalogues.

Chisel, Floor-Cutting (Board-lifting Chisel; Electrician's Chisel) *Fig. 208*
Solid steel spade-shaped Chisel with a thin blade $1\frac{1}{2}$–$2\frac{1}{4}$ in wide, tapered on both sides. Used for cutting across the grain of flooring boards in order to lift up a section.

Fig. 208

Chisel, Flutaroni: see *Chisel, Carving.*

Chisel, Framing (American Framing or Firmer; American Long Socket; Cant Chisel). *Fig. 198(a)*
A long, strong, Firmer Chisel with blade $\frac{1}{4}$–2 in wide, with a round or canted back. (In a canted back two bevels on the back meet in a ridge in the middle.) The handle is socketed and often hooped. These heavy Chisels, often referred to as 'American', are included in the Shelburne Museum's booklet *Woodworking Tools* (U.S.A., 1957) under 'Early Construction Tools'. Used for heavy carpentry and other framing work. (See also *Chisel, Slick.*)

Chisel, Gas Fitter's *Fig. 198(l)*
In solid steel $\frac{1}{2}$–1 in wide, with mushroom head.

Chisel, Gauge *Fig. 209*
A small cast-iron tool similar to a Chariot Plane, with a lever cap, designed to take a $\frac{1}{4}$-in Chisel as the cutter. Used for raising a shaving of any desired thickness for blind nailing, or for cutting grooves for inlaid decoration.

Fig. 209

Chisel, Glazier's
A name sometimes given to a strong Firmer Chisel with $1\frac{1}{2}$–2 in blade, often socketed, and with a very short handle. Used for hacking out hard putty and old glass from window frames. (See also *Knife, Hacking.*)

Chisel, Grafting: see *Knife, Grafting.*

Chisel, Gunstocker's: see also *Gunstocker. Fig. 198(f).* A short, heavy Firmer Chisel from $\frac{3}{4}$ to 3 in wide. To avoid the corners of the Chisel 'digging-in' and leaving a mark, the blade is sometimes made slightly gouge-shaped. Used for removing waste from the wood blank from which the gunstock is to be made.
Other Chisels used by gunstockers include:

(*a*) Dog-Leg Chisel: A cranked tool similar to the Carver's Chisel of this name. Used for clearing out the recesses in which the lock and other metal parts are fitted. Known as 'Shovels' in the gunstocker's shop, they are usually home made from old files, and consequently found with many different shaped blades both flat and gouge-shaped.

(*b*) Certain other Chisels are listed in nineteenth-century catalogues as 'Gunstocker's' but do not seem different from those used in other trades. These include a 'Gunstocker's Mortice Chisel' sized $\frac{1}{16}$–$\frac{1}{2}$ in, illustrated in Smith's *Key* (Sheffield, 1816), which appears to be the same as a Joiner's Mortice Chisel. 'Gunstocker's Gouges' sized $\frac{5}{8}$–1 in are recorded in the Sheffield *List* of 1888, but are not illustrated.

Chisel Handles: see under *Chisel. Fig. 197.*

Chisel, Hinge *Fig. 210*
Used mainly on the continent of Europe, these are thin Mortice Chisels designed for cutting the slots for receiving concealed hinge plates on doors and sashes. Types include:

(*a*) Slotted for side cutting, 1–$1\frac{3}{8}$ in wide.

(*b*) (*c*) With barbs for removal of chips, $\frac{7}{8}$–$1\frac{3}{8}$ in wide.

(*d*) Known also as a Square Gouge, Lipped Chisel, or Recessing Tool. A gouge-like tool in the form of a square trough $\frac{5}{8}$–$2\frac{1}{4}$ in wide, with cutting edges ground on the inside, with a solid steel head, or tanged for a wooden handle. It is sometimes described in continental catalogues as a Wagon-Builder's Driving Iron, but in others as a tool for cutting mortices for hinges.

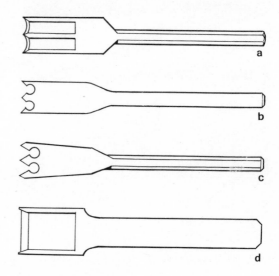

Fig. 210

Chisel, Hollow: see *Mortising Machine.*

Chisel, Hook: see *Chisel, Turning.*
In the U.S.A., the name Hook Chisel is given to a chisel blade with side-hook, mounted on a long handle and used for harvesting ice.

Chisel, Hoop: see *Cooper (3) Hooping Tools.*

Chisel, Iron Head
The name given in Smith's *Key* (Sheffield, 1816) to all-steel Chisels without a wooden handle, e.g. some types of *Chisel, Bruzz; Chisel, Shipwright's;* and *Chisel, Wagon-Builder's.*

Chisel, Lipped: see *Chisel, Hinge.*

Chisel, Lock Bolt: see *Chisel, Drawer Lock.*

Chisel, Lock Mortice (Swan-necked Chisel) *Fig. 198; Fig. 211*
A blade $\frac{3}{8}$–$\frac{5}{8}$ in wide, curving upwards at the sharpened end. Used for cutting the slots for mortice locks in doors and drawers, and other blind mortices, etc. Variants include:

(*a*) *Common or York pattern Fig. 198(p)* – with the blade of square section bent in a flat S-curve.

(*b*) *Swan Neck pattern Fig. 211* – with a long blade curving downwards with a sharp knee on the lower edge.

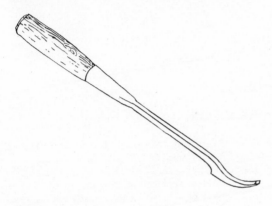

Fig. 211

In both types the curved and thickened lower edge of the blade acts as a fulcrum when levering out the chips at the bottom of a deep mortice or slot.

Chisel, Long etc.
'Long Strong' – a term applied by many of the older trade catalogues to Firmer, Mortice, or Turning Chisels of specially strong build and extra length.

'Long Thin' see *Chisel, Paring*.
'Long Pod' see *Chisel, Carving*.
'Long Cornered' see *Chisel, Turning*.
'Long Stalking' see *Chisel, Stalking*.

Chisel, Macaroni: see *Chisel, Carving*.

Chisel, Machine: see *Mortising Machine*.

Chisel, Millwright's *Fig. 196 No. 2471*
A very heavily made Firmer Chisel, $\frac{3}{8}$–2 in wide, the 1 in size being 10 in long to the bolster. The handle is often double-hooped. From illustrations in nineteenth-century catalogues this Chisel appears to have been longer than all the others except the Block-Maker's Chisel.
Used for trimming deep mortices and for trimming

wooden cogs, and other general millwright's work. (See also Chisels described under *Millwright*.)

Chisel, Moon: see *Millwright*.

Chisel, Mortising *Fig. 196; Fig. 198(h); Fig. 221*
A general term for various Chisels used for mortising. All have an extra strong blade, a stout handle to take the blows of the Mallet, and the blades are characterised by being thicker back-to-front than other Chisels, in order to resist bending when levering out the waste in a deep mortice.
Variants include:

(*a*) *Joiner's Mortice Chisel* (Common Mortice Chisel) *Fig. 196 No. 2411*
A blade $\frac{1}{4}$–1 in wide, with an oval bolster and a stout oval handle. Used by carpenters and joiners for general and heavy mortising. (Chisels of almost identical shape were used by the Romans.)

(*b*) *Sash Mortice Chisel Fig. 196 No. 2412*
A lighter blade, 8–9 in long and $\frac{1}{8}$–$\frac{3}{8}$ in wide. Used by joiners and cabinet makers for light mortising.

(*c*) *Wheelwright's Mortice Chisels Fig. 221*
Wheelwrights use the same Mortising Chisels as the carpenter but tend to employ the larger sizes. For instance, when taking out the core of the mortices in larger wheel hubs, the wheeler uses a socketed Mortice Chisel up to 20 in long overall, with a blade 1–2 in wide. The blades of these heavy chisels taper in thickness back-to-front from $\frac{1}{2}$ in at the cutting end up to 1 in near the socket in order to stand the strain when working in the bottom of a deep mortice. (See *Chisel, Wheelwright's*.)

(*d*) *Socket Mortice Chisels Fig. 198(h)*
A name given to the stout Mortice Chisels that are provided with a socket for the handle instead of a tang.

(*e*) *Machine Mortice Chisels:* see *Mortising Machine*.

(*f*) See also *Chisel, Lock Mortice*.

Chisel, Nail Cutting: see *Chisel, Ripping*.

Chisel, Paring (Long Thin Paring Chisel: some modern tool lists include this tool under the general heading of Bevel-edged Chisels.) *Fig. 196 Nos. 2405 and 2420; Trowel Shank type Fig. 198(n)*
A long, thin blade $\frac{1}{4}$–2 in wide and from 9 in long from edge to bolster, either plain or with bevelled edges, often fitted with an octagon-shaped handle. Used by patternmakers, cabinet makers, joiners, and other tradesmen for hand paring and fitting without the use of a Mallet.

A variant known as a Trowel Shank Paring Chisel has a cranked shank and is used for paring deep trenches or recessing, and to facilitate paring over wide, flat surfaces by keeping the hand clear of the work.

Chisel, Parting

A name given to various Chisels usually V-shaped, used for grooving. See *Chisel, Carving; Chisel, Turning.*

Chisel, Pattern Maker's

A name sometimes given to a Paring Chisel, particularly one with a cranked (trowel) shank.

Chisel, Plane Maker's: see *Plane Maker.*

Chisel, Plugging *Fig. 198(m)*

An all-steel Chisel with a skewed end to the blade, 8–12 in long and $\frac{1}{8}$–$\frac{1}{4}$ in thick. Used by carpenters, plumbers, and other tradesmen for removing the mortar from brickwork joints to take plugs or wedges for fixing skirtings, picture rails, door and window frames, pipe hooks, etc. Now largely superseded by modern masonry boring and fixing devices.

Chisel, Pocket: see *Chisel, Butt; Chisel, Sash Pocket.*

Chisel, Print and Block Cutter *Fig. 212* see *Print and Block Cutter.*

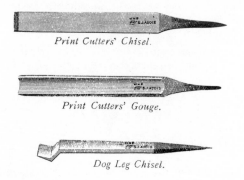

Print Cutters' Chisel.

Print Cutters' Gouge.

Dog Leg Chisel.

Fig. 212

Chisel, Pruning *Fig. 213*

A stout socketed Chisel 2–2$\frac{1}{2}$ in wide, sometimes provided with a sharpened side hook. Used with a long handle for pruning small branches.

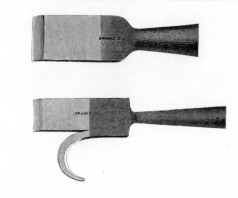

Fig. 213

Chisel, Purfling: see *Violin Maker.*

Chisel, Recessing: see *Chisel, Hinge.*

Chisel, Registered (Registered or Shipwright's Chisel) *Fig. 214* and *Fig. 196 No. 2428*

A Firmer Chisel with a strong blade and originally a square bolster, replaced in the modern tools by a round-necked bolster, fitted to a hardwood handle, usually decorated with two groups of scored lines. They are listed from $\frac{1}{8}$–2 in wide and with a blade up to 15 in long.

The iron ferrule next to the bolster is longer than usual, with one end solid in which there is a square hole to take the tang. A leather washer is often provided to soften the impact of the bolster on the end of the ferrule which, in the case of an ordinary Chisel, is taken by the end grain of the handle. These are the features which were probably 'registered' in the first place, but when and by whom is not known. The other end of the handle is fitted with another iron ferrule (or hoop) to keep the wood from splitting.

According to the late Mr. A. Collier (London, 1950) the original 'Registered Chisel' was used by bargebuilders. From a study of the lists, the tool appears to have come into use about 1870 (it does not appear in Gilpin's list of 1868). Up to the present the first known mention is in a list of Turner, Naylor and Marples (Sheffield, *c.* 1870) where the handles are described as 'Registered or Shipwrights' Chisels – Double-hooped, one hoop with solid end for tangs'. Howarth (Sheffield, 1884) and Ward & Payne (Sheffield, 1900) list the Registered Chisel next to the Iron-Headed Shipwright's Chisel. Mr. Collier's suggestion is further confirmed by Mathieson (Glasgow, 1900) who calls them 'Strong Ship or

Registered Chisels'. In recent years the Registered Chisels in the smaller sizes from ¼ to ½ in have become popular among joiners for mortising.

Fig. 214

Chisel, Right-Hand Corner: see *Chisel Turning* (Scraping Tools).

Chisel, Ripping *Fig. 219; Fig. 198(i)*

(*a*) An all-iron Chisel used for rough work of all kinds, including cutting through nails when necessary, and also as a wedge, e.g. for separating planks nailed together. See also *Chisel, Wagon-Builder's*.

Moxon (London, 1677) writes of it as follows: 'The *Ripping-Chissel* . . . is a Socket-Chissel, and is about an Inch broad, and hath a blunt Edge. Its Edge hath not a *Basil*, as almost all other Chissels have, and therefore would more properly be called a *Wedge* than a *Chissel* But most commonly Carpenters use an old cast off *Chissel* for a *Ripping Chissel*.

Its Office is not to cut Wood, as others do, but to *rip* or *tear* two pieces of Wood fastened together from one another, by entering the blunt Edge of it between the two pieces, and then knocking hard with the Mallet upon the head of the Handle, till you drive the thicker part of it between the two pieces, and so force the power that holds them together (be it Nails, or otherwise) to let go their hold: For its blunt Edge should be made of Steel, and well tempered, so that if you knock with strong blows of the Mallet the *Chissels* Edge upon a Nail (though of some considerable Substance) it may cut or brake it short asunder.'

(*b*) Another Chisel known as a Ripping Chisel is illustrated under *Upholsterer*.

Chisel, Roller Cutter's

A short firmer-type blade, and according to Ward and Paynes' Catalogue (Sheffield, *c.* 1911) made in sizes from 4 to 7 in wide. Purpose unknown.

Chisel, Rubber Cutting: see *Knife, Rubber Tapping*.

Chisel, Sash Pocket (Pocket Chisel) *Fig. 198(d)* and *Fig. 727* under *Window Making*

A very thin, parallel-sided blade up to 5 in long and 1½–2½ in wide, tapered in thickness and ground on both sides. Used for cutting, or completing the cut started by the Saw, when making the pockets in a pulley-style in the frame for sliding sashes. (See also illustration and information under *Window Making*.) It is thin enough to enter the saw kerfs without enlarging them.

Chisel, Shipwright's *Figs 196; 198; 215*

Shipwrights used the heavier varieties of Chisels and Gouges, usually those described under *Chisel, Socket* or *Chisel, Registered*. Some were specially made with very long blades, from 24 to 36 in overall, with 1½–2 in wide cutting edges. They were used for such heavy work as making lock gates or for cutting the socket holes in the head of a capstan. Special Chisels include:

(*a*) *Boat Builder's Chisel* (Barge builder's Chisel; Shipwright's Sharp Iron Chisel) *Fig. 198(c)*

A strong all-iron Chisel with a parallel blade from ¾–2 in wide. Used for heavy work in ship and boat-building.

(*b*) *Ship's Slice* (Ship Slicer; U.S.A.: Slick) *Fig. 215; Fig. 198(e)*

A very broad, flat Chisel, with a blade 2–4 in wide, up to about 30 in overall length, and with a slightly rounded cutting edge. The socketed handle is offset to keep the hand clear when paring a flat surface, and sometimes terminates with a knob for pushing. Used for fairing and removing waste from the deck and elsewhere, particularly in places where the Adze cannot reach. At H.M. Dockyard, Portsmouth (1969) a Slice with a 14 in handle was used for clearing tar and dirt from the sheave-slots of ship's blocks.

Fig. 215

Chisel, Sinking Down: see *Plane Maker*.

Chisel, Skew: see *Chisel, Firmer; Chisel, Plane Maker's (Cheeking); Chisel, Turning*.

Chisel, Slick (Slice) *Fig. 216*

A long, broad chisel about 30 in overall with a blade up to 3½ in wide. More common in the U.S.A. than in Britain, it is used for cleaning up the sides of large mortices in construction work, and for levelling surfaces. The top of the handle is often swollen for pushing from the shoulder. (See also *Chisel, Shipwright's* (Slice) and *Chisel, Framing*.)

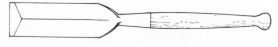

Fig. 216

Chisel, Slice: see *Chisel, Shipwright's*.

Chisel, Socket *Fig. 195; 196 No. 2429*
A method of fixing a handle on a Chisel or Gouge. Tanged blades are unsuitable for heavy work, for when struck heavily the tang is liable to split the handle. To overcome this disadvantage, a tapered iron socket is forged solid with the blade for the reception of the handle. This method, albeit more expensive, gives a long life to a good handle, especially when also fitted with a hoop at the top end – known as 'ring-head'.

Chisel, Spade: see *Chisel, Carving*.

Chisel, Spoon-Bent: see *Chisel, Carving*.

Chisel, Stalking (Long Stalking Chisel) *Fig. 198(g)*
A strong Socket Firmer Chisel from ¼ to 2 in wide. From catalogue illustrations the blade appears to be *c.* 14 in long; it is described in Goldenberg's 1875 Price List as 'extrafort'.

Use unknown, but Mr. R. Paterson (U.S.A., 1968) writes that he heard the name Stalking used by a shipwright for a 1¼ in Framing Chisel used for removing heavy waste from large rebates. It has been suggested by a cabinet maker that it was used for digging out the centre-hole in tables whose centre legs were known as 'stalks'. Messrs Robert Sorby & Sons, who were established 200 years ago in Sheffield, inform us that they believe the Stalking Chisel was sold only to the cotton industry and was probably used for the shaping of cotton bobbins etc.

Chisel, Trenail (Trenail Cutter)
An all-iron, cranked Chisel used by railway trackmen and platelayers for cutting off the wooden pegs (known as trenails) when re-gauging etc. Trenails are used for securing the chairs (which hold the rails) to the sleepers beneath.

Chisel, Turning (also Turning Gouges) *Figs. 217; 218.*
Wood-turning tools are distinguished from the usual carpenter's Chisels and Gouges by their plain tangs with no bolster, and by being in general longer in both blade and tang. They are also fitted with a characteristically long handle, often turned with a bulb near the ferrule. Moxon (London, 1677) writes that the handles are made 'tapering towards the end, and so long that the *Handle* may reach (when they use it) under the Arm-pit of the *Workman*, that he may have more stay and steddy management of the Tool'.

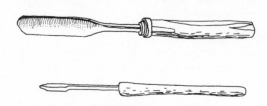

Fig. 217   Home-made examples

(a) (b) Gouge and Chisel Fig. 218
For turning softwoods, and for ordinary work in hardwoods, the tools range in width from ¼ to 2 in; the overall length of the 1 in size varies from 20 to 24 in. The Gouges are used for removing wood quickly and for smoothing hollows; the Chisels are used for paring a smooth surface. Gouges are bevel-ground on the outside (out-cannel); the Chisels are ground with a slightly rounded bevel on both sides when new, but are usually ground flat afterwards by the wood turner himself. Their cutting edge is either square or skewed; if skewed they are sometimes known as Long-Cornered Chisels.

Home-made Turning Gouges, as sketched below, are commonly found in wheelwright's shops where they are used for turning wheel hubs.

(c) Hook and Side Tools
This type of Turning Chisel has a sharpened hook or a sharpened angular blade, and a straight or cranked shank. It is used for hollowing, undercutting the sides and bottoms of deep works, and for screw-cutting. (The tools called 'Grooving Hooks' by Moxon (London, 1677) are scrapers and belong to the next category.)

(d) Scraping Tools
For accurate turning in hardwoods special tools with only one obtuse bevel are used, shaped according to the profile to be worked, and used with a scraping action. The simplest forms, with a blade ⅜–¾ in wide,

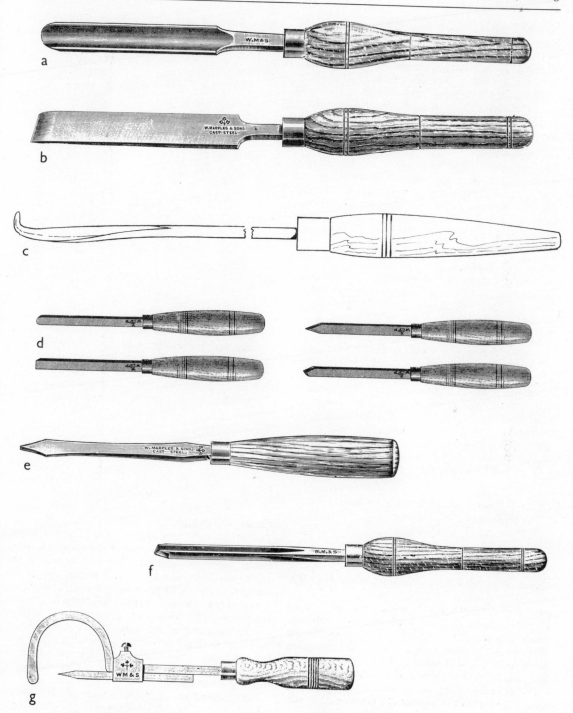

Fig. 218 Factory-made examples of Turning Chisels

145

such as the Right and Left-hand Corner Chisel, the Round and Square Nose Chisel and the Diamond Point, are sometimes referred to as Patternmaker's Turning Tools, and sometimes as Wood-Turning Scrapers. Recently two other useful shapes have been designed for school work: the Side-cutting Diamond and the Side-cutting Round. Nineteenth-century catalogues often illustrate more than 60 shapes for these tools.

(e) Parting Chisel

The end of the blade tapers to a point, leaving a cutting edge $\frac{1}{8}$ in wide, which is bevel-ground on both sides. The sides of the Chisel at the cutting end are often relieved, giving a flared appearance; this is done to avoid binding against the wood being turned. It is used for parting off the work on the lathe when the turning is finished.

(f) V-Tool (Bruzz)

Although sometimes called a Bruzz, it is much lighter than that tool, and is ground on the outside instead of the inside. Used for marking where details are to be worked.

(g) Sizing Tool (Wood-Turner's Gauge)

A small chisel-shaped cutter is fixed to the shank of the tool with a screwed clamp, and may be adjusted to any required distance from the hook end. Used for sizing repetition work on the Lathe. A series of grooves is cut with the tool, giving the finished diameter of the work at various points, the rest of the contour being turned afterwards. In use the tool is pulled away from the work so that the end of the hook bears against the revolving wood and the handle is then pressed downwards.

Chisel, V-Shaped: see *Chisel, Bruzz; Chisel, Carving; Chisel, Parting; Chisel, Turning.*

Chisel, Veiner: see *Chisel, Carving.*

Chisel, Wagon-Builder's (Wagon Chisel) *Fig. 219; Fig. 198(i)* and *(k)*
An all-steel Chisel, usually octagonal like a Cold Chisel in appearance but sharpened with only one bevel. The shank is sometimes offset to enable the blade to be used close to a surface. Though often home-forged, it was listed in sizes from 8 to 9 in long, and from $\frac{1}{2}$ to 2 in wide. It was also occasionally listed in gouge form as a Wagon Gouge. (The entirely different Coachmaker's Chisel is occasionally listed as a Wagon-Builder's Chisel.)

Used by wheelwrights for cutting off the tops of wedges when boxing wheels, and as a Chisel for rough work in carpentry, military work, railway wagon building, etc. See also *Chisel, Ripping,* and *Chisel, Trenail.*

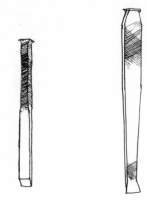

Fig. 219

Chisel, Wedging, (Wheeler's) *Fig. 220*
A blacksmith-made all-steel round-backed Chisel used for opening the grain in the hub to start the wooden wedges which centralise and secure the bearing-boxes. The blade is slightly splayed so that after being driven into the hub it can be loosened for withdrawal by tapping on each side with a hammer.

Fig. 220

Chisel, Wheelwright's *Fig. 221*
The wheelwright uses ordinary Carpenter's Chisels, but in addition there are a few special types, often rather large and rugged. The heavier kinds are commonly made with a deep iron socket to take the wooden handle, or they are made in steel throughout. To resist hard blows from the Mallet, the handle top is often hooped. They were often made by a local smith, but many of the factory-made specimens to be found in England bear the name of the Sheffield firm of I. Sorby who appear to have specialised in making these heavy but handsome Chisels during the nineteenth century.

Chisels used in this trade include: *Chisel, Bent; Chisel, Bruzz; Chisel, Coachmaker's; Chisel, Mortising; Chisel, Wagon Builder's; Chisel, Wedging* (also *Gouge, Boxing* and *Gouge, Coachmaker's*).

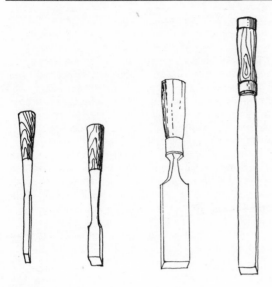

Fig. 221

Chit: see *Froe*.

Chiv (Chive): see *Plane, Cooper's Chiv*.

Chop
Part of a holding appliance, e.g. the jaws or cheeks of a Vice (see *Bench Chops*).

Chopping Block: see *Block, Chopping*.

Chuck
A device for holding the work or tool in a Lathe, or the device at the foot of a Brace or Drilling Machine for holding the Drill or Bit.

Cigar Box Opener *Fig. 222*
A knife-like tool. One edge of the blade has a small notch near the end, and there is a protruding metal 'hammer' on the back.

Its use is described as follows by J. Leon & Co. (wholesale tobacconists, London, 1968): 'The general purpose of the tool is for opening wooden cigar boxes, the flat lids of which are usually hinged along one side and fastened by a single nail or pin at the centre of the opposite side. The whole box is

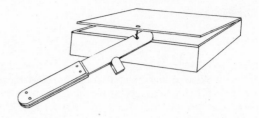

Fig. 222

usually sealed with a thin layer of paper. The purpose of the knife-like blade is to slit the paper and lever open the lid. The slot in the blade is fitted round the nail immediately below the lid, to avoid the blade accidentally entering the box and damaging the cigars. It can also be used for extracting the nail if it should become bent. The protruding boss is normally used for hammering the nail back into place when the box is closed.'

Circular Cutting Gauge (Circular Router) *Fig. 223*
A stock of hardwood, 18–20 in long overall, with a handle at each end. The middle of the stock is plated at the top and bottom, with a vertical slot ⅜ in wide down the centre. In some patterns the slot runs nearly the whole length of the stock, in others about half way along. A vertical pin can be fixed at any desired position in the central slot by means of a thumbscrew. A vertical cutter is fixed by means of a screw or wedge at one end of the stock, clear of the slot.

This unusual tool is said to have been used for cutting out circular holes in the tops of wash-stands – an essential article of bedroom furniture before the introduction of hand basins and separate bathrooms. The lighter tools of this type were also used for cutting the circular openings and veneers of barometer cases. A pilot hole was bored in the centre to take the pin, and the spur cutter adjusted to describe a circle of the radius required.

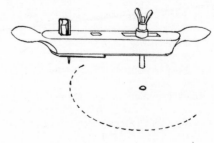

Fig. 223

Circular Square: see *Square, Radial.*

Clamp: see *Cramp.*

Clave *Fig. 224*
A heavy block of wood standing on four legs about 2 ft 6 in high, with a sunken top in which a work-piece can be wedged. Used by makers of Ship Blocks and others. Diderot (Paris, 1763) illustrates a Clave in use by sabot makers. (See *Ship's Block-maker* for *Clave Board.*)

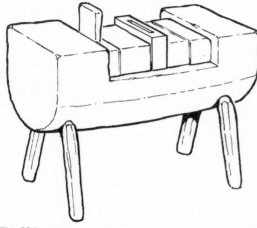

Fig. 224

Clavestock
An obsolete term for an Axe or Chopper for cleav-ing wood. 'A clavestock and rabetstock, carpenter's craue, and seasoned timber, for pinewood to haue'. (Thomas Tusser, *Five hundreth pointes of good husbandrie*, 1573.)

Claw
Various tools, sometimes called Claws, will be found under *Box and Case Opening Tools; Hammer, Claw; Nail Extracting Tools.*

Cleaver or Cleaving Iron: see *Basket Maker; Bond Splitter; Cleave; Froe; Knife, Cleaving;* Splitting Tools under *Woodland Trades.*

Cleck or **Cleek:** see *Chime Hook* under *Cellarman.*

Clench Tools are mentioned under the following entries: *Hammer, Clench; Hammer, Ship Clench; Staple Clincher.* (A Clench Cutter is a farrier's tool.)

Clip: see *Carver's Clip.*

Clip Wrench: see *Wrench, Coachbuilder's.*

Clog-Sole Maker *Fig. 399* under *Knife, Clogger's*
There are two distinct tradesmen concerned with clog making: the itinerant clog-sole maker who normally works outdoors; and the clog maker proper, who, working in a town or village workshop, re-shapes the rough soles, fits the leather uppers, and produces the finished clogs.

The usual wood for clog soles is alder or sycamore. The timber is sawn into short lengths and cleft and trimmed with an Axe into rough blocks. They are then shaped with a straight-bladed Bench Knife and stacked for seasoning in a pile shaped like a bee-hive.

For tools used in clog-sole making and finishing, see *Knife, Clogger's.*

Note on Sabot Tools
Unlike the English clog which is a wooden sole surmounted by a leather upper, the European sabot is made entirely of wood. Some of the special tools used for this trade are found occasionally in English workshops. They include:

Augers. With a spoon-shaped nose. One of these has a flat, shield-shaped nose, with one edge up-turned and sharpened, and provided with a screw lead.

Chisels. A trough-shaped Chisel rather like the Lipped Chisel illustrated under *Chisel, Hinge.*

Knives. Various hooked knives are used for paring the insides of the sabot.

Close Chamfer Howells Iron: see *Plane, Cooper's Chiv (e).*

Club A name given to various Mallets, Cudgels, or Mauls. See List under *Maul.*

Coachbuilder (Coachmaker) *Fig. 225*
(*Note:* Since coachbuilder's work has much in common with that of the *Wheelwright,* reference should also be made to that entry.)

Heavy springless coaches were being made in England in the sixteenth century. Spring suspension was introduced about 1670 in the form of the C Spring. According to Philipson (London, 1897) this is the oldest and yet most perfect method of suspending carriages on springs: 'The wheels and underworks may go over the roughest roads, yet the passenger (in the body of the carriage) can ride without discomfort, as the leather braces change the short jerky motion caused by the wheel striking against obstacles and uneven places in the roadway into a pleasant swing.' This method necessitated a perch – a bar connecting the fore and after carriage, which was expensive and rather heavy. The invention of the elliptic spring by Obadiah Elliot in 1805, led to the lighter type of carriage with the body connected directly to the springs – without the need for a perch – such as the ordinary Landau or Victoria. The idea spread to most other light carts and carriages, including, eventually, the 'horseless carriage' or motor car. (The suspension of the coach in *Fig. 225* is of transitional design with both C and elliptic spring and also a perch.)

Many of the carriages of the nineteenth century are of great beauty. Their design gradually became standardised, for unlike the wheelwright and wainwright, the coachbuilder was able to consult textbooks and standard patterns, although his workshop methods remained largely traditional until the end. This event came a little sooner for the coachbuilder than for the wheelwright and wainwright, for horses continued to work on the farms for some twenty years or more after their use on the roads had come to an end.

The Coachbuilder's Work
Whereas it was customary for a wainwright to construct the whole vehicle himself, the coachbuilder's

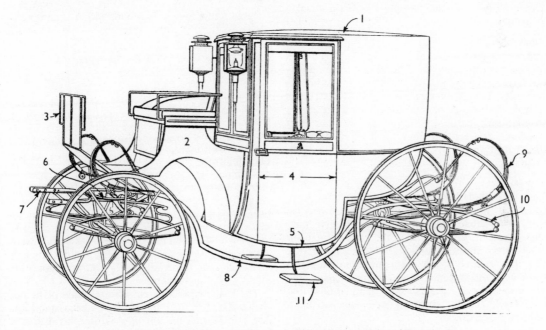

—SINGLE BROUGHAM, HUNG ON C AND UNDER-SPRINGS WITH PERCH.

Fig. 225 Parts of a Brougham Carriage (From 'The Art and Crafts of Coachbuilding', J. Philipson, 1897)
  1. Cant rail* and roof
  2. Boot
  3. Dasher
  4. Door pillars*
  5. Rocker*
  6. Wheel plate on which the fore-carriage rotates
  7. Futchell ends to take shafts (a splinter bar and pole is fitted when driving a pair of horses)
  8. Perch
  9. C-spring
  10. Under-spring
  11. Step

  * Name of underlying part of frame

work tended to become separated into specialised trades, e.g. the bodymaker, undercarriage maker, wheelwright, coachsmith, carriage trimmer, and carriage painter. The workshop was often divided into separate compartments in which these separate operations were carried out.

Like the wheelwright's and wainwright's trade, a basic feature of the coachbuilder's work was strength. Joints had to be very strong and tough hardwoods were essential. But although strength was important, the work was on a lighter scale, and had to be brought to a much higher finish than was needed for, say, a farm cart. Coachbuilder's tools were made on light and graceful lines. Many were specially designed for shaped work, for a coach had few straight lines in its framing; the body viewed in any direction is a combination of elliptical or compound curves.

Materials

The woods used were hard and well seasoned; ash for the frames, mahogany and other hardwoods for the door and body panels, and elm for footboards and seats. Writing of ash, J. Philipson (1897) says: 'For the framework of a carriage body we require a strong, tough, hard and fibrous wood without great heaviness, and we find those qualities combined in the greatest perfection in the hedgerow Ash for which there is a great demand in England.'

The Workshop

The coachbuilder's shop contained much the same equipment as the wheelwright's and wainwright's, but it was often designed on more modern lines. One important difference was the need for a trimming shop, in which upholstery was carried out, and also for a completely separate paint shop free from dust or shavings. (See *Painting Equipment*.) These special departments were often placed at first floor level, and the carriages were winched up a ramp to reach them.

The smithy was large in order to handle the quantity of iron work used for undercarriages, the folding hoods (known as heads), the brake mechanisms, and the iron plates used to strengthen the framing of the body of the carriage itself.

Tools

For general tools used by coachbuilders, see *Wheelwright*. For some of the tools used by the coachsmiths and undercarriage makers, see *Wheelwright's Equipment* (3) *Tyring Tools*; the tools used in coach trimming include saddlers' tools, which are not included here, but see *Upholsterer*.

The tools developed more exclusively for coach work are described under the following entries:

Cramp, Coachmaker's
Cramp, Carriage Trimmer's
Gauge, Panel
Gauge, Wheelwright's and *Coachbuilder's* (including Coachmaker's Sweeps, Wheelwright's Mortice Gauges, Rim Gauge, etc.)
Hammer, Coach Trimmer's
Lap Set
Plane, Coachbuilder's (including Door Check Plane, Tee Rabbet Plane)
Router, Coachbuilder's (including Grooving, Beeding, Boxing, Moulding, Jigger and Pistol Routers)
Wrench, Coachbuilder's (including Axle Key, Cap Wrench, Clip Wrench, Step Wrench)

Coach Screw

A large screw for use in wood, with a square head driven by turning with a Wrench.

Coach Wrench: see *Wrench, Coachbuilder's*.

Coak

A heavy wooden peg used for joining timbers in shipbuilding etc. (See *Trenail; Auger, Coak Boring*.)

Coffin Maker and Undertaker *Fig. 226*

(The following notes are based on an article kindly contributed by Mr. Charles Hayward.)

In country districts the local builder or wheelwright was also the undertaker, and coffins were usually made only to special order on a death taking place. In towns, however, undertaking was a trade in itself, and a number of 'boxes' (as coffins were called), of various sizes, were kept in stock, though in winter time, especially during an epidemic, this might be rapidly exhausted.

The vast majority of coffins are made of home-grown elm, 1 in thick for top, sides, head, and foot, and $\frac{3}{4}$ in for the bottom. Elm is used because it is usually plentiful and one of the cheapest hardwoods; it bends easily at the kerfs and will take nails well without splitting. For best quality work oak is used.

After a death has occurred, the undertaker's first job is to take measurements. Normally the only sizes needed are the length from the head to the foot, and width at the shoulders.

Making the coffin

The sides are 2 in wider at the head than at the foot. 'Kerfing' is done at the shoulder to facilitate bending; this consists of about six kerfs sawn through to within $\frac{1}{8}$ in at the top edge and nearly through the whole thickness of the board at the bottom edge.

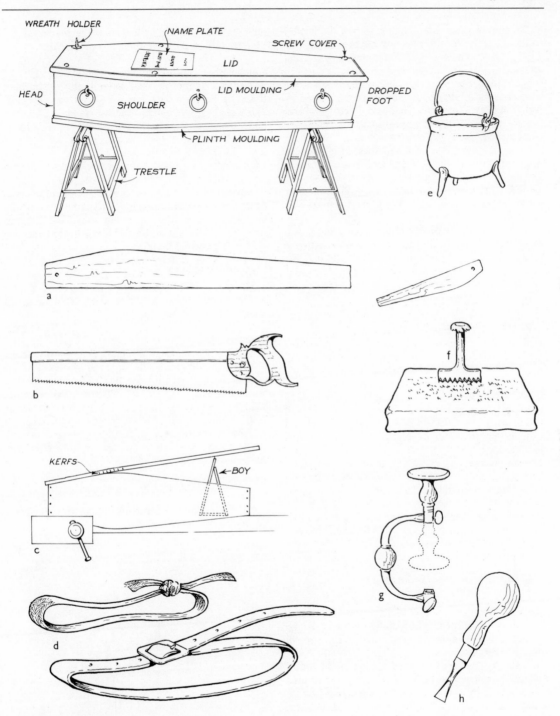

WREATH HOLDER
NAME PLATE
SCREW COVER
LID
HEAD
SHOULDER
LID MOULDING
DROPPED FOOT
PLINTH MOULDING
TRESTLE

a

b

KERFS
BOY
c

d

e

f

g

h

Fig. 226 Coffin maker's tools and equipment

(*a*) Templets
(*b*) Kerfing Saw for kerfing the shoulders
(*c*) 'Boy' supporting loose side
(*d*) Strap and Web, for carrying a coffin up or down stairs
(*e*) Pitch Pot, for pitching the inside of the coffin
(*f*) Pinking Tool for decorating the edge of the calico lining
(*g*) Undertaker's Brace ⎫
(*h*) Coffin Screwdriver ⎬ for screwing down the lid

The first stage of assembling is to nail the bottom to the head and foot to make all edges true with the Trying Plane. The head portion is nailed down first, and a contraption known as a 'Boy' (c) is used to support the loose end of the side while this is done. This consists of three odd pieces of elm nailed together in triangular form. To give the necessary drop (angle) the foot is pulled outwards about 2 in before being nailed.

Paupers' coffins were of the plainest kind. General construction was normal, but there was no finish of any kind. The wood was straight from the Saw, unplaned, nails were not punched in, joints were not levelled, and there were no mouldings; insides were unlined.

In some cases especially in summer time, it was necessary to pitch the inside. The pitch was heated in a Pitch Pot (e) and poured inside. The whole coffin was then lifted and rolled so that the molten pitch ran along all round the joints.

After completion
Handles, name plate, and inside lining are fixed when the coffin is actually needed. Calico is largely used for lining; this and the inside sets and frilling are fixed with tacks. The edges of the frilling are cut out in a dog-tooth pattern with a Pinking tool (f). When an awkward staircase has to be negotiated, a Strap and Web (d) is used to give the bearers a means of gripping the coffin. The strap is tightened just beneath the shoulders of the coffin; the web (an ordinary length of upholsterer's chair web) is tied near the foot.

Some undertakers carry a special stubby Screwdriver (h) or folding Brace (g) which goes easily into the pocket or a small bag. This is used to screw down the lid before removing the coffin from the house.

Some undertaker's equipment and tools are illustrated but see also *Brace, Undertaker's; Saw, Kerfing;* and *Screwdriver: (2) Special.*

Comb: see *Brushmaker; Graining Comb.*

Comb Maker
Though not regarded as a woodworking trade the following comb maker's tools may crop up in other workshops:

Floats. A Float with a curved blade used in comb-making is called a Graille; Floats of Vee-shaped cross section, in ascending widths, are called Found, Carlet, and Topper. A float for abrading flat surfaces, made like a Baker's Rasp, is called a *Quannet.*

Stadda. A double bladed saw for cutting the teeth of the comb (see *Saw, Stadda*). The Languid (Lan-

guet) is a spacing piece between the blades of the Stadda Saw to preserve their distance apart.

Combination Tools ('Universal' Tools)
A term sometimes applied to tools that have more than one function e.g. *Box and Case Opening Tools; Fencing Tool; Hammer-Pincers; Plane, Combination;* Universal Spoke Shave under *Shave, Spokeshave (Metal).*

Commander: see *Beetle; Basket Maker.*

Compass (Small compasses are often called Dividers) In its simplest form (which has remained practically unchanged since Roman times), the Compass consists of two straight and equal legs connected at one end by a movable joint. Its main purpose is to take measurements and to describe circles. Size of leg varies from about 4 to 30 in.

The types of Compasses follow closely those of their counterparts which are described, sometimes in greater detail, under *Callipers.*

Compass, Beam (Trammel) *Fig. 227*
A wooden or metal bar of rectangular section, about 2–5 ft long, and two heads, of wood or metal, which slide along the bar and can be fixed in any desired position by means of wedges or screws. The trammel heads are usually pointed, but one may carry a pencil holder instead.

Used by millwrights, shipwrights, carpenters, and others to describe large sweeps or circles, or for marking out large work-pieces.

(a) A home-made example
(b) Trammel heads

Another example is illustrated under *Wheelwright's Equipment: (2) Spoke Tools.*

a

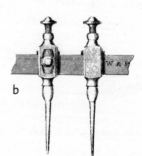

b

Fig. 227

Compass, Bow: see *Compass, Spring*.

Compass, Cooper's *Fig. 228*
The following types are commonly found in the cooperages. One of their chief uses is to find the correct radius for the cask head. This is done by 'stepping' the tool around the croze groove until the points meet after six steps.

(*a*) An iron Wing Compass with legs 7–8 in long. Though called a Cooper's Compass in the tool lists, other Compasses of this type and size would serve as well. They were usually provided with a series of ornamental serrations on the shoulders.
(*b*) Made of wood with tapered legs up to 18 in long, fitted with iron points.
(*c*) Also of wood, in the form of a bow 9–12 in overall, which is made from a piece of springy ash, steamed and bent into the shape of a 'U'. The legs are connected by a wooden cross-piece with screws of opposite hand at the ends, tapped into the thick part of the bow. This enables the user to open or close the bow by merely twisting the cross-piece. The example illustrated is probably French, but according to Coleman (Ireland, 1944) the same type is used in Ireland, and is often made by the cooper himself.

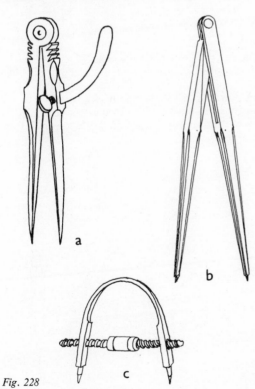

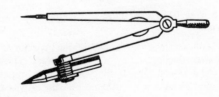

Fig. 228

Compass Dividers: see *Compass, Spring*.

Compass, Lancashire
For the meaning of this term see *Calliper, Lancashire*.

Compass, Millwright's *Fig. 229*
Wing Compass of the Lancashire type, but with a screw adjustment on the root of the wing for fine setting. Coarse adjustment is made in the usual way where the wing passes through a slot in the leg and is locked with a thumbscrew. The legs, of tapered square section with a sharp point, are from 6 to 12 in long. The legs are often slightly bowed, enabling the tool to be used more easily on a cylindrical surface, and also, when required, as a Calliper. Used for dividing and setting out the wooden cogs on mill gears and for similar work. See *Millwright*.

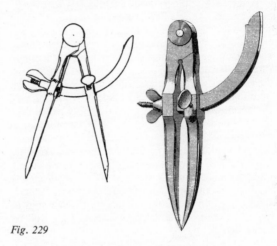

Fig. 229

Compass, Pencil *Fig. 230*
One type is the familiar 'school' drawing instrument about 4 in long overall, one leg ending with a pointed steel pin, the other with a fitting to take a pencil. These are used in the workshop for marking out small circles, but a much larger type is also used, usually made of wood throughout, with legs 15 in or more long, with one leg ending in a holder to take a crayon.

Fig. 230

153

Compass, Rack Wing

A Wing Compass of the Lancashire type. Teeth are cut on the outer circumference of the wing to form a rack which appears to engage with an additional thumbscrew in the leg. It is not clear whether this thumbscrew provides a rack-and-pinion action or merely acts as a fixing screw. Long-established makers of Callipers and Compasses such as Messrs Peter Stubs of Warrington (1970) have been unable to enlighten us on this point.

Compass, Ship Builder's *Fig. 231*

Shipwrights used similar drawing and measuring instruments to other woodworkers. The Beam Compass is used for much of the circular marking-out in the mould loft. Very large wooden compasses are usually home made, and are called Sweeps.

In the 1904 catalogue of Goldenberg (Alsace) there is an illustration of a Shipbuilder's Compass made in sizes up to about 10 in long. The points are turned at right angles to the legs, and the points are chisel-shaped. A similar compass is illustrated by Rålamb (Stockholm, 1691). We are not certain of its purpose. According to Sir Flinders Petrie (London, 1917), when describing a similar Compass made in Roman times, the points are bent to avoid the difficulty of the legs being oblique when wide open; he is assuming, presumably, that the compass is being used for measuring distances and not for drawing circles. Another explanation is that this tool combines in some degree the functions of both Compass and Calliper. We have not come across this type of compass in any English shipyard, but since Goldenberg's tools were sold in most countries of the world, examples may well be in use here.

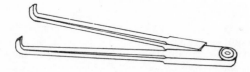

Fig. 231

Compass, Spring (Bow Compasses; Dividers) *Fig. 232*

Like Spring Callipers, these tools are made in two patterns:

(*a*) *'Lancashire' Pattern Dividers* with characteristic shaped forging and graceful chamfers. The legs are 5–10 in long, square in cross-section, taper to sharp points, and are connected at the top with a circular spring forged in the piece. The distance be-

tween the points is regulated by a screwed rod passing through one leg and pivoted to the other, fixed by a winged nut.

(*b*) *'American' Pattern.* In the modern version, the legs are separate but connected at the top by a flat spring bent to a circular shape, fitting tightly into notches at the top of each leg, just above the centre of the loose stud which acts as a fulcrum. The width of the opening is regulated by a turnscrew.

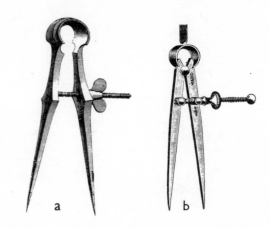

a   b

Fig. 232

Compass, Wheelwright's *Fig. 233*

All kinds of Compasses are to be found in the older wheelwright's shops, including some with very long legs used for marking out the frame of carts and waggons. Examples, some obviously home made, are shown below. The length of the legs varies from about 6 to 30 in.

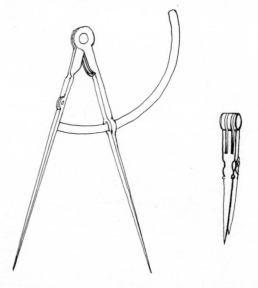

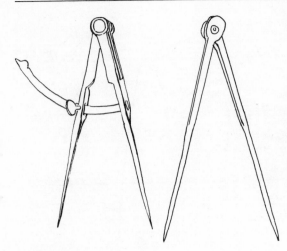

Fig. 233

Contraction Lath: see Pattern Maker's Contraction Rule under *Rule*.

Conversion
The sawing up of timber into usable sizes.

Cooper: (1) THE TRADE *Figs. 235*
The cooper makes wooden vessels formed of staves which are bound together by hoops of wood or iron. The art is one of great antiquity and was most probably developed from basketmaking. Though well established in Roman times, the trade appears to have declined in England after their departure, but was

Compass, Wing (Wing dividers) *Fig. 234*
(*a*) Like its counterpart, the Wing Calliper, the legs are held in position by means of a thumbscrew acting on a narrow steel strip shaped to form a quarter circle and fixed to one of the legs. In some patterns the end of the wing is ornamented with an ogee contour.

(*b*) The so-called 'American Pattern Wing Calliper' is a modern version of the traditional pattern. The legs are rectangular in section for about half their length, then rounded and tapered to a point. Fine adjustment is provided by a knurled nut engaging on a screw at the root of the wing and tensioned by a flat spring inside the leg.

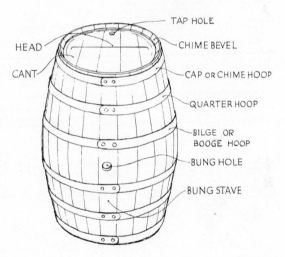

Fig. 235 (*a*) A typical beer cask

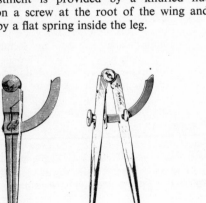

Fig. 234

re-introduced by the Anglo-Saxons. The shape of the cooper's main product, the cask, is a cylinder with bulging sides. How or when this peculiar shape was invented is not known; but it is this bulge (or 'bilge' as coopers call it) which gives the cask its strength. For when confined by hoops, the staves, with their tapered ends, behave like the stones of an arch.

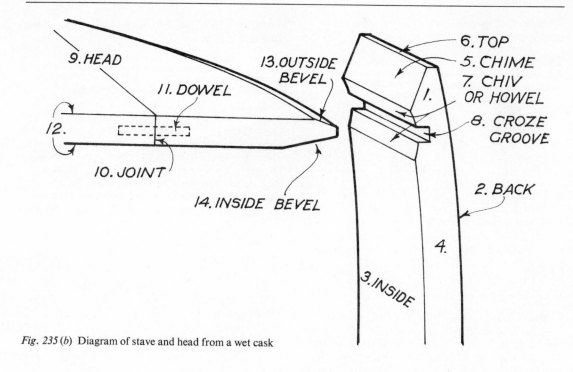

Fig. 235 (b) Diagram of stave and head from a wet cask

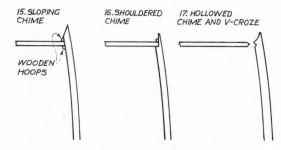

Fig. 235(c) Diagram of stave and head from a dry or 'dry-tight' cask

As other people have a sign,
I say – just stop and look at mine!
Here, Wratten, cooper, lives and makes
Ox bows, trug-baskets, and hay-rakes.
Sells shovels, both for flour and corn,
And shauls, and makes a good box-churn,
Ladles, dishes, spoons, and skimmers,
Trenchers, too, for use at dinners.
I make and mend both tub and cask,
And hoop 'em strong, to make them last.
Here's butter prints, and butter scales,
And butter boards, and milking pails. . . .

Today, few cooperages remain, and most of those that do are maintained by the brewers. Though the beer cask is now giving way to the metal container, wooden casks used for the storage and transport of wines and spirits are likely to survive for many years to come.

The Cooper's Work
Today the trade is divided into two main branches, wet and dry.

Wet. This includes the coopers who make beer and spirit casks for the breweries and distilleries, and

Until the early years of this century, cooperages could be found in many English towns and in a few of the larger villages. Unlike the cooper of today, they made other things besides casks – as may be gathered from the often-quoted cooper's sign of the early nineteenth century from Hailsham in Sussex:

156

the wine coopers who make the casks for wines and spirits and repair those coming from abroad.

A branch of the wet trade which has now almost disappeared is that of the 'white cooper', who made innumerable vessels needed for farm and household, such as buckets, milk churns, wash-tubs, cattle-feeding tubs, and vessels for home-brewing and general laundry and dairy work.

The making of casks for holding liquids demands the greater skill, for they must be perfectly tight to avoid the slightest leakage, and they must stand the strain of rough handling during transport over great distances.

Dry. This includes the maker of the rougher type of cask for provisions such as butter, flour, apples, meat, or fish, and for a vast variety of dry chemicals, for breakable goods such as glass and china, and for loose goods such as nails and bolts. The dry cask was the principal method of containing and transporting these materials before the advent of the mass-produced metal drum, crate, or cardboard carton.

Though less exacting than wet coopering, the dry cooper's trade is complicated by the great variety of types. Casks for holding nuts and bolts do not need to be watertight, but must be very strong. Barrels for herrings must be cheap yet watertight – they are sometimes known as 'dry-tight'.

Materials

The staves and heads for wet coopering are normally made of oak, but chestnut is sometimes used for transporting inferior wines. For whisky, the distillers prefer oak casks previously used for sherry, from which the spirit acquires its characteristic colour.

Dry work is done in beech, sycamore, pine, ash, poplar, chestnut, elm, or from second-hand woods from old casks. The hoops, if not of iron, are made from split hazel and willow. Dried rushes ('flags') are used for making liquid-tight joints between the head-boards, in the croze, and sometimes at the stave ends.

Machine-made casks

Towards the end of the nineteenth century machines were invented to perform almost every part of the cooper's work. These included stave-sawing, cutting, dressing, and jointing machines; chiming, howeling and crozing machines; a setting-up and hydraulic bending machine; wooden hoop riving, cutting, and dressing machines, and many more. Yet casks continued to be made by hand, at least in the smaller cooperages, right up to the present time.

(*Note*: The tools of the cellarman – the man in charge of the cellars where the wines and other liquors are stored – are often found in cooper's shops. His tools are listed under *Cellarman*.)

Historical note

From the evidence of the apprentice indentures of Bristol, Norwich, and Great Yarmouth, of 1554–1646, (W. L. Goodman 1972), a typical kit of cooper's tools included the following:

Addes or Thixtell (Adze); Howel (Howel Adze); Barge Axe (probably a Side or Broad Axe); Hatchet; Heading Knife (a Drawknife); Shave; Crowes or Creves (Croze); Compass; Wimble (Brace); Saw; Jointer; and Lave (presumably a Lathe, but for what purpose is not known). In one or two instances, there was a Froward (Froe), and a Knape (identity unknown).

The Howel at this time was a small Adze with a gouge-shaped blade, used for cutting the shallow concave surface on the inside top of the staves, on which the croze groove for the head of the cask is subsequently cut out (see diagram). It should be explained that this shallow concave surface is called by the same names as the tool that makes it; and two different words are used for it – howel and chiv. Before *c.* 1800 the surface appears to have been called a howel, and it was cut with a Howel Adze.

At some unknown date, probably after 1800, this surface came to be called the chiv, and the plane-like tool that cuts it is called a Chiv. But in the U.S.A., and sometimes in Ireland, the surface is still called the howel and the Chiv Plane that cuts it is also called a Howel.

It is of interest to note that, according to Coleman (Ireland, 1944), Irish coopers were at that time cutting the surface either with a Howel Adze, or by what he calls a 'Stock Howell', which was their name for a Chiv Plane. It may be inferred, therefore, that the Chiv Plane is a direct development of the Howel Adze, i.e. an adze-blade bedded in a wooden plane-stock. This term 'Stock Howel' may explain the curious name for a small Chiv in France – the 'Stockholm' which, presumably, has nothing to do with the capital of Sweden.

For a description of the Howel and Chiv, see *Adze, Cooper's Howel; Plane, Cooper's Chiv*.

Principal Tools used in making a Cask

(*Note:* (1) The numbers in brackets relate to the diagram *Fig. 235* (*b*). Tools marked thus * are described under *Cooper* (3) *Hooping Tools*.)

Entries for tools used

(*a*) *Shaping the staves*

The ends of the stave (1) are 'listed' (tapered off) with an axe.

Axe, Cooper's

157

The back of the stave (2) is rounded with a 'Backing Knife'. *Drawing Knife, Cooper's Backing*

The inside of the stave (3) is hollowed with a 'Hollow Knife'. *Drawing Knife, Cooper's Hollowing*

The edges of the stave (4) are planed on the Jointer and given the correct 'shot' (bevel). *Plane, Cooper's Jointer*

(b) Assembling the staves and hooping

The Staves are assembled in a cone-shaped formation inside an iron hoop.

Steam is used to soften the staves, which are kept hot by a fire from wood shavings contained in a Cresset. *Cresset**

The splayed staves are drawn together by driving on successively smaller wooden hoops known as Truss Hoops. *Truss Hoops**

or by the action of tightening a rope. *Hooping Windlass* etc.*

The iron hoops are joined by riveting them on a Cooper's Anvil. *Cooper's Anvil** (Bick Iron)

Hoops are driven on with a Hammer and Hoop Driver. *Hoop Driver**

Wooden hoops (if used) are cut, joined, and nailed with a special Adze, *Adze, Cooper's Nailing*

and sometimes with the help of a knife known as a Cooper's Whittle. *Cooper's Whittle**

(c) Cutting the chime

This is the bevelled edge (5) at the top and bottom of the cask. In wet casks it is cut with a Rounding Adze *Adze, Cooper's Rounding*

and finished off with a Sharp Adze. *Adze, Cooper's Sharp*

The top of the chime is 'topped' to give a level surface (6) on which the fence of the Chiv and Croze will subsequently bear. This is done with a special Plane. *Plane, Cooper's Sun*

The chime of dry and very small casks is cut with a small chiv-like tool called a Sloper or Flencher. *Plane, Cooper's Chiv*

(These tools cut a sloping bevel or shoulder which allows the head to be dropped in and secured by a wooden hoop or by other means, with or without a croze groove (15, 16, and 17). They are suitable for softer or thinner woods and could not

be used for cutting the oaken chime of beer casks.)

In the U.S.A. the chime is sometimes cut with a jigger-shaped Drawknife. *Drawing Knife, Cooper's Chamfering*

(d) Cutting the chiv and croze

i.e. smoothing and grooving the inside surface at both ends of the cask to receive the heads.

The chiv (U.S.A. 'howel') is the surface (7), just below the chime, which is levelled to take the croze groove. This is cut with a Chiv. *Plane, Cooper's Chiv (U.S.A.: Howel Plane)*

When the right size of Chiv is not available, or when doing repair work, this surface is cut with a type of Drawknife. *Drawing Knife, Cooper's Jigger (U.S.A.: Howelling Knife)*

Formerly, and occasionally today (e.g. in Ireland) the chiv surface was cut with a small hollow-shaped Adze. *Adze, Cooper's Howel*

The croze is a groove (8) round the inside ends of the cask in which the heads are fitted. It is cut with the Croze. *Plane, Cooper's Croze*

In repair work, the croze groove is sometimes cut with a curved Saw. *Saw, Riddle*

(e) Making and fitting the heads

The edges of the pieces (10) that make up the head (9) are trued on the Jointer. *Plane, Cooper's Jointer*

The edges are drilled for dowels (11) with a Brace. *Brace, Cooper's*

After assembly, the head is planed level (12) with a large type of Shave. *Shave, Cooper's Heading Swift*

After marking out with a Compass, the corners of the head are sawn off and rough-rounded with the Cooper's Axe. *Saw, Cooper's Heading* *Axe, Cooper's*

Finally, the edges of the heads (13 and 14) are bevelled with a large Drawing Knife (sometimes preceded by the Cooper's Axe). *Drawing Knife, Cooper's Heading*

In the U.S.A. this bevel is sometimes cut with a 'Heading Trammel'.

Knife, Cooper's Chamfering

Split rushes are inserted between the boards of the head, and also in the croze groove into which the head is fitted, to prevent leakage. This is done with the help of the Flagging Iron.   *Flagging Iron*

(f) Cleaning and finishing
Before the heads are inserted, the inside joints between staves are levelled.

Plane, Cooper's Stoup; Shave, Cooper's Inside; or Drawing Knife, Cooper's Round Shave

The outside of the cask is finished.

Shave, Cooper's Downright; Scraper, Cooper's Buzz

(g) Boring the bung and tap hole
The tools used are described under:

Auger, Cooper's Bung; Bit, Cock Plug

List of tools used by coopers.
See under the following entries:

Adze (Cooper's):
  Bunging
  Chimney Pipe
  Howel
  Nailing
  Notching
  Rounding
  Sharp
  Trussing
Anvil – see *Cooper (3) Hooping Tools*
Auger, Cooper's Bung Borers:
  Burners
  Bushing Tools
  Continental
  Cylindrical
  Taphole
  Thief
Axe, Cooper's
Bellows
Bick Iron, see Anvil under *Cooper (3) Hooping Tools*
Bit, Cock Plug
Bit, Cooper's Dowelling
Block, and Block Hooks, see *Cooper (2) Furniture*
Brace, Cooper's
Branding Iron, see *Burn Brand*

Bung Borers and Burners, see *Auger, Cooper's Bung Borer*
Bung Chain, see *Cellarman*
Bung Removers:
  Bung Chisel and Pick, see *Cellarman*
  Bung Tickler, see *Cellarman*
  Flogger, see *Cellarman*
  Shive and extractors or vice, see *Cellarman*
Cellarman's Tools, see *Cellarman*
Chime Hook, see *Cellarman*
Chincing Iron, see under *Flagging Iron*
Compass, Cooper's
Cork Removers and Corking Tools, see *Cellarman*
Cresset, see *Cooper (3) Hooping Tools*
Dip (Diagonal)
Drawing Knife, Cooper's:
  Backing
  Chamfering
  Crumming
  Heading
  Hollowing
  Jigger
  Round
Flagging Iron
Flogger, see Bung Removers under *Cellarman*
Furniture, see *Cooper (2) Furniture*
Gauge, Stave
Gimlet, Brewer's ⎫ see also *Cellarman*
Gimlet – Wine Fret ⎭
Groper, see under *Plane, Cooper's Chiv*
Hammers, Cooper's (Hand, Sledge, and Set)
Heading Board, see *Cooper (2) Furniture*
Heading Swift, see *Shaves, Cooper's*
Hooping Tools, see *Cooper: (3) Hooping Tools:*
  Anvil (Bick Iron)
  Chime Maul
  Chisel
  Cresset
  Hoop Driver
  Hoop Knife
  Hooping Dog
  Punch
  Truss Hoop
  Windlass
Howel, see under *Adze, Cooper's Howel*
Jumper
Knives (see also Drawing Knife):
  See *Knife, Cooper's Whittle; Cooper (3) Hooping Tools*
  For *Stuffing Knife*, see Chinching Iron under *Flagging Iron*
Pincer, Cooper's see *Cellarman*
Plane, Cooper's:
  Backmaker's Jointer

Chiv (also *Groper, Sloper,* and *Flincher*)
Croze
Jointer
Stoup
Sun (*Topping Plane*)
Saw, Cooper's:
   Bilge and *Concave Saws* under *Saw, Crown*
   Head
   Riddle
Saw Tub, see *Cooper* (2) *Furniture*
Scraper, Cooper's:
   Buzz
   Plane, Box Scraper
   Round Shaves, see *Drawing Knife, Cooper's*
Server, see *Truss Hoop* under *Hooping Tools*
Shave, Cooper's:
   Downright (*Outside Shave*)
   Heading Swift (*Plucker*)
   Inside Shave (*Tub Shave*)
   Pail Shave
Shaving Horse
Steam Chest and *Steam Bell*
Timber Scribes
Velincher, see *Sampling Tubes* under *Cellarman*
Vice, Cooper's (*Raising Iron*)

Fig. 236 (a)

(b) *Heading Board* (Head Stand; Scouring Board)
A board measuring about 3 ft long and 9 in wide, with a ledge or narrow shelf about 12 in from the foot. The head of the cask is laid on the Heading Board with its edge resting on the ledge; the Board is then stood on the ground or Block and while its top leans against the cooper's body, he shaves the head smooth with a Heading Swift. The operation is sometimes called 'Swifting the head'.

Cooper: (2) FURNITURE *Fig. 236*
Each cooper works near his own Block and Saw Tub. The Shaving Horse on which staves are trimmed stands nearby, and the long Jointer Planes for trueing the stave edges are stood up against the wall or one may be ready for use with its end propped up on the jointer legs. The Bench serves as a shelf for tools and Shaves, and Drawing Knives are hung along its front edge. Burn Brands, Truss Hoops, Hoop Iron, and Flags (Rushes) are suspended from nails on the walls behind. In the bigger cooperages there may be a separate chamber, steam chest, or steam bell used for softening the staves before hooping.
   Other furniture includes:

(a) *Cooper's Block*
A tree stump about 2 ft 6 in high. This is used both as a Bench and a Chopping Block; and when fitted with Block Hooks it is also used for shaping staves, as an alternative to the Shaving Horse, especially for longer staves. The ends of the staves are hooked under the grip of the toothed hooks; the upper hook is used when hollowing the stave, and the lower hook when backing the stave.

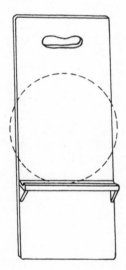

Fig. 236 (b)

(c) *Saw Tub* (Saw Stool)
Made from a sawn-off cask and standing about 18 in high. It is used as a Sawing Stool.

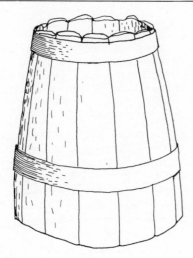

Fig. 236 (c)

Other cooper's furniture is described under the following entries:

Anvil and Cresset under *Cooper* (3) *Hooping Tools*
Shaving Horse
Steam Bell
Steam Chest

Cooper: (3) HOOPING TOOLS *Figs. 237; 238*
The following tools are used by coopers for hooping a cask. The process involves the softening of the staves by heat, bending them into cask form, and then forcing iron or wooden hoops over the outside of the cask.

(a) Adze Fig. 237 (a)
   1. For jointing and fixing wooden hoops. (See also *Adze, Cooper's Notching.*)
   2. For beating truss hoops over the cask to bend the staves. (See also *Adze, Cooper's Trussing.*)

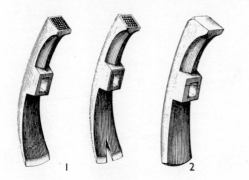

Fig. 237 (a)

(b) Anvil (Bick Iron; Beck Iron; Cooper's Stake; Studdie; T-Anvil) *Fig. 237 (b)*
   1. A T-shaped Anvil or Stake about 30 in high overall, set upright in a block of wood. The shank is usually square in section, and the slightly rounded top has two or more holes to receive a punch when punching the rivet holes in hoops. It is also used for hammering over the rivets when joining the hoops.
   2. An Anvil with double horns is occasionally used in cooper's shops.

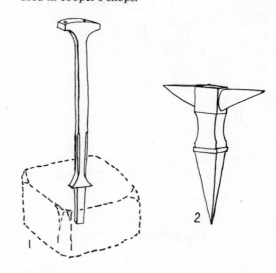

Fig. 237 (b)

(c) Chime Maul (Beater) *Fig. 237 (c)*
A heavy bar of iron or hardwood, about $2\frac{1}{2}$–3 ft long. Used for knocking on the chime hoops, i.e. the hoops surrounding the head of a cask.

Fig. 237 (c)

(d) Cresset (Crisset; Scots: Lummie) *Fig. 237 (d)*
A brazier, usually made from old hoop iron, in which shavings and bits of wood can be burnt. An open-ended cask is placed over the burning Cresset to warm up the wood and so make it more pliable for bending into its final barrel-shape form by means of the Truss Hoops. During the process the cask is mopped over with water and the Cresset sometimes splashed to produce steam. Older coopers declare that the Cresset is superior to the more modern steam-oven because after using it the staves tend to *stay* bent. The Cresset is usually employed in any case

to dry out the moisture from the cask after steaming, and this is said to shrink the fibres on the inside of the cask which helps to set the staves in barrel form.

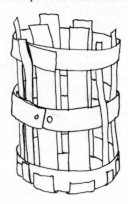

Fig. 237 (d)

(e) Chisel Fig. 237 (e)
Cold Chisels for cutting iron hoops. One with a flat blade for cutting hoops to length, and one with a curved blade for cutting the round nose of the hoop.

Fig. 237 (e)

(f) Hooping Dog (Cooper's Crank; U.S.A.: Lever Hook. Goldenberg (Alsace, 1875) calls it a Cooper's Tang. In France it is called a Traitoire or Tiretoir.) *Fig. 237 (f)*
A wooden lever about 21 in long, with a Vee-shaped iron cap at one end, above which is pivoted a hooked iron bar.

Depicted by Diderot, it is still used by coopers in Continental Europe for stretching or levering a wooden hoop over the top of the staves. It can also be used for pushing a recalcitrant stave inside a Truss Hoop. It is included here because it is seen occasionally in English cooperages, e.g. when repair work is done on casks having wooden hoops. At the Port of London Authority's vaults (1967) a length of cartspring, operated like a Tyre Lever, was used for the same purpose.

Fig. 237 (f)

162

(g) Hoop Driver (Drift; Driver; Shoe Driver; Scots: Hose Driver) *Fig. 237 (g)*
A wedge-shaped tool used by coopers for driving hoops over the outside of a cask. Most of them are grooved at the nose to prevent them from slipping off the hoop. The wooden handles are often ringed with iron to prevent splitting under the heavy blows from the Cooper's Hammer. Varieties include:

1. *Burton Driver* (English Driver). A wedge-shaped steel shoe, grooved at the bottom, and socketed to take a wooden stock, usually of beech, ringed at the top. (The sketch shows a home-made handle in a Burton-type shoe.)

2. *Scotch Driver*. Similar to the Burton but the steel shoe is necked to make removal and replacement of the shoe or stock easier.

3. *Bent Driver* (Solid Driver; Scots: Harn Driver). Solid iron throughout, with a tapering handle. The head is struck on the upper flat face with a Hammer.

4. *Straight Driver*. A T-shaped tool in solid iron, with the bottom edge concave and the side hollowed out.

5. *Wood Driver*. Similar in general outline to (1) above, but all wood except for a metal ring at the top. Used for driving wooden hoops.

6. *Socketed Driver*. Socketed like a Chisel and designed for light work on small hoops or wooden hoops.

7. *Hammer Driver* (U.S.A.: Nantucket Driver). A Hammer with a straight pane grooved along its edge which serves as a driver. Can be used for hammering down the hoop; or it can be held on the hoop and struck with another Hammer.

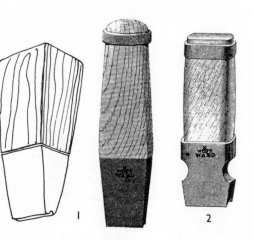

1   2

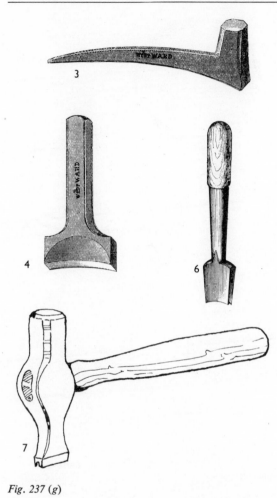

Fig. 237 (g)

(h) Hoop Notching Knife Fig. 237 (h)
Illustrated in Smith's *Key* (Sheffield, 1816), this has a blade similar in appearance to a Cleaving Knife. It was used, we believe, to cut Vee-shaped notches in both ends of a split hazel rod so that the notches interlock with one another to form a complete wooden hoop. Tool makers' catalogues of the later nineteenth century often list a 'Cooper's Whittle' – a similar looking Knife (illustrated here) which is probably also used for the same purpose. Another tool used for notching wooden hoops is described under *Adze, Cooper's Notching*.

Fig. 237 (h)

(i) Hooping Windlass (Capstan; Dutch Hand; Stave Cramp) *Fig. 237 (i)*
As an alternative to the use of Truss Hoops, various cramping devices are used by coopers for drawing staves together into barrel form. This applies particularly to lighter casks and to dry work.

1. One pattern, called a Dutch Hand, consists of a length of rope which surrounds the top of the cask, and a wooden bar which acts as a lever for drawing the rope tight.

2. Another, sometimes known as a Capstan, is operated by means of a capstan-like drum which tightens a wire loop. (A similar device is illustrated by Diderot (Paris, 1763) under 'Tonnelier'.)

3. Like the Dutch Hand (1), the staves are drawn together by a rope loop, but the rope is tightened by twisting the T-handle.

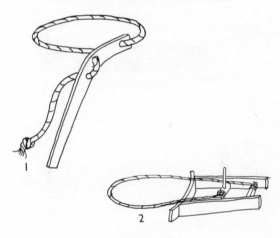

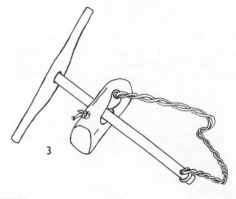

Fig. 237 (i)

(j) Cooper's Punch Fig. 237 (j)
Two kinds are used:

    1. A steel bar, tapered to a blunt point at one end. Used for punching holes in hoops to take the rivets. A hole is punched in the outer flap of the hoop, and the rivet is then driven through the lower flap without previous punching.

    2. A Punch used for small hoops and ornamental work. It contains a hole bored vertically upwards from the face. This is driven over the stem of the rivet, after penetration, to force the flat ends of the hoop tightly together. A variant has a protrusion at the foot containing a semi-circular recess known as a 'mould'. It is used to 'dome' the tail of the rivet.

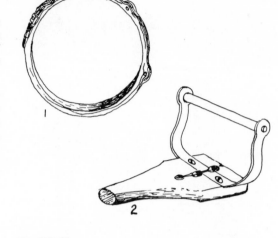

Fig. 237 (k)

Fig. 237 (j)

(k) Truss Hoop (Gathering Hoop) *Fig. 237 (k)*
    1. A strong hoop of bent ash, splayed on the inside face. A series of hoops of decreasing size is driven down over the cask to bend the staves to shape, which are then held by the iron hoops.

    2. When the Truss Hoop becomes worn, or when one of a slightly smaller size is required, the hoop is wound with spun yarn. This reduced the diameter. The yarn is wound on the hoop with a 'Server'. This is a bat-shaped piece of wood provided with a stirrup-shaped fitting to carry a reel of tarred yarn. In use the yarn passes through one of three holes bored through the middle of the bat.

    3. *Passer* (Passing Hoop; Parson Hoop). According to Mr. David Murison (Editor, Scottish National Dictionary, 1966) a Passer in cooper's language refers to a thick ring of steel used both as a Trussing Hoop when setting up the barrel staves, and as a Gauge. When a barrel is finished the Passer is run up and down the sides which it should just touch, i.e. 'pass' the inner circumference and no more.

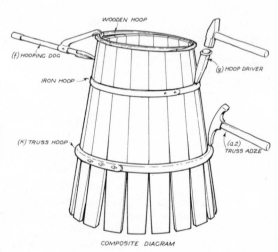

Fig. 238 Composite diagram of some hooping operations

Cooper's Fret: see *Gimlei, Wine Fret.*

Coping: see *Scribing.*

Coppice Trades: see *Woodland Trades.*

Corking Tools: see *Cellarman.*

Corner Tools
A term applied to certain tools including the following: *Brace, Corner; Chisel, Bruzz; Drill, Hand; Saw, Angle.* See also *Chamfer.*

Cornering Tool *Fig. 239*
A strip of flat steel, 6 in long and about $\frac{1}{2}$ in wide, with each end bent round in opposite directions. Each end has a small hole with sharp edges which act as a cutter.

The tool is laid on the sharp corner ('arris') of the wood and pushed or pulled according to the grain, taking off a thin shaving and leaving a rounded corner.

Fig. 239

Counter-Bore
A recess made to take a screw or bolt-head below the surface, e.g. in the deck of a ship, or in the handle of a Saw. See *Auger, Deck Dowelling; Auger, Hub Counterboring; Bit, Deck Dowelling.*

Countersink: see *Bit, Countersink.*

Cove: see *Plane, Moulding.*

Cramp (Clamp; Scots: Glaun) *Fig. 240. Overleaf*
Holding and tightening devices for holding work together during assembly or when being glued. Unlike some other members of the 'holding' family, such as the Vice, Cramps are portable and can be taken to the work in hand. Most of them have two jaws, one or both of which can be drawn together by a screw.

Cramp, Bar: see *Cramp, Joiner's; Cramp, Sash,* etc.

Cramp, Buhl: see *Cramp, Fretwork.*

Cramp, C. or Circle: see *Cramp, Gee.*

Cramp, Carriage Trimmer's *Fig. 241*
Made from two pieces of wood, hinged at the top, usually with leather. The two jaws were drawn together by a screwed nut and bolt; layers of leather were often sewn together to serve as a nut. The Coach Trimmer would have a dozen or more of these Cramps to hold the leather or other covering material while it was being sewn on the metal frames of the wings, dash boards, and elsewhere.

Fig. 241

Cramp, Coachmaker's *Fig. 240(d)*
A strong wrought iron G-Cramp made in sizes 8–22 in and used when framing-up the body and undercarriage.

Cramp, Cooper's: see cramping devices under *Cooper (3) Hooping Tools.*

Cramp, Corner (Picture Framing Cramp) *Fig. 242*
These devices are made to grip the corners of mitred joints used in making and glueing picture frames, mirrors, and other framing work (continued p. 167).

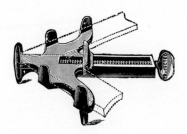

Fig. 242

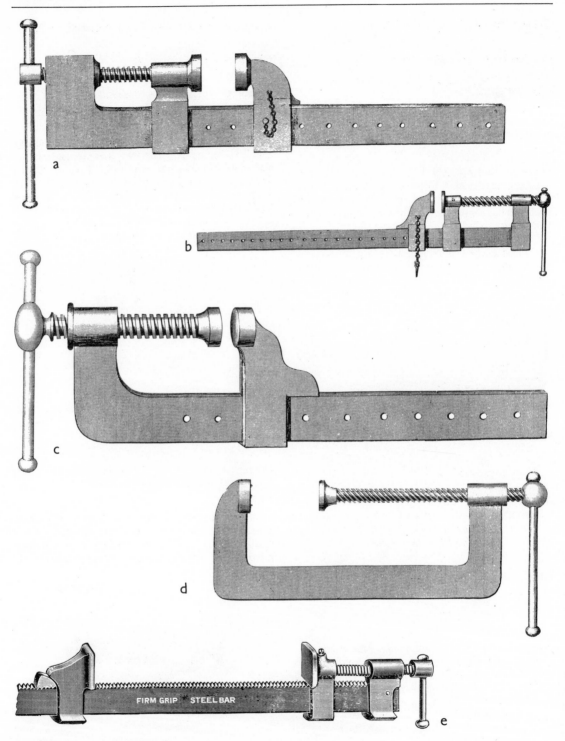

Fig. 240 Group of large-sized Cramps
(*a*) Joiner's Cramp   (*b*) Joiner's Sash Cramp   (*c*) Ship Cramp   (*d*) Coachmaker's Cramp   (*e*) Joiner's Quick-Grip Cramp

(a) Individual Corner Cramps
A metal casting, usually triangular, with two arms at right angles and a central screw with a loose jaw for tightening up. Used for cramping the mitred corners of picture frames, and similar work, when glueing and nailing.

(b) Four-Corner Cramps
Made in various patterns, this Cramp consists essentially of four corner-pieces connected by bars, with either separate or simultaneous cramping action. It can be used for round, oval, or square work. A home-made version consists of four identical corner-pieces of hardwood. A groove is made round the outside of each to take a cord. In use, the cord is tightened by wedging away from the frame, bringing all four corner-pieces into action at the same time.

Cramp, Drill: see *Drill, Cramp.*

Cramp, Flooring *Fig. 243*
There are two main types, both are used for 'cramping-up', i.e. forcing floor boards closely together before nailing.

Fig. 243 (b)

(a) Lever type
A long iron lever arm, forked at the lower end, with serrated cam-like jaws to grip the sides of the joists. A bar with a serrated end is hinged about one third the height of the lever arm to act as a prop. In use the sides of the fork are forced against the edge of the flooring board by pressing the lever forward.

(b) Screw type
A metal base plate is fitted with a serrated lever-cam which fixes it at the required position on the joist. A plunger is then forced against the edge of the flooring board by means of a worm-and-bevel gear or by a lever operating a rack-and-pawl device.

Cramp, Fretwork, Buhl, or Carver *Fig. 244*
Small iron G-Cramps carrying a screw with a shoe at one end and a thumbscrew at the other. Made in several sizes to take work from about 2 to 5 in wide, and used for holding small workpieces in cabinet, Buhl, carving, and other light work.

Fig. 243 (a)

Fig. 244

167

Cramp, Gee (Circle-End Cramp; G-Clamp) *Fig. 245* and see also *Fig. 240(d)*

A general term for Cramps made in the form of a letter G with a screw passing through one jaw and terminating in a loose circular cap which protects the workpiece from the revolving end of the screw. They are used for all kinds of holding and assembling.

The cap or shoe on the screw end is often ball or hinge jointed and is thus capable of taking workpieces at an angle. These Cramps are made in many sizes and patterns from the tiny Buhl or Fretwork Cramp up to the giant sized Ship Cramps. Some modern G-Cramps have a screw known as a 'Springgrip' fitted to the boss through which the main screw passes. This allows the cramp to be tightened by finger and thumb, leaving the other hand free to hold the work.

Fig. 245

Cramp, Glueing Screw: see *Brushmaker; Violin Maker.*

Cramp, Hand Screws *Fig. 246*

Two hardwood blocks 10–18 in long and about 1½–2 in square are connected by two wooden screws, one in the centre and the other near the back of the jaws. The central screw passes freely through a plain hole in one block, and through a tapped hole in the other. The outer screw is tapped through the same block, and its shouldered end fits loosely in a blind hole or seating in the other block. The work to be cramped together is placed in the jaw opening, the two screws adjusted to bring the blocks nearly parallel, and the

final cramping is done by tightening the outer screw, thus forcing the jaws together.

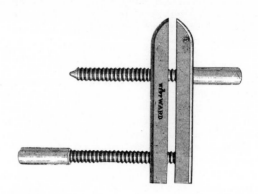

Fig. 246

Cramp Heads

A pair of cramp jaws provided with metal straps or other means of attachment so that they can be fitted to any suitable length of timber. When not in use the heads can be carried in the tool bag, and serve as a kind of portable Joiner's Cramp.

Cramp, Hexagon: see *Cramp, Universal.*

Cramp, Hoop: see various cramping devices under *Cooper (3) Hooping Tools.*

Cramp, Joiner's (Bar Cramp; Sash Cramp)

In metal Fig. 240(a) (b)

Similar in principle to the G-Cramp, but with an adjustable span ranging from about 2 to 7 ft effected by means of a bar (Scots: 'sword'), along which one or both of the jaws or heads can slide. One head is adapted to take a vice-screw and tommy-bar, this screw being made to impinge on the workpiece itself or on to a separate sliding head. The other jaw can be fixed at suitable distances along the bar by a pin passing through it. In some patterns the bar is serrated to form a rack on which the sliding jaw can be quickly secured in any desired position (*Fig. 240(e)*).

The bar may be flat or of T section, with or without feet, and lengthening pieces (known in north England and Scotland as 'Ekes') may be attached if necessary by means of a sleeve.

The Sash Cramp is a somewhat lighter version of the Joiner's Cramp (*Fig. 240(b)*); the Ship Cramp is similar but much heavier (*Fig. 240(c)*).

These tools are used for 'cramping-up' and holding

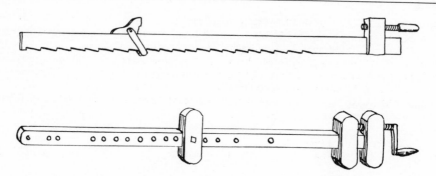

Fig. 247

large frames in joinery and cabinet work such as doors or window frames, when nailing and glueing.

In wood Fig. 247
A wooden bar with a short upright arm at one end carrying a wooden screw, or an iron screw with tommy-bar. The bottom edge of the bar is racked and the top edge grooved. The sliding jaw, made from a single solid block, runs in the groove and is held in position by an iron strap pivoted to the block and fitting in the appropriate slot of the rack, which is made to take work up to 5 ft.

Cramps of this type made by the craftsman himself are often found in country workshops, but they are still factory-made in France and elsewhere on the Continent of Europe. A simpler, home-made version, with sliding cheeks and iron screw, is also illustrated.

Cramp, Picture Framing: see *Cramp, Corner.*

Cramp, Quick Action: see *Cramp, Sliding Gee Pattern.*

Cramp, Saddle
A Sash or Joiner's Bar Cramp with deep jaws, used for curved work or for framing when there are projecting members.

Cramp, Sash: see *Cramp, Joiner's.*

Cramp, Saw: see *Saw, Sharpening: Holding Devices.*

Cramp, Ship *Fig. 240(c)*
A very heavy and strong variety of the Joiner's Cramp made in various sizes e.g. to take 3 ft, the bar measures $4 \times \frac{3}{4}$ in and the Cramp weighs 115 lb; to take 4 ft the bar measures $4\frac{1}{4} \times \frac{3}{4}$ in and the Cramp weighs 135 lb. A very heavy version of the G-Cramp is also used for ship work. Used for bending ship's

planks and holding them in place while they are secured.

Cramp, Sliding Gee Pattern (Glueing Cramp; Quick Action Cramp) *Fig. 248*
A light G-shaped Cramp consisting of a steel bar with a fixed jaw at one end and a sliding arm running along the bar carrying a steel thumb screw. Some models have the inside edge of the bar serrated so that the arm is held rigid as soon as the pressure is applied by the vice screw. Used for all types of cramping. The bar is made long enough to take work from 4 to 24 in wide.

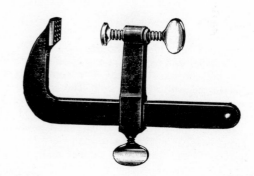

Fig. 248

Cramp, Spoke: see Spoke Bridle under *Wheelwright's Equipment (1) Spoke Tools.*

Cramp, Stave: see Hooping Windlass under *Cooper (3) Hooping Tools.*

Cramp, Strake: see Strake Tools under *Wheelwright's Equipment (3) Tyring Tools.*

169

Cramp, Universal (Combination Cramp; Hexagon Cramp) *Fig. 249*

Two double-cranked arms with swivel shoes, connected together by a system of levers actuated by a central threaded rod which also carries a round shoe at its end. It can be used for corner-cramping large mitred frames or for holding a plank while pressing on its edge, e.g. when lipping doors.

Fig. 249

Cramp, Violin: see *Violin Maker.*

Cramp, Wood *Fig. 250* (see also *Fig. 247*)

A term applied to a wooden cramp of sturdy make with square corners which are often comb-jointed and reinforced with an iron bolt. The end of the screw is sometimes provided with a projecting metal point to hold a protective piece of waste wood in place against the workpiece when the Cramp is tightened. Made in many sizes to span from 3 to 18 in or more. Used mainly by cabinet makers for holding workpieces.

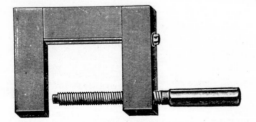

Fig. 250

Crank

A double bend in a bar or rod designed to produce rotary motion, e.g. in a carpenter's Brace, and (later) for transforming horizontal into rotary motion, as in a steam engine.

Cresset (Crisset): See *Cooper* (*3*) *Hooping Tools.*

Creves (Crevis; Crevice; Cravice)

A term appearing in the Apprentice Indenture Records, Bristol, 1549–1574. Possibly a Croze.

Cringle: see Hand Fid under *Sailmaker.*

Crocodile: see *Wrench* (*2*) *for Gripping Rounds.*

Croom Iron: see *Drawing Knife, Cooper's Round Shave.*

Crotch Grabs: see Span Dogs under *Timber Handling Tools.*

Crow Bar (Gavelock; Pinch Bar; Tilly Iron; Wrecking Bar) *Fig. 251*

A strong steel rod, often clawed at one end and having a spike at the other. Used for levering up or moving heavy objects.

Fig. 251

A special Crow Bar used for setting hurdles, known as a Shepherd's Bar, Poll Prytch, or Fold Shore, is described under *Hurdlemaker*.

Crown Wheel
The larger gear wheel of a bevel reduction gear; a wheel with cogs or teeth set at an angle to its plane, like a crown.

Croze or **Crowe**: see *Plane, Cooper's Croze*.

Crumm: see *Drawing Knife, Cooper's Crumming*.

Cuckoo: see Fangle Iron under *Millwright*.

Cudgel: see *Mallet, Woodland Worker's*.

Curette
A name sometimes given to certain cooper's Drawing Knives used for trimming the inside of casks, such as the Jigger. (See *Drawing Knife, Cooper's Jigger*.)

Curfing Iron: see *Print and Block Cutter*.

Cutter (Iron) of a Plane: see under *Plane*.

Cutter and Chisel Grinder: see *Grinding Appliances*.

Cutting Gauge: see *Gauge, Cutting*.

Cutting Lip
The inbent cutting edge of a Shell Auger and Nose Bit. (See diagram under *Auger, Shell*.)

Cyma Recta or **Reversa**: see *Plane, Moulding*.

D

Dader: see Mortising Axe under *Hurdle Maker.*

Dado
Panelling or moulding fitted around the lower part of a wall in a room.

Dannocks
Gloves made in tough leather, worn by hedgers and other tradesmen when handling rough materials.

Deck River or **Iron:** see Reaming Iron under *Caulking Tools.*

Depthing Router (Depthing tool): see *Plane, Router.*

Devil
A name applied to various scarifying and other tools, including the ordinary Scraper and the various shave-type Scrapers used by chairmakers. See also Devil under *Millwright*; Go Devil under *Plane, Cooper's Chiv*; and Devil's Tail under *Jumper.*

Die Stock and **Dies:** see *Screw Dies and Taps* and *Metal Working Tools.*

Dip (Diagonal; Dipping or Gauging Rod) *Fig. 252*
These wooden rods, usually home made, each marked according to cask size, were used by coopers and others as a rough test for capacity. They were made in pairs, the chisel-shaped ends being usually plated with iron or brass.

Their method of use is not entirely clear, but judging from the more modern Dip – an iron rod hinged at the centre like a pair of dividers – one rod was laid diagonally within the cask, a scratch was made at the upper end, and the other rod laid from the scratch across to the opposite side of the cask. If the top rod reached to an agreed distance from the croze at the top of the cask, the size of cask was accepted as correct.

Fig. 252

Dish or Dished
A term applied to wheels. 'Dish' is the inclination of the spokes in relation to the axis of the hub. See *Wheelwright.*

Dividers: see *Compass; Froe.*

Doctor's Web: see *Web, Calico Printer's.*

Dog
A name given to many different tools, including Cant Dog, Ring Dog, and Span Dog under *Timber Handling Tools;* Hooping Dog under *Cooper (3) Hooping Tools;* Sawing Dogs under *Sawing Horse;* Spoke Dog under *Wheelwright (2) Spoke Tools;* Tyring Dog and Strake Dog under *Wheelwright (3) Tyring Tools.*

Dog, Joiner's (Joint Cramps) *Fig. 253*
A U-shaped forging, 2–4 in along the back, with wedge-shaped points. Used for driving into the ends of two boards to keep them together while glueing the joint etc.

Fig. 253

Dog Leg
A term meaning bent into the shape of a dog's hind leg, as in Dog-Leg Carving Chisel.

Dog, Timber (Hook Nail) *Fig. 254*
Iron rods, varying from about 6 to 18 in long, with the ends turned down at right angles and sharpened, often to a chisel-shaped point. Sometimes the bar is twisted so that the ends point in different directions, when they are said to be 'at mock' with one another. Timber Dogs are driven into heavy timbers to hold them in position when being used as shores, e.g. on buildings or in shipyards. They are also used by sawyers for holding a log steady on the saw pit. One end is driven into the log, and the other into the surrounding timbers or sill of the pit.

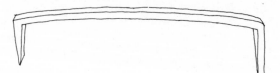

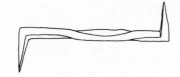

Fig. 254

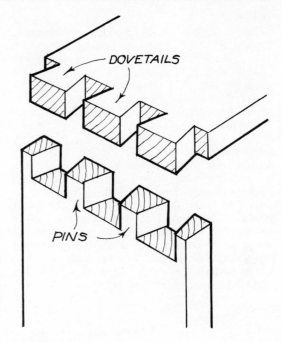

Fig. 256

Dolly: see Bending Block under *Wheelwright's Equipment (3) Tyring Tools.*

Donkey *Fig. 255*
Various work-holding devices are known as Donkeys. (See *Horse* for list of appliances called Horse, Mare, or Donkey.) The example illustrated is the Donkey used by makers of Windsor chairs. This is a curved block of timber designed to be held in a Vice, and used to hold curved material such as Windsor tops and chair splats while being finished with Shave and Scraper.

Fig. 255

Donkey's Ear Board: see *Mitre Appliances.*

Dotter: see *Measuring Sticks.*

Dovetail Joint *Fig. 256*
A name given to one of the best and strongest joints for joining wooden pieces together at right angles to each other and end to end. It is widely used for boxes, drawers, and the carcases of cabinets of all kinds. The pins (or tenons) on one piece are fan-shaped like a dove's tail, and fit into sockets of corresponding shape in the other piece. (See *Plane, Dovetail; Saw, Dovetail.*)

Dovetail Marker *Fig. 257*
The Bevel can be used for marking dovetails (see *Square, Bevel*), but special templets in wood or metal are sometimes made to save having to set the angle each time. One type is of thin hardwood with projecting shoulders; another is of sheet brass.

Fig. 257

Dowel

A round peg used to join wooden parts together, e.g. the felloes of a wheel, or parts of a chair. See also *Trenail*. Tools used for making and fitting dowels include: *Auger, Deck Dowelling; Bit, Cooper's; Bit, Deck Dowelling; Bit, Dowel; Bit, Dowel Trimmer* (and *Rounder); Dowel Centre; Dowel Plate; Plane, Rounder;* Tine Former under *Rake Maker.*

Dowel Centre *Fig. 258*

A short brass cylinder with an overhanging top provided with a spike in the centre. The body has a diameter of $\frac{1}{4}$ in rising in sixteenths to $\frac{1}{2}$ in.

Used to locate the exact position of the dowel (or dowels) in a butt joint, e.g. in chair or cabinet making. The hole for the dowel is bored in the desired position on one member, and the appropriate Dowel Centre is inserted. The other member is then held in position and a smart blow with the mallet marks the corresponding hole.

Fig. 258

Dowel Plate (Dowel Cutter; Peg Cutter) *Fig. 259*

A steel plate about $\frac{1}{2}$ in thick with up to six holes in it, decreasing in size from about $\frac{5}{8}$ in down to $\frac{1}{4}$ in, and often home-made.

Cleft pegs of square-sectioned wood are driven successively through the holes down to the size of the pin or dowel required. It is commonly used by wheelwrights for making the dowels which join the ends of the felloes. The factory-made plates sometimes provide a small projecting vee-tooth or serration on the inner circumference of each hole. This forms grooves on the dowel which allow air and surplus glue to escape.

Fig. 259

174

Dowelling Stock: see *Brace, Cooper's.*

Downright: see *Shave, Cooper's.*

Draft Shave: see *Drawing Knife.*

Drag Shackle: see Span Dogs under *Timber Handling Tools.*

Draw Bore Pin (Draw Borer; Draw Irons; Drift Pin or Hook; Drift Bolt; Hook Pin.) *Figs. 260; 261.*
A tapering steel pin used for drawing up mortice-and-tenon joints, and made in different patterns, as follows:

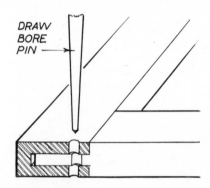

Fig. 260 Diagram showing Draw Bore Pin in action

(*a*) A tapering steel pin about $\frac{3}{8} - \frac{5}{8}$ in. in diameter at the thick end and 8–12 in long. It is either tanged into a wooden handle like a Butcher's Steel, or made in steel throughout with a thickened head which is sometimes eyed. Another all-iron type, sometimes made in large sizes, has a flattened triangular head which is undercut to form a 'chin' which can be struck with a Hammer to help removal.

(*b*) The type used by coachbuilders is a tapering steel pin fitted with a wooden cross-handle like a gimlet. The coachbuilder had several of them, the common sizes being $\frac{1}{4}$, $\frac{5}{16}$ and $\frac{3}{8}$ in. in diameter and about 8 in long.

These pins are used for drawing up mortice-and-tenon joints in framed work which cannot be pulled together with a Cramp. A hole is first bored in the mortised workpiece, the tenon inserted as tightly as possible, and the position of the hole marked by pricking with an Awl or with the Twist Drill, if used. The tenon is then removed, and a hole bored through it, slightly nearer the shoulder; the distance is a matter of judgement, depending on the hardness of the timber used. When the joint is re-assembled, it is drawn up tight by inserting the Draw Bore Pin, later to be replaced, if required, by a dowel.

Moxon (London, 1677) describes the above process but uses the dowel pin itself to draw up the joint. (See also *Drift*.)

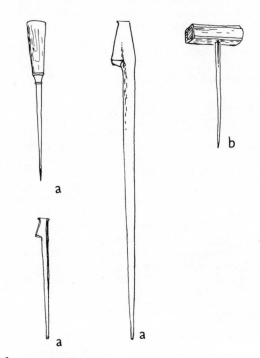

Fig. 261 Examples of Draw Bore Pin

Drawing Knife (Draw Knife; Draw Shave; Draft Shave; Shaving Knife) *Figs. 262; 263 overleaf*
The common form consists of a flat or curved blade made in sizes from 8 to 18 in long and up to about $2\frac{1}{2}$ in wide. The blade is normally chisel-shaped in section and bevel-ground on its front edge. Tapering tangs at both ends of the blade are bent at right angles to the cutting edge and are fitted with wooden handles, usually turned, with the end of the tang clenched or riveted over.

These tools are used in many different trades for the removal of surplus wood and for rounding and chamfering. In operation the work may be held between the bench and the user's chest, but more often in a Shaving Horse, Brake, or Vice, and the tool drawn towards the user.

A Drawing Knife is included in a group of Viking Shipwright's Tools *c*. A.D. 100 in the State Historical Museum in Stockholm. Tools of this type were widely used in medieval Russia for smoothing a surface after using the Axe or Adze, but no examples have been noted in medieval illustrations in the West. The modern form is depicted by Moxon (London, 1677) among carpenter's tools, and by Diderot (Paris, 1763) among cooper's tools.

Drawing Knife, American Pattern *Fig. 262 Shape C; Fig. 263 (a)* and *(j) overleaf*
A flat blade, 8–14 in long, the edge of which is curved towards the user. The blade therefore appears 'saddle'-backed with a bellied (convex) cutting edge. Unlike most other patterns, the sharpening bevel is ground all the way from the back down to the cutting edge. This is described in some lists as 'razor blade', in others as 'fluted blade.' The handles are often splayed slightly outwards.

Most of these features are present in the 'Planes, Façon Souabe' (Swabian Pattern Drawknives) in the list of Goldenberg of Alsace (1875). It seems probably that the type was taken to the U.S.A. by European settlers, and later back to Europe as the 'American Pattern'.

Drawing Knife, Barking
A Drawing Knife of large or medium size and of varying shapes used for de-barking logs. It is held with the bevelled edge of the blade underneath, in which position it is less liable to be drawn into the wood.

Drawing Knife, Carpenter's *Fig. 262 Shape A* and *Fig. 263 (b)* and *(d)*
A straight blade 8–18 in long and up to about 2 in wide. It will be noted that the 'London Pattern' has a moulding on both ends of the blade. This is the only Drawing Knife showing any sort of decoration among the innumerable variants illustrated in nineteenth and twentieth century tool catalogues.

Drawing Knife, Chairmaker's
The normal Carpenter's patterns were used. Mr. L. J. Mayes writes as follows on their use in the Windsor chair trade (High Wycombe, 1950):

'The most universal tool in chairmaking. The turner used it to shave chair legs, stretchers, and other turned parts, before going into the lathe. He also used it to 'tune' the pole of a pole lathe by shaving away the underside until the right degree of spring was achieved. In general chair work it was used for almost every kind of preliminary shaping, and the accuracy achieved by a skilled craftsman is of the kind that needs to be seen to be believed.

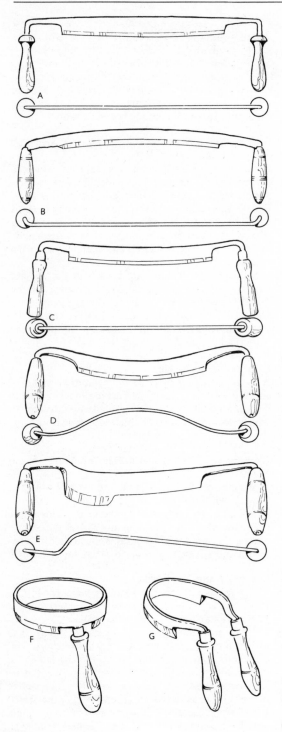

Drawknives were sometimes 'lined and steeled' by the user himself. The original cutting edge was ground off, the blade softened by heating, split and a piece of old rasp placed in the split and hammer-welded in, then tempered and ground. The tool would have the qualities of the tool-steel of the rasp, backed up by the softer and more resilient metal of the original blade. A delicate example of the smith's craft, a lined-and-steeled Drawknife was much valued and would last almost a lifetime.'

Drawing Knife, Chamfering: see *Drawing Knife, Cooper's Chamfering; Drawing Knife, Stop Chamfer; Drawing Knife, Wheelwright's.*

Drawing Knife, Cooper's *Figs. 264; 265 overleaf*
These tools, which are known simply as 'Knives' by the Coopers themselves, tend towards the heavier and larger sizes with blades measuring up to 16 in long and fitted with wooden handles usually plain turned without ferrules.

They are used for taking off unwanted timber from the back and inside of the staves, and for paring the bevel surrounding the heads.

There is some confusion both in nomenclature and use as between the Cooper's Crumming Knife, Chamfering Knife, and Jigger. An attempt at differentiation has been made under these entries, but the reader should also refer to the list of alternative methods and tools under *Cooper.*

Fig. 262 Drawing Knife: Typical shapes

              Example

(*A*) *Flat and straight*

        Drawing Knife, Carpenter's

(*B*) *Flat and 'circular backed'*
(also known as arched, hog-backed, circular-edged, or fish-backed)

         Cooper's Heading

(*C*) *Flat and 'saddle-backed'*
(also known as American, Hollow-backed, or Bellied)

         American

(*D*) *Bent*
to give a hollowing cut

        Cooper's Hollowing

(*E*) *Part flat and part bent*
A combination of shapes (A) and (D)

        Cooper's Crumming

(*F*) (*G*) *Round or half round*

        Cooper's Round Shave

Fig. 262

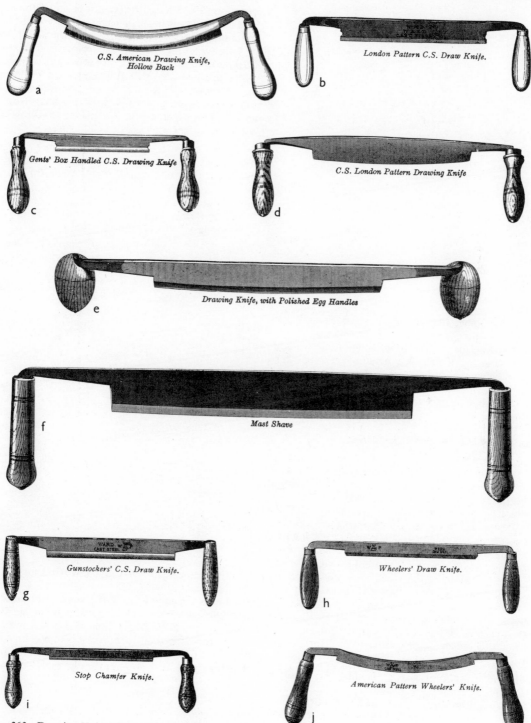

Fig. 263 Drawing Knives for various trades (see entries)

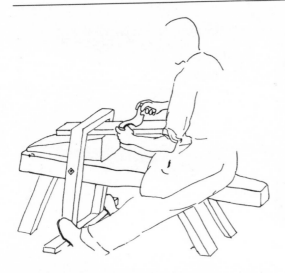

Fig. 264 Cooper's Shaving Horse in action

Drawing Knife, Cooper's Backing (Backing Knife; Cooper's straight Stave Knife) *Fig. 265 (a)*
A straight blade 10–18 in long, similar to the Carpenter's Drawing Knife. The handles are often bent down below the cutting edge to encourage a 'greedy' cut.

Used to trim and give a slight convexity to the outside of the stave.

Drawing Knife, Cooper's Chamfering *Fig. 266*

(a) Jigger type
The *Chronicle of the Early American Industries Association* (U.S.A., September 1965, Vol. XVIII No. 3) illustrates a tool of this name which in England might be called a Cooper's Jigger except that the blade appears to be straight or only slightly curved. The *Chronicle* declares that the tools were used for cutting the chime, but Brombacher (U.S.A., 1922) lists them as 'Cooper's Chamfering and Howeling Knives', thus indicating that they were used for cutting both chiv and chime. An illustration of a workman using this tool to cut the chime bevel is given in Eric Sloane's *Museum of Early American Tools* (U.S.A., 1964). (English coopers used an Adze for cutting the chime bevel – see *Cooper* for alternative methods and tools.)

(b) Radial Knife (U.S.A.: Heading Trammel or Chamfering Knife)
This American tool, which we have not seen in English cooperages, resembles a Beam Compass (Trammel) with a cutter fitted at one end instead

of a point. It is designed to cut a bevel on the edge of a cask-head instead of using a Draw Knife; or for cutting the chime bevel instead of using an Adze.

A tool working on a similar principle, but fitted with a square-shaped knife and used for cutting the chime bevel of a beer cask, is illustrated in Brombacher's List (U.S.A., 1922). The compass point is supported on an adjustable three-pronged 'frog' – an appliance which can be wedged into position just below the top of the cask.

(c) A tool called a *Chamfer Plane* ('Rabot a chanfrein'), and used for the same purpose, is illustrated by Legros (Liège Museum, 1949). Made entirely of wood, it consists of a shave with cutting-iron bedded and wedged like a plane; one of the handles is extended to form a long arm which carries an adjustable compass point to act as a pivot.

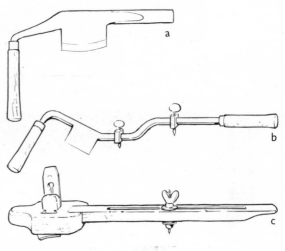

Fig. 266

Drawing Knife, Cooper's Crumming (Crumm Knife) *Fig. 262 Shape E; Fig. 265*
The name Crumming Knife is given to a number of different Cooper's Drawing Knives. It has a blade which combines both a straight and hollowing section in the same tool. Its purpose is to combine the function of backing and hollowing a stave without changing tools. It is not, apparently, a very popular tool.

There are three types:

(a) Common Crumming Knife Fig. 265 (b)
A wide blade 12–16 in long. For just over half the

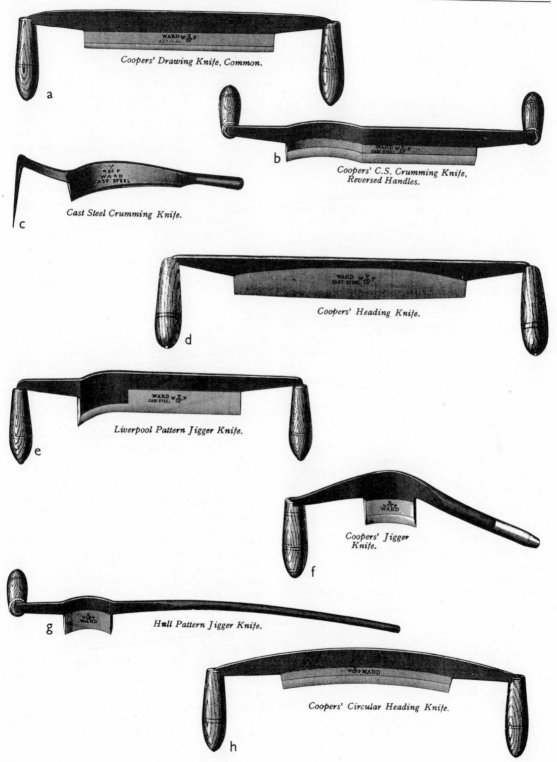

Coopers' Drawing Knife, Common.

a

Coopers' C.S. Crumming Knife,
Reversed Handles.

b

Cast Steel Crumming Knife.

c

Coopers' Heading Knife.

d

Liverpool Pattern Jigger Knife.

e

Coopers' Jigger
Knife.

f

Hall Pattern Jigger Knife.

g

Coopers' Circular Heading Knife.

h

Fig. 265 Cooper's Drawing Knives (Ward & Payne Ltd., Sheffield, 1911)

length the blade is flat; the remainder is bent to a shallow hollow and is bevelled on the inside. The wooden handles are sometimes turned down at right angles to the blade on the side opposite the bevel, a position known as 'reversed'.

(b) Liverpool Crumming Knife (sometimes called a Jigger Knife) *Fig. 265 (e)*
A blade 7–10 in long, of which two-thirds is flat, the remaining portion being bent round in a smooth hollowing curve. The handles are turned down in the same plane as the blade.

(c) Belfast Pattern Crumming Knife Fig. 265 (c)
A form of Jigger but with a long curved blade which tapers from right to left. The handles are like those of a Jigger, i.e. a solid-iron extension at one end and a wooden handle bent in line with the blade at the other. This tool can also be used for cutting the chiv surface. For alternative methods and tools see *Cooper*.

Drawing Knife, Cooper's Heading (Heading Knife) *Figs. 262; 265*
A large, flat blade up to 2¼ in wide and 16 in long. There are two types, 'straight' and 'circular backed'. One has a straight back and straight cutting edge (*Fig. 262 Shape A* and *Fig. 265 (d)*; the other has a 'circular' blade (also known as 'hog backed') and a slightly concave cutting edge which helps to prevent the knife from slipping off the work (*Fig. 262 Shape B*, and *Fig. 265 (h)*). After rough-cutting the bevel round the heads with the Cooper's Axe, the Heading Knife is used for smoothing and finishing. For alternative tools and methods, see *Cooper*.

Drawing Knife, Cooper's Hollowing (Belly Knife) *Fig. 267 and Fig. 262 Shape D*
The blade is bent in a shallow hollowing curve. It is made in sizes up to about 12 in long and 2¼ in wide. Used to trim and give a slight concavity to the inside of the staves.

Fig. 267

Drawing Knife, Cooper's Jigger (U.S.A.: Howeling Knife; Jigger Knife; Runcorn Jigger) *Fig. 265*
A short, bent, hollowing blade 2½–4 in wide. The tang on the left is wood-handled for holding outside the cask. The handle on the right is an extension of the back of the blade, forged solid with a round or square section, for grasping inside the cask.

Used for paring down the chiv surface on repair work, or when the correct size of Chiv is not available. Much of the cooper's work is very hard, but the use of a jigger for this operation appears to be the hardest work of all.

For alternative methods and tools see *Cooper*. Variants include:

(a) London Pattern (Fig. 265 (f)) with a blade 2½–3½ in long, usually bevelled on the inside. The handle on the right is of solid round steel, bent slightly downwards and tapering to a point.
(b) Scotch or Wick Pattern (sometimes called a Crum Knife) with a blade 2½–6 in long, usually bevelled on the outside. The solid steel handle on the right is rectangular in section, in line with the back of the blade or bent downwards.
(c) Hull Pattern (Fig. 265 (g)) with a blade 2¾ in wide, bevelled on the outside. The all-iron handle on the right is about 18 in long, long enough to pass right through a small cask so that both handles can be grasped outside it. Like the French Curette it may also be used for cleaning the inside of a cask.
(d) Liverpool and Belfast Pattern. See *Drawing Knife, Cooper's Crumming.*
(e) See also *Drawing Knife, Cooper's Chamfering* for jigger-like tool used in the U.S.A. for cutting the chime.

Drawing Knife, Cooper's Round Shave (Erasing Iron; Croom Iron; Inshave; Scorper) *Fig. 268; Fig. 262, Shapes F and G.*
A round blade, curved into the form of a complete or part circle about 2–6 in. in diameter.

These tools are used by Coopers for reaching down inside a cask to level the joints between staves (particularly when the grain is too variable for a Stoup Plane or Inside Shave to be used) and also for cleaning the inside of a cask if it becomes foul. They are also in common use for erasing brands, marks, and painted letters from casks and boxes.

(a) The *Two-Hand Round Shave* has a semicircular blade about 4 in wide, fitted with tangs bent round to bring the wooden handles in line with the blade. Both home and factory made examples are illustrated.
(b) The *One-Hand Round Shave* (U.S.A.: Closed Scorp) has a small blade formed on a closed ring

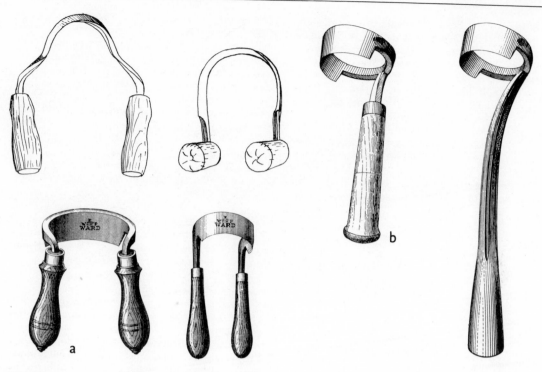

Fig. 268

2½–4 in. in diameter, to which is forged a tang to take a wood handle. In another pattern the ring blade is forged on to the neck of a long, curved steel socket into which the handle is fitted. Diderot (Paris, 1763) illustrates a similar tool which has an open-ended blade like the knife used by spoon or bowl makers – see *Knife, Hooked. (Note:* A very small version of this tool, called a Peg Knife, is used in the Shoemaker's trade for cutting off the wooden pegs that protrude inside the shoe.)

Drawing Knife, Egg Handled *Fig. 263 (e)*
A term used when the tool is fitted with egg-shaped handles. They give a good grip for pulling because there is room for the fingers to surround the back of the handle.

Drawing Knife, Gentlemen's (Gent's Drawing Knife) *Fig. 263 (c)*
A small version of the Carpenter's Drawing Knife, with a blade from 5 to 7 in long and turned handles, often of boxwood.

Drawing Knife, Gunstocker's *Fig. 263 (g)*
A strong Drawing Knife with a straight blade 1½ in wide and 8–11 in long. Used for paring and removing waste wood when making gunstocks. See *Gunstocker.*

Drawing Knife, Handle Maker's *Fig. 269*
A Drawing Knife about 8–10 in long overall, with the back bent upwards in the middle to form a semicircle 1¾–2 in. in diameter, bevel ground on the outside of the curve. The tangs are bent at right angles to bring the handles into the same plane as the blade in the usual way.

We have seen this tool being used by the makers of wooden shovels, Messrs Lusher & Marsh (Norwich, 1949), for cleaning the inside corners of the 'D' handles (which were made solid with the rest of the shovel) and also for cleaning the root of the shaft where it meets the wooden blade. At the St. Albans City Museum there are several of these knives in different sizes, said to have come from a local maker of Barn Shovels. These tools are shown hanging on the wall of the Shaft (Handlemaking)

181

Shop of Ward & Payne, Sheffield, in a photograph in their 1911 catalogue, with a pile of D-handle Spade shafts nearby. The mystery is that although these tools were probably made in Ward & Payne's works, there is no mention of them in their list, which gives 30 different patterns of Drawing Knife among the 7,300 items listed. (For other tools used in this trade see *Spade and Shovel Maker*.)

In *Gwerin* (Vol. I, No. I 1956) this tool is included among those used by a Bedfordshire wood turner and rake maker. In the Shelburne Museum Handbook (U.S.A., 1957) a similar tool is described as a 'Coachmaker's Draw Knife', but we have never seen or heard of it used in this country by coachmakers.

Fig. 269

Drawing Knife, Hoop Maker's *Fig. 270*

A tool used for smoothing the inside surface of split-hazel and other rods, used until recently for making hoops for casks. Some Drawing Knives made for this purpose had their handles bent at right angles to the plane of the blade, as illustrated. The reason for this is not clear, but it may have been done to make the tool suitable for splitting the rods (like a Froe) as well as shaving them.

Fig. 270

Drawing Knife, Howeling: see *Drawing Knife, Cooper's Jigger*.

Drawing Knife, Mastmaker's (Mast Shave) *Fig. 263 (f)*

A large and heavy variety of the Carpenter's Drawing Knife, with a straight, flat blade $2\frac{1}{4}$–$3\frac{1}{2}$ in wide and 10–20 in long. Draw Knives used in the mast and spar shop of H.M. Dockyard, Portsmouth (1969) were 24 in long overall with a $2\frac{1}{2}$ in wide blade. We were told that the handles were bent slightly downwards after purchase to prevent the tool from 'digging in'.

Used for trimming masts and spars (see *Mast and Spar Maker*).

Drawing Knife, Router Type

A Router blade mounted on the back-bar of a Drawing Knife. See Boxing Router under *Router, Coachbuilder's*.

Drawing Knife, Saddle-Tree Maker's

Illustrated by Gilpin (Cannock, 1868), this appears to be like the common carpenter's Drawing Knife, $9\frac{1}{2}$ in long, but with a slightly hollow back.

Drawing Knife, Shipwright's

No special Drawing Knives are listed by the tool makers for this trade (except the Mastmaker's), but a continental list (Goldenberg, 1875) illustrates a Shipbuilder's Drawing Knife 11–15 in long with a flat blade slightly arched and with the handles made in the form of round wooden discs set at right angles to the blade.

Drawing Knife, Spanish

Illustrated by James Cam (Sheffield, *c.* 1800). They differ from English patterns in that the blade runs the full length of the back of the knife and the handles are longer.

Drawing Knife, Stop Chamfer (Stop Chamfer Knife) *Fig. 263 (i)*

A narrow, flat, straight-bladed Drawing Knife 5–14 in long. Used for working stopped chamfers on joists, wagon frames, etc. (See *Drawing Knife, Wheelwrights'* for a description of the work when applied to carts and wagons.)

Drawing Knife, Straight Stave: see *Drawing Knife, Cooper's Backing*.

Drawing Knife, Wheelwright's *Fig. 263 (h)* and *Fig. 271*

A narrow, flat, straight blade, 8–14 in long, used for trimming spokes and shafts, and for working chamfers on wagon and cart framing. George Sturt (Cambridge, 1958) describes the operation as follows:

'The object was to relieve the horses of every ounce possible. To this end the timbers were pared down, here and there, to a very skeleton thinness. But wherever strength was essential – where a mortice or a bolt-hole was made, or where a bearing was wanted for another timber, there nothing was shaved away; the squared timber was left square. The result was a frequent shaving out of short curves, which indeed had some look of being meant for beauty, but their usefulness came first. At a rough guess I should say the eighth part at least of the weight of squared waggon timbers was taken away with the draw-shave. Shafts were reduced even more than that . . . .'

Goldenberg (Alsace, 1875) illustrates a pattern 'for felloes', i.e. for the sections of the rim. This has a short blade ground on both sides, which is offset from the back of the knife so that it can cut at a level below the handles. This may have been intended to keep the hands above the work when trimming the side of a felloe.

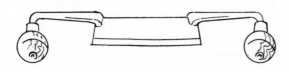

Fig. 271

Drawing Knife, Wheelwright's Radial

This unusual tool is described by Richardson (U.S.A. 1903). It operates in a similar way to the Radial Knife described under Drawing Knife, Cooper's Chamfering. It consists of a straight knife with a handle at one end; at the other there is a spigot which is pivoted from a hole in a perforated bar fitted across two spokes, leaving the knife lying across the felloe of a completed wheel. When the knife is moved in radial fashion it is made to pare down the side of the felloe.

Dressing Stick: see *Plumber's Tools*.

Drift

A name given to several different tools used in woodworking and other trades. The Drift is a rod, often tapered, used for clearing out holes, driving a bolt out of a hole, holding drilled objects in position while fitting them together, clearing out chips from the square chisel of a Mortising Machine, or for enlarging holes, e.g. in a metal plate.

The term is also applied to a wooden Bobbin used by Plumbers for smoothing out a dented lead pipe, to a Cooper's Hoop Driver (see *Cooper (3) Hooping Tools*), and to a Shepherd's Bar (see *Hurdle Maker*). There is also some overlapping in use and nomenclature with the Draw Bore Pin.

Drill

This term is used loosely for Drilling Machines, Hand Drills, and sometimes also for the Bits which do the actual drilling.

Drill, Archimedean (Archimedean Brace) *Fig. 272*

Varying in length from about 6 to 15 in, these tools consist of a head, usually of wood, a stem cut or twisted into the form of a slow spiral, a driving (or 'travelling') handle containing a nut cut internally to engage with the spiral, and a screw chuck or pad to take Bits up to about ⅛ in. in diameter. The rotating action is obtained by sliding the handle up and down the spiral stem so that the bit rotates alternately in opposite directions; for this reason the V-shaped bits are ground on both sides (see *Bit, Archimedean*). The driving handle is usually in the form of a metal nut, often wood covered.

These drills were in common use until recent years for boring small holes in thin wood, metal, and other materials, and they were useful for working in confined spaces where a Brace could not be operated.

Though apparently unknown before the nineteenth century and now largely superseded by the Hand Drill, the Archimedean principle led to the Spiral Drills (and other tools) which produce continuous motion in one direction (see *Archimedean Drill, Double Spiral*) and later to the group of modern tools made on this principle, including Screwdrivers, Nut Wrenches, and even Taps for cutting threads. Variants include:

(a) Plain Archimedean Drills
As illustrated, including two of somewhat primitive design probably locally made.

(b) Centrifugal type
A stock 8–12 in long, with a metal bar fitted across it near the chuck. Small metal weights fitted at each end of this bar act as a fly-wheel which helps to promote steady motion. The driving handle is fitted with an internal ratchet and so drives only on the push stroke.

(c) Side-handle type (Persian Drill)
The stem is of the type known as twisted clock-pinion, and is operated by a separate side handle. Many of this type are very finely made, and were used for delicate work, including dentistry.

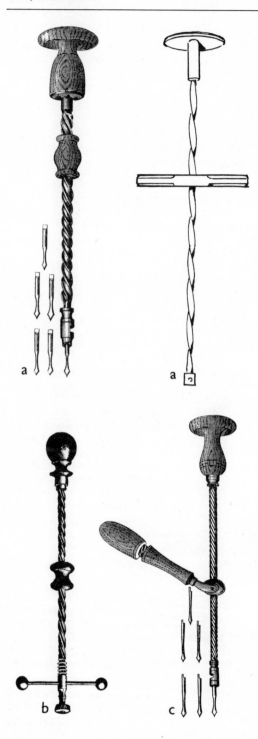

a

b   c

Fig. 272

(d) Twisted Wire Pattern [Not illustrated]
In this pattern, the spiral stock is made from a length of twisted wire.

Drill, Archimedean, Double Spiral Pattern (Reciprocating Drill; Spiral Ratchet Drill; Yankee Drill) *Fig. 273*.
In spite of its commonly used alternative name, this early twentieth century version of the Archimedean Drill does not reciprocate – at least not the Drill itself – and this is its great advantage over previous patterns. By means of a double spiral cut on the stem, and a reversing device within the travelling handle, a constant forward motion is obtained in whichever direction the handle is moved. Moreover, as its early advertisers truly stated, the slow spiral – not above 20° – 'minimised friction and magnified power'. It could be used in places where a Brace or Hand Drill could not easily be operated. The length is 16–18 in overall and the chuck is usually three-jawed and self-centering.

Though this Drill is rarely used today the principle of the double spiral is applied to many modern tools. (See, for example, *Screwdriver, Spiral Ratchet.*)

Fig. 273

Drill, Automatic: see *Drill, Push.*

Drill, Beam: see *Drill, Press.*

Drill, Bench (Drilling Machine) *Fig. 274*
This is a metal-working tool but is often found in woodworking shops. Made in many different types, it usually consists of a strong cast-iron frame for bolting to the bench, on which is mounted a vertical drill-spindle driven through bevel gears by a large hand wheel. Pressure on the work to be drilled is exerted by a hand-operated rack or by automatic screw feed. A more primitive version of the Bench Drill, often home-made, is described under *Drill, Press.*

Fig. 274

'bib' or breast-plate. This is usually a small metal plate curved to fit the human body to which it is secured by straps. On the outer surface of the plate is a pad of metal containing a shallow hole which receives the blunt point on the upper end of the revolving steel rod (*Fig. 275*).

The breast plate depicted by Bergeron (Paris, 1816) is shaped like a small violin as if to match the fiddle-like bow which is illustrated nearby. Moxon (London, 1677) writes: 'The *Drill-Plate*, or *Breast-Plate*, is only a piece of flat Iron, fixt upon a flat Board, which Iron hath an hole punched a little way into it, to set the blunt end of the Shank of the *Drill* in, when you drill a hole: Workmen instead of it, many times use the *Hammer*, into which they prick a hole a little way on the side of it, and so set the *Hammer* against their Breast.'

Fig. 275　A Bow Drill in use, with breast-plate

Drill, Bow (Fiddle Drill) *Fig. 275; 276*
These time-honoured instruments for drilling holes are rotated by the string of a bow which is wound round the bit-stock.

The Bow Drill as we know it was in use in Egypt about 2500 B.C. This remarkable invention (whose earliest form may have been an arrow rotated by a bowstring twisted around it) is widely distributed, and it was used in Europe for many drilling operations until supplemented by the Brace in the Middle Ages. The Pump Drill, which may be a development of the Bow Drill, is also of ancient origin, but there is no evidence of its use before Roman times.

The Modern Bow Drill
In its simplest form, the body of the drill consists of a cylindrical or bobbin-shaped stock (round which the bow-string is wound) mounted on a steel rod of which the lower end holds the bit, and the upper end carries a head (or nave) by which the stock is held and pressed against the work. Alternatively, for work on harder material such as iron or stone, the stock is pressed against the work by a

The Bow Fig. 276
The stock is rotated by the back-and-forth movement of a bow which imparts a reciprocating motion to the Bit which is consequently designed to cut equally well in both directions. The bow is normally made of wood (*b*); the cord is attached to one end, is given a single turn round the stock and is then secured to the other end of the bow.

The bow used in ancient times may have resembled the weapon of that name. The modern version is an almost straight stick resembling a fiddle bow, but with a slack cord. The bow used by clockmakers,

185

often made of whalebone, is bow-shaped. Straight metal bows are sometimes referred to as 'Swords'; these were rapier-like strips of steel (*a*), with a hook at the top end and a turned handle at the other. The cord is attached to the hook, and after passing round the drill-stock is tightened by a ratchet-and-pawl device near the handle.

Use

Bow or fiddle drills are suitable for comparatively light work, such as the boring of small holes in wood, metal, and stone. They are still used (or were until very recently) in piano making, clock making, and lettering on stone, and by china repairers and cutlers. (See *Drill, Passer.*)

Bow Drills used in China Fig. 276 (c)

A Chinese pattern has a system of strings which may help to increase the torque. Two strings run from the forward tip of a straight bow back to the stock on which they are twisted, one at the bottom of the stock and one at the top. These rotate the drill on the forward stroke of the bow; the two strings, being widely spaced, are capable of imparting a stronger drive without slipping than the usual single cord. A third string runs from the handle end of the bow to the central part of the stock. On the backward stroke this string rewinds the two driving strings. The operator would thus press on the stock while pushing the bow forward, and release the pressure on the return stroke.

Another drill used in the Far East is described as a 'Shipbuilder's Thong Drill' by Hommel (U.S.A., 1937). It is made on a similar principle to the Bow Drill, except that instead of being operated by one man using a bow, the thong is pulled by an assistant. One man holds the head of the stock while another pulls the leather thongs alternately to impart a reciprocating motion. This must be the Drill referred to in Homer's Odyssey Book IX: 'Like a man boring a ship's timber with a drill which his mates below him twirl with a strap they hold at either end, so that it spins continuously'.

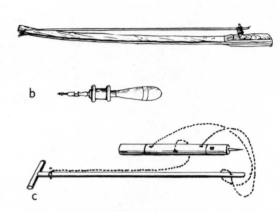

Fig. 276　Examples of Bow Drill
(*a*) With a metal bow or 'sword'
(*b*) With a wooden bow
(*c*) A Bow Drill from China

Drill Brace: see *Brace, Corner.*

Drill, Breast (Drill Stock; Corner Drill) *Fig. 277*
A drilling tool larger and heavier than the Hand Drill, with the bevel gear carried on a steel pillar or cast-iron frame.

(*a*) The early mid-nineteenth-century forms were fitted with a large iron saucer-shaped head, which was later superseded by one with a concave head in the form of a breast-plate. In these early models, sometimes known as English, Glasgow or Registered Breast Drills, the bits were held by friction in a tapered square socket, or by means of a screw.

(*b*) The later patterns were fitted with the Barber screwed chuck, the driving bevel-gear was adjustable in two positions to give alternative speeds, and a spirit level was often fitted to guide the operator.

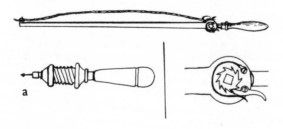

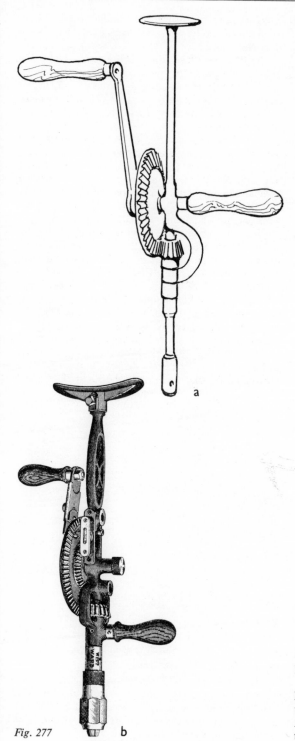

Drill, Centrifugal: see *Drill, Archimedean.*

Drill, Corner: see *Drill, Hand; Brace, Corner.*

Drill, Cramp (Smith's Drill) *Fig. 278*
A drilling appliance with the general shape and appearance of a G-Cramp without the swivel cap on the screw. This is replaced by a hollow seating to take the pointed head of an iron crank, which blacksmiths often refer to as a Wimble. This was an early method of putting pressure on a drill when boring metal. The Cramp is normally held in a vice during this operation.

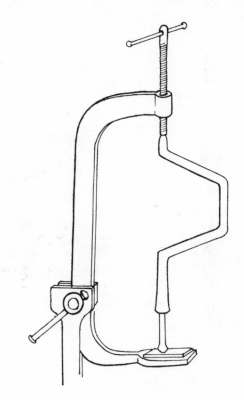

Fig. 278

Drill, Fiddle: see *Drill, Bow.*

Drill, Hand (Wheel Brace; see also alternative names under (*b*) below) *Fig. 279*
Note: The term Hand Drill is also sometimes applied to an Engineer's or Smith's Bench Drill.

Fig. 277

b

187

(*a*) The modern form of Hand Drill was an American innovation of about 1870 which reached this country about the turn of the century. It usually had a light openwork cast-iron frame to take the gears, a three-jaw chuck and a long, plain turned handle in line with the stock, often with a removable knob for a side handle. Later the design was simplified by borrowing the cylindrical pillar from the Breast Drill to take the gears and side handle.

(*b*) An earlier and very handsome-looking Hand Drill, sometimes found in British workshops, was most probably imported from the Continent. It is named in tool catalogues and elsewhere as an Angle Brace, Gearing Brace, Drill Brace, or Corner Drill.

It is about 11 in long overall, with a stock consisting of a stout U-shaped forging carrying two bevel gear wheels, driven by a crank, the bit being held in a square, tapered hole in a simple chuck and fixed with a thumbscrew. In the early forms the gear had a ratio of 1:1, but this was raised later to 3:1. A similar tool is illustrated by Bergeron (Paris, 1816).

which is normally driven by a fiddle bow and kept in contact with the work by means of a breast-plate. (See *Drill, Bow.*)

The Passer bit is a double (bifurcated) drill made of two thin pieces of steel rod, welded together at the head and sharpened at the other to form flat cutters. The tool is an early form of hand-operated Router, doing similar work to that of the modern Machine Router.

The process is as follows. A steel template is pierced with the shape of the hole or recess required and is held against the material to be worked. The bifurcated drill is made to rotate within the shaped hole in the template. As it does this, the springy legs of the Passer follow the inside outline of the template, in which they are confined, and by their eccentric movement cut, or rather rout out, the required shape. The depth of the hole is regulated by shoulders, which are cut near the tip of the drill legs, bearing on the inside edge of the template.

One of its chief uses is to pierce or recess ornamental shapes (often square, shield, or oval) in the handles of pen and pocket knives in the Sheffield cutlery trade; hence the legend that the Sheffield Cutler can drill a square hole. It is also used for recessing high-quality hardwood instrument cases and boxes to take the brass mounts, and for routing the recesses to take the brass washers and 'diamonds' in carpenters' rosewood Squares.

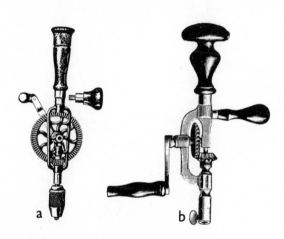

Fig. 279

Fig. 280

Drill, Hollow: see *Bit, Annular.*

Drill, Jobbers Twist: A term used for a Twist Drill of cheaper quality.

Drill, Morse: see *Drill, Twist; Drill, Straight Fluted.*

Drill, Passer (Parcey; Parsa; Parcer; Breast-plate-and-Parsee) *Fig. 280*
An early name for a Gimlet or Bow Drill, but now more often applied to a remarkable and special bit

Drill, Persian: see *Drill, Archimedean.*

Drill, Press *Fig. 281*
Mainly used for metal working but often found in the older woodworking shops. Variants include:

(*a*), (*b*) *The Beam Drill*
Made almost entirely of wood, this is a cumbrous but effective means of applying pressure to the top of a hand-operated iron crank (often called a Wimble) by means of a weighted beam and without any screwing-down mechanism.

In both the examples illustrated the pressure is increased by adding to the number of weights hung on the free end of the beam. This pressure can be released by raising the beam with a rope or lever.

(*a*) Courtesy of the Halifax Museum; (*b*) from a coachbuilder's workshop in Luton, Bedfordshire.

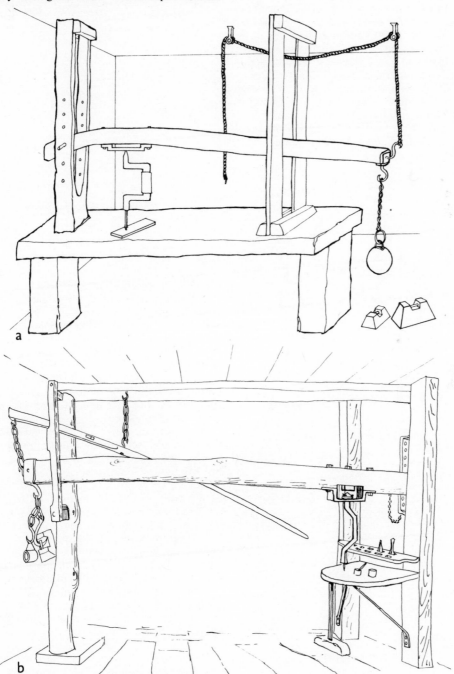

Fig. 281 (*a*)–(*b*) Typical Beam Drills

(c), (d) *Bench Press Drills* (Smith's Drill)
These easily adjusted and efficient machines were mounted on the bench itself, or on the wall above, with the swinging jib overhanging the bench. The jib or arm can be moved radially or extended telescopically towards the operator, and is fixed in position by a locking screw. The Bit is hand driven by means of the iron crank or Brace on which pressure is applied from above by a vertical screw. This screw must be given a turn from time to time in order to keep the drill pressed down on the work. (c) comes from a Hertfordshire smith's shop; (d) from the workshop of a Cambridgeshire laddermaker.

Note: Other Drills operated by mechanical pressure include *Drill, Cramp*; Ratchet Brace under *Brace, Corner.*

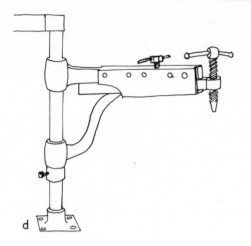

Fig. 281 (c)–(d) Typical home-made Bench Drills

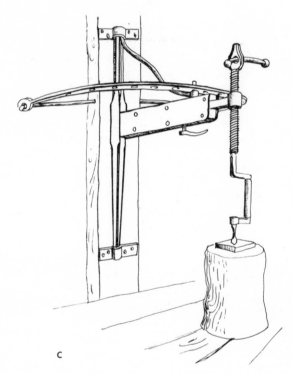

Drill, Pump (Up-and-down-drill) *Fig. 282*
Like the Bow Drill, this ingenious tool is driven by a cord wound round its spindle which imparts a reciprocating motion. But unlike the Bow Drill, the cords are operated by a cross-arm which is moved up and down with a pump-like action. The tool consists of a vertical spindle fitted into a stone or metal flywheel which is situated just above the bit-holder at the lower end. The cross-arm can be moved freely up and down the spindle which passes through its centre. Reciprocating motion is imparted to the spindle by a cord which passes from one end of the cross-bar up to a hole in the top of the spindle, and then down to the other end of the bar. The hole in the cross-arm acts as an outside bearing to the spindle, and so enables the drill to be operated with one hand.

The action begins with a slight twisting of the spindle with the fingers, the cross-bar being held in one hand; the two cords are thus given a spiral twist round the upper part of the spindle. When the cross-bar is sharply depressed, the spiral is unwound and this rotates the spindle. At its lowest point, pressure on the cross-bar is relaxed and the momentum of the flywheel is sufficient to wind up the cord in the opposite direction, and as the cords shorten, to draw the bar upwards ready to be pushed down again and start turning the Bit in the opposite direction. Thus, the drill can be kept in constant, but reciprocating, motion.

This tool, which was known in Roman times, is used for drilling hard substances such as stone or metal. It cannot be used on soft wood because the Bit would bind and prevent the 'flywheel' from re-winding the cord; but it could drill shallow holes in very hard wood. It is employed by masons and jewellers and, until recently, by itinerant china re-pairers, who used it for drilling broken china to take the rivets. For this purpose it could be worked with one hand while holding the broken object in the other.

Fig. 282

Drill, Push (Automatic Drill; Push Brace; Yankee Push Brace) *Fig. 283*
A name given to a drill of Archimedean type enclosed in a tube or sleeve and driven by pressure on the handle. This, like that of the Spiral Ratchet Screw Driver, contains an enclosed spiral and a spring which brings the handle back ready for the next stroke. A set of bits are often provided in a 'magazine' handle.

Originally designed to take Brace Bits, it is still being made for drilling small holes and has the ad-vantage of needing only one hand to operate it.

Fig. 283

Drill, Reciprocating: see *Drill, Archimedean, Double Spiral Pattern.*

Drill, Sculptor's: see *Drill, Bow.*

Drill, Smith's: see *Bit, Smith's Drill; Drill, Bench; Drill, Cramp; Drill, Press.*

Drill, Spiral Ratchet: see *Drill, Archimedean, Double Spiral Pattern.*

Drill, Stock: see *Drill, Breast.*

Drill, Straight Fluted (Morse Straightway Drill) *Fig. 284*
A Drill Bit for machine or hand use with a straight flute. Made in fractional sizes from $\frac{1}{16}$ to $\frac{1}{2}$ in. (See *Drill, Twist* for the Morse Drill.)

Fig. 284

Drill, Stub
A name given to very short Twist and other Drills.

Drill, Twist (Jobber's Drill; Morse Drill) *Fig. 285*
Boring Bits of high-grade steel, circular in cross-section, with a spiral flute either at a constant angle or with increased twist towards the point. The main variations relate to the shape of the shank, which is straight or tapered for holding in a machine chuck as in (*b*); or square-tapered for holding in a Brace, as in (*a*). Two types of nose are shown:

(*a*) With spurs and a point lead, designed for wood only.

(*b*) With a nose ground to a point at an angle of 60°, for metal, wood, and general use.

Both straight and spiral flute Drills are often known in the workshops as Morse Drills, but the catalogue description 'Genuine Morse' indicates that they were manufactured by Morse Twist Drill Machine Co., founded by Stephen A. Morse in 1861 in East Bridgewater, Mass., U.S.A. Morse also devised the table of taper sizes which are known as 'Morse tapers'.

Fig. 285

191

Drill, Up-and-Down: see *Drill, Pump.*

Drill, Wall (Jumper) *Fig. 286*
Iron tools about 9 in long used for boring a hole in brickwork or masonry to take, for example, a wooden fixing-plug. The nose end is made in the form either of a smith's blunt-pointed Drill Bit, a star-shaped Chisel, or tubular Crown Saw. Operated by hammering and turning the tool between blows, this tool is now superseded by power-driven tungsten-carbide tipped Drills. (See also *Chisel, Plugging.*)

Fig. 286

Drill, Yankee
A term sometimes applied to Archimedean Drills of the double spiral type.

Drilling Machines
For hand-operated Drilling and Boring Machines, see *Drill, Bench; Drill, Cramp; Drill, Press; Hub Boring Engine; Wood Boring Machine.*

Driver: see Hoop Driver under *Cooper* (3) *Hooping Tools;* Stake Driver under *Beetle.*

Driving Stool: see *Rake Maker.*

Dumcraft: see *Jack, Lifting.*

Dust Bellows: see *Pianoforte Maker* and *Tuner.*

Dutch Hand: see Hooping Windlass under *Cooper* (3) *Hooping Tools.*

Dutchman: see *Jumper.*

Dwang
A Scots name for certain tools which are used as levers or wrenches, e.g. an upholsterer's web strainer (see *Upholsterer*).

E

Eatche (Each)
A Scottish form for Adze. 'Let me hae a whample at him wi' mine eatche – that's a'.' (Sir Walter Scott, *The Bride of Lammermoor*, chapter XXV.)

Edge Mark
A V-shaped mark denoting the face edge of timber indicating that it is straight and square to a previously trued face side.

Edge Tools
A term used to describe tools with a sharpened edge or blade. Light Edge Tools include the Chisel, Gouge, Plane Iron, Drawing Knife, and other types of Knife. Heavy Edge Tools include the Adze, Axe, Stock Knife, etc.

Eke
A Scots and northern England term for a lengthening bar or extension piece for a Cramp. (See *Cramp, Joiner's*.)

Elsin (Alsene)
An obsolete term, except in northern dialect form, for *Awl*.

End Grain
The grain exposed in wood by a cross cut made at right angles to the direction of the grain.

Engine
A name given to certain tools including the brush maker's Comb and Hackle (see *Brush Maker*); the *Hub Boring Engine*; and Stail Engine (see *Plane, Rounder*).

English Vice
A term applied to a Vice with the cheek vertical and extending from the floor to the bench surface. See *Vice, Woodworker's; Bench, Woodworker's*.

Engraver: see *Wood Engraver*.

Equipment: Including workshop furniture. See list under *Workshop Equipment*.

Erasing Iron (Scraper)
A name given to a number of different tools for removing marks and brands from boxes, including: *Drawing Knife, Cooper's Round Shave; Plane, Box Scraper; Scrapers*.

Expanding Boring Tools: see *Auger, Expanding; Bit, Expanding*.

Extension Appliances
For examples see Extension Holder Bit under *Brace, Corner;* Eke under *Cramp, Joiner*.

Extractors: see Spoke Extractor under *Wheelwright's Equipment (2) Spoke Tools;* Nail and Screw Extractor under *Bit, Annular*. Also see list under *Nail Extracting Tools*.

Eye
Of tools – a hole, e.g. the eye of a Hammer or Axe in which a handle is inserted. 'Square or Adze Eye' indicates a tapered socket of square or oblong shape; 'Axe Eye' indicates socket of egg-shaped cross-section.

F

Face
Of a tool – the working surface of a tool. Of wood – the trued side or edge from which all subsequent marking and testing is done. (Known also as the fair side or face edge.)

Falconer's Plough: see *Plane, Plough, Circular.*

Fangle Iron: see *Millwright.*

Fantail (Flared; Splayed)
A term used to describe a shape which spreads out (like a fan), e.g. the blade of some Axes or a Chisel with a tapered blade.

Fawcet Tool: see *Pipe and Pump Maker.*

Fawn-Foot Handle *Fig. 287*
A specially designed handle for Felling Axes, with an oval cross-section and smooth double curve, swelling at the foot with the end shaped like the foot of a young deer. See note on handles under *Axe.*

Fawn's Foot Hickory Felling Axe Handle.

Fawn's Foot Hickory Hatchet Handle.

Fig. 287

Feather Edge
Tapering to a thin edge. The term is used to describe, for example, Twist Augers or Bits which have a thin outer edge to the twist.

Feeler: see Spoke Set under *Wheelwright's Equipment* (2) *Spoke Tools.*

194

Felloe
A segment forming the rim of a wooden wheel. See *Wheelwright.*

Felloe Patterns: see *Gauge, Wheelwright.*

Fence
A fixed or adjustable guide, usually vertical, fitted to some Planes, Routers, and other tools. The fence is held against the face of the work in order to regulate the width of a mould, the size of a rebate, or the position of a groove.

Fencing Tool
A name given by various makers to a combination type of cutting plier, 9 in long, with a hammer on one side of the head and a pick or single claw on the other.

Ferrule
A metal ring or cap, e.g. on a tool handle to prevent it from splitting. See diagram under *Chisel.*

Fid: see *Sailmaker.*

Fiddle: see Fiddle Drill under *Drill, Bow;* Fiddle under *Wheelwright's Equipment* (2) *Spoke Tools;* also tools under *Violin Maker.*

File *Figs. 288; 289*
A metal bar, usually of hardened steel, having one or more of its surfaces covered with a series of raised cutting edges or teeth, designed to cut by abrading. The teeth are indented by Chisels before the metal is hardened; until recently this was a highly specialised hand process.

Known since the Bronze Age, files are now chiefly used for metal work. Those included here are used in woodworking shops for various smoothing or fitting operations, and for Saw and Bit sharpening. Reference may also be made to the other members of the file family – the *Rasp, Float,* and *Rifler.*

The 'Cut' Fig. 288
The range of cuts shown in the Sheffield Illustrated list of 1888 runs from Smooth with 64 teeth to the inch through 2nd Cut, Bastard, and Middle Cut to Rough, with 11 teeth to the inch. The typical cuts illustrated are called:

(*a*) Float or single Cut
(*b*) Double Cut
(*c*) Rasp Cut

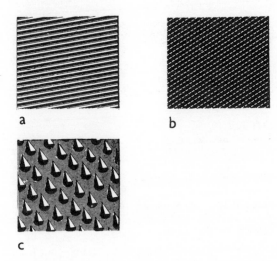

a   b

c

Fig. 288

File Shapes Fig. 289
Those illustrated show the most common types.

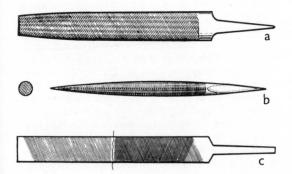

a

b

c

d

e

Fig. 289   (*a*) Half-round
(*b*) Round (or 'rat tail')
(*c*) Flat (or 'ward')
(*d*) Three-square (or 'saw')
(*e*) Fusiform (or 'cigar-shaped')

File Brush (File Card; File Cleaner)
A rectangular mat of fine spring steel wire, each point bent at an angle, mounted on canvas or leather, and attached to a shaped wooden pad. Used to clean the teeth of clogged-up Files.

File, Cabinet *Fig. 290*
Flat or half-round, tapering towards the end, from 4 to 16 in long. Used by cabinet makers and other tradesmen for smoothing.

Fig. 290

File, Pottance Chequering (Potence File)
A name given to a flat, parallel-sided file of rectangular section, about $\frac{9}{16}$ in wide and 4 in long, used by Gunstockers for producing grooves when chequering the stock. (Pottance or Potence is the lower bearing of the verge in a watch.) See *Gunstocker.*

File, Rat Tail (Round File) *Fig. 289 (b)*
A round tapering file, 4–20 in long, used by woodworkers for enlarging holes or grooves, etc.

File, Saw Handle: see *Rasp.*

File, Saw Sharpening: see *Saw Sharpening: Files.*

File, Straw *Fig. 291*
A term which appears to refer to Files which are wrapped in straw rope, presumably as a method of packing.

195

Fig. 291

File, Warding *Fig. 289 (c)*
A thin, tapered, flat file, from 4 to 6 in long, used
for cutting keys for locks.

Fillister: see *Plane, Fillister;* Fillister Router under
Router.

Fipple: Scots name for a *Nose Bit.*

Fire Fork: see *Wheelwright's Equipment (3) Tyring
Tools.*

Fishing Rod Maker *Fig. 292*
Early fishing rods were simple lengths of hazel, bamboo, or ash, but hickory or greenheart was sometimes used for the more expensive rods. About 1900,
the split-cane rod was developed – a rod built up
radially from triangular sections. The bamboo cane
is first split into six or more sections with a knife,
the pith removed, and after planing to shape, the
sections are glued together. Since the cane is not
always straight, the split pieces are immersed in hot
silver sand to make them sufficiently pliable to be
straightened. One method of trimming the sections
is to lay the triangular pieces in tapered grooves on
a block of wood, and to plane off the surplus cane
until flush with the surface of the block. A later
development is the rod made of steel tube. This was
first stepped but later tapered. A still later development is the rod made of fibre glass.
Special tools used include:

(*a*) *Trap.* When making the solid rods, the square
wood is held in the lathe, and rounded by a Rounding
Plane (known as a Trap) which is made in hinged
form. By passing the tool up and down the rod, and
by regulating the pressure on the handles, the wood
is rounded in section and tapered in length.

(*b*) *Spoon.* The joints between the lengths are of
the cone-and-socket type, and the female half of the
joint is tapered with a Reamer known as a Spoon.
The joint was later strengthened by metal tubes, and
later still, screwed joints were introduced.

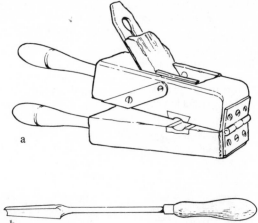

a

b

Fig. 292

Fish Tail
A term used to describe the shape of a fan-shaped
tool, e.g. Fish Tail Carving Chisels. See also *Fantail.*

Fitter: see *Wheelwright's Equipment (2) Spoke Tools.*

Flagging Iron and Chincing Iron *Fig. 293*
These tools are used together and are both included
below:

(*a*) *Flagging Iron* (Caulking Iron; Prying Rod; Rush
Iron; Stave Wrench; U.S.A.: Rushing Lever)
An all-iron forging about 2 ft or more long, forked
at one end with the prongs ('teeth') bent over at
right angles. The lower end of the handle is flattened. The ends of the prongs are T-shaped, but in
the Scottish pattern are bent over, sometimes in
opposite directions.
It is used by coopers in the process of flagging.
In order to prevent leakage, dry rushes ('flags') are
inserted into the croze-groove and between the separate parts of the head; and after repairs, between the
staves at their chime ends.
The flags are put into the croze-groove with the
fingers, or with the help of a Chincing Iron (see
below). The Flagging Iron is needed mainly when
making repairs, when for example it is desired to fit
a new stave; in such a case, the chime hoop is
knocked off, and the Flagging Iron used to force
the new stave outwards. In operation it is held horizontally with one prong inside the cask and one
outside. When the handle of the tool is pushed sideways by the cooper's hip, one prong forces the top

of the new stave outwards. By this means, both hands are left free for inserting the flags into the croze-groove.

The flattened tail is used for slipping between staves or head-pieces to 'knock them along' during assembly. It can also be used to help in removing a head.

(*b*) *Chincing Iron* (Bolster; Chintz; Rush Knife; Stuffing Knife)
A chisel-like tool, all iron or wooden-handled. Used for forcing the dried rushes into the croze-groove.

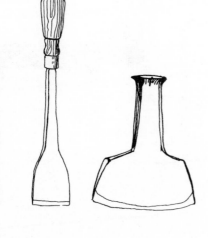

a

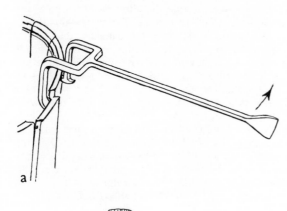

b

Fig. 293

Flammard or **Flamming iron**: see *Froe*.

Flared (Dovetail; Fantail; Fish-Tail; Splayed)
Terms used to describe a shape which spreads out like a flame or fan, e.g. the blade of some Axes or Chisels.

Flat Iron: see *Veneering and Marquetry.*

Flattener, Veneer: A name sometimes given to a Veneering Hammer. See *Veneering and Marquetry.*

Flatter: see *Saw Sharpening; Files.*

Flencher (Flincher): see *Plane, Cooper's Chiv.*

Flexible Saw: see *Saw, Chain.*

Flit Plough: see *Plane, Plough.*

Float *Fig. 294*
A very coarse single-cut File or Rasp with parallel teeth, spaced 4–5 to the inch, and often home made from an old File. The narrow patterns have the teeth cut along one edge; the wider, tapering Floats have teeth cut on the flat; square, tapering Floats are sometimes made with teeth on two adjacent faces, for working into corners.

Used occasionally for cleaning out mortices, and also used extensively by plane makers for shaping the bed and throat of a Plane (see *Plane Maker*). Workers in horn use various Floats; one with a curved cutting edge is known as a 'Graille'.

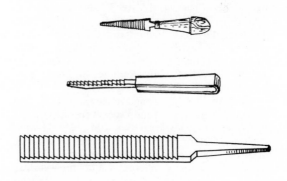

Fig. 294

Flogger: see Bung Removers under *Cellarman.*

Flute
A groove cut for decoration or as a means of producing a cutting edge on a tool.

197

Fluter and Fluteroni Tool: see *Chisel, Carving.*

Fold Shore (Drift; Pitcher): see Shepherd's Bar under *Hurdle Maker.*

Follower: see Traveller under *Wheelwright's Equipment (3) Tyring Tools.*

Footprint: see *Wrench (2) for gripping rounds.*

Fore Check: see Moving Fillister under *Plane, Fillister.*

Forester: For some of the tools used in this trade see *Tree Feller; Timber Handling Tools.*

Forging: see *Iron and Steel.*

Forkshaft (Fork Staff Rounder): see *Plane, Forkstaff; Plane, Rounder.*

Frail: see *Tool Basket.*

Frammer: see *Froe.*

Fret: see Gimlet under *Cellarman.*

Froe *Fig. 295*

This tool has many alternative names including those listed below. (The word Froe is also used as an adjective for dry or brittle.) Chit; Cleaving Iron; Dill Axe; Divider; Flammard; Flamming Iron; Frammer; Fromard; Fromward; Frow; Froward; Frummer; Helpmate; Lath Axe; Lath River; Pole Axe; Ramhead; Rending Axe; Riving Axe; Side Knife; Split Axe; Thrower.

A blade about 6–12 in long, wedge-shaped in cross-section provided with a round socket for the handle, which is set at right angles to the cutting edge. Though sometimes factory made, those found in country workshops are nearly always blacksmith made.

It is used in many different trades for splitting timber lengthwise into boards, segments, or billets. This operation is also called cleaving, rending, or riving. Splitting oak logs for making wheel spokes is known as 'spoke-cleaving'. The finished spoke is often described as 'rent', and is considered vastly superior to a sawn spoke: it is asserted that flaws and weaknesses in the grain remain hidden in the sawn article. Furthermore, cleaving is a much quicker method than sawing for producing lengths of wood for making chair legs, tent pegs, laths, spokes,

hurdles, and many other products of the woodland worker or country workshop. The Froe is driven by striking the back of the blade with a Mallet or Club (see *Mallet, Woodland Worker's*).

When cleaving timber for components such as wheel spokes or cask staves, the log or block of wood is placed upright and the edge of the Froe laid across the end of the grain. The split is begun by driving in the Froe like a wedge. It is then used as a lever which, when pushed or pulled sideways by its handle and then struck again when necessary, opens the split and finally separates the two parts.

When cleaving longer material for making palings, hurdles, fork or rake handles, etc., the poles are held horizontally in a Brake. The Froe is first driven into the end of the pole; the handle is then rocked from side to side while the blade is gently but firmly guided down the middle of the rod, sometimes helped with a mallet blow, until the two halves fall apart. This operation of 'turning away' the wood may perhaps give a clue to one of the early names for this tool – Fromward.

H. C. Edlin (1949) relates that 'In most forms of cleaving it is the 'feel' of the wood rather than the sight of it that guides the craftsman's hand, and one worker in hazel has told me that once a cleft has been started he can keep it running with his eyes shut.' (See also *Lath Maker; Woodland Trades.*)

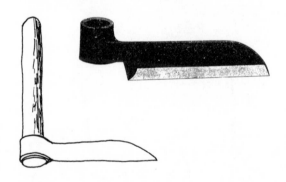

Fig. 295

Froe Club: see Mallet under *Woodland Trades.*

Froe, Cooper's Curved

A Froe with a blade curved like a large Gouge. According to E. Sloane (U.S.A., 1964, p. 31) it was used in the United States for riving staves from the billet. It produced a stave with the inner face already roughly hollowed and the outer rounded. This saves time in shaping straight staves for pails, shallow vats, etc. as well as saving timber.

Froe, Lath Maker's (Lath Maker's River) *Fig. 296*
A Froe with a sharpened bevel at the end of the blade as well as along the lower edge. See *Lath Maker*.

Fig. 296

Frog
The bed for the cutting iron in a metal Plane. See diagram under *Plane*.

Fromward (Fromard; Froward): see *Froe*.

Frow: see *Froe*.

Frow Horse: see Cleaving Brake under *Woodland Trades (2) Brakes*.

Frummer: see *Froe*.

Fulcrum
The point of support, or pivot, of a lever.

Furniture, Workshop: see list under *Workshop Equipment*.

Furrowing Strips (Furrow Spline): see *Millstone Dresser*.

Fusiform
Spindle-shaped or cigar-shaped, e.g. a file which tapers towards each end.

G

Gathering Hoop: see *Cooper (3) Hooping Tools.*

Gauge

Many tools and appliances are loosely called Gauges, but we have confined the entries which follow to those which are (*a*) used for marking lines, (*b*) act as a standard or measure to which material or work-piece must conform, or (*c*) are commonly referred to as Gauges, e.g. Cutting or Rabetting Gauges.

But see also *Measuring Sticks; Patterns, Templets, and Jigs; Plane Maker (Patterns and Jigs); Rule; Square and Bevel.*

Gauge, Bit *Fig. 297*
A small Clamp fitted to the body of the Bit by means of a thumbscrew. This carries either a fixed or adjustable stop which acts as a depth gauge when boring holes.

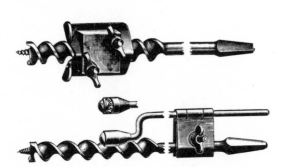

Fig. 297

Gauge, Butt *Fig. 298*
A modern Gauge with spurs mounted on the ends of sliding rods, used to mark the seatings for butt hinges on doors. When the spur at one end of the longer stem is set for gauging on the edge of the door, the spur at the other end is automatically set for marking from the back of the jamb. The shorter stem marks off the thickness of the butt. It can also be used for other purposes such as marking along a rebate.

200

Fig. 298

Gauge, Circular: see *Circular Cutting Gauge.*

Gauge, Clapboard or Siding
An adjustable cranked rod mounted on a base and used for regulating the amount of overlap between the clapboards. (Listed by the Stanley Rule & Level Co., U.S.A., 1902.)

Gauge, Coachbuilder's: see *Gauge, Panel; Gauge, Wheelwright's and Coachbuilder's; Square, Coach-builder's.*

Gauge, Combination *Fig. 299*
A name given to Mortice Gauges which combine the functions of a Marking and Mortice Gauge. The stem has adjustable spurs on one side for marking mortices, and a single marking or cutting spur on the other.

Fig. 299

Gauge, Cutting (Slitting Gauge) *Fig. 300*
Two variants are illustrated:

(*a*) This tool is identical with the Marking Gauge, except that it is fitted with a small pointed knife or cutter instead of the spur. The knife is held in position by a wedge, so that it may be taken out for re-sharpening. The fence is often faced with narrow brass strips to resist wear. Used for deep scoring parallel to the edge, especially across the grain when marking the shoulders of joints; or for the stringing and banding for veneers. It can also be used for cutting thin wood into strips, working from both sides.

(*b*) Another variety, known as a Slitting Gauge, is made on heavier lines. A fence like that of the Panel Gauge is mounted on a graduated stem about 18 in long, with a cutter near one end, and next to it is fitted a Jack-plane type handle with a flat base in which is sometimes fitted a roller. Used for cutting out thin boards such as drawer bottoms. (See also *Circular Cutting Gauge.*)

a

a

b

Fig. 300

Gauge, Depth
Various devices for checking the depth of mortice holes, dowel holes, etc., usually home made, but use may be made of a Marking Gauge with the spur removed. See also *Gauge, Bit.*

Gauge, Grasshopper (Handrail Gauge) *Fig. 301*
The stem acts as the fence, and a wooden cross-bar passing through the end of it at right angles holds a pencil or metal point at each end. A similar Gauge is illustrated in the Shelburne Museum booklet No. 3 (U.S.A., 1957) with an extension arm hanging down from the end of the cross-bar to mark at various depths. Used for straight and circular gauging over obstructions, e.g. over the top of a stair rail.

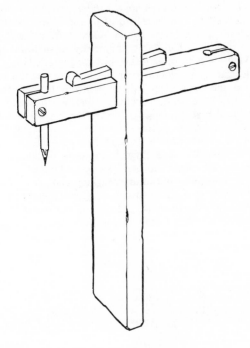

Fig. 301

Gauge, Handrail: see *Gauge, Grasshopper.*

Gauge, Jointer
An adjustable metal fence which can be attached to a wood or metal Plane. Used for guiding a Plane at the desired angle when planing a bevelled or square joint.

Gauge, Marking *Figs. 302; 303 overleaf*
The factory-made Marking Gauge is usually of beech-wood and consists of two parts, a fence about $2\frac{1}{2} \times 2$ in and a stem about 9–10 in long, sometimes graduated, and carrying a pointed steel spur at one end. The head is fixed in any required position by means of a wood thumbscrew or by wedge.

Tradesmen and apprentices often made their own

Marking Gauges which may still be seen in the work-shops in a great variety of patterns, some of which are both ingenious and beautiful. The home-made method of fixing the fence is almost invariably by captive wedge.

Marking Gauges are used for marking lines parallel to the face side or edge when planing workpieces to size, and for similar parallel lines when marking rebates, joints, etc.

The Romans do not appear to have used this tool, and the first known record of its use comes from a picture *c.* 1600 (Hieronymus Wierix's *Holy Family*). At that time there was apparently no method of fixing the adjustable fence. But the joiner's Gauges illustrated by Diderot (Paris, 1763) are secured by long, narrow wedges passing vertically through the depth of the head, a method which is still used in modern French Gauges. The earliest known use of the thumbscrew occurs in Smith's *Key* (Sheffield, 1816). This became the standard method for most commercially made English Marking Gauges, although a wedge is still used for Panel Gauges.

Early eighteenth-century Marking Gauges do not always have a spur. Instead the workman held a spike or pencil against the end of the stem, as is done today with a Thumb Gauge.

d

e

Fig. 302   Marking Gauges: factory-made examples
(*a*) A common type in plain beechwood
(*b*) Plated to resist wear
(*c*) Patternmaker's type with captive wedge
(*d*) 'London' pattern often made in ebony
(*e*) An all-metal example with a round stem

Gauge, Millwright: see *Millwright; Millstone Dresser.*

Gauge, Mortice *Fig. 304 overleaf*
A Marking Gauge with two spurs instead of one, used for marking the double parallel lines showing the position of a tenon or mortice or similar joint, thus avoiding the need to scribe two lines separately. Variants include:

(*a*) Two separate stems locked in the fence by a wedge. A type commonly made on the continent of Europe since the eighteenth century.

(*b*) An American version with two metal stems. Small cutting-wheels are provided instead of a spur in some cases.

(*c*), (*d*), (*e*) Single stemmed Gauges have been preferred in Britain. The earliest factory-made Mortice Gauges we know of are depicted in Smith's *Key* (Sheffield, 1816). Similar ones are shown here from the *Sheffield List* of 1888, with different means of adjustment. The better quality Gauges are finely made from ebony or rosewood, with brass inserts.

(*f*) For long runs of standard joints, Gauges with fixed fences and spurs permanently set in the required position are preferred, especially for work with twin or double mortice-and-tenon joints. Similar special Gauges, larger and of cruder workmanship, often with nails for spurs, were used for marking tusk-tenon joints in trimmers and trimming joists, where three and sometimes four parallel lines were required. The example illustrated is home made, the stem solid with the fence. See also *Gauge, Wheelwright's and Coachbuilder's.*

a

b

c

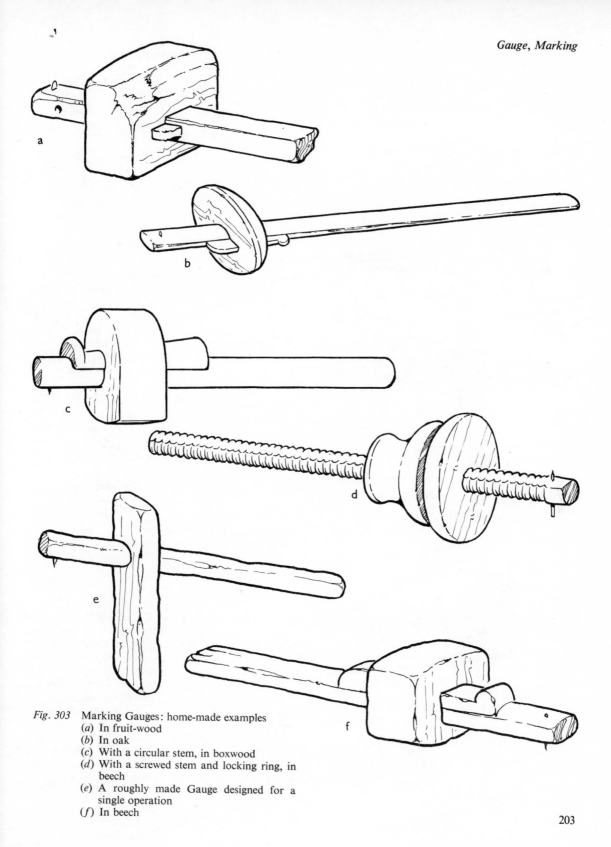

Fig. 303 Marking Gauges: home-made examples
  (*a*) In fruit-wood
  (*b*) In oak
  (*c*) With a circular stem, in boxwood
  (*d*) With a screwed stem and locking ring, in
  beech
  (*e*) A roughly made Gauge designed for a
  single operation
  (*f*) In beech

(*g*) A home-made version of the adjustable Gauge (*c*) above, with the inner spur fitted on a wooden slide, and with a captive wedge.

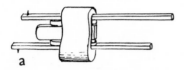

a

b

c

d

e

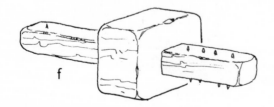

f

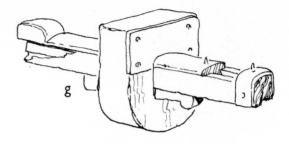

g

Fig. 304

Gauge, Mullet *Fig. 305*

A short length of wood used for testing the thickness of panels. A groove is cut on one edge of a piece of wood with the same tool that was used to work the grooves in the stiles and rails; or an offcut from one of these grooved members is used. Its purpose is to test the edges of the panels brought down to the required thickness without having to try them in the separate members of the framing.

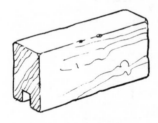

Fig. 305

Gauge, Panel (Coach Panel Gauge) *Fig. 306*

A large wooden Gauge, often made in mahogany, with a wide fence. The stem, up to about 30 in. in length, which carries the marking spur at one end, is held in the fence by a captive wedge.

In the more usual form the fence is asymmetrical, shaped to give a thumb-hold from the side; in other patterns the fence is symmetrical, sometimes ornamented with mouldings. This Gauge is used for marking out wide panels or boards. The lower edge of the fence is rebated on one side to enable the fence to be held down firmly on the edge of the board to be marked, thus holding the stem and spur at a fixed height above the surface of the board.

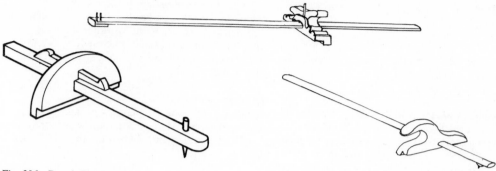

Fig. 306   Panel Gauges

Gauge, Rabbet (Rebating Gauge) *Fig. 307*
A brass stock fitted at an angle to the end of a
handle and carrying a cutting spur and adjustable
fence. Made for rebating small straight or curved
work, e.g. picture frames and the like.

Fig. 307

Gauge, Ramshorn: see Coachmaker's Sweeps under
Gauge, Wheelwright's and Coachbuilder's.

Gauge, Saw
A fence or stop mounted by adjustable screws to the
side of a Saw to control the depth of cut.

Gauge, Side
The term is used in the List of James Cam (Sheffield,
c. 1800) as follows:

'No. 481 Side Gauges, Single slide 10/- per dozen
482   „   „   Double „   13/- per dozen'

The price is about the same as that of the Marking
Gauges; only the Mortice Gauges are appreciatively
dearer. The term is also mentioned in Thomas
Sheraton's *The Cabinet Maker & Upholsterer's
Drawing Book* (1791). We do not know its purpose
nor what it looked like.

Gauge, Skew Mortice: see *Plane Maker (Patterns and
Jigs).*

Gauge, Spider: see *Square, Coachbuilder's.*

Gauge, Spoke: see *Gauge, Wheelwright's; Wheelwright's Equipment (2) Spoke Tools.*

Gauge, Stave
A home-made wooden bar, slightly hollowed and
graduated along one side, and with a stop or foot
at the end. Used by coopers for testing the length
and width of cask staves.

Gauge, Thumb *Fig. 308*
One type consists of a rectangular strip with a notch
at one end to hold a pencil while the fingers act as a
fence. An adjustable type consists of a flat fence with
a narrow slide running tight in a dovetailed groove
at right angles to the edge of the fence, the working
end usually chamfered down. These tools are useful
for marking guide lines for chamfers etc., where the
score made by the marking gauge would be objec-
tionable.

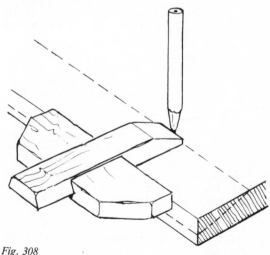

Fig. 308

Gauge, Turner's: see Sizing Tool under *Chisel, Turner's.*

Gauge, Tyring: see Traveller under *Wheelwright's Equipment (3) Tyring Tools.*

Gauge, Violin Maker's (Fiddlemaker's Calliper or Thickness Gauge) *Fig. 718* under *Violin Maker*
Various Gauges are used by violin makers for measuring the thickness of the front and back plates of the instrument. All are of the Calliper-Gauge type and have one feature in common – a deep U-shaped frame (like the frame of a Fret Saw), to enable the points of the Gauge to reach the centre of the violin. One type consists of a U-shaped metal frame with a moving jaw connected to a dial for registering the thickness. A simplified form of this Gauge is often made in wood by the violin maker himself.

Gauge, Whalebone: see Spoke Set under *Wheelwright's Equipment (2) Spoke Tools.*

Gauge, Wheelwright's and Coachbuilder's *Fig. 309*
A number of different Gauges and Patterns are used when making spokes and felloes, for marking out mortices, and for the many other operations which go to the making of wheels, carts, wagons, or ploughs. A selection of those most frequently to be found are described below. More specialised Gauges were used by the coachbuilders. Mr. John O. H. Norris (Manchester, 1966) explained the reason for their use:

'A coach, chariot, brougham, landau, or similar body has few straight members in its framing. The body viewed in any direction is a combination of curves, elliptical or compound. This means that all joints were at an angle with the vertical and horizontal planes, so adjustable gauges are needed to get the surfaces to be joined at the correct angle.'

(a) Angle Gauge
These little 'crossed sticks' are riveted at the centre to a predetermined angle and used to check the inclination of a blind mortice, e.g. a spoke mortice hole in the wheel hub. This inclination is measured so that the spokes will leave the hub at an angle sufficient to give the wheel the required amount of 'dish'.

(b) Spoke Gauges: see *Wheelwright's Equipment (2) Spoke Tools.*

(c) Coachmaker's Sweeps (Coach Templates)
Thin, flat strips of curved wood, often of mahogany, are marked out from the working drawing, cut as required, and used by the coachbuilder as templates for marking out the pillars, rails, etc. which make up the frame.

(d) Felloe Patterns
Quantities of these templates were kept hanging in the workshops. Made from thin board, they were used for marking out felloes on the plank before sawing out.

(e) Mortice Gauge (Wheeler's and Coachbuilder's)
Movable pegs are fitted at right angles to each other in the same stock which is usually thickened at one end to take them. So far as we know, it is always home made.
Used for marking out the two sides of a mortice hole. A peg is adjusted by tapping the end lightly on the bench. There are no wedges, the pegs being fitted tightly in the stock.

(f) Another version consists of a short piece of wood with indentations cut to provide several separate fences each with its corresponding fixed marking spur. Used for marking out mortices and tenons of standard size in the frame, pillars, etc. without resort to an adjustable Gauge.

(g) Rim Gauge (Boxing Bevel; Boxing Try)
Invariably home made. A short arm, often made from an old spoke, is hinged to a longer arm of about 2–3 ft in length.
Used for trueing the axle-box within the wheel-hub while the wheel is laid flat, as an alternative to trueing a wheel by swinging it vertically and measuring its clearance as it turns. The short arm of the Gauge is set in the axle-box and the long arm is revolved to see where it touches the rim of the wheel. The box is centred by means of wedges (see *Chisel, Wedging*) which are inserted when the Rim Gauge shows that the box needs to be edged over to make the wheel run true.

(h) For other Gauges used in these trades, see Spoke Fitter, Spoke Set, Spoke Trammel under *Wheelwright's Equipment (2) Spoke Tools;* Traveller under *Wheelwright's Equipment (3) Tyring Tools;* Spider Gauge, Horizontal Square etc. under *Square, Coachbuilder's;* Panel Gauge under *Gauge, Panel* (a large version of the Marking Gauge often found in coachbuilding shops).

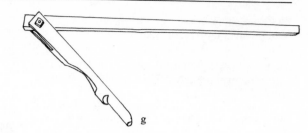

Fig. 309

Gauging Rod: see *Dip*.

Gavel *Fig. 310*
This small wooden mallet is made in various ornamental forms and can be mistaken for a tradesmen's tool. It is used by chairmen and auctioneers.

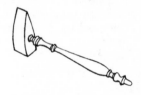

Fig. 310

Gavelock (or **Gablock**)
A dialect term for a Crow Bar.

Gee Throw
A stout wooden Crow Bar with a pointed, curved iron shoe. Used by lumbermen for shifting heavy logs etc. (See *Timber Handling Tools*.)

Geelum
Scots joiner's term. See *Plane, Rebate*.

Gentleman's Tools (Gents Tools)
Nineteenth-century trade catalogues include many tools, usually lighter than the ordinary, which are described as 'Gentleman's', and are intended for the use of amateurs. (Still lighter tools, such as very small hammers, are sometimes listed as 'Lady's'.)

207

The Gent's Saw is an instance where this term has been retained until recent years.

Gimlet (Gimblet; Nailsin; larger Gimlets 9–12 in long are sometimes called Spout, or Gutter Gimlets.) *Fig. 311*
A miniature Auger with a spiral 'twist' or shell body and a screw point; the handle, usually in beech or boxwood, forms a 'T' with the shank. Used for boring small holes from $\frac{1}{8}$ to $\frac{3}{8}$ in. in diameter, as pilot holes for nails, screws, etc. Unlike the Awl, which makes a hole by squeezing the material apart, the Gimlet starts by squeezing, but finishes the hole to size by side-cutting.

Gimlet, Auger Type *Fig. 311, No. 1941*
A twisted body like a Twist Auger from $\frac{1}{8}$ to $\frac{1}{2}$ in. in diameter. The nose is usually of the Scotch pattern with a screw lead.

Gimlet, Bell Hanger's *Fig. 311, No. 1951*
A very long Gimlet $\frac{1}{4}$–$\frac{3}{8}$ in. in diameter and 18–36 in long, usually provided with an egg-shaped handle.

Formerly used by bell hangers for boring holes for the passage of the strings or wires installed for pulling domestic bells; now used by electricians when laying electric and telephone wires.

Gimlet, Boat *Fig. 311, No. 1957*
A miniature version of the Ship Gimlet and used for a similar purpose on small boats.

Gimlet, Brewer's (Spile Gimlet) *Fig. 311, No. 1958 and 1960*
A short Shell or Twist Gimlet, but with a stouter body. The handle is often turned in the shape of a barrel, and sometimes decorated with hoop-like rings. Used for boring a tapered vent-hole in the shive (a flat bung) which is afterwards stoppered with a tapered peg known as a spile. See also *Cellarman*.

Gimlet, Farmer's *Fig. 311, Nos. 1950, 1926, 1927*
A name given to a large strong Shell Gimlet with a plain handle.

Gimlet, Gutter (Spout Gimlet)
A name given to a strong Gimlet of twist or shell type but 9–12 in long. Used presumably when fixing gutters.

Gimlet, Pod: see *Gimlet, Swiss.*

Gimlet, Ring Handle
A name given to Gimlets, usually imported, made from a length of steel wire which is bent round to form the handle. See illustration under *Gimlet, Swiss.*

Gimlet, Shell *Fig. 311, Nos. 1921 and 1946*
A shell side-cutting body, terminating in a short tapered screw point. This variety of the Gimlet usually cuts at its full width just above the screw point, after which the body tapers off very slightly towards the handle. This avoids unnecessary friction in the hole.

Gimlet, Ship *Fig. 311, No. 1955/56*
A strong Shell Gimlet about 10 in long and from $\frac{1}{4}$ to $\frac{5}{8}$ in. in diameter, with screw lead; either eyed for a wooden cross-handle, or left plain for an extended shank to be welded on. Used for boring holes for spikes or rivets in a ship's planking etc.

Gimlet, Skate *Fig. 311, Nos. 2483 (b) and 1961*
A very short Shell or Twist Gimlet. Sometimes made with a hollow brass tubular sheath to cover the blade when carried in the pocket; when unscrewed, the sheath is inserted through an eye in the shank to act as a cross-handle. Used for boring the sole and heel of the boot when fixing on skates.

Gimlet, Spike *Fig. 311, Nos. 1946 and 1948*
A name given to strong Twist or Shell Gimlets from $\frac{1}{4}$ to $\frac{5}{8}$ in diameter. Used for boring holes to take the large nails known as spikes.

Gimlet, Spile: see *Gimlet, Brewer's.*

Gimlet, Spout: see *Gimlet, Gutter.*

Gimlet, Sprig (Cabinet Gimlet)
A term used for fine Shell or Twist Gimlets up to $\frac{1}{8}$ in. in diameter. (A sprig is a small nail with a very small head.)

Gimlet, Swiss (Scots: Wilk Bit) *Fig. 312*
Like the *Bit, Swiss Gimlet*, this has a podlike shell body for about two-thirds of its length, after which the pod is given a half twist, and tapers off into a pointed screw lead.

Fig. 312

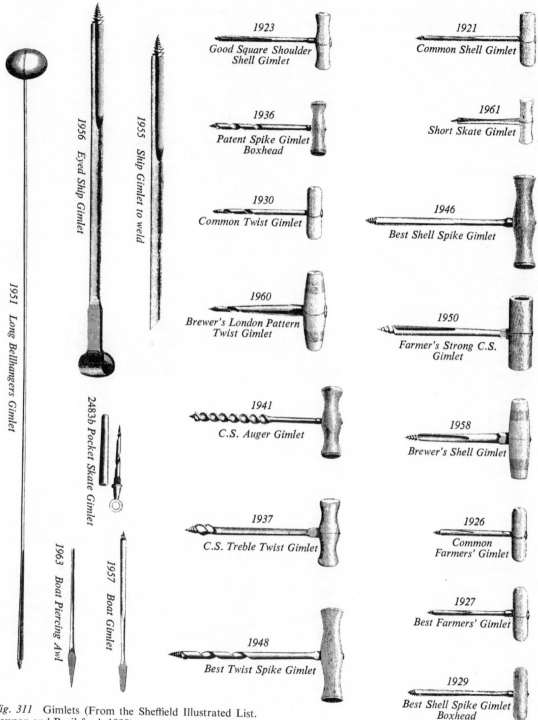

Fig. 311 Gimlets (From the Sheffield Illustrated List. Pawson and Brailsford, 1888)

1956 Eyed Ship Gimlet

1955 Ship Gimlet to weld

1951 Long Bellhangers Gimlet

2483b Pocket Skate Gimlet

1963 Boat Piercing Awl

1957 Boat Gimlet

1923 Good Square Shoulder Shell Gimlet

1936 Patent Spike Gimlet Boxhead

1930 Common Twist Gimlet

1960 Brewer's London Pattern Twist Gimlet

1941 C.S. Auger Gimlet

1937 C.S. Treble Twist Gimlet

1948 Best Twist Spike Gimlet

1921 Common Shell Gimlet

1961 Short Skate Gimlet

1946 Best Shell Spike Gimlet

1950 Farmer's Strong C.S. Gimlet

1958 Brewer's Shell Gimlet

1926 Common Farmers' Gimlet

1927 Best Farmers' Gimlet

1929 Best Shell Spike Gimlet Boxhead

Gimlet, Treble Twist *Fig. 311, No. 1937*
A name given to a strong Gimlet with a short twist of only three turns.

Gimlet, Twist *Fig. 311, No. 1948*
A name given to Gimlets in which a spiral groove has been cut and which tapers slightly toward the screw point. This spiral groove has a sharp edge which has some side-cutting action. Owing to the slight taper, the body cuts at full width only about half-way up the spiral. Made in sizes up to $\frac{1}{4}$ in.

Gimlet, Wheeler's
A name given to a long Gimlet similar to a Bell Hanger's or Ship Gimlet.

Gimlet, Wilk: see *Gimlet, Swiss.*

Gimlet, Wine Fret (Brewer's Fret; Cooper's Fret) *Fig. 313*
A short boring tool resembling a Twist Gimlet save that the body is a slender cone with a shallow twist and a screw lead, and the upper part is shouldered and sometimes fitted with a brass collar. The Pocket Fret is smaller, and takes apart so that it can be carried inside the pocket. It is used for boring holes of $\frac{1}{4}$ or $\frac{5}{16}$ in. in diameter, usually through the head of a cask, in order to take a sample of the liquor. This is normally done by means of a Tasting Tube (see Sampling Tube under *Cellarman*).

The cone-shaped body and shallow twist ensure a fairly smooth hole which can be safely plugged afterwards; the shoulder and collar prevent the escape of liquid until the sampling apparatus is ready.

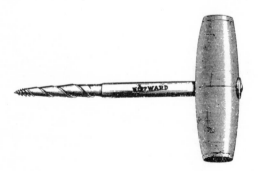

Fig. 313

Ginnet
A variant of *Jennet* – an early term for a Carpenter's or Shipwright's Adze.

Girthing Strap: see *Timber Girthing Tape and Sword.*

Glasspaper (Sand Paper) *Fig. 314*
Finely crushed and graded glass or other abrasive glued to kraft paper. The grades normally used in woodworking are called Flour, Fine, Middle, and Strong.

Used for smoothing wood, Glasspaper is normally folded round a cork or wooden block known as a Sandpaper Block or Cork Rubber. For cleaning up round holes or hollows, a round piece of wood of the appropriate diameter is sometimes used, with a saw kerf cut down the middle in which to insert the sheet of glasspaper, which is then folded round the block.

(*Note:* In most woodworking trades the term 'Glasspaper' is used rather than 'Sandpaper'. Curiously, however, the machine which performs a rougher though similar task is called a Sanding Machine, and the machinist speaks of the general operation as 'sanding'.)

Fig. 314

Glass Cutter *Fig. 315*
Until the latter part of the nineteenth century sheet glass was cut with a diamond. With the advent of special steels, the glass-cutting wheel was introduced; such wheels were listed as early as 1885 and are now used by most tradesmen. The diamonds were set in a metal head mounted on a slim wooden handle on which it is free to swivel a few degrees to left and right. The cutting wheels were originally mounted in the same way, but today are mostly set on a metal handle and are usually provided with 'racks' or 'gates', i.e. grooves of different widths cut in the side of the head and used for breaking off narrow strips (or rough edges) which cannot be grasped in the fingers.

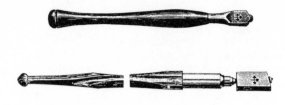

Fig. 315

Glaun
Scots joiners' term for *Vice* or *Cramp*.

Glazier
This trade is often combined with that of the painter or joiner. The glazier fixes glass in the frames of windows and elsewhere. The glass was cut with a diamond but more recently a hardened steel wheel has been used instead. The panes are usually secured in a rebate with putty. In leaded-light work the glass is held by strips of lead. The tools used include: *Glass Cutter; Hammer, Glazier's; Hammer, Sprig; Knife, Chisel; Knife, Glazier's; Knife, Hacking or Chipping; Larrikin* (for lead lights)*; Pliers, Glazier's; Rule, Glazier's; Square, Glazier's.*

Glue Plate
A metal plate used for transferring glue to the surface of a fretted workpiece. The plate is covered in glue and the work placed upon it to pick up the glue. This method prevents the glue from escaping over the edges of the work – as usually happens if a brush is used.

Glue Pot and Brush (Glue Kettle) *Fig. 316*
A double metal pot with wire handles. It is usually of cast iron, the inner pot being tinned to avoid the glue becoming discoloured by rust. The outer pot contains water which, on the principle of the double saucepan, prevents the glue from overheating. Traditionally, the Glue Pot is heated on the workshop stove, or on a gas ring, but nowadays the outer pot often contains a built-in electric heater. Multiple sets with a row of inner pots mounted in a single heating tank are made for use in large workshops.

Glue brushes generally have stiff bristles bound to the thick end of a tapering round handle, but the glue brush used by chairmakers was commonly made by taking a piece of round cane and beating one end of it so that the fibres frayed out to form a brush.

Moxon (London, 1677) writes: 'The *Glew-pot* is commonly made of good thick lead, that by its

Substance it may retain a heat the longer, that the *Glew Chill* not (as Work-men say when it cools) when it is to be used.'

Further extracts from Moxon show that he did not suggest that the initial boiling should be done in the lead pot: 'The clearest, driest and most transparent Glew is best .... When you boil it, put it into a clean Skillet or Pipkin, by no means greasie, for that will spoil the Clamminess of the Glew.... When it is well boiled, pour it into your Glew pot to use....'

Fig. 316

Glueing Screw: see *Brush Maker; Violin Maker.*

Glut
A wedge of wood or iron.

Goat's Foot
A name sometimes given to tools provided with a nail-pulling claw. See *Nail Extracting Tools.*

Gouge *Figs. 317; 318*
Gouges are made on similar lines to Chisels, but the blade is hollow in cross section for cutting curved surfaces. They are normally made in widths from ¼ to 2 in, and in sets of eight standard curves, from 'flat' through 'middle' and 'scribing' to 'fluting'. Reference may be made to the diagram under *Chisel* for nomenclature, and to the appropriate chisel entry for information on Gouges that are related by usage.

There are two main kinds of blade: those with the grinding bevel outside – known as 'out-cannel'; and those with the grinding bevel on the inside and known as 'in-cannel'.

In general, out-cannel Gouges are used for the hollowing-out of curved, saucer-like depressions (Fig. 317(*a*)); the in-cannel gouge is required when it is necessary to make a cut in a straight line, e.g. for scribing a moulding (Fig. 317(*b*)), or when boxing a wheel hub.

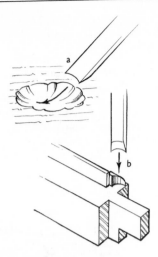

Fig. 317 (a) Out-cannel
         (b) In-cannel

Gouge, Blockmaker's *Fig. 318 (g)*
A strong Gouge with a socketed handle, 16–20 in long overall, ground in-cannel, and from $\frac{1}{2}$ to 2 in wide. It is used for cutting the score and other parts when making ship's blocks. Those seen at H.M. Dockyard, Portsmouth (1969) were only 10 in long with handles rounded at the top as if for pushing. The blades were $\frac{3}{4}$–1 in wide, $\frac{3}{8}$–$\frac{1}{2}$ in thick, and bevel ground in-cannel. See also *Chisel, Blockmaker's; Ship's Blockmaker.*

Gouge, Boxing: see *Gouge, Wheelwright's.*

Gouge, Bushing: see *Gouge, Wheelwright's.*

Gouge, Carving: see *Chisel, Carving.*

Gouge, Chinese *Fig. 318 (b)*
Made for export, this is a short out-cannel Gouge, $\frac{1}{4}$–2 in wide, with a flared blade tapering back sharply from the cutting edge.

Gouge, Coachmaker's *Fig. 318 (f)*
Made with either in- or out-cannel, with the blade often fitted with a double-hooped handle. Used for rough paring in both coachmaker's and wheelwright's work.

Gouge, Entering: see *Chisel, Carving* (Front Bent).

Gouge, Firmer *Fig. 318 (a); Fig. 196, Nos. 2452, 2474, 2449 on page 131*
A tool similar in appearance to the Firmer Chisel but with a hollow blade, $\frac{1}{16}$–2 in wide, shaped to standardised curves. In ordinary workshop practice the term Firmer Gouge is reserved for the out-cannel form, but this Gouge is also made with the grinding bevel inside. Used for all general hollowing work.

Gouge, Fluting
A name given to Gouges shaped to cut deep channels.

Gouge, Gunstocker's: see *Chisel, Gunstocker's.*

Gouge, Millwright's *Fig. 318 (e)*
An extra strong blade, 10–12 in long, from $\frac{1}{4}$ to 2 in wide, and usually ground in-cannel. See also *Chisel, Millwright's.*

Gouge, Nave
A name sometimes given to a Boxing Gouge. See *Gouge, Wheelwright's.*

Gouge, Paring *Figs. 318 (h) and 196*
Lighter than the Firmer Gouge and made from $\frac{1}{4}$ to $1\frac{1}{2}$ in wide ground in-cannel. Normally used without a mallet for hand paring work. The common type is often called a Scribing Gouge. Variants include:

> Long Thin Paring Gouge – *Fig. 196, No. 2448* under *Chisel*
> Trowel Shank Paring Gouge – *Fig. 318 (h)*

Gouge, Pattern Maker's *Fig. 318 (i) and (j)*
There are several types of Gouge made for pattern makers, including those illustrated.

Gouge, Registered
A strong Firmer Gouge, out-cannel, with the characteristic 'Registered' double-hooped handle and special ferrule with a leather washer. See *Chisel,*

Fig. 318 Gouges: Typical examples. *Opposite*
         (a) Firmer Gouge, double hooped
         (b) Chinese Gouge
         (c) Socket-Type Gouge
         (d) Scribing Gouge
         (e) Millwright's Gouge
         (f) Coachbuilder's Gouge
         (g) Blockmaker's Gouge
         (h) Paring Gouge with trowel shank
         (i) }
         (j) } Patternmaker's Gouges

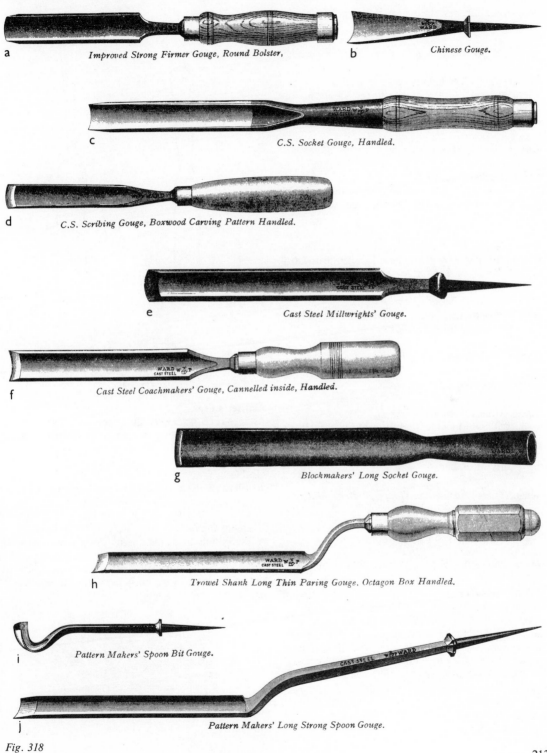

a   *Improved Strong Firmer Gouge, Round Bolster,*

b   *Chinese Gouge.*

c   *C.S. Socket Gouge, Handled.*

d   *C.S. Scribing Gouge, Boxwood Carving Pattern Handled.*

e   *Cast Steel Millwrights' Gouge.*

f   *Cast Steel Coachmakers' Gouge, Cannelled inside, Handled.*

g   *Blockmakers' Long Socket Gouge.*

h   *Trowel Shank Long Thin Paring Gouge, Octagon Box Handled.*

i   *Pattern Makers' Spoon Bit Gouge.*

j   *Pattern Makers' Long Strong Spoon Gouge.*

Fig. 318

213

Registered for description of this type of handle. Used for extra heavy work.

Gouge, Rubber Tapping: see *Knife, Rubber Tapping.*

Gouge, Sash (Scribing Gouge) *Fig. 318 (d)*
Light in-cannel Gouges made in sets of six or nine sizes from ⅛ to 1 in. Used for scribing sash or door stuff. See *Window Making; Scribing or Coping.*

Gouge, Sinking Down: see *Chisel, Plane Maker's.*

Gouge, Socket *Fig. 318 (c)*
A strong Gouge with a socket forged integral with the blade to receive the handle. See also *Chisel, Socket.*

Gouge, Sowback
A name given to a short Gouge with a gentle upward-curved blade, ⅛–1 in wide. Used for large-scale hollowing work.

Gouge, Spoon: see *Gouge, Pattern Maker's; Chisel, Carving; Gouge, Wheelwright's.*

Gouge, Square: see *Chisel, Hinge.*

Gouge Stone: see *Oilstone Slip.*

Gouge, Timber Testing
A short gouge ¾ in wide, slightly bent like a Carver's Curved Gouge, and provided with a leather sheath. Used by timber merchants and others for testing the quality of timber.

Gouge, Trowel Shank: see *Gouge, Pattern Maker's; Gouge, Paring.*

Gouge, Turning: see *Chisel, Turning.*

Gouge, Wagon Builder's: see *Chisel, Wagon Builder's.*

Gouge, Wheelwright's (Bushing Gouge; Boxing Gouge or Chisel; Nave Spoon) *Fig. 319*
A very heavy Gouge, usually fitted with a socketed handle and often blacksmith-made. The length varies from 14 to 18 in overall, and the shallow curved blade varies in width from ¾ to 3 in. Used for excavating wheel hubs to take the axle-bearing which is known as the 'box'.

Gouging Machine and Guillotine: see *Woodwind Instrument Maker.*

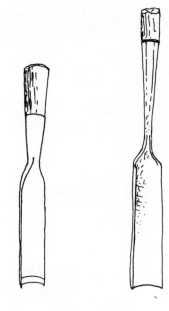

Fig. 319

Graft
Of carpentry – to lengthen a piece of wood, e.g. a beam; or to replace a part, e.g. the rotted foot of a door post.

Grafter's Froe
Mercer (U.S.A., 1929) illustrates this small Froe. It was used for splitting the parent stock of fruit trees to insert the graft.

Graille
A Float with a curved cutting surface, used by workers in horn and by comb makers.

Graining Comb *Fig. 320*
A comb-like instrument with long, thin, parallel teeth. It is used for imitating in paint the marks and grain of various woods. Though rarely practised today, graining was, until recently, an important part of the painter's work.

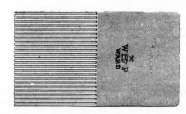

Fig. 320

Grannie's Tooth
An alternative name for a Router described under *Plane, Router.*

Grapnel (Grappling iron; Snudgel) *Fig. 321*
A hook with three or four prongs joined at their bases to a ring, and measuring from 10 to 15 in overall. Often found in country workshops and used for recovering a lost bucket or a dead animal from a well.

Fig. 321

Graver: see *Wood Engraver.*

Grease Pot (Grease Box; Grease Horn) *Fig. 322*
A box or pot to hold the grease, oil, or tallow used for lubricating Brace Bits, Augers, or Saws, and for dipping screws before insertion.

Wheelwrights and joiners often used a home-made pear-shaped box carved from the solid wood, about 4½ in long with the lid rotating about a screw at the narrow end. Another type, seen in a Cambridge-shire laddermaker's shop, was simply a short log hollowed out at the top to take the grease, with a sloping handle stuck in the side. This was used for lubricating Augers which were in almost continuous use.

Other methods include the following:

(*a*) *Planes.* For lubricating the soles, tradesmen often have a small open-topped box or block provided with a piece of felt or other material kept soaked with linseed oil. Like the Oilstone box, this has two spikes in the bottom to hold it still on the bench as the plane is drawn backwards over it. A stub of wax candle is also used for this purpose.
(*b*) *Sawyers* kept an oil pot and brush in the wall recess of the pit close to the bottom sawyer, from which oil could be applied to the saw.

(*c*) *Chair Makers* fill a depression in the bench itself with grease. A saw kerf is made to run up to this depression into which the saw can be inserted for lubrication. (See Framing Block under *Chair Maker* (3) *Workshop Equipment.*)

(*d*) *Other tradesmen* often used a cow-horn for holding grease, e.g. see *Sailmaker.*

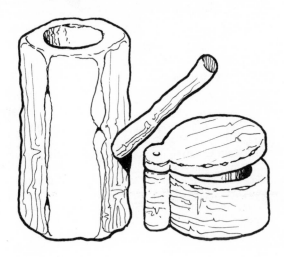

Fig. 322

Grifon
A metal scraper curved in a semi-circle, with a ser-rated edge, used by sculptors and others for paring or scraping down plaster, clay, or wax.

Grinding Appliances (Tool Holders) *Fig. 323*
These include various devices designed to hold a plane iron, Axe, or Chisel against a grindstone or oilstone at a proper angle.

(*a*) *Attachment to the Grindstone* [not illustrated]
Two parallel bars carrying a clamp to hold a cutter or chisel are pivoted from a point above and behind the stone. The operator can thus lower the tool against the stone at a constant angle.

(*b*) *Tool Grinding Rest*
A metal holder for plane irons etc. with a small roller mounted beneath it. This roller revolves against the circumference of the grindstone, thus keeping the edge at the same angle, and enabling the user to hold it steady. This is used mainly for holding plane irons or Chisels when sharpening on the oilstone. It is particularly useful for beginners who have not acquired the knack of keeping the

tool at the same angle throughout the sharpening stroke.

Fig. 323 (b)

(c) Home-made Tool Holders

For Plane and other cutting irons: a wooden holder about 6–8 in long with a saw kerf in the end into which a cutter can be inserted. It was specially useful for holding small cutters which it would be very difficult to hold against a grindstone with the bare hands. For Axes: a curved piece of wood, one end of which is inserted in the eye of an Axe to hold it when grinding the cutting edge.

Grindstone *Fig. 324*

A cylinder of fine-grained natural sandstone, from about 12 to 40 in wide and up to 5 in thick, mounted on a spindle of which one end is extended to form a crank. Factory-made Grindstones are usually mounted on a cast-iron or stout wooden stand, and the lower half of the stone is often enclosed in a metal or wooden trough containing water. Some Grindstones were provided with metal arms containing a clamp to hold the tool while grinding (see *Grinding Appliances*).

Driven by hand-crank or treadle (and later by motor), Grindstones are used for grinding the cutting bevels of edged tools, such as Chisels, Gouges, and plane irons. In the older workshops the spindles ('gudgeons'), extending from both sides of the stone, were often run in plain iron or wooden bearings mounted on the wooden posts set on each side of the stone.

In bodger's work, out in the woods, the gudgeons were grooved ('necked') and secured by staples driven over them into the side posts, one of which might be fastened to a growing tree. A tin can with a hole in it allowed water to drip on to the stone when in use.

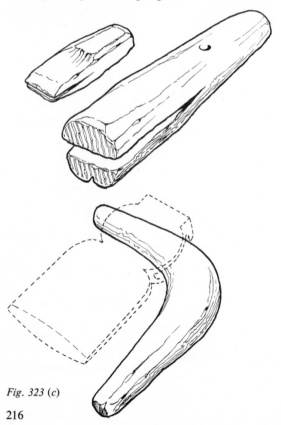

Fig. 323 (c)

Fig. 324

Grip: see Woodman's Grip under *Woodland Trades* (*1*).

Grocer's Tools: see *Box and Case Openers; Butter and Cheese Sampler; Hammer, Grocer's Warehouse; Nail Extracting Tools; Rasp, Baker's.*

Grommet: see Hand Fid under *Sailmaker.*

Grooving Tools: For examples see list under *Plane, Grooving;* also under *Routers; Scratch Stock.*

Groper: see *Plane, Cooper's Chiv.*

Groping Iron

A fifteenth-century term for a Gouge (Halliwell, 1847).

Grove: see *Plane, Cooper's Croze.*

Grozing Iron: see Plumber's Iron under *Plumbing Tools.*

Grubbing Mattock: see *Axe, Grubbing.*

Gullet

Of Saws – the spaces between the teeth which carry the sawdust. See diagram under *Saw.*

Gun Metal

A name given to bronze. A common naval specification is 88% copper, 10% tin, and 2% zinc, but certain specifications allow a limited replacement of tin by lead. Much used for ship's fittings, propellers, etc., owing to its strength and resistance to corrosion by sea water. Occasionally used for the stocks of nineteenth-century metal Planes.

Gunstocker

Gunstockers make the wooden stock or butt of a gun, which carries the lock mechanism and the gun barrel itself. Though special lathes can be used, stocks for the best guns are made by hand.

Most stocks are made of walnut, preferably French, which owing to its reliability and beautiful grain is considered to be the best wood for the purpose. Its virtues are described as follows in A. L. Howard's *A Manual of the Timbers of the World* (London, 1920). ' Although walnut requires some time to season, and shrinks considerably during the process, yet when subsequently exposed to drying or moistening influences it stands excellently, and it is exceedingly difficult, if not impossible, to find another wood possessing this attribute to the same degree.'

The rough billet of walnut is held in a vice and the waste wood pared off with a broad Chisel. An old gunstock is often used instead of a mallet. This is followed by shaping with Drawknife, Spokeshave, and Rasp, and finally with glass paper and a Rubbing Stick (a wooden bar faced with leather).

The recesses for the lock and other metal parts are first cut in the stock with an ordinary Mortice Chisel, and are then cleaned out with a cranked Dog-Leg Chisel which gunstockers call a 'Shovel'. Marking out the recess for the lock etc. is done by holding the metal part concerned in the flame of a paraffin lamp to 'smoke' it. It is then placed on the partly recessed stock, where it leaves a mark on the wood which must be cut away.

The cross-hatching on the butt is done with a Graver which has a saw edge and is known as a Checkering Tool, or with a Pottance Checkering File.

Some of the special tools used will be found under the following entries: *Checkering Tool; Chisel, Carving (Dog-leg); Chisel, Gunstocker's; File, Pottance Chequering; Float; Plane, Gunstocker's; Rasp (Gunstocker's); Rubbing Stick.*

H

Hack
Scots joiners' term for a small *Adze*.

Hackle: see *Brushmaker*.

Hacksaw: see *Saw, Hack*.

Haft (Helve)
A handle, e.g. of a Hammer, Axe, or Mallet.

Halflin (Hafflin)
A Scots term. See *Plane, Trying*.

Hammer *Fig. 325 opposite*
The first Hammers were hard stones held in the hand; later, a handle of wood or antler was fastened to the stone head with thongs; later still, the head was perforated and the handle wedged into the hole. Some Bronze Age Hammers were cast, with sockets into which a knee-shaped handle was fixed, as in the socketed Axes. The Romans sometimes bifurcated the pane to form a claw for extracting nails. A plain hammer with a square head is shown in medieval pictures of tools and is still used on the Continent; we call it the French Hammer. With a round, necked head it is called the London, Exeter, or Warrington Joiner's Hammer. The 'strapped' hammer with languets to hold the head firmly on the handle appears in Durer's 'Melancholia' of 1514, and is still popular today as exemplified by the Canterbury Hammer. Strapping is done in two different ways. In one, shown in (*b*), the straps are forged on to the head itself, which may be eyed to take the handle or left solid. The other method (*c*) is to fit separate plates which pass through the eye and are bent or stepped to fit over the top edge of the hammer-head and thus hold it securely. This is the method which appears to have been illustrated in Moxon (London, 1677) and is favoured today by toolmakers in Continental Europe.

Hammer, Adze Eye (Adze Eye Nail Hammer) *Fig. 326*
A Claw Hammer with a round face and tapered neck; the eye is tapered and usually extended downwards like that of an Adze, thus giving the tool its usual name.

It was developed in the U.S.A. in 1840 by a blacksmith called David Maydole, as a means of avoiding the weakness of the ordinary eyed Claw Hammer; and it has now been almost universally accepted in Britain and the U.S.A. as a general-purpose Carpenter's Hammer. The deep adze eye effectively prevents the head from being wrenched off the handle when the claw is used for extracting nails. Writing about this Hammer, H. R. Bradley Smith (U.S.A., 1966) related that 'Six carpenters had come to Norwich to build a house. One lost his hammer. He went to Maydole who offered him his latest creation. This man immediately fell in love with his new tool where "the head and the handle were so united that there was never likely to be any divorce between them". Each of his five fellow carpenters had to have one. Their contractor ordered two more; then a local storekeeper stocked two dozen; next a New York dealer in tools bought a large supply. This sudden demand unexpectedly put David Maydole into the business of making hammers, and his blacksmith shop materialized into a factory.'

Fig. 326

Hammer-axe
Various tools with a hammer face on one side of the head and an axe blade on the other, e.g. *Hammer, Grocer's; Hammer, Pincer; Hammer, Scaffolder's*.

Hammer, Bath: see *Hammer, Centre Pane*.

Hammer, Bench (Joiner's Bench Hammer) *Fig. 327*
An octagon face, plain pane, and flat underside to the head.

Fig. 327

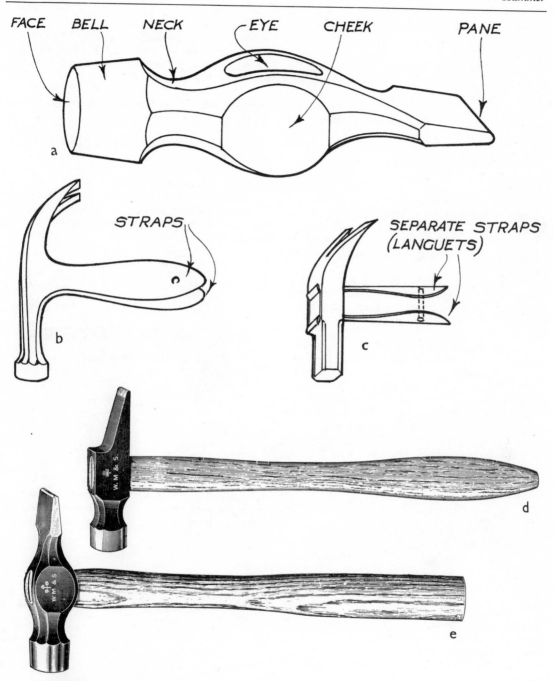

FACE BELL NECK EYE CHEEK PANE

a

STRAPS

b

SEPARATE STRAPS
(LANGUETS)

c

d

e

Fig. 325 The Hammer
   (*a*) Parts of a common Joiner's Hammer
   (*b*) Claw Hammer with forged straps
   (*c*) German Claw Hammer with separate straps
Typical Joiner's Hammers:
   (*d*) Exeter
   (*e*) Warrington

Hammer, Benwell: see *Upholsterer; Hammer, Upholsterer.*

Hammer, Block Point (Mounter's Hammer)
A light Warrington or similar type of hammer with a short handle. Used by wigmakers for inserting block-points to hold the foundation galloons (ribbons) and bracings in position on a wooden block. (Block-points are thin headless nails.)

Hammer, Boat Punch
A small version of the *Hammer, Ship's Maul,* weighing 13 oz.

Hammer, Box and Case Opening: see *Hammer, Grocer's and Warehouse; Box and Case Openers.*

Hammer, Box-Maker's (Packing-Case Maker's Hammer) *Fig. 328*
From 12 oz to 18 oz in weight, with a circular face, chamfered neck, and a flat claw extended on one side to form a sharp spike. It is described in manufacturers' lists as a 'Box-maker's Hammer with Dabber'.

A very similar Hammer is described by continental tool-makers as a Lathing Hammer or a Carpenter's Roofing Hammer.

Used in box and packing-case making. The spike is used for piercing the metal hoops or straps before nailing, and for 'dabbing' a pilot hole in hard woods for nail or screw. A London boxmaker informed us that the spike could also be used for twisting together wire hoops on dry casks. See also *Hammer, Manchester Spike.*

Fig. 328

Hammer, Brick: see *Builder's Tools; Scutch.*

Hammer, Bristol Pattern *Fig. 329*
A Joiner's Hammer with a head similar to the Exeter pattern.

Fig. 329

Hammer, Bush
A bundle of 6–8 hardened steel plates with sharpened ends bolted together to form a heavy hammer-head. Used for dressing stone and sometimes for levelling millstones. See *Millstone Dresser.*

Hammer, Butter: see *Hammer, Grocer's and Warehouse; Butter and Cheese Sampler.*

Hammer, Cabriolet: see *Hammer, Coach Trimmer's Cabriolet.*

Hammer, Canterbury *Fig. 330*
A Claw Hammer with two straps ('languets') secured by rivets to the handle. A carpenter's general purpose Hammer which is still in common use. For information on straps see *Hammer.*

Fig. 330

Hammer, Carpet *Fig. 331*
A light Hammer with a round tapered head, a plain cross pane, and a claw fitted to the heel of the handle. Used by carpet layers. See also *Upholsterer.*

Fig. 331

Hammer, Carrick *Fig. 332*
A Joiner's Claw Hammer with straps. See also *Hammer, Scotch.*

Fig. 332

Hammer, Cask: see *Hammer, Grocer's and Ware-house*.

Hammer, Caulking: see *Caulking Tools*.

Hammer, Centre Pane (Bath Head Hammer) *Fig. 333*
A Joiner's Hammer with an 'Exeter' type head, but with the cross pane tapered at both top and bottom – making the head symmetrical.

Fig. 333

Hammer, Chairmaker's *Fig. 334*
A double-faced Framer's Hammer, often smith made. The faces are sometimes 'mushroomed', i.e. of slightly larger diameter than the body, and slightly convex.

It was used by chairmakers for operations such as 'legging-up', i.e. when legs were driven into the seat and stretchers into the legs. Mallets were not used, the framers claiming that the smooth face of the Hammer had less tendency to mark surfaces or split end-grains.

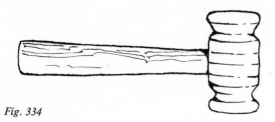

Fig. 334

Hammer, Clavot (Clough; Clove; Clotte)
A name given in the Apprentice Indenture Records of 1621 (Goodman, 1972). It is probably a Ship-wright's Claw Hammer.

Hammer, Claw (Scots: Kluvie) *Fig. 335*
Hammers with a bifurcated pane were used in Roman times and are frequently shown in medieval pictures, especially in scenes of the crucifixion. In modern times special types have been developed for general use by carpenters, farmers, gardeners, and others, and for particular trades such as upholsterer, saddler, clogger, shipwright, etc.

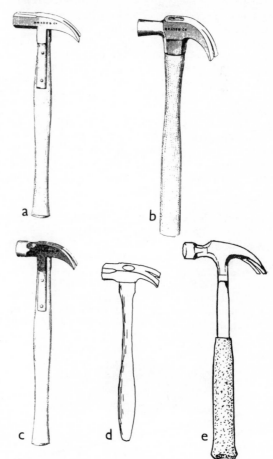

Fig. 335 Typical Claw Hammers
   (*a*) Kent Pattern with straps
   (*b*) Adze-Eye Pattern
   (*c*) Canterbury Pattern with straps
   (*d*) A home-made example
   (*e*) All-steel example of the 'Estwing' or 'Stanley' type with a plastic-coated handle

Owing to its use for extracting as well as driving nails, the Claw Hammer was never entirely satisfactory with its handle wedged into the usual round or oval eye, because the handle was always working loose or breaking off altogether. Various ways of avoiding this weakness have been tried, either by:

1. Deepening the eye itself
2. Fixing the head to the handle by means of straps ('languets')
3. Making head and handle of one solid forging.

The most successful example of method (1) is the 'Adze-eye' Claw Hammer, invented in 1840 and now practically universal as a general Carpenter's Hammer both in the U.S.A. and in Britain. Method (2) is shown in Albert Durer's famous picture 'Melancholia' (c. 1514) and also in Moxon's Plate 8 of carpenter's tools (1677). It has come down to us in the various types of Kent, Canterbury, and Scotch Claw Hammers and Saddlers' and Upholsterers' Hammers. For methods of strapping, see under *Hammer*.

The solid forged type (3) is represented by the various patterns of grocers' and packers' hammers; and in a more sophisticated form by the all-steel Claw Hammer patented in the U.S.A. by Ernest Estwing in 1926.

Hammer, Clench (This name is sometimes given to hammers used for riveting.) *Fig. 336*
A heavy Hammer with an octogonal or round face, chamfered neck, and a cross pane with a thick rounded edge. A general-purpose Hammer which can be used as an anvil when held behind the head of a rivet or nail to hold it in position while its point is hammered over. Clench hammers used by shipwrights are illustrated under *Hammer, Ship Clench*.

Fig. 336

Hammer, Clogger's *Fig. 337*
A Claw Hammer with a round face and long chamfered neck. Used for nailing the leather uppers of clogs to the wooden soles.

222

Fig. 337

Hammer, Coach Framing: see *Hammer, Framing*.

Hammer, Coach Trimmer's *Fig. 338*
A light Hammer with a long head, octagonal or round face, and a long, tapering cross pane provided with a claw at one side. The head is normally eyed and not strapped. The flat cross pane is used for driving tacks in confined spaces. See also *Hammer, Lightweight Types*.

Octagon Coach Trimmers Hammer Head.

Round Coach Trimmers Hammer Head.

Fig. 338

Hammer, Coach Trimmer's, Cabriolet (French Upholsterer's Hammer) *Fig. 339* and *Fig. 712* (Upholsterer)
A light Hammer with a long strapped head, round face, and a clawed cross pane.

A special feature of this pattern is a general roundness in all directions and the smallness of the face which is adapted for driving tacks on very fine work. The French pattern is similar, but perhaps even more rounded, and it is often provided with a handle in rosewood or ash, decorated with rings. Used by coach trimmers but is often listed with the tools of the upholsterer by whom it is also used.

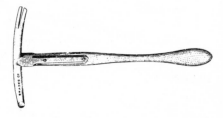

Fig. 339

Hammer, Cooper's (Flue Hammer) *Fig. 340*
Hammers used by coopers include the following:
Hand Hammer
A heavy, short-handled hammer, weighing from 2 to
4 lb.

(*a*) The London or Burton pattern is similar to a
Smith's Hand Hammer, with an octagonal face, flat
cheeks, and a flat-sided straight pane with or without short chamfers.

(*b*) The Scotch pattern has a round face with
chamfered neck. The face is used for beating Truss-
Hoops on and off, for striking Hoop Drivers when
driving on the hoops, and for riveting hoops. The
straight pane is used for flaring hoops ('making
splay') to follow the bulge of the cask.

(*c*) Flue Hammer: (not illustrated) a name given
to the Cooper's Hand Hammer because of its use
for flaring cask hoops. (Flue means to expand or
splay.)

Two-Handed type Hammers
(*d*) Sledge: (not illustrated) similar to the Cooper's
Hand Hammer (*a*), but 8 lb in weight and mounted
on a longer handle for swinging with both hands.
Used for finishing off the cask, i.e. beating on the
chime hoops; also for flaring very wide hoops.

(*e*) Set: a heavy Hammer with a flat, square face.
Probably used as an anvil when riveting vat hoops
too large to be riveted on the Cooper's Anvil.

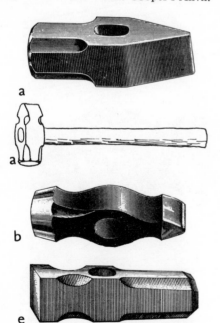

Fig. 340

Hammer, Cooper's Driver (U.S.A.: Nantucket
Driver. See *Hoop Driver* under *Cooper* (3) *Hooping
Tools*.)

Hammer, Coppering: see *Hammer, Ship's Coppering
and Sheathing*.

Hammer, Dog Head: see *Saw Doctor and Sharpener*.

Hammer, Dolly
A heavy head weighing about 18 lb and mounted on
a 3 ft handle. See *Wheelwright's Equipment* (3)
Tyring Tools.

Hammer, Double-Clawed *Fig. 341*
A Hammer provided with two sets of claws. The
extra claw was presumably used for pulling long
nails, i.e. it lifted higher. The origin and indeed the
purpose of this extraordinary tool is doubtful. It is
illustrated by Sloane (U.S.A., 1969) who calls it a
'Double-clawed Shaker Hammer'. The Chronicle of
the Early American Industries (U.S.A., 1969, Vol.
XXII) records that the date November 4th, 1902
appears on one of the Hammers.

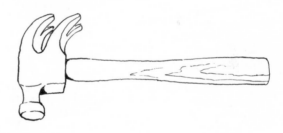

Fig. 341

Hammer, Double-Faced: see *Hammer, Framing*.

Hammer, Edinburgh: see *Hammer, Scotch*.

Hammer, Engineer's *Fig. 342*
Of the many types of metal worker's hammers used
in woodworkers' shops those illustrated are perhaps
the most common. They are used for all kinds of
general work, for driving heavy nails, plugging walls,
and for riveting. (*a*) With ball pane. (*b*) With cross
pane. (*c*) With straight pane.

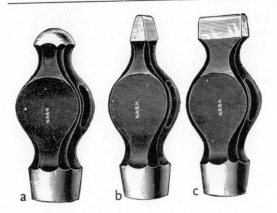

Fig. 342

Hammer, Flooring and Roofing (Octagon-faced Clench Hammer) *Fig. 344*
A heavy Hammer with octagonal face and thick, chamfered cross pane. Used by carpenters and others for heavy work on floors etc.

Fig. 344

Hammer, Exeter *Fig. 343*
Two kinds are described below. Many toolmakers' lists make no distinction between the London and Exeter pattern Hammer, but some show the pane of the Exeter as chamfered along the edges on both top and bottom while the London pattern is left square.

(*a*) A medium-weight Joiner's Hammer, and still in common use. The central part of the head is square in section with flat cheeks; the cross pane is tapered from the top only, and ends in a fairly narrow edge suitable for riveting.
(*b*) The 'Exeter Claw' has the square head of the Exeter Hammer, but with a claw instead of the cross pane. Being without straps, it is a cheaper Hammer than the Kent or Canterbury Claw patterns.

Hammer, Flue: see *Hammer, Cooper's*.

Hammer, Framing (Coach Framing; Double-faced Hammer; Scots: Mash or Massie; Railway Carriage or Wagon Builder's Framing Hammer) *Fig. 345*
A large double-faced Hammer usually with deeply chamfered necks. One of the faces is sometimes domed. The handle is usually short but a long-handled home-made version used by a Cornish wheelwright is also illustrated. It has an unusual flat, rectangular head. These Hammers are used for framing up coaches, railway carriages, carts and wagons, and similar work.

Fig. 343

224

fruit trees, such as peaches, plums, cherries, etc. were trained. When the nails have to be removed, owing to changes in the shape of the tree, the spike acts as a fulcrum, keeping the head clear and preventing damage to young growths and buds.

(*c*) The so-called Horticulturist's Hammer has a strapped head with a square poll turned down at its lower edge to form a claw, and an axe blade on the other side.

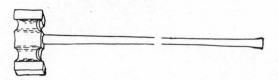

Fig. 345

Hammer, French Pattern *Fig. 346*

A name given to a Hammer with a square face and square body, flat along the underside, and a cross pane with no side chamfers. It is illustrated in medieval pictures and is still used as a Joiner's Hammer on the Continent of Europe. (Another so-called French Hammer is described under *Hammer, Coach Trimmer's, Cabriolet*.)

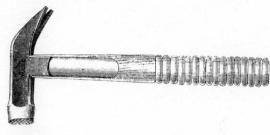

a

Fig. 346

Hammer, Fruiterer's: see *Hammer, Grocer's and Warehouse*.

Hammer, Gardener's (Garden Hammer; Wall Hammer) *Fig. 347*
Variants include:

(*a*) A Claw Hammer of the Kent, Canterbury, or Scotch pattern, with a checkered face and a round handle, bulbous or ring-turned, or with a plain round knob at the end.

(*b*) The spiked pattern was used by gardeners when fixing and removing nails from the wall on which

b

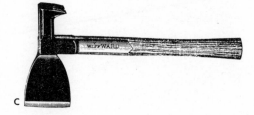
c

Fig. 347

Hammer, Gentlemen's and Ladies' *Fig. 348*
A light Claw Hammer with a strapped head, round face and straight claw, similar to an Upholsterer's Hammer. The so-called Ladies' Hammer is the same except that the claw is curved. It would be hard to say what this difference implies.

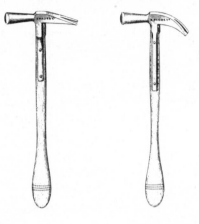

Fig. 348

Hammer, Glasgow: see *Hammer, Scotch.*

Hammer, Glazier's *Fig. 349*
A strapped Claw Hammer with a round face and tapering neck. A Continental Glazier's Hammer is in steel throughout and has an octagonal face and straight claw. The handle is drawn down to form a blunt chisel-like edge at the heel. (For another Hammer used by Glaziers see *Hammer, Sprig.*)

Fig. 349

Hammer, Grocer's and Warehouse *Fig. 350*
Hammers used for packing, box opening, cask opening, etc. include the following (see also *Box and Case Opening Tools*):

(*a*) *Butter Firkin Hammer* (Butter Hammer; Cask Hammer)
An eyed Hammer with a square face and a long, tapering pane, plain or clawed. Near the end of the pane there is an upward 'kink'. Used by butter makers and grocers for taking out the lid of a butter cask (e.g. when taking a sample), without cutting the hoops. The long pane of the Hammer is pressed down between the lid and the inside of the cask, and the lid loosened by pressing the hammer-handle downward, while the 'kink' at the end of the pane holds the tool in position over the top of the cask. This 'kink' also acts as fulcrum when extracting nails. (See also *Butter and Cheese Sampler.*)

(*b*) *Fruiterer's Hammer*
A light Hammer with a checkered face and a long, flat claw. The head may be eyed or strapped. Used for opening fruit boxes.

(*c*) *Grocer's Hammer* (Warehouse Hammer)
An all-steel head with an octagonal face on one side and an axe blade on the other. The long iron handle terminates in a claw or chisel blade. Another pattern has a claw on the head. It is intended for general use in the shop and warehouse including case opening.

(*d*) *Orange Chest Hammer* (Orange Case Hammer)
A strapped Hammer with square face, the neck often ornamented with mouldings, and an adze-like blade. There is a nail-pulling slot in the side of the blade, or a claw extends from the side of it. The earliest illustration we have seen of this tool is in Richard Timmins' catalogue, *c.* 1800. Used for opening and closing cases of oranges.

(*e*) *Packer's Hammer*
All-steel with a solid claw head. The steel handle terminates in a chisel or claw. Used for opening and closing cases and casks.

Hammer, Hammond Claw: see *Hammer, Adze Eye.*

Hammer, Joiner's
A general term for a Hammer of small to medium size which is used by joiners and other tradesmen for general purposes. They all have a cross pane which tapers to a fairly narrow edge, which enables pins and small nails to be 'started' between finger and thumb, and which is also suitable for spreading the end of small rivets. In modern tool catalogues,

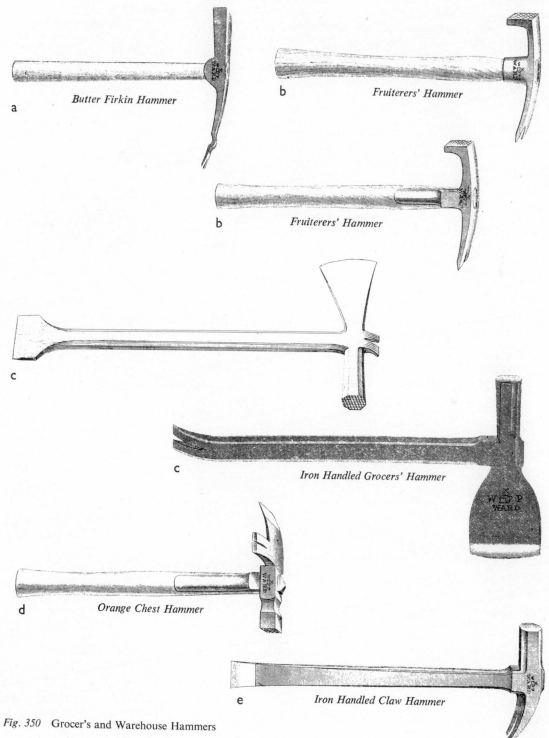

a   *Butter Firkin Hammer*

b   *Fruiterers' Hammer*

b   *Fruiterers' Hammer*

c

c   *Iron Handled Grocers' Hammer*

d   *Orange Chest Hammer*

e   *Iron Handled Claw Hammer*

Fig. 350   Grocer's and Warehouse Hammers

227

the Warrington, Canterbury, and Exeter Joiner's Hammers survive as the most usual patterns in use today.

Hammer, Kent *Fig. 351*

A Claw Hammer with an octagonal face and neck and with the head strapped to the handle. A small protruding piece of metal is often found at the centre top of the head. It is not known whether this originates in some earlier method of fixing the head or is provided to increase the leverage by raising the height of the fulcrum.

The earliest illustration we have encountered is in Smith's *Key* (Sheffield, 1816). It still survives as one of the few remaining regional types. An apparently identical tool is illustrated in Gilpin's list of 1868 and is called a Shoebridge Hammer.

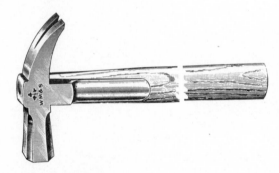

Fig. 351

Hammer, Ladies': see *Hammer, Gentlemen's and Ladies'.*

Hammer, Lancashire Pattern *Fig. 352*

In Gilpin's list (Cannock, 1868) this name is given to what appears to be a Warrington Joiner's Hammer. In Wynn Timmins' list (Birmingham, 1892) the term is applied to a Hammer of the Exeter type.

Fig. 352

Hammer, Lathing: see *Hatchet, Lathing.*

Hammer, Lightweight Types

Long-headed, lightweight Hammers are designed for light work such as the fixing of leather and cloth covers to furniture and coaches, and for saddlery. Their main function is to drive or extract small tacks, often in very confined spaces. Though differentiated according to the following trades, many of them are interchangeable, and may be so found in use.

Hammer	Face	Pane	Claw	Strapped Head
Coach Trimmer's	Square (occasionally round)	flat cross	side	No
Coach Trimmer's Cabriolet	Round and very small	curved clawed	end of pane	Yes
Saddler's	Round	flat cross	sometimes at side	Yes
Upholsterer's	Round	flat clawed	end of pane	Yes

Note: Other lightweight Hammers include *Hammer, Gent's; Hammer, Pattern Maker's; Hammer, Tack.*

Hammer, London Pattern *Fig. 353*
In some lists the London and Exeter pattern Joiner's Hammers are identical, but the consensus of opinion seems to be that the pane of the London was perfectly plain and that of the Exeter chamfered.

Fig. 353

Hammer, Lowe's Pattern *Fig. 354*
A Claw Hammer with a strapped head and deeply chamfered neck.

Fig. 354

Hammer, Lump: see Dolly Hammer under *Tyring Tools.*

Hammer, Magnetic
A fairly light hammer with a small round face, and either a claw or a long peg-shaped pane. The face is magnetised, so that a nail can be picked up by its head by the hammer itself, and driven home, often at one stroke, without using the other hand to hold it. Used by boxmakers and others when a very large number of nails have to be driven into the work.

Hammer, Manchester Pattern *Fig. 355*
A Joiner's Hammer with a round face, flat above and below the eye, and with cross pane flared out wider than the head.

Fig. 355

Hammer, Manchester Spike (Packing Case Manchester Spike Hammer) *Fig. 356*
A Manchester Pattern head with a pointed spike in the middle of the flat cross pane. Probably used for the same purpose as the *Hammer, Boxmaker's.*

Fig. 356

Hammer, Marking *Fig. 357*
This tool is in the form of a Hammer with letters cut on one or both faces, or a small Adze or Axe with the letters cut on the poll. The blades are used for cutting a 'blaze' (removing a piece of the bark) and the poll is used for imprinting initials. This is done to a standing tree intended for felling or on any other timber to record measurements or possession. According to Mr. H. L. Edlin, these tools are not common in Britain, but are used extensively in Germany. In the French State Forests special marks are used by a senior official and only trees so marked may be cut. This process is called 'martellage' (hammering) and when not in use the official Hammer is kept locked up.

The Government mark of the broad arrow was used in the North American Colonies about 1750 to mark trees to be requisitioned from any colonist's land for use in naval building – a practice which gave rise to considerable resentment.

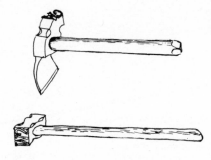

Fig. 357

Hammer, Mash or Massie: see *Hammer, Framing.*

Hammer, One Handed: see *Rampin.*

Hammer, Orange Chest: see *Hammer, Grocer's and Warehouse.*

Hammer, Packer's: see *Hammer, Grocer's and Warehouse.*

Hammer, Packing-case Maker's: see *Hammer, Box Maker's.*

Hammer, Pattern-maker's *Fig. 358*
A light Warrington type head with either a cross or ball pane and a comparatively long, slender handle. Weight of head 4–8 oz.

Used for light nailing and other work when making wooden patterns for the foundry.

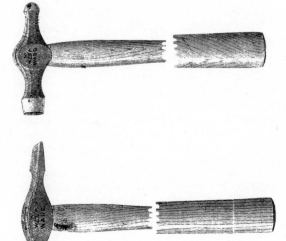

Fig. 358

Hammer, Perth: see *Hammer, Scotch.*

Hammer, Pin
A name sometimes given to lightweight Hammers used for driving small nails or pins. See also *Hammer, Ship's Spike Set.*

Hammer-Pincers (Combination Tool; Hatchet Pincers) *Fig. 359*

Pincers with hollow serrated jaws which are extended on one side to form a hammer and on the other a blunt axe. The lower end of one handle sometimes serves as a screwdriver, and the other as a claw. These tools were commonly imported from Continental Europe. A similar tool was occasionally used by grocers and confectioners to combine shop pincers with a sugar or toffee breaker.

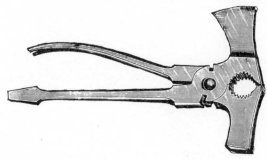

Fig. 359

Hammer, Plough *Fig. 360*
An all iron Hammer often smith made, with a square face and a pane in the form of a nut wrench. Another wrench is forged on the heel of the handle. Used for undoing the nuts which secure the coulter and share, and for hammering the latter into place. This tool is usually carried on the plough beam.

Fig. 360

Hammer, Railway
A name given by Richard Timmins (Birmingham, *c.* 1800) to a large Claw Hammer with solid strapped head, octagonal face, and a strong claw.

Hammer, Ripping
A general term in the U.S.A. for Hammers with a comparatively shallow-curved claw. Used by carpenters for ripping off old cladding or match-board, and for general purposes.

Hammer, Riveting

A general term for a Hammer of small to medium size made in a number of different patterns. All have a cross pane which tapers to a narrow edge suitable for hammering over or spreading the end of small rivets. Some modern Joiner's Hammers are sometimes described as Riveting Hammers.

In the seventeenth and eighteenth centuries rivets were more commonly used by joiners than today, e.g. for fixing fittings such as hinges.

Hammer, Roofing: see *Hammer, Flooring and Roofing.*

Hammer, Roseneath: see *Hammer, Scotch.*

Hammer, Saddler's *Fig. 361*

A light Hammer with a long strapped head similar to the Common Upholsterer's Hammer, but with a solid cross pane instead of a claw. (A claw is sometimes provided at the side of the pane.) The pane is tapered to a narrow edge and can thus be used for driving tacks etc. into places otherwise inaccessible.

Fig. 362

Hammer, Scaffolder's *Fig. 363*

A round face, chamfered neck, broad cheeks, and a round-shouldered axe blade instead of a pane. Used by the erectors of wooden scaffolding.

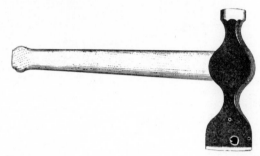

Fig. 363

Hammer, Scotch *Fig. 364*

A solid strap-headed Claw Hammer distinguished from the Kent or Canterbury types by having a somewhat longer neck. The handles often have a bulbous grip. Within the group there are several regional variations which differ only in the detail of strap or shape of chamfering on the head. See also *Hammer, Carrick.*

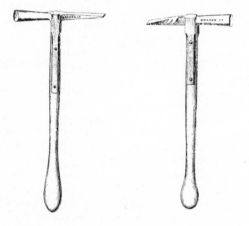

Fig. 361

Hammer, Saw Maker's

Of the several special Hammers used in this trade, one, the Dog Head Hammer, is sometimes found in woodworking shops where it is used for straightening and tensioning Saws. See *Saw Doctor and Sharpener.*

Hammer, Saw-Setting *Fig. 362*

The special Hammer used for setting Saws is described under *Saw Setting Tools.*

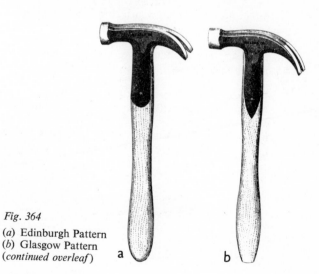

Fig. 364

(*a*) Edinburgh Pattern
(*b*) Glasgow Pattern
(*continued overleaf*)

a   b

231

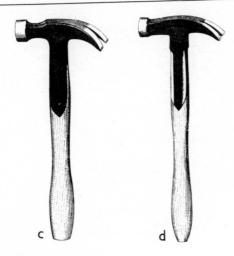

Fig. 364

(c) Perth Pattern
(d) Roseneath Pattern

Hammer, Screw

A name sometimes given to an adjustable Screw Wrench. See *Wrench*.

Hammer, Scrim: see *Upholsterer*.

Hammer, Shackle Pin: see *Hammer, Ship's Maul*.

Hammer, Sheathing: see *Hammer, Ship's Coppering and Sheathing*.

Hammer, Ship's (According to Mr. David D. Murison, Editor of the Scottish National Dictionary, the following names are given to heavy hammers used in northern shipyards: Bilfie; Mundy; Peltie.)
Shipwrights use many of the ordinary Carpenter's Hammers, and in addition the special Hammers and Mauls described in the following entries.

Hammer, Ship's Carpenter's *Fig. 365*

A Claw Hammer with an octagonal face, tapering neck, and strong claw. Probably listed under this name to give Ship Chandlers a distinctive hammer to offer to their sea-going customers.

Fig. 365

Hammer, Ship Clench *Fig. 366*

Typical examples are illustrated. The Scotch Pattern has a round face and cross pane like a Warrington Hammer. For purpose, see *Hammer, Clench*.

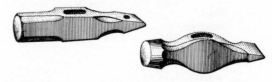

Fig. 366

Hammer, Ship's Coppering and Sheathing *Fig. 367*

(a) *Coppering Hammer*
An eyed Claw Hammer with a long head and 'mushroom' face. In the London pattern the neck is round in section; in the Scotch pattern the neck is deeply chamfered. Weight from $1\frac{1}{2}$ to 2 lb.

This special Hammer is used for fixing copper sheets on the ship's bottom. A small block of wood (or a wad of lead, known as a 'Beater', fitted with a rope handle) is held over the sheet, and this is hammered to shape the sheeting to the curves of the ship. The wide face of the Hammer does not dent the sheet when nailing. The long claw is used for ripping off old sheeting. The exceptional length of the head enables it to be used on sharply concave surfaces where an ordinary Hammer head would not reach. (See also *Punch, Coppering*.)

Copper sheeting was fitted in order to prevent the growth of marine vegetation on the ship's bottom which impeded its way through the water. This was done on both wooden and iron ships. In the case of the latter, wooden sheathing was put on first and the copper sheet nailed over it.

(b) *Sheathing Hammer*
These were Claw Hammers with a round face. Their special use was for fixing wooden sheeting over iron ships as a foundation for the copper sheet which was nailed over it. And when a ship was built for service in northern waters a run of wooden sheathing was sometimes fitted along the water-line to protect the hull from ice.

a

Shoe Hammer

London Pattern Shoe

Cramping Hammer

Fig. 367

Fig. 369

Hammer, Ship Maul (Pin Maul; Spike Maul; U.S.A.: Top Maul) *Fig. 368 overleaf*

A round-faced Hammer with chamfered neck, oval eye, and a heavy, tapering pin at the other end, mounted on a straight handle about 2 ft 9 in long, and made in sizes weighing from $1\frac{1}{2}$ to 8 lb to meet the varying needs of a shipwright's work. The face is used for driving trenails and spikes (large nails). If there is trimming to be done afterwards, the Maul is turned, and the spike driven below the surface of the wood with the pin. The Maul is also used for general purposes, including the releasing of the iron Dogs when launching.

Smith's *Key* (1816) and James Cam's price list records a smaller 13 oz version of the Ship Maul called a Boat Punch. Two larger sizes have checkered faces; one weighing 2 lb is called a Boat Sticking Hammer, and one weighing $3\frac{1}{2}$ lb a Nailing Off Hammer. We do not know the meaning of these terms.

Hammer, Ship's Spike Set (Pin Hammer)

According to an American shipwright, Mr. Robbin Paterson (1966), when spikes had to be driven below the surface, they used a modified type of Ship Maul, called a Spike Set. One man held the tool by its handle with its pin on the head of the spike, while another struck the 'face' with a Beetle. It was usually made by the yard smith; the pin was cupped at the tip to hold the head of the spike or bolt.

Hammer, Shoe *Fig. 369*

These 'mushroom'-faced Hammers, with down-turned panes, are sometimes found in woodworking shops because of the need to repair boots or shoes on the spot.

Hammer, Shoebridge: see *Hammer, Kent.*

Hammer, Sledge *Fig. 370*

A heavy Hammer with flat cheeks and octagonal face and neck, either double-faced or with a cross or straight pane. These heavy smith's Hammers, sometimes weighing as much as 14 lb, are used by tree fellers for driving wedges into the saw cut, by shipwrights when driving wedges to lift the hull in the process of launching, and by wheelwrights for driving spokes into the hubs of heavy cart and wagon wheels. This last work is described by George Sturt (1923): 'He picks up [the spoke] in one hand, and, with sledge-hammer in the other, lightly taps the spoke into its own mortice. Then he steps back, glancing behind him belike to see that the coast is clear; and, testing the distance with another light tap (a two-handed tap this time) suddenly, with a leap, he swings the sledge round full circle with both hands, and brings it down right on top of the spoke .... This way of driving spokes was probably very antique, and, being laborious and costly, it had died out from my shop before I had to retire myself. Hoop-tyres, superseding strakes, had indeed made

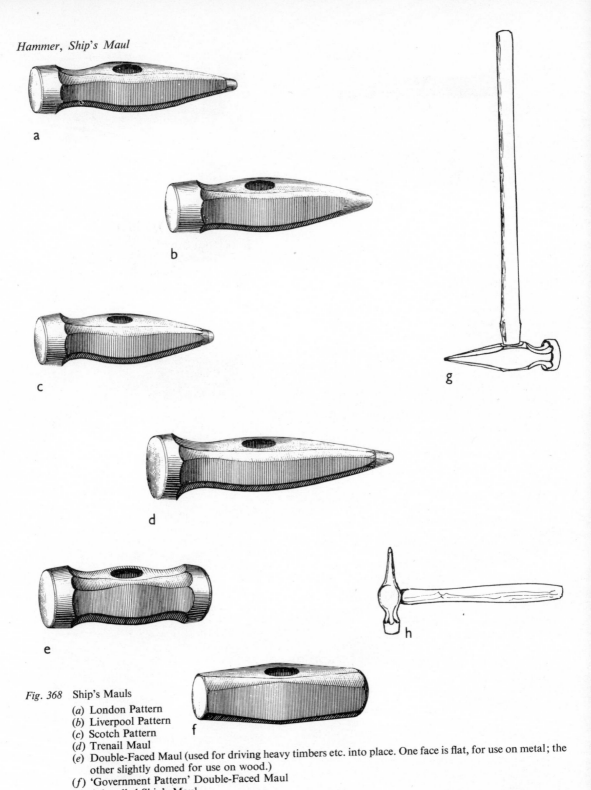

Hammer, Ship's Maul

Fig. 368   Ship's Mauls

    (*a*) London Pattern
    (*b*) Liverpool Pattern
    (*c*) Scotch Pattern
    (*d*) Trenail Maul
    (*e*) Double-Faced Maul (used for driving heavy timbers etc. into place. One face is flat, for use on metal; the other slightly domed for use on wood.)
    (*f*) 'Government Pattern' Double-Faced Maul
    (*g*) A handled Ship's Maul
    (*h*) Shackle-Pin Maul. A small version of the Ship Maul used for 'breaking' a shackle, i.e. knocking the pin out to open it.

such strenuous arm-work less necessary; and the lighter wheels for spring-vans ... did not otherwise need putting together so strongly.' (See also *Wheelwright's Equipment* (2) *Spoke Tools*.)

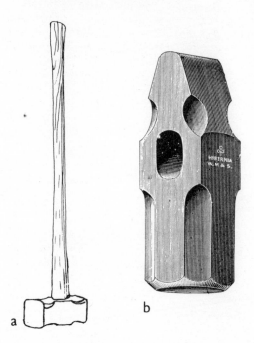

Fig. 370 (*a*) A double-faced home-made Sledge Hammer
(*b*) A straight-pane factory-made Sledge Hammer

Hammer, Spike: see *Hammer, Box-maker's; Hammer, Manchester Spike; Hammer, Ship's Spike Set.*

Hammer, Spinet: see *Pianoforte Maker.*

Hammer, Sprig (Glazier's Flat Head Sprig Hammer; Tilter) *Fig. 371*
The double faces are rectangular and one side of the head is entirely flat, and therefore adapted for driving sprigs (small cut nails) close to the glass in picture framing and similar work.

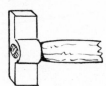

Fig. 371

Hammer, Stanley Nail
A term sometimes applied to all-steel and other modern Claw Hammers made by the Stanley Tool Co. See, for example, *Hammer, Adze Eye; Hammer, Claw.*

Hammer, Staple: see *Staple Puller.*

Hammer, Strapped Head
A term for hammers attached to the handle by side straps. See *Hammer; Hammer, Claw.*

Hammer, Tack *Fig. 372*
A light Hammer of the Exeter, Warrington, and other types, provided with either a claw or cross pane. The handle is often bulbous at the heel end. Used for driving small nails.

Fig. 372

Hammer, Telephone (Telephone Engineer's Hammer) *Fig. 373*
A very light Warrington-pattern Hammer with cross pane and a long, thin handle. Used for driving staples over telephone wires and for similar work.

Fig. 373

Hammer, Tuning: see *Pianoforte Maker.*

Hammer, Upholsterer's *Fig. 374*
Very light strapped hammers with a claw. See *Upholsterer* for further particulars. Those illustrated here are the London and Benwell patterns.

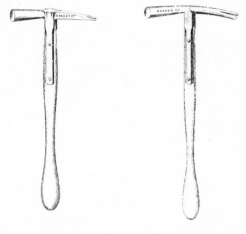

Fig. 374

Hammer, Veneering *Fig. 375*
Not used for striking but for smoothing and squeezing out excessive glue when fixing Veneers. See *Veneering and Marquetry*.

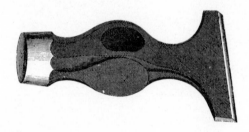

Fig. 375

Hammer, Wall: see *Hammer, Gardener's.*

Hammer, Warehouse: see *Hammer, Grocer's and Warehouse.*

Hammer, Warrington (Joiner's Hammer; Lancashire Pattern Riveting Hammer; Scots: Pin Hammer) *Fig. 376*
A Hammer made in a full range of sizes (5–33 oz) as well as in the lighter tack and pattern makers' sizes. It has a round face with a neck chamfered on

each side of the rounded cheeks. The cross pane is symmetrical and tapers down on both sides to a rounded tip for starting 'pins' and for riveting. Used as a general-purpose Hammer by joiners, carpenters, cabinet makers, and other tradesmen.

A study of the sources and enquiries in Warrington itself have failed to establish when and why this type of Hammer came to be known as the 'Warrington'. It is so named in the Sheffield *List* of 1862, the earliest reference so far encountered. The handled example shown here was made in *c.* 1854 by John Constable, a blacksmith at Hardwick in Norfolk.

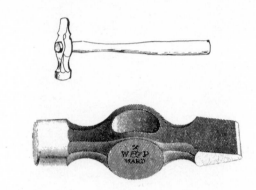

Fig. 376

Hammer, Wheelwright's: see *Hammer, Framing; Hammer, Sledge;* Dolly Hammer under *Wheelwright's Equipment* (3) *Tyring Tools.*

Hand Protector or **Leather:** see *Brush Maker; Sailmaker* (*Palm*).

Hand Screws: see *Cramp, Hand Screws.*

Handed Pairs
Tools such as Planes and Routers made in pairs to work with the fence on the left or right side of the work according to the direction of the grain. (See also *Matched.*)

Handle Maker *Fig. 377*
Today many tool handles are turned on special lathes. The following notes refer to the practice of making handles from cleft timber which is afterwards trimmed by hand.

The best handles for tools such as Axes, Hatchets, and Pickaxes are usually made from cleft ash and shaped with the Drawing Knife and Spokeshave. While being shaped, the work is sometimes held on a 'Fiddle', a device which is illustrated under *Wheel-wright's Equipment* (2) *Spoke Tools*.

The long straight handles of rakes, hoes, and pitchforks are called 'stails' or 'tails'; the curved handles for scythes are called 'snaithes' or 'sneads'. They are round in section, and are made from ash poles cleft into segments and then trimmed with a Drawing Knife and finished with a Rounder Plane.

Setting Pin (Bending Brake; Bending Horse)
A stout post with horizontal pegs of wood or iron protruding from it. The handles are first made pliable by steaming and are then bent to shape on this device. They can be either made straighter, as for pitchfork handles, or more crooked, as in the case of scythe handles. After bending, the poles are left to dry while strained between pegs on a frame, in order to retain their shape.

Tools used for making handles include the *Plane*, *Rounder* and a special Drawing Knife, illustrated below and described under *Drawing Knife, Handle Maker's*. It is used for cleaning up the inside corners of spade handles.

Fig. 377

The terms 'caulked handle' and 'fawn-foot handle' are explained by the diagram under *Axe*. Notes on special handles will be found under *Adze; Auger; Axe; Chisel; Spade Maker*.

Handles: see under various entries for the tools concerned, e.g. *Adze; Auger; Axe; Chisel; Hammer*.

Hat-Block Maker
A maker of wooden moulds for shaping hats. Metal blocks are often used for men's hats; but to meet the innumerable shapes demanded by fashion, most women's blocks are made in wood, usually sycamore. Ordinary woodworking tools are used, including large Gouges, Rasps, and Carving Chisels.

Hatchet (Chopper)
Hatchets are light Axes, from 1 to 3 lb in weight. They are made in the more usual patterns: Kent, Scotch, Wedge, etc., for general use, and also for special trades and purposes such as Lathing and Shingling. Note: If the reader cannot find a particular Hatchet under its usual name, reference should be

made to the entries under *Axe*, where a diagram of a typical Axe or Hatchet head will also be found.

Hatchet, Australian *Fig. 378*
The blade has a waving curve in outline, and there is a claw on the poll.

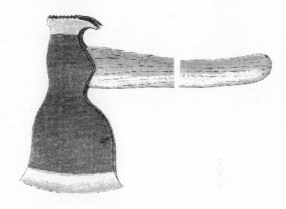

Fig. 378

Hatchet, Broad: see *Hatchet, Side*.

Hatchet, Canadian: see *Hatchet, Wedge*.

Hatchet, Claw *Fig. 379*
A name given to Hatchets provided with a claw for extracting nails. The lugs are usually round and the claws protrude either above or below a heavy square poll. Sometimes used for shingling (see Hatchet, Shingling).

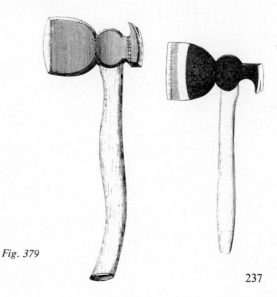

Fig. 379

237

Hatchet, Cooper's: see *Axe, Cooper's.*

Hatchet, Fantail *Fig. 380*
A flared, straight-sided blade with no poll.

Fig. 380

Hatchet, Fireman's: see *Axe, Fireman's.*

Hatchet, Forester's: see *Hammer, Marking.*

Hatchet, Gardener's: see *Hammer, Gardener's.*

Hatchet, Hewing
A term used by Mercer (U.S.A., 1929) for small Axes used for rough-surfacing, splitting, chopping, and nailing.

Hatchet, Hunter's
A name sometimes given to a light Hatchet of the Kent or wedge type (see *Axe, Wedge*). Another form, illustrated by Arnold & Sons (London, 1885), has a scale handle into which folds a small tapering saw (see *Saw, Hunter's*). Used by hunters and others as a multi-purpose portable tool.

Hatchet, Kent (Carpenter's Hatchet) *Fig. 381*
The usual form of bench or general-purpose Carpenter's Hatchet. The head has the same shape as that of the Kent Axe, but is thinner and lighter.

Fig. 381

238

Hatchet, Lathing (Lathing or Lath Hammer; Latch Hatchet) *Fig. 382*
The blade is always flat along the top, with the lower edge usually notched for pulling nails. The poll may be polygonal, square, or occasionally round, and is usually checkered on the face. The Lathing Hammer is the same tool, but its head is fixed to the handle with straps instead of being eyed. Some continental Lathing Hammers are similar to the Boxmaker's Hammer – with one claw short and the other long and pointed. Gilpin (Cannock, 1868) illustrates what appear to be Lathing Hatchets but have longer blades. They are listed in roughly ascending order of blade length, as Brick, Plastering, and Sheathing Hammers. Only the Plasterer's Hammer has a nail-pull in the lower edge of the blade.

Used by plasterers for cutting to length and nailing plaster laths. The flat top enables the tool to be used close to a ceiling. There are many variants of which those illustrated are typical.

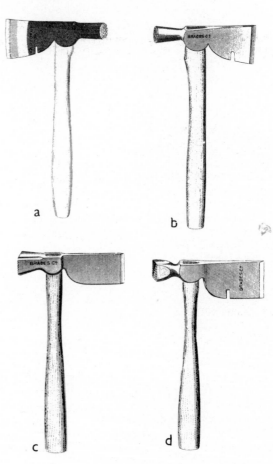

a

b

c

d

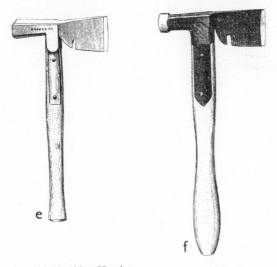

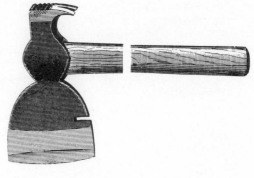

Fig. 383

e

f

Fig. 382   Lathing Hatches
   (*a*) Polygonal poll
   (*b*) Round poll
   (*c*) Square poll
   (*d*) Square poll, necked

   Lathing Hammers
   (*e*) Square poll
   (*f*) Scotch pattern with round poll

Hatchet, Side (Broad Hatchet) *Fig. 384*
A Kent pattern with the blade ground on one side only.

Fig. 384

Hatchet, Scotching
A name mentioned in the Apprentice Indenture Records (Bristol, 1532–1646) which may relate to the process known as 'scutching' – a term used by Scottish coopers for levelling the joints in the head of a cask by reducing the thickness of one of the pieces with an Adze (*cf. Scutch*).

Hatchet, Shingling *Fig. 383*
A round-shouldered blade usually provided with a notch in the lower edge for nail pulling. Among several variants there are two chief patterns – one with a hammer-poll (often checkered), the other with a claw on the poll, usually below it. Used for trimming and nailing roof shingles. See *Shingle Maker*.

Hatchet, Splitting Out: see *Axe, Chairmaker's*.

Hatchet, Sugar (Sugar Cleaver) *Fig. 385*
This tool might be mistaken for a Lathing Hatchet but for the fact that the blade is sometimes decorated with a moulding on the shoulders. Another pattern, known as a Sugar Cleaver, looks like a small Butcher's Cleaver, some all-iron, others with a turned handle. Used up to about 1900 for breaking pieces off a loaf (or cone) of sugar. The domestic Sugar Nippers were used for breaking these pieces into smaller pieces for the table.

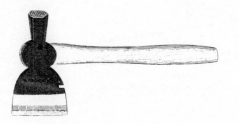

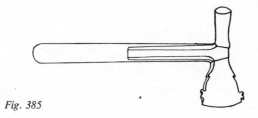

Fig. 385

239

Hatchet, Tomahawk *Fig. 386*

A name given by Gilpin (Cannock, 1868) to the following two Hatchets, presumably made for export:

(*a*) A Wedge Hatchet listed as an 'American pattern Tomahawk'.

(*b*) A long bladed Hatchet listed as a 'Colonial pattern Tomahawk'.

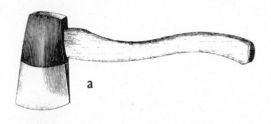

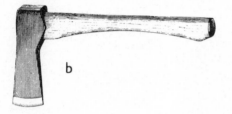

Fig. 386

Hatchet, Wedge (Canadian Hatchet; Hunter's Hatchet) *Fig. 387*

A light wedge-type head fitted to a short handle.

Fig. 387

Hawk: see *Builder's Tools.*

Hawksbill

Term for a hooked cutter. See, for example, *Plane, Cooper's Croze.*

Heading Swift or **Plucker:** see Heading Swift under *Shave, Cooper's.*

Heading Tools

Tools used by coopers when making the heads of casks. These include: *Axe, Cooper's; Brace, Cooper's; Drawing Knife, Cooper's Heading;* Heading Board under *Cooper* (2) *Furniture; Saw, Head; Shave, Cooper's* (*Heading Swift*)*; Vice, Cooper's.*

Heel

Of tools – the rear part, e.g. of a Saw or Plane.

Helpmate: see *Froe.*

Hoe (Shipwright's): see *Caulking Tools.*

Hog-Backed

Convex; rising in the middle like a hog's back.

Holdfast *Fig. 389*

The simplest form was an iron bar in the shape of a figure-of-seven. A more convenient factory-made tool consists of a stout iron pillar 10–18 in long with a short arm set at right angles at one end of which a curved lever is pivoted, with a screw at one end and a swivel-shoe at the other. The pillar is inserted in a slightly larger hole in the bench top, the work placed under the shoe, and when the screw is tightened up the work is held fast to the bench. The figure-of-seven type had to be driven tight by hammering from above and released by knocking behind the head. Holdfasts of this type were known to the Romans and are in use today, not only on the bench, but also, for example, on the platform at one end of a saw pit to hold smaller pieces, e.g. felloes, while being sawn.

The following extract from Moxon, p. 64 (London, 1677) is an interesting example of an early technical description:

'The Hold-fast, let pretty loose into round holes ... in the Bench: Its Office is to keep the Work fast upon the Bench, whilst you either Saw, Tennant, Mortess, or sometimes *Plain* upon it, etc. It performs this Office with the knock of an *Hammer,* or *Mallet,* upon the *head* of it; for the *Beak* of it being made crooked downwards, the end of the *Beak* falling upon the flat of the *Bench,* keeps the *head* of the *Hold-fast* above the flat of the *Bench,* and the *hole* in the *Bench* the *Shank* is let into being bored straight down, and wide enough to let the *Hold-fast* play a little, the *head* of the *Hold-fast* being knockt, the point of the *Beak* throws the *Shank* a-slope in the *hole* in the *Bench,* and presses its back-side hard against the edge of the *hole* on the upper Superficies of the *Bench,* and its fore-side hard against the opposite side of the under Superficies of the *Bench,* and so by the point of the *Beak*, the *Shank* of the *Holdfast* is wedged between the upper edge, and its opposite edge of the round hole in the *Bench*.'

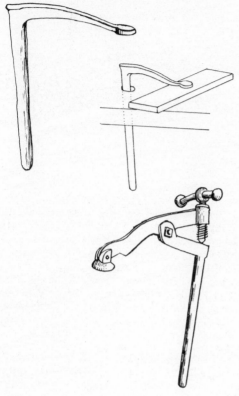

Fig. 389

Holding Devices: examples will be found under *Cramp; Holdfast; Plane Maker; Saw Sharpening: Holding Devices; Vice; Woodland Trades.*

Hollowing Tools: see list under *Rounding and Hollowing Tools.*

Holzaxt: see *Axe, Split.*

Hone
A smooth hard stone used to give a keen edge to a cutting tool.

Hook Rave: see Hoe under *Caulking Tools.*

Hook Tools
A name given to various tools, including those to be found under *Chisel, Turning; Knife, Hooked; Shave Hook; Timber Handling Tools* (*Woodhook*).

Hoop Maker
The making of wooden hoops for barrels and casks is now rarely seen in Britain. At one time hoops were used for hooping all kinds of dry casks; today they are used occasionally for replacing wooden hoops on imported wine casks. The raw material is usually hazel, but chestnut is sometimes used; the stronger Truss Hoops used by coopers for shaping casks are made of ash.

The first process is to cleave the rods. This is usually done with a small Bill Hook or Froe, but sometimes with a Bond Splitter, while the rod is held in a Brake. The bark is left on, but the inner wood is shaved off with a Drawing Knife.

The preliminary bending is done when the wood is green or, if later, after soaking to make the rods pliable.

The split rods are bent on a Hoop Bending Easel. This is a wooden frame consisting of eight radiating spokes with a series of holes bored in each, into which wooden pegs are inserted. The hoop is bent round inside the pegs to a circle of the required size.

For other tools used in hoop making, see *Drawing Knife, Hoop Maker's; Woodland Trades.*

For tools used for fitting the hoops, see *Cooper (3) Hooping Tools.*

Hoop Tightener (Strap or Band Tightener)
A strong iron bar with one end turned at right angles to form a short head which is slotted to receive a steel strip. The tool is used by box and case makers for stretching steel or plastic straps over a box or case before nailing it on. A modern version is designed to tighten and fasten in one operation.

Hooping Grate: see Tyre Oven under *Wheelwright's Equipment (3) Tyring Tools.*

Hooping Tools: see *Cooper (3) Hooping Tools.*

Hopper
A form of trough, diminishing in diameter towards the base.

Hoppus Measure
A set of tables used by sawyers and timber merchants for calculating cubic and superficial size etc. They were originally published by E. Hoppus in the eighteenth century (see Bibliography and References).

Horizontal Square: see *Square, Coachbuilder's.*

Horse, Donkey, or Mare
Many kinds of workshop equipment are given one of these names including those described under the following entries:

Bending or Setting Horse (under *Handle Maker*)
Broom Horse (under *Broom Maker*)
Brush Maker's Donkey (under *Brush Maker*)
Buhl Horse (under *Veneering and Marquetry*)
Chairmaker's Donkey (under *Donkey*)
Donkey's Ear Shooting Board (under *Mitre Appliances*)
Marquetry Horse (under *Veneering and Marquetry*)
Painting Horse (under *Painting Equipment*)
Saw, Donkey
Saw Sharpening Horse (under *Saw Sharpening: Holding Devices*)
Sawing Horse
Tining Horse (under *Rake Maker*)
Wheel Horse (under *Wheelwright's Equipment (1) Furniture*)

Horsing Iron: see *Caulking Tools* (*b*).

Hot Rasp: see *Rasp, Two-Handed.*

Housing
A trench or groove in a piece of wood to hold or support another piece, e.g. the end of a shelf.

Howel (Howelling tool)
A term seldom used in Britain, but in the U.S.A. it is the name given to the special cooper's Plane we call a Chiv or Chive, and also to the smooth surface produced by the Chiv on the inside ends of the staves in which the croze groove is subsequently cut. In the sixteenth and seventeenth centuries the Howel was almost certainly an Adze and not a Plane.

The word appears in the following entries: *Adze, Cooper's Howel; Adze, Chairmaker's Howel; Cooper* (Historical note)*; Drawing Knife, Cooper's Chamfering; Drawing Knife, Cooper's Jigger; Plane, Cooper's Chiv.*

Hub Boring Engine (Boxing Machine; Box Engine; Bushing Engine; Boxing Knife; Nave Borer) *Fig. 390*
This tool is designed to bore out a wheel-hub to make room for an iron bearing known as the 'box'. It is used as an alternative to the wheelwright's Chisel and Gouge. It works on the same principle as the machines employed to bore out the interior of an engine cylinder or gun. A screwed boring-bar, on which small cutters are mounted, is passed through a hole previously bored through the axis of the hub. When the boring-bar is turned the cutter revolves, and being self-feeding, it is drawn slowly through the hub, cutting off an internal shaving as it goes. The bar is held centrally within the hub by the dogs (or 'spiders') which are driven (or nailed) on to each end of the hub. The bar passes freely through one of the dogs but is threaded into the other, so that the cutter is drawn along as the cross-handle is turned. Three variants are illustrated.

Hurdle Maker (1) THE TRADE
Hurdles are portable wooden frames or screens mainly used as temporary fences. There are two types: gate hurdles made from cleft poles, and wattle hurdles which are made basket-like from interwoven rods.

Both types are employed for folding sheep and other animals. Wattle hurdles have the longer history for they are known to have been used before the Christian era for making huts or cart sides, and from then onwards for all kinds of hut and house construction, including the still surviving 'wattle and daub' infilling of wooden framed houses.

Gate hurdles are usually made in the village, often in the hurdle maker's own house or yard; wattle hurdles are more often made in the woodlands where the material is harvested.

Hurdle Maker (2) GATE HURDLES *Figs. 391; 392*
The material used is coppice poles of oak, ash, elm, willow, or chestnut, according to locality, cleft with the Froe in the Brake, and finally trimmed and pointed with Axe and Drawing Knife. The gate hurdle is 6 to 8 ft long and 3 to 4 ft high and consists of six or seven horizontal bars or 'slotes', two heads or end pieces, and two diagonal braces. The bars fit into mortices in the heads, made by boring two $\frac{1}{2}$ in holes at each end, and removing the core between by means of a Mortice Chisel or Mortice Knife. Sometimes a jig or frame is used on a low bench for assembly, but often it is done on the ground, the rails being fitted into the heads and pinned with a nail or oak peg through the projecting tenon. The braces are fastened with long nails clenched over. The example illustrated is a 'pig' hurdle and was made at Langley in Hertfordshire. It is more heavily built than a sheep hurdle. But the method of construction is the same, and also the names of the parts.
The *Shepherd's Bar* (also called a Poll Prytch, Fold Shore, Fold Drift, or Fold Pitcher) *Fig. 391 (f) overleaf*

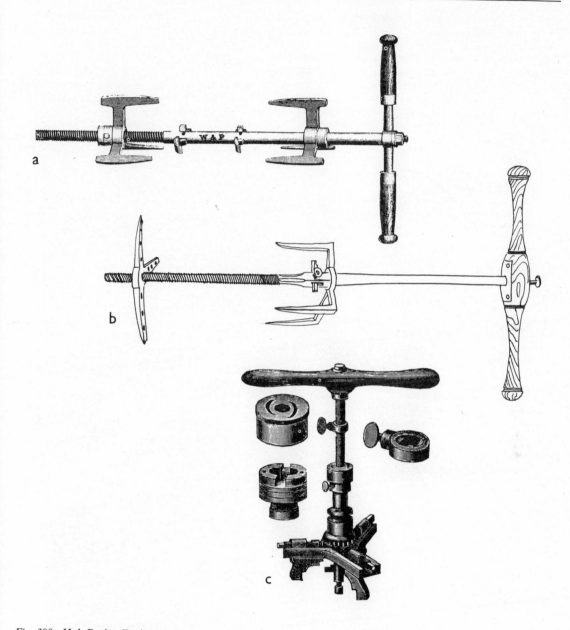

Fig. 390 Hub Boring Engine

    (*a*) A typical factory-made example
    (*b*) Made by the village smith
    (*c*) A late nineteenth-century example with a self-centring chuck

This is a heavy iron Crow Bar used for making a hole in the ground to take posts for hurdles and for 'dibbling' holes to take the feet of the hurdles themselves. The point is swollen to prevent its sticking in the ground. The hollow depression above the point is used for hammering the tops of stakes.

The tools used include some of those listed under *Woodland Trades*, including the splitting tools and Brakes, but also the following more specialised tools (*Fig. 392*):

Fig. 391   A typical Pig Hurdle

(*a*) Head
(*b*) Slotes (Ledges; Cross bars; Spars; Slats)
(*c*) Braces (Stretchers; Stays)
(*d*) Upright (Right-up)
(*e*) Coupling or Shackle
(*f*) Shepherd's Bar (or Fold Shore)

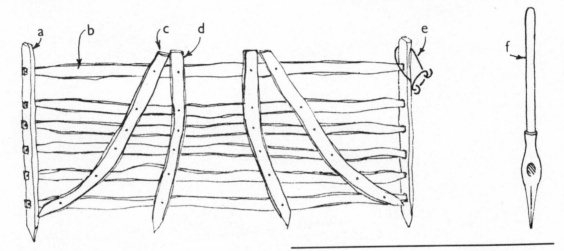

1. *Mortising Stool*

A home-made stool about 5 ft long and 2 ft high. That illustrated, from Oxfordshire, is designed to hold heads when drilling and mortising.

2. *Mortising Axe or Knife* (Tomahawk; Dader; Two-bill; Twybill; Twyvel)

A double-bladed tool, 11–15 in across the head, with a short handle. One end of the blade is a pointed triangular knife with a thick back; the other end is chisel shaped. It is used for cutting mortices in the heads, as an alternative to the use of a Chisel. Holes are bored at each end of the mortice and the wood between them cut with the knife along the grain; the other end picks out the wood. In a Midland pattern the knife and hook are carried at the extremities of a U-shaped forging. In another form the 'twybill' shape is abandoned and the knife is held in a turned handle, another handle being fitted to the side of the tool. There is therefore no chisel 'pick' in this pattern.

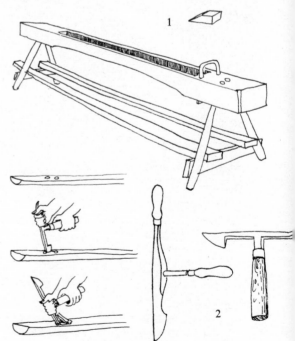

Fig. 392   Hurdlemaker's Stool and Mortising Knife

3. *Hurdlemaker's Brace*

Hurdlemakers use ordinary carpenter's Braces, but a special pattern was made in wood or metal with a specially long head, and an extra long foot to carry the thumbscrew chuck. This extra length enables the hurdlemaker to work without so much stooping when the work is laid on the ground or on a low bench.

Hurdle Maker (3) WATTLE HURDLES

Wattle hurdles are made from coppice hazel or willow, some of which is cleft and some left in the round. The hurdle is started by fixing upright 'sails' or 'shores' in a Hurdle Mould, the two end pieces being left in the round and those in between being of cleft wood. Weaving is started with the round rods and finished with several rows of round rods braided to give a firm top edge. The finished hurdle is usually 4–6 in long and about 3 ft high, but larger hurdles are made for special purposes. Willow hurdles are made in a similar manner, but the sails or shores are usually held upright in a vice formed by two stout square timbers bolted together.

The tools used, including Bill Hook and Maul, are described under *Woodland Trades*. The only specialised tool is the *Hurdle Mould*. This is a half-log or square balk about 7 ft long and 6–8 in wide, slightly curved in plan. Ten holes are bored in the top, the end holes being about 6 ft apart, the others spaced equally between them. The vertical sails are stuck into these holes which enables the horizontal rods to be woven in and out of them. When built in a curved mould the hurdles are said to be less liable to twist or wind when the material dries.

Hutchit: see *Boat Grip*.

I

In Cannel
The inside grinding bevel of a Gouge. See explanation under *Gouge*.

Inlay; see *Veneering and Marquetry*.

Inshave (Inside Shave): see *Drawing Knife, Cooper's Round Shave; Shave, Cooper's*.

Intarsia: see *Veneering and Marquetry*.

Iron
Of a plane or similar tool – the cutter (see *Plane*).

Iron, Bottoming: see *Shave, Chairmaker's*.

Iron, Caulking: this category of tools includes the following irons: Bent, Blunt, Butt, Cleaning, Crease, Deck, Dumb, Horsing, Jerry, Making, Meaking, Reaming, Reaping, Set, Sharp, Single or Double, Spike, Trenail or Trunnel. See *Caulking Tools*.

Iron, Cleaving: see *Froe*.

Iron, Erasing: see *Drawing Knife, Cooper's Round Shave; Plane, Box Scraper; Scraper*.

Iron, Flamming or Flammard: see *Froe*.

Iron, Flat: see *Veneering and Marquetry*.

Iron and Steel used in tool making
The following notes relate to the different kinds of steel used in the making of edge and other hand tools, and to some of the metallurgical terms encountered in the catalogues of the tool makers. (For non-ferrous metals used see *Brass* and *Gun Metal*.)

Annealing. A heat treatment applied to steel and other metals in order to relieve internal strains and improve strength, elasticity, and ductility to meet the stresses to which it is subjected in service. It is usually done by heating and holding at a certain temperature, followed by slow cooling.

Blister and Shear Steel. These terms relate to an early method of converting iron into steel by impregnating wrought iron with carbon in a charcoal furnace. This method has long been discontinued.

Carbon Steel. A general term for a good quality steel which is usually made in small crucibles or, to-day, in electric arc furnaces. The amount of carbon in the steel determines its properties. For instance, a comparatively high proportion of carbon (1·2%) would be suitable for the production of a fine cutting edge, as is required in razors, wood-working cutting tools, and files. A carbon content of not more than 0·7% is used when a tougher type of steel is needed, e.g. for the production of heavy Chisels (such as those used by shipwrights and millwrights), for Screwdrivers and Crow Bars, and also for Plane Irons when a harder quality of steel is welded on to the cutting edge.

Case Hardening. A process for hardening the surface of mild steel to resist wear, while retaining a softer core for toughness and resistance to shock. The articles are packed and heated in carbonaceous materials and subsequently quenched and tempered.

Cast Iron. An alloy of iron with up to about 4·0% carbon. It can be readily cast into shapes, but has limited strength and ductility and cannot be forged. A good quality cast iron is used for the body of modern metal planes, and for the base-plates of machines.

Cast Steel. The expression 'cast steel' was used in nineteenth-century toolmakers' catalogues to emphasise the fact that it was made in crucibles and then cast into moulds before rolling, instead of the more primitive method of impregnating wrought iron with carbon (see Blister Steel above). The process was invented by Benjamin Huntsman in Sheffield *c*. 1740. The superiority of this steel for cutting tools was so outstanding that the description 'cast steel' was taken to signify a tool of high quality.

Forging. The operation of shaping hot metal by means of hammers. This is the basic method of manufacturing iron and steel hand tools. In more recent times the process of hand-hammering has declined with the introduction of drop-forging, the Steam Hammer, and the Rolling Press.

'High Speed' and Special Steels. High Speed Steel was introduced about 50 years ago as an improvement on the old carbon steel. This was also made in crucibles and it was found that by the addition of certain alloying metals, such as chrome, vanadium, tungsten, and molybdenum, a very tough and heat-resisting cutting edge could be obtained. High Speed Steel is used when the cutting surface acquires a high temperature through friction – for example, for metal-cutting lathe tools and certain wood-working machine knives.

Today steel is made with many different constituents for various purposes. One of the most outstanding is the chrome-vanadium steel used where toughness and strength (and not a fine cutting edge) are required, e.g. for motor car springs, Crow and Wrecking Bars, Wrenches, and Screw-driver blades. The use of tungsten carbide-tipped tools is described below.

Malleable Iron. A type of cast iron in which the structure has been modified by heat treatment to impart a measure of ductility.

Mild Steel. A good quality low carbon steel mainly used for structural work, and for a multitude of manufacturing uses including motor-car bodies. Owing to its low carbon content it cannot be hardened, but it can be readily forged or bent. It is used by toolmakers for parts of tools or equipment not required for cutting.

Silver Steel. A soft and ductile steel with a low carbon content. It is supplied ground to a bright finish in small-diameter rods. It is useful for model makers and in laboratories. It cannot be hardened owing to its low carbon content.

Steeling. Until the mid-nineteenth century, steel was used sparingly since its production was expensive. Hardness required for edge tools was obtained by either (*a*) carbonisation: the edge was heated in a charcoal fire and then rapidly quenched in cold water – a method known in pre-historic and Roman Britain, or (*b*) by the equally old method of 'steeling', i.e. thin strips of good quality steel were let into the edge of a wrought-iron blade and joined by forge-welding in the fire.

Tempering (Drawing). The reheating of hardened steel to reduce its hardness. Steel tools are hardened by heating to about 760°C followed by quenching in oil or water. They are tempered by reheating to a predetermined temperature (according to the future requirements of the tool) and subsequent cooling in air. This renders the tool less brittle yet still hard enough to maintain a cutting edge.

Tool Steel. A general term for a hard, tough steel from which good cutting tools are made.

Tungsten Carbide. This is an exceedingly hard material which can be cemented to the tips of cutting tools such as Twist Drills or Circular-Saw teeth. This enables the tool to be used on hard materials for a much longer period before blunting.

Wrought Iron. A tough, ductile iron which contains a very small amount of other elements, and which in the process of manufacture retains a proportion of slag. Subsequent working produces the characteristic fibrous structure which gives the appearance of 'grain'. Used for making gates, forks, chains, and many earlier tools not requiring a cutting edge. Now often replaced by mild steel.

Iron, Veneering: see *Veneering and Marquetry.*

J

Jack
A name given to several different tools including the following: Jack Fork Stail, under *Plane, Rounder;* Jack Staff under *Millstone Dresser; Lifting Jacks; Plane, Forkstaff; Plane, Jack.*

Jamb Duster: see *Brush; Workshop.*

Jarvis, Barrel: see *Shave, Cooper's (Downwright).*

Jarvis, Wheeler's: see *Shave, Jarvis.*

Jennet (Jénet; Ginnet)
An early term for a Carpenter's or Shipwright's Adze.

Jenny: see *Calliper, Jenny Leg.*

Jerry Iron: see *Caulking Tools.*

Jig: see *Patterns, Templates, and Jigs.*

Jigger
A name given to a number of different tools, including *Drawing Knife, Cooper's Jigger; Router, Coachbuilder's Jigger.*

Jim Crow
A name given to a number of tools, including a Screw Cramp used for straightening iron rails or bars.

Jock
Scots term for Callipers with straight legs.

Joiner: see *Carpenter.*

Joint Cramp: see *Dog, Joiner's.*

Jointer: see Jointing Tools under *Pipe and Pump Maker; Plane, Cooper's Jointer; Plane, Jointer.*

Jumper (Billy; Devil or Devil's Tail; Knocking-up Iron; Raising Iron) *Fig. 393*
A heavy round-iron rod about 3 ft 6 in long, curved round at one end to nearly a right angle. Introduced through the bung hole of a cask, it is used by coopers to lever the head into position if it sticks below the level of the croze channel.

Another tool for raising heads if they become jammed is known as a Dutchman. Invariably home made, it consists of a 10–12 in length of hoop-iron, turned up at one end and studded with three or four protruding rivet heads. The hooked end is forced underneath the edge of the head where it rests inside the cask, and one end of a Flagging Iron is caught under one of the protruding rivets and then levered upwards. (See also *Vice, Cooper's.*)

Fig. 393

Jumper, Wall: see *Drill, Wall.*

248

K

Kerf
The narrow slot made by a saw.

Kevel
A northern term for a heavy hammer.

Kluvie
Scots term for a Claw Hammer.

Knape
A cooper's tool mentioned in early Apprentice Indenture Records (W. L. Goodman, 1972). Identity unknown.

Knee
Of timber – a naturally crooked piece. Such pieces were sought out and kept for making into curved parts of wagons or ships.

Knee Vice: see Besom Grip under *Broom Maker*.

Knife *Fig. 394.*
The ordinary Knife is not much used by the woodworking trade, but the term 'knife' is applied to many other cutting tools. For instance, coopers refer to Drawing Knives as 'Knives', e.g. Backing Knife, Hollowing Knife, Heading Knife, and so on. (These will be found under *Drawing Knife, Cooper's.*) Glaziers refer to their special tools as Chipping, Hacking, and Putty Knives; woodland workers have a splitting tool which they call a Riving Knife; the long, guillotine-like instruments used by clog-sole makers, rake-tine makers, and others are called Bench or Stock Knives; and certain scribing tools are referred to as Marking or Race Knives.

By cross-reference we have endeavoured to assist the reader in finding a tool which has acquired the title of Knife in addition to some other name; and we have entered below a description of knife-like tools used in the woodworking and allied trades.

Knife, Backing: see *Drawing Knife, Cooper's Backing.*

Knife, Basket-Maker's *Fig. 395*
Knives used by basket makers are described under *Basket Maker*. A Picking Knife and Shop Knife are illustrated here.

Fig. 395

Knife, Belly: see *Drawing Knife, Cooper's Hollowing.*

Knife, Bench (Blocking Knife; Paring Knife; Peg Knife; Stock Knife; Sway Knife) *Fig. 396*
The blade is forged to an iron bar varying in length from 1 to 4 ft. This bar has a hook at one end, and a handle, often T-shaped, at the other. The hook engages with an eye-bolt fixed to a bench or stool to form a hinge (illustrated under *Knife, Clogger's*).

In operation, one hand holds the wood to be shaped, while the other, by moving the Knife up and down, pares away the unwanted wood, using the bar as a lever.

The larger of the two illustrated is used by cloggers and by brush and last makers; the smaller sizes are used for shaping rake tines, tent pegs, wooden spoons, etc.

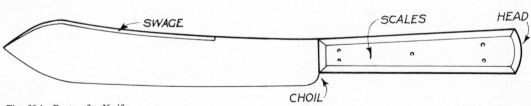

Fig. 394 Parts of a Knife

249

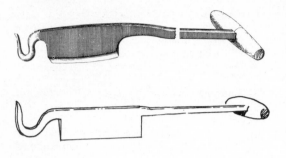

Fig. 396

Knife, Blocking: see *Knife, Bench.*

Knife, Chamfering: see list under *Chamfer.*

Knife, Chip Carving *Fig. 397*
Knives of various patterns: some with straight or skewed blades and others with the cutting edge inclined at 45° or 60° to the centre line of the handle. They are used to produce a form of decoration known as chip-carving. This consists of a series of recesses, usually of reversed pyramidal form.

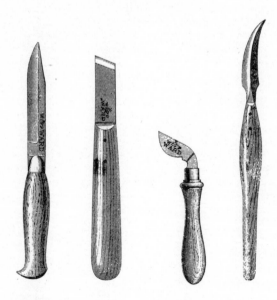

Fig. 397

Knife, Chipping: see *Knife, Hacking.*

250

Knife, Chisel (Stripping Knife) *Fig. 398(a)*
A blade with the end ground for scraping. Used by painters for removing old paint.

Knife, Circular: see *Drawing Knife, Cooper's Round Shave; Knife, Hooked.* (Not to be confused with the Circular or Moon Knife used by tanners and curriers.)

Knife, Cleaving
A form of small Bill Hook used for splitting lath and rods. See *Woodland Trades; Lath Maker.*

Knife, Clogger's *Fig. 399*
A Bench Knife is used for cutting the rough sole blanks for clogs, and until recently this work was carried out in the woodlands. The term Clogger's Knife applies to the following special Bench Knives which are used by the clog maker after the sole blanks are delivered to him. They are about 4 ft long, with differing blades. (See also *Clog-Sole Maker.*)

(a) Clogger's Paring Knife
A straight blade about 13 in long and 3½–4 in wide. Used for paring off waste and trimming the flat surfaces.

(b) Clogger's Hollowing Knife
A curved blade 4–6 in long and 4 in wide. Used for hollowing the upper side and edges of the sole.

(c) Clogger's Gripper
A narrow V-shaped tool, which is fixed in place of the blade. Used for cutting the channel round the edge of the sole into which the leather uppers are nailed.

(d) Clogger's Stool
The stool on which the above knives are mounted. The clogger uses one hand to hold the wood to be shaped, and with the other pares away the unwanted wood by 'levering' the knife up and down.

Knife, Cooper's
A name given to a number of cooper's tools including the following: *Drawing Knife, Cooper's;* Hoop Notching Knife under *Hooping Tools;* Chincing Iron under *Flagging Iron; Knife, Cooper's Whittle.*

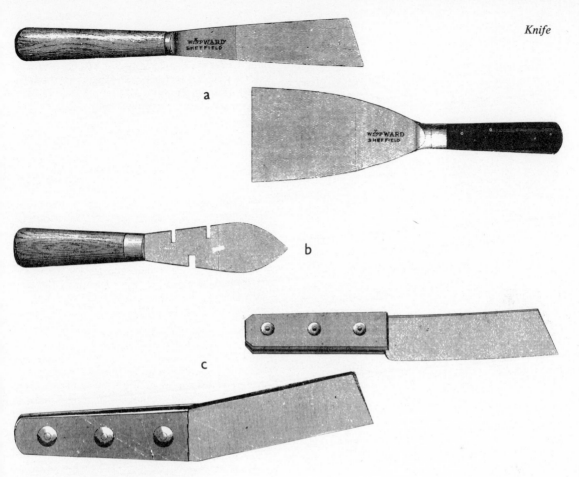

Fig. 398 Painter's and Glazier's Knives
    (*a*) Painter's Chisel Knives
    (*b*) Glazier's Stopping Knife
    (*c*) Glazier's Hacking or Chipping Knives

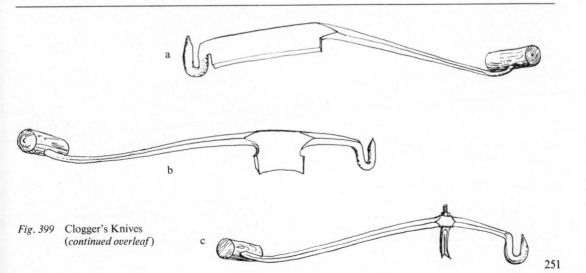

Fig. 399 Clogger's Knives
    (*continued overleaf*)

251

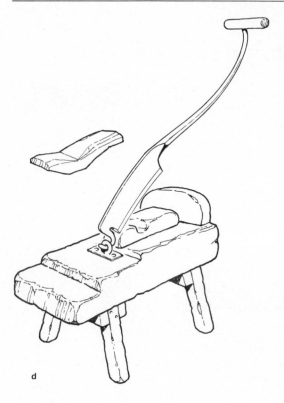

d

Fig. 399 Clogger's Knives and Stool

Knife, Cooper's Whittle (Hand Knife; U.S.A.: Cooper's Whitler)
A stout, pointed knife with a blade $2\frac{1}{2}$–6 in long, tanged into a wooden handle. The Scottish pattern has a spear-shaped blade with a central rib. Used for trimming, scarfing, and notching wooden hoops. See Hoop Notching Knife under *Cooper (3) Hooping Tools*.

Knife, Cratemaker's (Cratemaker's Bill)
A heavy knife used presumably for splitting and paring thin pieces of wood for making crates. See *Lath-Maker*.

Knife, Crumming or **Crumm**: see *Drawing Knife, Cooper's Crumming*.

Knife, Draw: see *Drawing Knives*.

Knife, Engraver's: see *Wood Engraver*.

Knife, Felt: see *Pianoforte Maker*.

252

Knife, Gilder's
A long blade used for cutting gold leaf on the gilder's cushion. It is not very sharp to avoid cutting into the leather of the cushion.

Knife, Glazier's Stopping *Fig. 398(b)*
A stiff-bladed Putty Knife with notches ('gates') cut in the edge for trimming glass sheets.

Knife, Grafting
Various forms of Knife, sometimes in the form of a small Bill Hook with a spike at the end bent down at right-angles, or in the form of a Chisel. Used in cleft-grafting for the propagation of fruit and other trees or plants. The stock is split with the knife and the cleft opened to receive a twig ('scion') of the new plant to be propagated.

Knife, Hacking (Chipping Knife) *Fig. 398(c)*
A parallel blade with a scale handle and a thick back to withstand hammering. Used by glaziers and others for hacking out broken glass from the frame before replacing a broken pane. The same knife with the blade set at an angle is called a Chipping Knife.

Knife, Heading: see *Drawing Knife, Cooper's Heading*.

Knife, Hollowing: see *Drawing Knife, Cooper's Hollowing*.

Knife, Hooked (Circular Knife) *Fig. 400*
A Knife-like tool with a blade in the form of a hook. (The same name is given to certain tools of similar shape used by shoemakers, including the Peg Knife for cutting off the protruding ends of pegs inside the shoe, and the Drag Knife, used for cutting leather.)

(*a*) *Chairmaker's Hook Shave* (Pulling-up Hook, Stick Hook; Shave Hook)
A steel hook forged flat in the plane of the handle. The inside of the hook is sharpened on one side. Used to 'fine-down' the ends of the sticks in a Windsor chair back to make them fit the holes bored for them.

(*b*) *Spoon and Bowl Maker's Hook Knife*
A curved Knife with its cutting edge set at right angles to a long handle. Used for hollowing wooden spoons and bowls. The handle is grasped with the hand close to the blade, while the upper end is held between the arm and body. (See *Spoon Maker*).

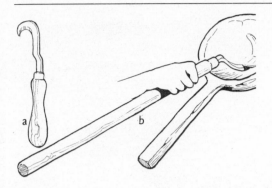

Fig. 400

Knife, Hoop Notching: see *Cooper (3) Hooping Tools.*

Knife, Howeling: see *Drawing Knife, Cooper's Jigger.*

Knife, Hurdle: see *Hurdlemaker.*

Knife, Jigger: see *Drawing Knife, Cooper's Jigger.*

Knife, Lath-Splitter's
A Knife made in the form of a small Bill Hook and used for splitting laths. See *Lath Maker; Woodland Trades.*

Knife, Marking: see *Marking Awl and Knife; Timber Scribe.*

Knife, Marquetry: see *Veneering and Marquetry.*

Knife, Mortising: see *Hurdlemaker.*

Knife, Notching: see *Cooper (3) Hooping Tools.*

Knife, Painter's: see *Knife, Chisel; Knife, Palette; Knife, Putty.*

Knife, Palette *Fig. 401(a)*
A thin blade, not sharpened but thinning towards the toe. Used by painters, coach painters, signwriters, and other tradesmen for mixing paints.

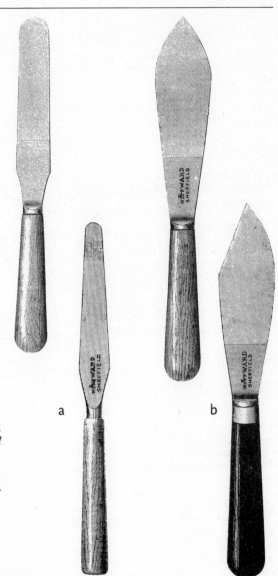

Fig. 401 Painter's Palette and Putty Knives
(*a*) Palette Knives
(*b*) Putty Knives

Knife, Paring: see *Knife, Bench.*

Knife, Peg: see *Knife, Bench; Knife, Hooked.*

Knife, Picking: see *Basket Maker.*

253

Knife, Purfling: see *Violin Maker.*

Knife, Putty (Stopping Knife) *Fig. 401(b)*
Blades of various shapes without a cutting edge, used by painters and other tradesmen for applying putty.

Knife, Rabbet: see Boxing Router under *Router, Coachbuilder's.*

Knife, Race (or Raze): see *Timber Scribe.*

Knife, Radial: see *Drawing Knife, Cooper's Chamfering; Drawing Knife, Wheelwright's Radial.*

Knife, Reed Making: see *Woodwind Instrument Maker.*

Knife, Riving (River)
A term for Knives made in the form of small Bill Hooks, and used for splitting. See *Lath Maker; Woodland Trades.*

Knife, Rounding: see *Drawing Knife, Cooper's Round Shave; Drawing Knife, Handle Maker's.*

Knife, Rubber Tapping *Fig. 402*
These tools are used in the rubber plantations in Malaya and elsewhere for 'tapping' trees for latex. This is done by cutting channels in the bark and collecting the latex in a pot as it drips downwards.
  (*a*) A V-shaped cranked gouge 13½ in long overall, used for cutting the vertical channels.
  (*b*) A gouge-like knife operated by either pushing or pulling and used for the oblique cuts on the tapping panel.
  (*c*) A special ring knife probably used for the same purpose as (*b*).
  (*d*) Timber – Scribes or Race Knives can be used as alternatives to (*b*).

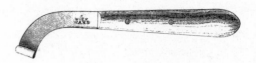

d

Fig. 402

Knife, Scriving (Scrive): see *Timber Scribe.*

Knife, Setting Out: see *Marking Awl and Knife.*

Knife, Side: see *Froe.*

Knife, Sluffing: see Chincing Iron under *Flagging Iron.*

Knife, Splitting: see *Lath Maker; Woodland Trades.*

Knife, Stave: see *Drawing Knife, Cooper's Backing.*

Knife, Stock: see *Knife, Bench.*

Knife, Stop Chamfer: see *Drawing Knife, Stop Chamfer.*

Knife, Stopping: see *Knife, Putty; Knife, Glazier's Stopping.*

Knife, Striking: see *Marking Awl and Knife.*

Knife, Stripping: see *Knife, Chisel.*

Knife, Stuffing: see Chincing Iron under *Flagging Iron.*
(The same name is given to a tool used by saddlers for stuffing straw into horse collars etc.)

Knife, Sway: see *Knife, Bench.*

Knife, Veneer: see *Veneering and Marquetry.*

Knife, Violin: see *Violin Maker.*

Knife, Voicer's: see *Organ Builder.*

a

b

c

254

Knife, Wheelwright's: see *Drawing Knife, Wheelwright's.*

Knife, Wood-Cutting: see *Wood Engraver.*

Knocker-Up (Knocker-Up Tool)
A term applied to a number of different tools used for raising parts during manufacture, e.g. *Jumper, Cooper's;* Knocker-Up under *Print and Block Cutter;* Knocking-Up Cups under *Organ Builder.*

L

Ladder Maker

Ladder sides are made from straight poles of ash, spruce, or Scots pine, sawn or cleft down the middle. The flat face is planed true and the holes marked on the centre line 9 in apart. Pilot holes are bored at intervals along the sides and then enlarged with a Taper Auger. The rungs are made of cleft oak or beech, shaped roughly with Axe and Drawing Knife, and the ends are then tapered in the 'Rung Engine' (see *Plane, Rounder*). The rungs are driven in and are often strengthened with iron rods under the bottom, middle, and top rungs, riveted over washers. (For other tools used see *Woodland Trades*.)

Ladder Rounder (or Engine): see *Plane, Rounder*.

Ladkin: see *Larrikin*.

Lancashire Tools and Steel Toys

(*a*) The term Lancashire Tool is still sometimes applied to a certain class of small steel tool which, at least until recent times, was put out to be made by small family businesses, many of which were situated in Lancashire. These 'little masters' were necessarily more or less anonymous. Today we buy 'Stanley' tools, 'Gillette' razor-blades, 'Remington' typewriters; they may be, and often are, made anywhere from China to Peru, but the brand name is the guarantee now, not the regional origin.

According to the *List of Prices* of W. & C. Wynn, of Birmingham, printed in 1810, Lancashire Tools included the following:

Bench and Hand Vices	Pliers
Callipers	Sliding (Watchmaker's)
Compasses	Tongs
Cutting Nippers	Spring Dividers
Pincers	Watch and Clock Turn-
	screws and Hammers

In the *Book Of Patterns of Lancashire Files and Tools* of the famous Warrington and Rotherham firm of Peter Stubs, *c*. 1845, the tools (apart from their speciality, the files) include:

Beck (Bick) Irons	Nippers
Broaches	Pincers (Carpenters)

Callipers	Pliers
Clock and Watch	Screw Plates
Hammers	Sculptor's Chisels
Dividers	Sliding Tongs
Gravers	Steel Bench Vices
Hack Saws	Stock and Hand Shears
Hand Vices	Wire and Nail Gauges

Many of these tools, whether included with the 'Steel Toys' in the Sheffield Lists, or in other lists as late as that of Buck & Hickman (London, 1935), are almost invariably qualified by the term 'Lancashire', and in the case of certain types of pincers, compasses, and callipers, this description is still valid to-day.

(*b*) The term 'Steel Toys' originally appears to have denoted various household appliances and small tools made wholly or mainly of steel, by Birmingham and other provincial 'little masters'. Judging from the W. & C. Wynn 1810 *List of Prices* (referred to above), by a process of elimination the 'Steel Toys' probably included:

Apple Corers	Muffin Toasters
Bed Wrenches	Nail Nippers
Bick Irons	Nut Cracks
Boot Hooks	Pastry Jiggers and
Carrot, Potato, and	Tweezers
Turnip Scopes	Pen Engines
Cheese and Butter	Sailors Palms
Tasters	Skates
Corkscrews	Spinnet Hammers
Curling, Pinching and	Steak Tongs
Craping Irons	Steel Cats and Dogs
Dog Couples	Steel Compasses and
Fruit Tongs	Bodkins
Gun Worms and imple-	Steel Tobacco Pipes
ments, and Mole Traps	Sugar Cleavers
Hand and Table (i.e.	Sugar Nippers
Bench) Vices	Toasting Forks
Key Swivels	Tuning Forks
Ladies Netting Vices	Tooth Drawers (i.e.
Larding Pins	Extractors)
Lemon Racers	Tweezers
Lobster Cracks	

In later Birmingham lists the distinction between 'Steel Toys' and other objects is still kept, but in the *Sheffield List* (1888) the 'Steel Toys' include some of the 'Lancashire Tools', principally the Pincers, Pliers, Callipers, Dividers, Screw plates, Hand and Bench Vices, Archimedean Drills, and Iron Braces, together with those of the domestic appliances listed above which were not by this time obsolete, such as Bed Keys. In the *Sheffield List* of 1910 the title 'Steel Toys' is given to pages described in the index as

'Lancashire Tools', while the surviving Corkscrews etc. are listed separately as such. The term now appears to be dying out.

Languet
Extensions to the cheek of a hammer-head which secure the head to the handle – such Hammers are known as 'strapped'. See under *Hammer*.

Languid
A term used for spacing pieces between the blades of a Saw used in comb making (*Saw, Stadda*); also a part of the mouth of an organ pipe (see *Organ Builder*).

Lap Board: see *Basket Maker*.

Lap Set *Fig. 403*
A metal wrench used for bending metal edge-plates that were fitted to motor carriage doors. The edge-plates were bent or set to conform to the curve of the carriage body. According to Mr. Arthur Collier (London, 1948) 'a large quantity were sold during the boom of 1904–14, the age of expensive wood and metal panel motor bodies.'

Fig. 403

Larrikin (Ladkin; Ladakin) *Fig. 404*
A strip of hardwood (often box) about 7–8 in long by $1\frac{1}{2}$ in wide and $\frac{3}{8}$ in thick, with blunt-pointed knife-like ends. Used by Glaziers for opening out the groove of the lead strips in leaded-light work to receive the glass. It is also used for pressing down the strip, to rub over uneven places and press them out, or to burnish off any rough or scratched surface.

Fig. 404

Latch
As applied to the chuck of a Brace, the term relates to a spring-loaded catch which drops into the notch provided in the tang of a Bit.

Lath: see Folding Lath etc. under *Rule*.

Lath-Axe (Lath River)
A name sometimes given to tools described under the following entries: Cleaving Knife under *Lath Maker* and *Woodland Trades; Froe; Hatchet, Lathing*.

Lath Maker *Fig. 405*
Riven (split) laths were in common use until recently for carrying plaster, or for supporting slates, thatch, or tiles. Owing to their unbroken grain, riven laths were stronger than sawn laths; but as H. L. Edlin pointed out (1949), riven laths involved rejecting all defective wood which did not cleave freely.

The materials and tools used were similar to those of the other woodland trades using riven material, but thin laths could be severed with a heavy knife instead of a Froe.

Tools used in this trade include the following:

(*a*) *Lath Maker's Froe* (Lath Maker's River)
A short blade made like a Froe but with a sharpened bevel at the square end of the blade as well as along its lower edge. Used for splitting out the laths. The sharpened bevel at the square end may have been provided to enable the tool to be used as a hatchet for trimming the ends of a lath.

(*b*) *Lath Splitter's Knife* (Cleaving Knife; Riving Knife)
Made in the form of a small Bill Hook with a straight or curved blade. The blade is often of wedge section, with the back thick enough for striking with a Mallet.

As well as being used for making laths, it is also used for splitting (as an alternative to the Froe) when making spelk baskets, hoops, and the like. Diderot (Paris, 1763, under Tonnelier) illustrates a similar tool being used by coopers for 'rough hewing' staves or paring hoops.
Note. For fixing laths, see *Hatchet, Lathing*. See also *Froe; Basket Maker; Woodland Trades*.

a

257

b

Fig. 405

Lathe

The Lathes described in the entries which follow are limited to those which have been adapted for particular woodworking trades. We have not included the range of pedal and power driven Lathes used for more general work.

Lathe, Boring: see *Brushmaker*.

Lathe, Pole *Fig. 190* under *Chairmaker*

The traditional Pole Lathe has a wooden bed on which are mounted wooden head-and-tail stocks which carry metal centres. The bases of these stocks pass through the gap in the bed and are secured from below with wooden wedges. From the thin end of a springy sapling (the pole) a cord is looped round the work-piece itself and thence to a treadle below. In the process of turning on an ordinary Lathe the cutting of the work-piece by the Chisel occurs when the work is rotating towards the operator; in the Pole Lathe this happens only on the down stroke of the treadle. When foot pressure is relaxed, the spring of

the pole winds the work-piece in the reverse direction and takes the treadle back to the 'up' position. The turning tool is withdrawn during this reverse phase.

Until about 1960, turners of chair legs for Windsor chairs – known as 'Bodgers' – set up Pole Lathes in the beech woods around the High Wycombe district of Buckinghamshire, protected by a temporary hut. The beech trees provided the raw material.

The reciprocating action of this Lathe was well suited to the bodgers because double-ended pieces (the stretchers, for example) could be treated at one setting, the cord drive being applied to the centre of the work and not at either end. But Pole Lathes were also used by other tradesmen such as makers of wooden bowls and greengrocers' measures.

See also *Turner; Chairmaker* (2) *the Bodger*.

Lathe, Wheelwright's *Figs. 406; 407*

Home-made examples, still to be found occasionally in Wheelwrights' shops, are made of wood throughout except for the steel centres which carry the work. These are mounted on wooden head-and-tail stocks which are set high enough to clear a hub blank about 18 in. in diameter. They are secured by wedges beneath the bed. The lathe is rotated by a rope from a hand-driven cart-type wheel mounted on the floor nearby or sometimes on the rafters above. The main purpose of these lathes was for turning wheel hubs, but their long beds enabled them to be used for turning land rollers and newel posts, repairing mill shafts, and similar work.

Fig. 406 A Wheelwright's Lathe

Fig. 407 A Wheelwright's Lathe driven from above

Lave
One of the tools listed in the early Apprentice Records.
Possibly a Lathe – see historical note under *Cooper*.

Lead:
The spike at the centre of the nose of an Auger or Bit.

Level (Plumb Bob; Spirit Level) *Figs. 408; 409*
This instrument is used to test whether a surface or
structure is perfectly level; another tool, known as a
Plumb Rule or Level, which is used for testing per-
pendicularity, is described here also. There are two
main types:

PLUMB BOB LEVEL *Fig. 408*
(*a*) *A-Level*
This tool was known in the early Egyptian period
and is still in use today. It consists of a wooden
frame joined together like the letter 'A' with a
plumb bob suspended from the apex. The two arms
being exactly equal, when the plumb line coincides
with a mark in the centre of the cross-arm, the surface
on which the tool stands must be level.

(*b*) *T-Level*
This was used from the fifteenth century until the
general adoption of the Spirit Level in the eighteenth
century. It consists of a vertical Plumb Board or
Rule rising at right angles from a horizontal straight
edge, with or without side braces.

(*c*) *Plumb Square*
A set square with a plumb line and bob near one
edge.

(*d*) *Plumb Rule* (*Plumb Board*)
The vertical member of a Plumb Level is used as a
Plumb Rule to test whether a post, wall, or other
structure is vertical. It usually consists of a piece of
wood about 5 ft long by 4 in wide, with the two long
edges perfectly straight and parallel. A line is marked
down the middle, and a hole large enough to take a
plumb bob is cut near the bottom. A saw kerf is cut
in the top, exactly in the centre, to hold the upper end
of the line.

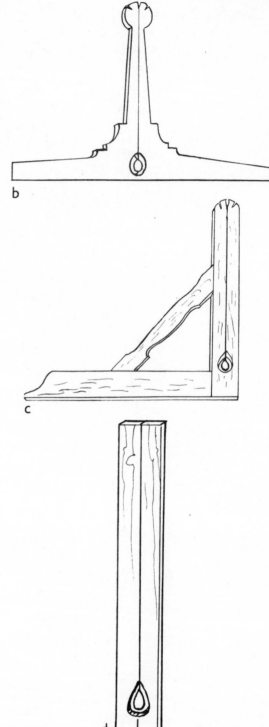

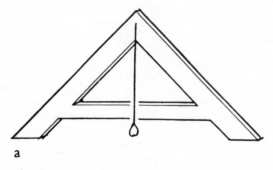

a

Fig. 408

260

a

Narrow Rosewood Spirit Level

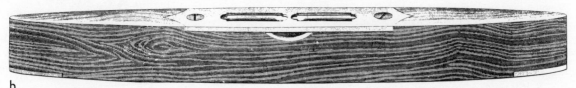

b

Rosewood Level, Taper Ends

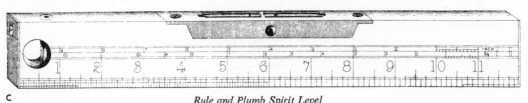

c

Rule and Plumb Spirit Level

d

Lancashire Pattern Rosewood Spirit Level

e

Brass Pocket Level, with Boxwood case

Fig. 409 Spirit Levels
    (*a*) With straight sides
    (*b*) A boat-shaped pattern
    (*c*) With graduated rule, and a Plumb Level for testing perpendicularity
    (*d*) With brass mounts
    (*e*) A pocket Level

A block of hardwood, usually ebony or rosewood, 6–12 in long and ¾–1 in wide, with a short glass tube, containing spirit with a bubble of trapped air, let into the upper surface of the block. A brass plate covers the tube, with a thin strip dividing it into two halves. When the bubble appears to be bisected by the strip, the tool is exactly level. A side window is sometimes provided so that the bubble can be seen when the tool is being used at eye-level.

Among the many variations are Levels with boat-shaped ends; ornamental brass inlays; rule markings along the stock; pocket sizes; and Levels made entirely in metal.

Spirit Levels were used in the eighteenth century by surveyors and others, but were not commonly adopted as a carpenter's tool until the nineteenth century. An early mention of the 'Spirit Level' occurs in Smith's *Key* (Sheffield, 1816).

Leveller (Cooper's Leveller): see *Plane, Cooper's Sun*.

Lever Cap
The wedge holding the cutting iron of a metal Plane. See diagram under *Plane*.

Lever Hook: see *Cooper (3) Hooping Tools*.

Lewis
Used when lifting heavy blocks of stone etc. See *Builder's Tools*.

Lifting Jacks (Scots: Dumcraft) *Fig. 410*
Three kinds of Lifting Jacks are still to be found in the older workshops:

(1) Those that lift by lever, e.g. the Cart and Carriage Jacks. A modern surviver is a wooden lever-jack used for lifting a grand piano in order to remove a leg.

(2) Those that lift by screw, e.g. Bottle Jacks.

(3) Those that lift by rack and pinion, e.g. the so-called Timber Jack.

Jacks for heavy lifting are now usually hydraulically operated.

(a) Cart Jacks
A wooden post about 3 ft high with a lever handle pivoted to the top. When the point is placed under the axle-bed, just behind the wheel hub, and the handle pressed downwards, the vehicle is raised sufficiently to free the wheel for removal. One side of the post is hollowed out to allow the lever to descend until its point is just past the vertical, when the weight of the vehicle holds the lever in position.

The other version illustrated is adjustable for height but in this case the handle must be secured while under load.

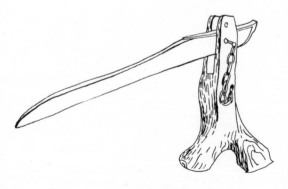

Fig. 410 (a)

(b) Carriage Jack

An all-iron Jack, mostly used by coachbuilders, consists of an iron lever pivoted at an adjustable height on a central pillar which is about 2 ft high. After lifting, the lever is either held in position by means of a pawl engaging a rack on the central pillar, or by a chain.

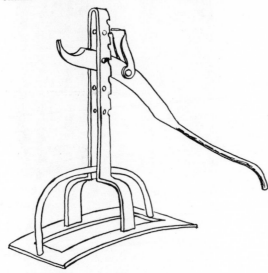

Fig. 410 (b)

(c) Carriage and Motor-car Jack

A later development of the lever-type Carriage Jack, which was also used for early motor-cars.

Fig. 410 (c)

(d) Bottle Jack (Screw Jack)

Early types consist of a short wooden pedestal about 1 ft 9 in high, usually iron bound at the head and foot, and containing an extendable screwed shaft with a forked or crown-shaped shoe on the head. The shaft is raised or lowered by turning a large nut which rests on the top of the pedestal. Used for lifting heavy weights such as a sunken barn, levelling the wind-shaft of a windmill, or lifting a heavily loaded wagon.

Nineteenth-century developments of the Screw Jack include a metal pedestal or tripod instead of the wooden 'bottle'; a tommy-bar operating through a hole in the top nut, sometimes with ratchet operation; and in the case of the so-called Halley's Screw Jack, the shaft is operated through a worm gear.

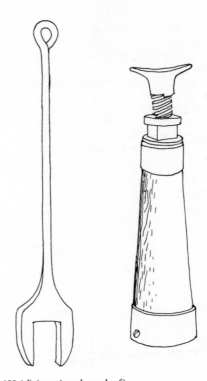

Fig. 410 (d) (continued overleaf)

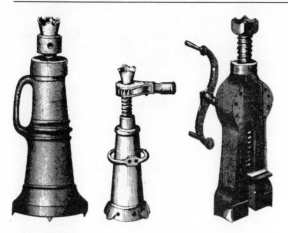

Fig. 410 (d)

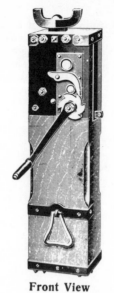

Front View   **Back View**

Fig. 410 (e)

(e) *Timber Jack* (Rack-and-Pinion Jack)
Illustrated by Moxon (London, 1677) and probably known well before that date, it consists of a stout wooden case, about 2 ft 6 in high, containing a racked iron post which is raised and lowered by a pinion operated by a handle on the outside of the case. The post has both head and footstep – the latter for jacking up objects when only 6 in off ground level. Used like the Screw Jacks, with the added convenience of the footstep. (The factory-made example illustrated was imported from Germany.)

Lighting, Workshop *Fig. 411*
The following are examples of candle holders and oil lamps, mostly home-made, which are often found in smithies and workshops.

(a) Sheet-iron candlestick for standing on a bench (sometimes with a spring ejector).

(b) Candle holder with spike for fixing into a wall.

(c) Candle holder with spike for sticking into the ground.

(d) Candle holder with slot for fixing to a shelf.

(e) Candle holder on an iron stem. The base is often made from an old horse-shoe and the candle bracket is sometimes made to slide up or down the stem.

(f) A wooden Candlestick found in saw pits and elsewhere where a greater height is required. It is about 4 ft 6 in high with holes bored at intervals on the stem. The candle holder is bat-shaped, tapered at one end to fit into one of the holes on the stem.

(g) Oil lamps. These are usually small cast-iron or sheet-metal pots with a narrow neck for the wick, with or without screw adjustment for the wick height. 'Bull's eye' lanterns and cart lamps (both oil and candle) were also used.

Line Pin: see under *Chalk Line and Reel*.

Line Reel: see under *Chalk Line and Reel*.

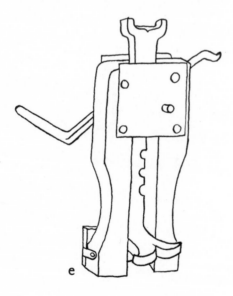

e

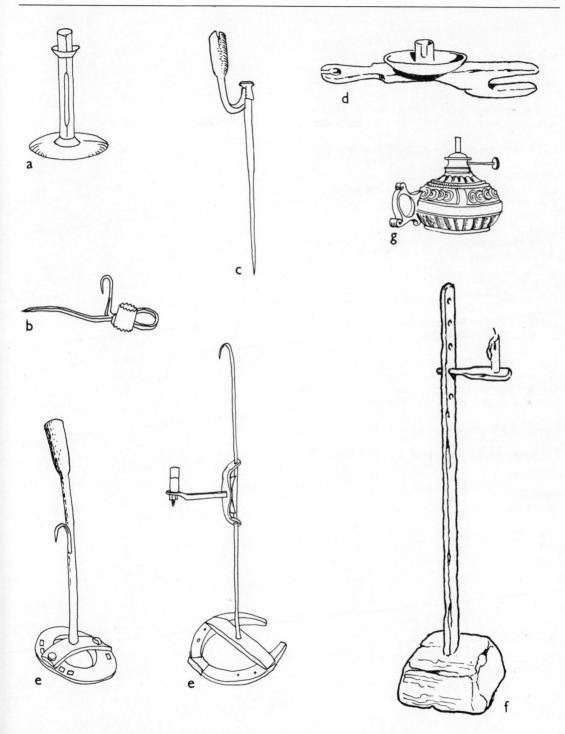

Fig. 411 Workshop Candle Holders and Oil Lamp

Lip Tool: see *Organ Builder.*

List (Listel)
A square wooden strip often used to separate mouldings.

Listing Tool: see under *Router, Coachbuilder's* and under *Axe, Cooper's.*

Log Hook: see Ring Dog under *Timber Handling Tools.*

Log Measuring Tools: see Log Stick under *Rule; Calliper, Timber.*

Loker
According to Halliwell (London, 1847) a Lincolnshire dialect name for a carpenter's Plane.

Long Arm
A term applied to various tools with a long reach.

Lubrication: see *Oil Can, Joiner's;* Oil Box under *Caulking Tools; Grease Pot.*

Lug
A small projection used for supporting an object or for attaching it to another part. Of Axes – the pointed extension below the cheek.

Lumber Stick: see Log Stick under *Rule.*

Lumbering Tools: see *Timber Handling Tools; Tree Feller.*

Lummie
Scots term for Cresset. See *Cooper (3) Hooping Tools.*

M

Macaroni Tool: see *Chisel, Carving*.

Machines (Hand Operated): see *Drill, Bench; Drill, Press; Grindstone; Hub Boring Engine; Lathes; Mortising Machine; Wood Boring Machine*.

Making Iron: see *Caulking Tools*.

Mallet
Mallets are Hammers made of wood. They are used for driving wooden objects which might fracture if struck with a metal Hammer, e.g. Chisels with wooden handles, wooden dowels or tent-pegs, etc. Mallet handles are often wedge-shaped at the top to prevent the head from flying off. (See *Maul* for list of other tools serving a similar purpose.)

Mallet, Carpenter's and Joiner's (Scots: Mell) *Fig. 412 overleaf*
The most common English carpenter's Mallet is made in hardwood, usually beech, and is almost always flat-sided. Some have a slightly curved head as if they were cut from a wheel felloe – as indeed they often were when made by a wheelwright for his own use. A common size for heavy work measures 1 ft 4 in overall with the head about 5 in wide. Some variants are illustrated.

Mallet, Carver's *Fig. 413*
The shape of the head could be described as a well-rounded truncated cone, mounted upside-down on a short handle. Home-made examples are more bun-shaped. This traditional shape enables the carver to strike the Chisel from any angle without altering the position of his hand on the handle. The Mallet head is made in beech, boxwood, or other close-grained hardwood.

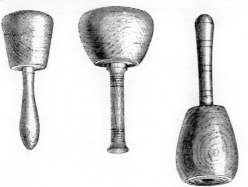

Fig. 413

Mallet, Caulking *Fig. 416 (b).* For description and use see *Caulking Tools*.

Mallet, Cooper's Flogger: see Bung Removers under *Cellarman*.

Mallet, Mason's
A wooden Mallet with a head similar to that illustrated under *Mallet, Carver's*, and *Builder's Tools*.

Mallet, Print and Block Cutter's: see Curfing Iron under *Print and Block Cutter*.

Mallet, Serving *Fig. 415*
A tool used for serving rope, i.e. binding a rope with yarn. For description and illustrations of other types see *Sailmaker*.

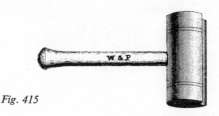

Fig. 415

Mallet, Shipwright's *Fig. 416*
Ordinary carpenters' Mallets are not much used in the shipyards. For light work shipwrights sometimes use a cut-down Caulking Mallet which they then call a Chiselling Mallet. In addition they use:

(*a*) For driving a Horsing Iron and similar work, a long-headed Beetle.

(*b*) For caulking, a Caulking Mallet. See also *Caulking Tools*.

(*c*) For heavy work, a Ship's Maul. For illustration see under *Hammer, Ship's*.

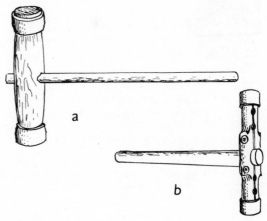

Fig. 416

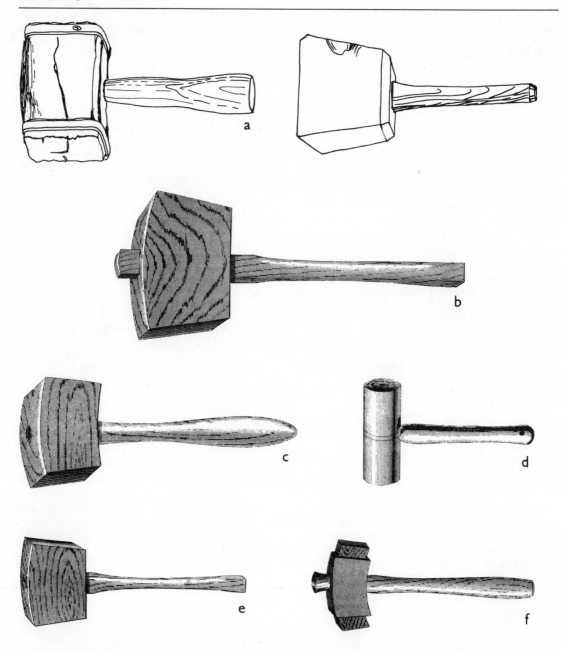

Fig. 412 Carpenter's and Joiner's Mallets

(*a*) Home-made examples (That on the left-hand is made from a wheel felloe.)

(*b*) With the handle fitted into a tapered mortice

(*c*) A factory-made copy of the felloe-type head

(*d*) With cylindrical head

(*e*) 'Gentleman's' or 'toy' Mallet measuring 3–4 in between faces

(*f*) An iron or brass head of square section with sockets at each end to take renewable hardwood faces

Mallet, Woodland Worker's (Batter; Beater; Beetle; Cudgel; Froe Club; Maul)

Mallets used by woodland and coppice workers are of two types. One is a conventional Mallet or small Beetle, with a round or square head 4–5 in. in diameter and with a 12–18 in handle. The other is a home-made cudgel or maul made in one piece from a small log some 3–6 in. in diameter, with one end shaved down to a convenient size for holding. More recently, this type is made in turned wood.

A conventional Mallet is mostly used for driving stakes and similar work. The cudgel or maul is used for driving the Froe and other tools. (See *Woodland Trades;* also *Beetle.*)

Mandrel

A cylindrical tapered rod of iron or wood designed to hold a hollow part while it is being forged or turned. A very heavy cone-shaped Mandrel is described under *Wheelwright's Equipment (1) Furniture.*

Mare: see *Horse; Shaving Horse.*

Marking Awl and Knife *Fig. 417*

Three main types are described below; they are all used for marking and setting out workpieces. The points are used for marking lines with the grain, or on end grain, and for picking out screw holes etc. The blade is used for cutting lines across the grain, as when marking the shoulders of tenons.

(*a*) *Marking Awl* (Chisel-End Marking Awl)

A round steel rod, 6–7 in long, tapering to a sharp point at one end and drawn down at the other to form a skewed chisel-like blade. The centre part is sometimes knurled to improve the grip.

(*b*) *Striking Knife*

A larger version of the Marking Awl. One type has wooden scale handles on the central part of the tool.

(*c*) *Handled Marking Awl* (Scribe or Scriber)

In its simplest form this is a metal spike set in a wooden handle. The factory-made tool usually consists of a round spike 4–5 in long, tapering to a sharp point, and fitted to a ball-shaped wooden handle with a sharply protruding collar just above the ferrule on which to press with the fingers.

a   *Marking Awl.*

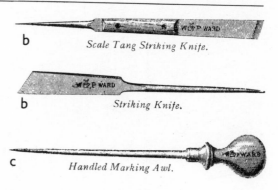

b   *Scale Tang Striking Knife.*

b   *Striking Knife.*

c   *Handled Marking Awl.*

Fig. 417

Marking Tools

For marking lines on timber, branding, etc. They include: *Awls,* see *Marking Awl; Adze* or *Axe* (Marking) under *Hammer, Marking; Burn Brands; Compasses; Gauge, Marking etc.; Hammer, Marking; Marking Awl and Knife; Pencil, Carpenter's; Purfling Tool* under *Violin Maker; Punch, Letter or Figure; Stencils; Timber Scribe* (including Scrive Hooks etc.).

Marline Spike: see *Sailmaker.*

Marquetry: see *Veneering and Marquetry.*

Massie (Mash)

Scots term for *Hammer, Joiner's* or *Hammer, Framing.*

Mast and Spar Maker *Fig. 418*

A Spar is a general term for a mast, yard, or boom used on a ship. A spar can range in length from 10 to over 100 feet, and may be round, oval, or pear-shaped in section. Spars are often made by the shipwright himself, but in the bigger yards a separate workshop is usually provided and manned by specialist mast and spar makers.

The timber used includes Scots pine and Norway spruce, which, for the largest masts, had to be imported from Scandinavia or the U.S.A. in ships specially constructed to load these long timbers. Some of the biggest masts are made of several separate pieces united by trenails.

The trade is described by, among others, David Steel in *The Art of Making Masts, Yards, Gaffs, Booms, Blocks and Oars (1797).*

In the Naval Dockyard at Portsmouth (1969) masts and spars are usually made from sawn timber.

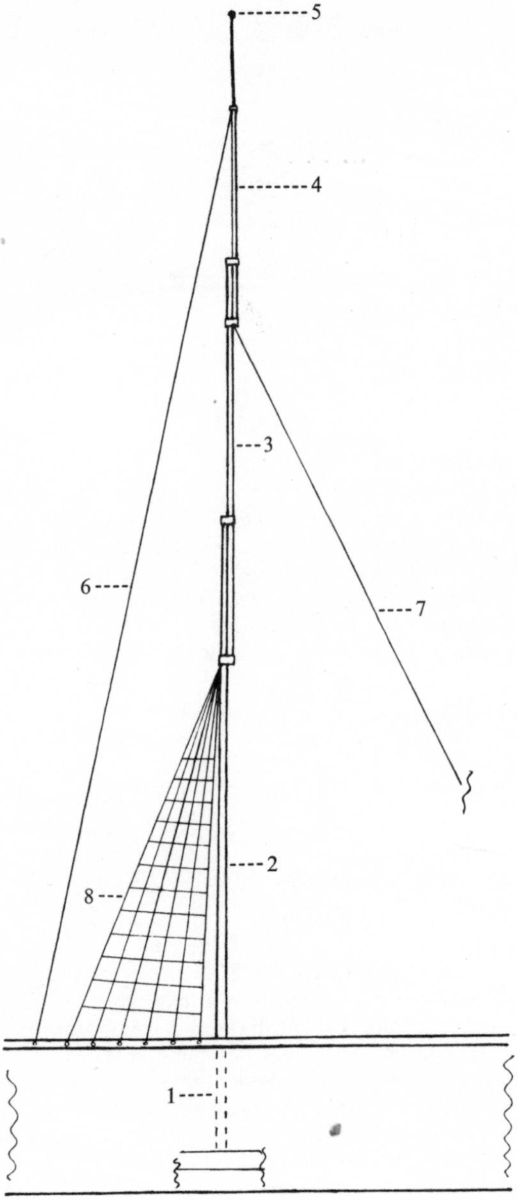

Fig. 418 Diagram of a main mast from a three-masted sailing ship (Height c. 200ft. Diameter at heel c. 3ft.
(1) Heel of the main mast stepped on the top of the keelson. (2) Main mast. (3) Topmast. (4) Topgallant and royal mast. (5) Truck. (6 & 7) Position of backstays and forestays which resist longitudinal wind pressure. (8) Shrouds which resist lateral wind pressure.

After marking with a chalk-line, the heavy waste is removed with a Mast Axe, followed sometimes by an Adze. The mast is then trimmed with a Draw Knife, levelled with a Jack Plane, and finally finished to a smooth taper with the hollow-soled Mast Plane followed by glasspaper.

In other yards, and notably in the U.S.A., large spars and masts are often built up from lengths of timber jointed together to form a square or circular shape, usually hollow.

Special tools used in mast making include: *Axe, Mast Maker's; Drawing Knife, Mast Maker's; Plane, Mast and Spar; Poker, Mast Maker's;* Mast Maker's Rule under *Rules.*

Note: For Spar-making for Thatch, see *Thatch-Spar Maker.*

Matched
A term used to describe a pair of Planes, of which one forms a groove and the other a tongue to fit into it. (See also *Handed Pairs* and *Plane, Matching.*)

Mattock: see *Axe, Grubbing.*

Maul (Batter; Beater; Beetle; Cudgel; Froe Club.)
A term for various tools used for driving, e.g. Maul under *Basket Maker; Beetle;* Chime Maul under *Cooper* (3) *Hooping Tools; Hammer, Ship Maul* (or *Pin Maul);* Mallet under *Woodland Trades.*

Maulstick: see *Painting Equipment.*

Measuring Sticks (Dotter; Length Rod; Size Stick; Obsolete term: Scanteloun.) *Fig. 419* (See also Log stick under *Rule.*)
Many tradesmen employ some form of measuring stick for marking the length of timber when the work involves cutting or making a large number of pieces of the same size. Some measuring sticks consist merely of a light rod a few inches longer than the length of the required component with a nail driven through one end to serve as a marker. Or the length is sometimes indicated by means of a notch formed an inch or two from the end of the rod – a method adopted when the measured pieces are sawn off without any pencilled or other marked line being made, the saw being started in the notch.

A typical example is the measuring stick used by chair makers and sometimes called a 'dotter'. This is a stick, about 16 in long, with a shoulder to act as a fence at one end, and a metal point (usually a wood-screw) at the other. It is placed over a leg or stretcher and given a sharp tap to mark the 'dot', i.e. the correct place at which a hole is to be bored.

Fig. 419

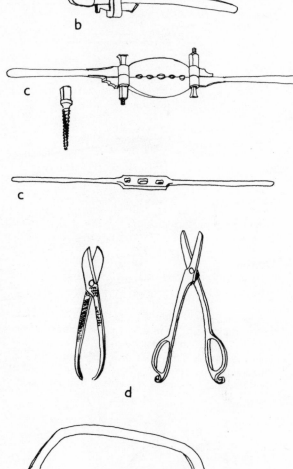

Measuring Tools include: *Boning Strips; Calliper; Compass; Gauge; Level; Measuring Sticks; Patterns, Templets and Jigs; Rule; Square; Templet, Sash.*

Mell

Scots term for a joiner's *Mallet*.

Metal-working Tools *Fig. 420*

The following metal-working tools are commonly found in woodworking shops. They are additional to those already included as belonging to such trades as coachbuilding. The specimens illustrated are mostly home made. Scale 1:8 approx.

(*a*) *Anvil.* Often situated in a separate part of the workshop to avoid danger from flying sparks.

(*b*) *Wrench.* An adjustable Spanner of the wedge type used for turning nuts. (See separate entry.)

(*c*) *Die stock, Tap Wrench, and Screw Taps.* Used for forming a thread on bolts and nuts. See also *Screw Dies and Taps.*

(*d*) *Hand Shears* (Tin snips). For cutting sheet metal. The scissor-handled example was made by a Sussex smith.

(*e*) *Hack Saw* for sawing metal. (See also *Saw, Hack.*)

(*f*) *Bolt croppers* for cutting off bolt ends etc.

(*g*) *Ratchet Brace Drill.* (See under *Brace, Corner.*)

(*h*) *Hot Rasp.* Usually home made. Operated by two men for rasping off hot metal. (See *Rasp, Two Handed.*)

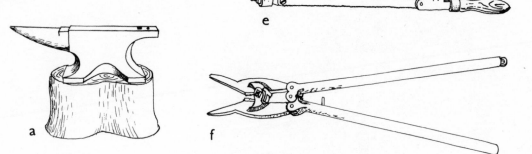

Fig. 420 Metal-working tools commonly found in wood-working shops (*continued overleaf*)

271

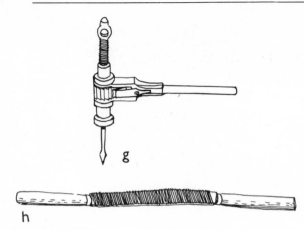

g

h

Fig. 420

Metals used in tool making. See *Brass; Gun Metal; Iron and Steel* (Cast, Mild, Carbon, High Speed, etc.)

Milled
A surface roughened by transverse grooves or ribs to afford a grip.

Millstone Dresser *Figs. 421; 422*
Since the dressing of millstones is sometimes carried out by millwrights, brief notes on their tools have been included. The work was also performed by itinerant stone dressers or by the miller himself. It was said that the dresser was able to prove his 'steel',

i.e. demonstrate his experience, by the black specks on the back of one hand. These are tiny steel chips which, springing from the Mill Bill as it strikes the stone, become embedded in the skin.

The process of dressing a millstone includes re-cutting the furrows and cracking, i.e. levelling, the 'lands' between. The furrows are cut on the working face of the stones to break up the grain and to guide the meal from the eye at the centre of the stone to its skirt, where it is collected as flour.

G. Ewart Evans (*The Farm and The Village*, Faber, 1969) quotes an East Anglian millwright. The following is an extract:

'For marking out a stone you need a land spline and a furrow spline, pieces of flat wood the exact size that you want your furrows to be. With one kind of dress these were $1\frac{7}{8}$ in and $1\frac{1}{8}$ in wide respectively. I lay the furrow spline on the stone and mark round its edge with a trimmed duck-feather dipped in a solution of soot and water (I always use a duck's or a goose's feather. An old stone man once told me that these were the best kind.) Then I place the land spline, which is the wider of the two, alongside the furrow and the land is marked out in the same way. You repeat this method until you have marked right round the stone . . . the furrow does not follow the true radius but is marked from the skirt to a point behind the centre of the stone. This amount of difference between the furrow and the radius is called the drift. This is the harp dress. But there is another method, called the sickle dress which has a curved furrow (hence the name) narrow at the eye but broadening out as it reaches the skirt. After the furrows have been cut the next job is cracking or chasing the lands . . . .'

An account of this work is given by S. Freese (Cambridge, 1957) and by J. Russell. (Newcomen, 1944).

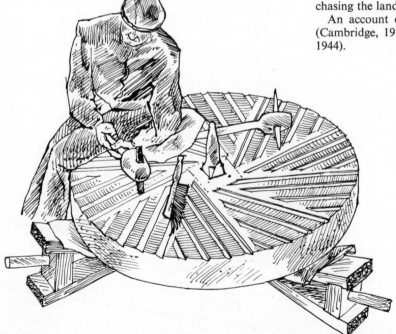

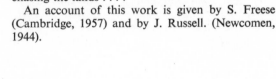

Fig. 421 A millstone dresser at work

Tools used include the following (*Fig. 422*):

(a) Mill Bill and Thrift

A hardened steel hammer-head, tapered off at each end like a wedge. This is held in a wooden handle called a Thrift, about 18 in long, with a club-shaped head, usually turned. There is a rectangular hole cut through the head in which the steel Bill is held. Used for dressing the working face of a millstone.

Variants include a Bill pointed at both ends (used for preliminary levelling); an iron socket on the end of a wooden handle for holding the Bill; and a Bill with an eye-hole to take a handle.

(b) Scraper

A steel plate, usually in a wooden handle, for cleaning the stone before dressing it.

(c) Jack Staff (Jackstick; Tracer Bar; Speech)

A wooden bar, with a square eye at one end to fit over the stone driving spindle (quant), and at the other end an upright quill. This is wedged in a small hole and can be set to touch the stone face when the spindle is slowly revolved. The bedstone level is checked by watching whether the quill just scrapes the stone at every point.

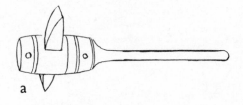

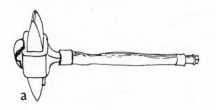

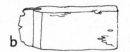

Fig. 422 Millstone dresser's tools

Millstone Gauges

These include:

1. Stone Staff (Proof Staff). A straight bar of hardwood used for checking the face of the millstone. Red iron-oxide paste, known as 'tiver', is applied to the surface of the staff, which is then rotated over the whole surface of the stone. If the stone is in good order the oxide will be distributed evenly over its surface.

2. Prover. A metal bar, kept in a wooden case, for testing the accuracy of the stone staff.

3. Furrow strips (Furrow and land spline). Strips of wood about $\frac{3}{8}$ in thick to gauge the width of the furrow and the 'land' between the furrows.

For other tools sometimes used for dressing millstones see *Hammer, Bush; Pritchet*. Spectacles or wire-mesh goggles were worn to protect the eyes.

Millwright *Figs. 423; 424; 424/1*

There is reason to believe that wheels driven by water were working in the Middle East in the first century A.D. Windmills were known in Persia in the ninth century A.D. and reached Europe some three centuries later. The carpenters who specialised in the construction of watermills and windmills have, since the fifteenth century, become known as millwrights. W. H. Pyne (London, 1845), writing on the work of the carpenter and wheelwright, says 'When he becomes a Millwright . . . we must admit his art assumes a high cast indeed.'

The Millwright's Work

The work of the millwright may be compared to that of the shipwright for the sheer size and weight of the timber to be shaped and jointed; to that of the wheelwright for the technique of making wheels of great strength; and to that of the engineer for his ingenuity and knowledge of mathematics. In a modern engineering works the men in charge of the installation and upkeep of the machine tools are referred to as 'millwrights'. And when some large machine, weighing perhaps many tons, has to be

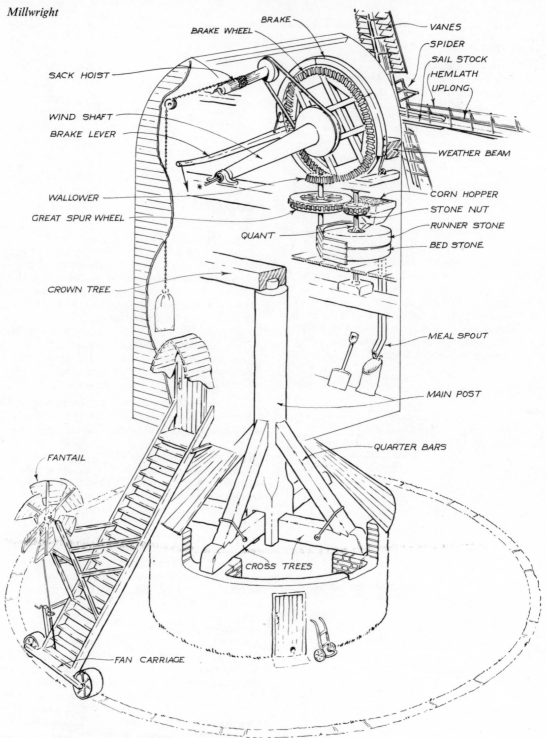

SACK HOIST

BRAKE WHEEL

BRAKE

VANES

SPIDER

SAIL STOCK

HEMLATH

UPLONG

WIND SHAFT

BRAKE LEVER

WEATHER BEAM

CORN HOPPER

STONE NUT

RUNNER STONE

BED STONE

WALLOWER

GREAT SPUR WHEEL

QUANT

CROWN TREE

MEAL SPOUT

MAIN POST

QUARTER BARS

FANTAIL

CROSS TREES

FAN CARRIAGE

Fig. 423 Diagrammatic view of a Post-Mill

* Striking rod for moving the spider which controls the angle of the vanes. (Operated by a chain outside the mill.)

274

dismantled or moved, the head millwright will direct the operation, just as his ancestors stood at the foot of the mill while some heavy piece of the internal machinery or a new stone was slowly hoisted into place.

Though simple in conception, the construction of a windmill or watermill is a complex process which demands a high degree of skill and experience. The work is described by several authors, notably by Wailes (London, 1954), and by Freese (Cambridge, 1951). Mill repairs are well described by Rose (Cambridge, 1937).

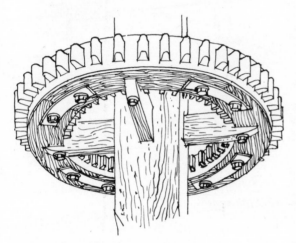

Fig. 424 A spur gear made of wood

Materials

The builder of wooden windmills needed large oak baulks of the highest grade. In a post mill, the central post must be strong enough to carry the whole weight of the structure, including the machinery, millstones, and the stock of corn and meal. As H. L. Eldin pointed out (London, 1949), 'It must face up to the thrust of a gale and the vibration of the machinery, and hold its burden so delicately poised that it can be swung around by the power of one man. An oak post thirty inches square was set to bear this heavy strain . . . .' Walter Rose (Cambridge, 1937) writes that 'My father would travel many miles to find suitable butts of oak of the right length and size, to cut for the [sail] stocks. I remember when they arrived on the timber carriage we had difficulty in turning it into the yard, so long were they.'

Various woods were used for the cogs which were fitted into the rims of the gear-wheels. Well-seasoned

apple and oak was much favoured, but also hornbeam and beech.

Water-wheels were framed in oak, but the slats, which made contact with the water, were commonly made of elm.

The Workshop

The few millwrights' workshops that are still recognisable as such look like large wheelwrights' establishments, together with the outdoor storage of a builder's yard. One such, at Hawkhurst in Kent, is described as follows by Wailes and Russel (Newcomen, 1953/55):

'On the ground floor there was a 12-in dead centre lathe with wooden shears; a large hand wheel provided motion mounted in a swinging frame capable of adjustment for belt tension and it could be moved endways to line up with work of varied length . . . .

Upstairs were two 9-in centre treadle drive lathes. Immediately opposite the doors, which were one above the other giving access to the ground and second floors, there was a removable joist and floor boards, these when taken up left an aperture where large wheels could be swung up on a mandrel for turning cogs, etc.

Behind the shop there was a small brass foundry and an arrangement for bending iron floats and buckets.

The sawpit was on the grass nearly in front of the 'Sawyers Arms'; timber was kept on this ground and sometimes large components were erected there – water wheel rings, etc. Very large holes were drilled with a large force brace used in a horizontal position long enough in the crank to be worked by two men; some of the drills were 1½ in. in diameter.'

Millwrights' Tools Fig. 424/1

The millwright uses the same tools as other carpenters engaged on heavy constructional work, including the indispensable Lifting Jack and Hoisting tackle.

The special tools listed below relate mainly to the fitting of wooden cogs and to repair work.

(a) Fangle Iron (Cuckoo; Millwright's Gauge or Scribe)

A scriber used for marking the pitch circle on gear wheels, and also for marking the length of the cogs, i.e. the face line on the diameter of the wheel. (The pitch circle is the imaginary line along which two gear wheels in mesh make a rolling contact, and from which measurements essential to the form and spacing of the teeth are taken.)

Two of the Fangle Irons illustrated are simple iron dogs, smith-made, with 6 in prongs and a fixed scribing head. The other more elaborate tool (which is given the unusual name of 'Cuckoo') came from William Warren, millwright of Hawkhurst in Kent. Made in iron throughout, it consists of two prongs 7 in long and 3½ in apart, joined at the top by a quadrant on which a moving arm can be fixed in any position by a thumb screw. At the end of this arm is a scriber set at about 60° and threaded for adjustment. The tool is decorated with stepped mouldings at the bosses and with stop chamfering in between. A thumb screw is missing from a socket in the centre of the arm; it probably held a strip of metal designed to act as a stay to prevent the cuckoo from shifting off its mooring.

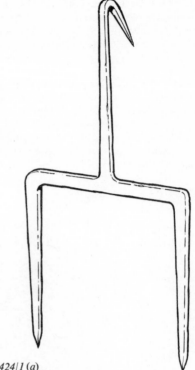

Fig. 424/1 (a)

(b) Striking Iron

An iron dog with spiked legs and a perfectly straight back 9 in long. Like the Fangle Iron, it is driven into a convenient beam in such a position that a gear wheel being re-toothed can be revolved with its cog ends in line with the back of the dog. As each cog passes, the iron back is used as a rule for scribing a line across its face to join the line previously made by the Fangle Iron on each side of the wheel at the circumference. This line will subsequently serve as a guide to the Chisel when paring down the head of the cog to shape.

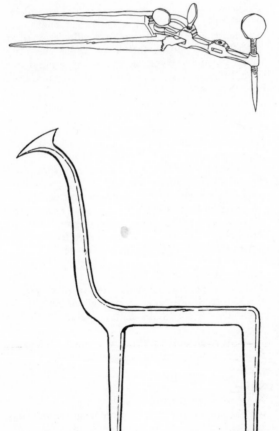

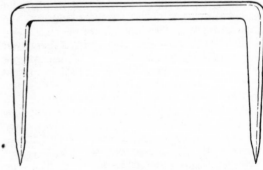

Fig. 424/1 (b)

(c) Moon Chisel

A crescent-shaped Chisel 3 in across mounted on the end of a metal rod about 14 in long. It is used for excavating a hole in the end of a heavy shaft (e.g. of a water wheel or spur gear). The hole is started with a Gouge and then continued by a levering movement with the Chisel. The purpose of the hole is to house the back end of an iron gudgeon. This is the heavy iron pin or spindle (sometimes called a trunnion) which protrudes from the end of the shaft to support the shaft in its bearing. The base of the gudgeon has four wings extending from its sides, which, when let into slots cut in the end of the shafts, hold the pin in place. It is centred by driving thin cleft oak wedges down on each side of these wings.

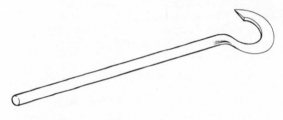

Fig. 424/1 (c)

(d) Devil

An iron rod about 2 ft 3 in long, the end flattened with a blunt squared edge on one side. It is used for levering out old wedges from the wings of a gudgeon ((c) above) when it is necessary to realign the wheel to make it run true.

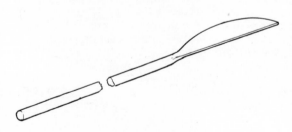

Fig. 424/1 (d)

Book

A name given to a board fixed in position near the gear wheel under repair. The Fangle Iron and Striking Iron were driven into it (in the absence of a handy beam near by) but its other purpose is to enable the millwright to record with a pencil the alignment of a wheel as it revolves.

Millwrights' Chisels

These are heavily built Firmer Chisels used for paring cogs, mortising, etc. (See *Chisel, Millwright's.*) Bruzz Chisels were also used for mortising work.

Compass

A Lancashire type of Wing Compass. See *Compass, Millwright's.*

Planes

Millwrights used the ordinary range of Planes, but the following more specialised Planes have been noted:

A small version of the Badger Plane with a stock only 6 in long × 2 in wide which is used for cleaning the corners of large, deep mortices.

A Cooper's Sun Plane was sometimes used for finishing the top of barrel-shaped articles, such as millstone-cases.

Mr. Jesse Wightman, an Essex millwright (1968), described planes used for making long wooden sail shafts: 'They were part of the tools and tackle of a millwright which were being cleared out in an after-decease affair like has so often happened. They were like two-in-one large wooden planes in one double length, two handles for two men, one to stand behind the other, length enough for this.'

Stone Dressing Tools: see *Millstone Dresser.*

Mining Carpenter

(The following notes are based on an article kindly contributed by Mr. F. A. Dixon of the National Coal Board, Gateshead, 1968.)

With the advent of intensive mechanical mining during the past twenty years the use of timber has declined and wood has been substantially replaced by steel. The use of woodworking tools has consequently diminished in recent years.

Timber was used for many years in all types of mining for the support of the underground roadways, working faces, and also for many deep mine shafts. Great strength was the main requirement and therefore heavy timber roughly hewn and strongly jointed was needed to withstand the enormous rock pressures encountered.

The woodworking tools required generally comprised an Adze, Auger, Saw, and several kinds of Axe.

The Axe had a short handle about 15 in long. It was used to cut notches to form the joints in the timber used to support either roughly arched or square shaped roadways. A long-handled Axe was at one time used for withdrawing the vertical supports by chopping into the base of the prop, but this practice has been replaced by mechanical devices.

Largely owing to the room and pillar method of mining practised for many years in the northern coalfields of Britain and the organisation of the miners in small groups under a junior official known as the 'Deputy', it became the practice for this official to undertake or supervise the erection of the coal-face supports, and generally he was responsible for the safety of the miners under his control.

The woodworking tools for the supports were a Saw and an Axe, locally known as the 'Deputies Axe', which was carried at all times as part of his standard equipment. This axe had a small notch in the lower edge of the blade for nail pulling, a poll for hammering, and a handle exactly two feet in length.

With these primary tools the Deputy could cut and notch timber as required, and erect or withdraw the timber props using the hammer end of the tool. The Deputy was also responsible for the installation and maintenance of the light railway track in his section. His Axe was used to measure the rail gauge of two feet, and the Axe's poll to fix the steel rails by nails or spikes to the wooden sleepers. The small notch in the cutting blade of the Axe was used to withdraw the nails or spikes when it became necessary to relay or repair the light railway track.

Miners' Axes are described under *Axe, Miner's Deputy; Axe, Miner's Welsh.*

Mitre Appliances *Fig. 425*
The following appliances are used to guide the Saw, Chisel, or Plane when cutting and trimming mitre joints; these are joints at the intersection, usually at 45°, between two workpieces. The appliances are often made by the tradesman himself and are designed for use in the vice or on the bench.

(a) Mitre Shooting Board
18–30 in long, with a rebate and dust groove along the front edge, and 45° stops on the top. Used with the Mitre or Shooting Plane laid on its side in the rebate for trimming the end of the workpiece or moulding to a mitred angle. (See also (h) below.)

(b) Donkey's Ear Shooting Board
A name given to a Shooting Board adapted for planing a mitre in the thickness of a board (e.g. along its edge) rather than across its end. The upper table, on which the workpiece is laid, is set at an angle of 45° to the lower table.

(c) Mitre Block
A block of hardwood, L-shaped in section, with two saw kerfs at 45° and one at 90° to the direction of

the length. Used for holding wood or mouldings of small dimensions to be sawn square or to a mitre.

(d) Mitre Box
A U-shaped block with saw kerfs cut as in (c) above. Adjustable metal guides are sometimes fitted to the upper edges of the saw kerfs to reduce errors due to wear. Used for holding large mouldings or wider strips when being sawn on the mitre. Modern, proprietary appliances for sawing mitres are designed on the same principle, but are made in metal and are provided with adjustable guides for the Saw and an adjustable protractor to regulate the angle. One example is illustrated under *Saw, Mitre-Box* and another under (g) below.

(e) Mitre Templet
A wood or metal templet, L-shaped in section, with both ends cut at an angle of 45°. Used for guiding a Chisel when cutting mitres worked in the solid moulding which, owing to their situation, could not be trimmed with the Plane.

(f) Mitre Shooting Block (German Mitre Shute; Mitre Jack)
A rectangular base on which is mounted a form of screw Cramp with the jaws shaped on one side at 45° to the base. In use the workpiece, which has been sawn to a mitre, is cramped between the triangular blocks, and the joint trued with the Plane. The 90° faces enable the ends of narrow stuff to be planed square.

(g) Adjustable Mitre Box
An all-metal version as described in (d) above. Another version is illustrated under *Saw, Mitre Box.*

(h) Adjustable Shooting Board
An all-metal version known as a 'Stanley Shute Board' with an adjustable angle setting. The special iron Plane, with skew cutter, is constructed to fit in the slide provided.

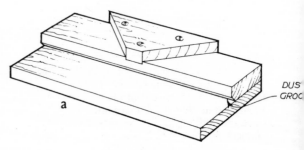

a

DUS
GROO

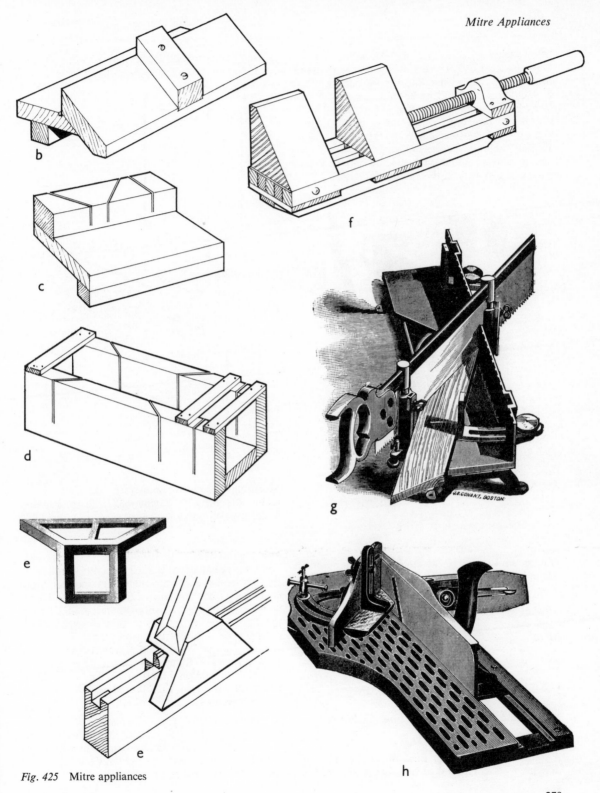

Fig. 425   Mitre appliances

Mitre Machine (Mitre Trimmer) *Fig. 426*
During the nineteenth century a number of hand-operated machines came into use for cutting mitres in mouldings and picture frames. Besides all-iron Mitre Boxes for guiding the Saw at any desired angle, a tool was developed which cuts a clean mitre at one thrust of the handle. The workpiece is held on an iron table at the desired angle while a strong handle carrying a double-edged knife is made to pass back and forth across the end of it.

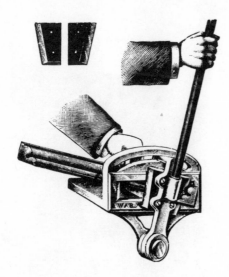

Fig. 426

Mitre Square: see *Square, Mitre.*

Mock
Objects are said to be 'at mock' with one another when they face in opposite directions, e.g. the spikes of a Timber Dog.

Monkey Wrench: see *Wrench, Monkey.*

Moot
A name given to a tool for rounding trenails. See *Plane Rounder, Trenail.*

Mortice Axe or **Knife:** see *Hurdle Maker.*

Mortice and Tenon Gauge: see *Gauge, Mortice.*

Mortice and Tenon Joint *Fig. 427*
A joint used in framing, furniture, buildings, and in many trades. The end of one part is prepared as

a tenon (or tongue) designed to fit into a mortice hole made in the other part.

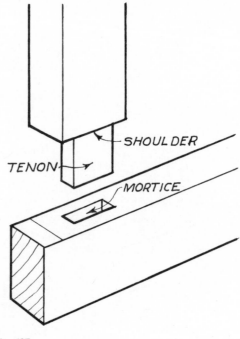

Fig. 427

Mortising Cradle or Stool: see Mortising Cradle under *Wheelwright's Equipment (1)*; Mortising Stool under *Hurdle Maker.*

Mortising Machine *Fig. 428*
Towards the end of the eighteenth century, and probably before, hand-operated machines were developed for making mortices. One of the simplest (still to be found in woodworking shops) consists of a sliding spindle, which carries the cutting chisel, mounted vertically on a heavy cast iron stand. To cut the mortice, the spindle is forced downwards with a long lever; it is drawn up again by a counterweight. The workpiece is held on a table below, and is advanced by a hand-screw feed. Chisels used on these machines were of several kinds, including:

(a) Machine Mortice Chisel [as illustrated]
A very strong Mortice Chisel of the ordinary shape, or channelled to cut a square corner, or with barbs so formed that on withdrawing the Chisel the chips are cleared from the mortice.

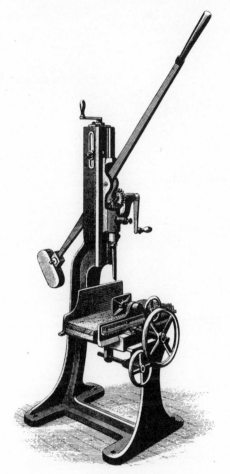

Mortising Machine.

Machine Mortise Chisel.

Self-coring Machine Mortise Chisel.

Fig. 428

(b) Hollow Mortising Chisel

Square in section, and hollow to allow an Auger Bit to pass through its centre. The Auger is made to revolve and travels in advance to bore a hole; the Chisel follows, and its four cutting edges transform the round hole into a square one. (See also, *Bit, Mortising*.)

(c) Double Chisel

Two sharpened edges which cut both ends of a mortice simultaneously. The chips gather into the depression between them.

(d) Chain Cutter

In more recent machines a continuous Chain Cutter is used instead of a Chisel.

Moulding and Turning Box (Fluting Box) *Fig. 429*

A box for holding workpieces such as columns or table-legs, while being moulded, fluted, or grooved for inlay.

The workpiece is held in position either by a screw or with wedges, and a Scratch Stock worked with its fence along the box side. The Scratch Stock can be operated in line with the axis of the workpiece or, by shifting one end of the workpiece, at any other angle. A registering device (known as a clock) can be fitted at one end. It consists of a disc with a number of holes spaced round the circumference according to the number of faces on the finished workpiece. (In engineering practice this device, when fitted to a Milling machine, is called an Index Plate.)

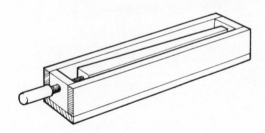

Fig. 429

Mouldings: see *Plane, Moulding*.

Moulds, Shipbuilder's

These are thin strips of wood cut according to the drawing made on the mould-loft floor (see *Shipwright*). They are used as a pattern when sawing out the ship's timbers, and some may be set up temporarily within a ship as a guide to the placing of the timbers.

Mouse (Thief)
A name given to a small weight attached to a thin string, used for piloting new sash cords over the sash pulleys when replacing a worn or broken cord. The free end of the line is tied to the cord, the Mouse passed over the pulley and allowed to drop as far as the sash pocket, when the cord can be pulled through and attached to the sash weight. Commonly made from a thin piece of lead or lead-covered cable, or better from a piece of fine roller chain. (See *Window Making*.)

Muller: see *Painting Equipment*.

Mullet: see *Gauge, Mullet*.

Mundy
Scots shipwright's term for a heavy Hammer.

Musical Instrument Maker: see *Organ Builder; Pianoforte Maker and Tuner; Violin Maker; Woodwind Instrument Maker and Reed Maker*.

N

Nail Bag
Made of leather or canvas and provided with a belt. Worn by lathers, carpenters, and others for holding nails.

Nail Extracting Tools (Goat's Foot; Nail Pull)
These include: *Adze, Cooper's Nailing;* Broken Screw Borer under *Bit, Annular; Box and Case Opening Tools; Hammer, Claw; Hatchet, Lathing; Hatchet, Shingling; Nail Puller; Tack Lifter or Claw;* Strake Nail Drawer under *Wheelwright's Equipment (3) Tyring Tools.*

Nail Passer
A name sometimes given to a *Gimlet.* See also *Piercer.*

Nail Puller *Fig. 430*
A stout iron pipe fitted with a pair of sharp jaws at the end: one fixed, the other pivoted and extended to form a lever. A round rod fits loosely in the tube and is provided with a head in the form of a knob or a hammer. Its purpose is to drive the jaws over an embedded nail when forced smartly down.

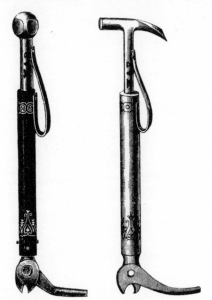

Fig. 430

Nail-Pulling Bar *Fig. 431*
A very heavily made iron bar about 3 ft long with the ends bent to form a fulcrum (like those of a Crow Bar); on each end an iron loop is pivoted. In operation the loop is dropped over the nail and the bar is lowered. This action grips the nail and levers it upwards. Used for drawing spikes out of ship's timbers and similar work.

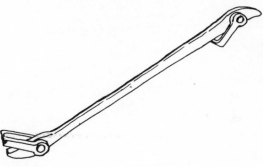

Fig. 431

Nail Set: see *Punch, Nail.*

Nailsin
A dialect term for Gimlet (Halliwell, 1847).

Nave Tools: see *Gouge, Wheelwright's; Hub Boring Engine; Lathe, Wheelwright's.*

Navegar (Navegor; Naforgar; Nafegar, etc.)
An obsolete term for *Auger.*

Neck Tongs: see *Cellarman.*

Needle: see various needle-like tools under *Pianoforte Maker and Tuner; Sail Maker; Upholsterer.*

Nelson: see *Shave, Nelson.*

Nib
As applied to tools, a small protuberance, e.g. the nib on the toe end of a Saw. See also Sneck under *Plane* (Cutting Irons).

Nicker: see *Spur.*

Nicking Tool: see *Organ Builder.*

Nippers: see under *Pincers* or *Pliers.*

Nog (Nogg)
A wooden plug, e.g. a plug inserted into a wall to provide a fixing. Also an alternative name for *Plane, Rounder*.

Norris (Tool Makers): see *Plane, Spiers and Norris Types*.

Notch
A recess, usually (but not always) V-shaped.

Obsolete Tool Names

Among those included are the following (of which some may still be found in dialect use):

Addes (Adze)
Clavot or Clove Hammer (Claw Hammer)
Creves or Crowes (Croze)
Elsin (Awl)
Groping Iron (Gouge)
Inbowing Plane (Moulding Plane)
Jennet (Adze)
Lave or Side Lave (Probably a Lathe)
Loker (Carpenter's Plane)
Navegar, Nafugar, etc. (Auger)
Sage (Saw)
Thixel (Adze)
Twybil (Mortising Axe)
Wimble or Wymble (Brace)

Ogee

A form of moulding. See *Plane, Moulding*.

Oil Box or Pot

Boxes or pots made in earthenware, wood, or metal used in many trades for holding grease or oil for lubrication. For examples, see Oil Box under *Caulking Tools; Grease Pot;* Grease Horn under *Sailmaker*.

Oil Can, Joiner's *Fig. 432*

A small conical tin can with a side handle, and a screwed nozzle which sometimes contains a spring-loaded valve to prevent escape of oil if knocked over. Used mainly for applying oil to the oilstone or slip when sharpening tools.

Fig. 432

Oilstone (Whetstone) *Fig. 433*

A natural or artificial stone about 8 in long, 2 in wide, and 1 in thick. The natural stones from quarries in Britain include: Charnley Forest, Rag, Tam o' Shanter, Welsh, and Water of Ayr; from North America – Canada, Washita or Lilywhite, and Arkansas. Artificial stones of carborundum (silicon carbide) are often made with a coarse grit on one side and a medium or fine grit on the other. Most joiners made their own wooden cases, hollowed out from the solid. Small pins driven through the bottom project sufficiently to hold the stone firm on the bench; or a strip of leather is glued beneath each end to serve the same purpose. The stones are kept oiled and used for sharpening Plane Irons, Chisels, etc. The late Mr. Arthur Collier (a Brixton ironmonger and tool merchant) related that, when first trying to sell artificial oil stones in 1901, carpenters would not take them. So he used to throw one of the stones against the wall, and when the carpenter saw that it did not break he sometimes agreed to try one.

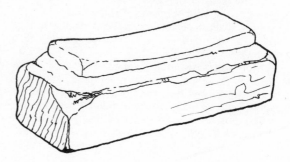

Fig. 433

Oilstone, Slip (Slip Stone; Gouge Stone; Gouge Slip) *Fig. 434 overleaf*

A small Oilstone made in various shapes, the most usual size being from 4 to 6 in long and 2 to 2½ in wide, with one long edge thicker than the other, and both rounded. They are used for sharpening in-cannel Gouges and the curved cutting edges of moulding plane irons etc., and for removing the burr after honing out-cannel Gouges. Square or triangular shaped slips are used for sharpening V-shaped Chisels.

Fig. 434

Old Woman's Tooth: see *Plane, Router.*

Organ Builder *Fig. 435*

Organs were being built for churches and monasteries in England as early as the tenth century. They consisted (as they do today) of metal and wooden pipes supplied with wind by bellows. The pipes are sounded when the organist presses the appropriate keys or pedals which admit air to the pipes. An organ built for Winchester Cathedral *c.* 900 is said to have contained 400 pipes and 26 bellows and required 70 men to work it.

The greater part of the instrument is made of wood. The only parts often made of metal are the pipes, which are made of a tin-lead alloy which is soft enough to be worked with woodworking tools. Ordinary woodworking tools are used for building the case of the organ, except for the rather unusual form of the 'Old Woman's Tooth' Plane used for cutting shallow depressions (scores) on the surface of the soundboard on which the slides operate. (See below).

More specialised tools belong to the 'voicer' – the man who is said to 'give the organ pipe its voice' – for the quality of the sound depends largely on the adjustments he makes to the metal or wooden pipes.

The tools described were seen at the Organ Works of Messrs J. W. Walker & Son, Ltd., (Ruislip, 1968).

(a) Parts of a metal organ pipe

The illustration shows the lower part of a metal mouth or flue pipe – the name for those pipes which have a wind-way and mouth, as against the reed pipe in which the tone is produced by a vibrating metal tongue inside the pipe. The orifices are regulated by the voicer with the tools described below. Similar principles but different methods are used for voicing wooden pipes. Reed pipe mechanisms

are regulated with several simple tools including a 15 in long steel strip used for adjusting the reed stop and a 6 in rod used for reducing the curvature of the vibrating brass reed.

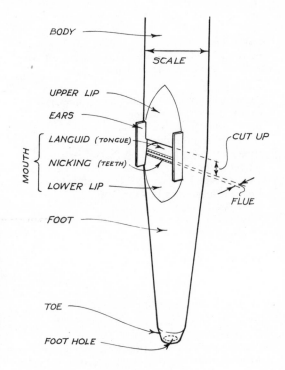

Fig. 435 (a)

(b) Voicer's Knife

A short, home-made knife with the blade ground at an angle and set in a wooden handle. Used for trimming and cutting away the sides of the mouth-piece when voicing the pipes.

(c) Voicer's Nicking Tool

A graver-like tool used for nicking the edge of the languid and lower lip at the 'flue'. This controls the development of 'edge tones' by splitting up the stream of air. Nicking tools for wooden pipes have a saw edge.

(d) Voicer's Knocking-up Cups

Brass bell-shaped tools made in different sizes and used for tapering the tip of the metal pipe at the foot where it rests in the hole in the soundboard, in order to regulate the amount of air entering the pipe. This is done by tapping the tip of the pipe with the hollow part of the Cup. This is also done to wooden pipes when the tips are metal covered.

(e) Voicer's Tuning Cones (Tuning Brasses)
Hollow brass cones used for tuning the smaller pipes. To sharpen the note, the top of the pipe must be slightly opened, i.e. made wider. This is done by lightly tapping the orifice at the head of the tube with the outside of the Cone held downwards within the pipe, thus forcing the soft metal slightly outward. The opposite effect – a flatter note – is brought about by a slight closing of the pipe; this is done by a light tap with the open end of the cone which covers the top of the pipe like a hat. Wooden pipes are tuned by means of an adjustable piston at the top.

(f) Other Voicer's Tools [not illustrated]
Languid Wires are metal rods of varying lengths, which are introduced through the opening at the foot to make adjustments to the languid. The Lip Tool, which looks like a square-ended moulder's tool ground thin at the ends, is used for adjusting the size of the opening of the 'flue'.

(g) Thicknessing Plane
A wooden plane similar in appearance to a Jack Plane but with an iron leaning only a degree or two backwards from the vertical and ground square with the sole. Length: 13 in; width: $2\frac{3}{4}$ in; height: $2\frac{1}{2}$ in; width of iron: 2 in.

Used for thinning down or flattening the surface of the tin-lead alloy sheets that are used for most of the metal organ pipes up to 4 ft in length. The edges of the sheets are trimmed by 'shooting' with a steel Mitre Plane with a low-angle cutting iron. (cf. *Plane, Metal Cutting.*)

(h) Scoring Plane
The sole is 8 in long and only 1 in wide, with the sloping front hollowed out to the same width as the cutting iron, and as far back as the mouth. The cutting iron is gouge-shaped to cut a shallow groove; a straight edge is used to guide the Plane when doing this. Its purpose is to cut grooves ('scores') on the surface of the soundboard on which the slides operate. These slides regulate the amount of air reaching any particular range of pipes and consequently the volume of sound that will come from them. The wind is apt to leak out between soundboard and slide and blow into a different hole from that intended, thus causing 'the wrong pipe to speak'. This 'running of the wind' is prevented by cutting scores in the top of the soundboard running zig-zag between the outlet holes; these conduct the leaking air to the end of the slide and away.

The only mention of this plane we have found in a trade catalogue is in a price list issued by G. Buck (London, *c.* 1860), not illustrated but described as an 'Organ Builder's Grooving Plane' and priced at 7/–.

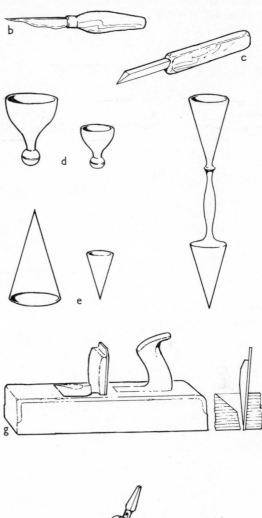

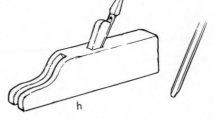

Fig. 435 (b)–(h)

Overrunner: see Truss Hoop under *Cooper* (*3*) *Hooping Tools*.

Out-Cannel

An outside grinding bevel; see explanation under *Gouge*.

Ovolo

A form of moulding; see *Plane, Moulding*.

P

Pad
A name given to a handle designed to hold a number of different tools – see *Awl Pad*; also to a separate bit-holder in which a Bit is permanently secured, and which fits into a socket at the foot of the Brace.

Painter *Fig. 436*
Certain tools of the painter's trade will be found under *Painting Equipment* below; also under *Knives* (*Chisel, Pallette, and Putty*); *Graining Comb*. An apron sold for use by painters is illustrated.

Fig. 436

Painting Equipment (Wheelwright's and Coach-builder's) *Fig. 437*
The colours used for wagons and carts depended on the tradition of the neighbourhood. Ronald Blythe (*Akenfield*, Allen Lane, 1969) quotes a Suffolk wheelwright: 'There was red lead and vegetable black, white lead, which was like thick distemper, and there was Chinese red and Venetian red, all these were the old colours used by the wagon-makers. The body-work was all painted blue. Always blue. The blue rode well in the corn. The wheels were done in Chinese red and lined-out with Venetian red, which was marvellously expensive – about £1 an ounce. We mixed all the paints here. Paint for small jobs was ground on a little stone but if we had a lot to do we ground it in a paint-mill. . . .'

Carriage painting became a separate trade, and used a great variety of colours both natural and made up, varnishes, japans, lacquers, and sizes. Pumice stone and other abrasives were used for rubbing down; many coats were applied and rubbed down before the final finished coat and varnish were put on.

John Philipson (London, 1897) writes of the work of the carriage painter as follows: 'A carriage owes much of its character and its power to please the eye, to the manner in which it is painted. It is in the painter's power to bring into prominence all that is graceful in the outline of a carriage, so that when the carriage is finished, and placed in the showroom, it may attract the attention and admiration of intending buyers.' Mr. P. Clarke, coachbuilder (Manchester, 1967), relates that the coach painters had a saying: 'Putty and paint will turn the Devil into a Saint.'

The tools and equipment to be found in the paint shop of a wheelwright or coachbuilder include the following:

(*a*) *Muller and Slab*. In many paint shops the pigments were ground on a stone slab of about 1 ft 6 in square; on this the raw colours, supplied in cakes, were ground by means of a Muller. This is a large pebble cut in half and ground flat on one side. It was held in the palm of the hand and worked in circles to grind and mix the paints.

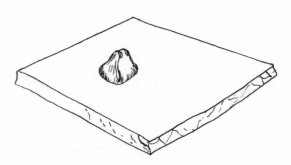

Fig. 437 (a)

(*b*) *Paint Mill*. Heavily constructed in cast iron, this is made in the form of a giant coffee mill with the same type of internal grooved grinding cone driven through bevel gears by a cranked handle. It superseded the Muller and Slab.

(*d*) *Brushes*. Those used on cart and coach painting include the ordinary painter's 'sash' tools, varnish brushes, and the more specialised sable 'liners' and 'writers' set in goose quills for lining, lettering, and heraldic painting.

Fig. 437 (b)

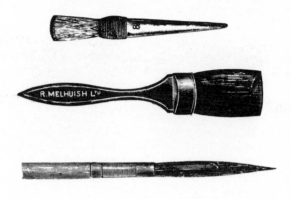

Fig. 437 (d)

(*c*) *Painting Horse*. A wooden post mounted on a heavy base with an old axle arm fitted at the top on which to hang a wheel while being painted.

(*e*) *Maulstick* (Mallstick; Mahlstick). A light stick (sometimes made in three pieces to be screwed together) with a small pad at the upper end. The pad is usually made of cotton wool covered with soft leather. The stick is held by sign writers and heraldic coachpainters with its padded end against the work to act as a support to the hand holding the brush.

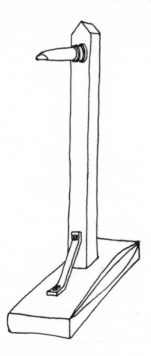

Fig. 437 (e)

(*f*) *Painter's Knives*. See *Knife, Chisel; Knife, Pallette; Knife, Putty*.

Paired: see *Handed*.

Palm: see *Sailmaker*.

Pane (pein, peen, pin, etc.)
The part of a hammer-head which is opposite to the striking face (see diagram under *Hammer*). Ball pane: a rounded ball on the end of the pane. Cross pane: a flat pane at right angles to the handle. Straight pane: a flat pane in line with the handle.

Parallel Strips: see *Boning Strips*.

Fig. 437(c)

Parting Tools: see *Chisel, Parting.*

Passer (Parsa; Parcey; Parcer): see *Drill, Passer;* or Truss Hoop under *Cooper (3) Hooping Tools.*

Pattern Maker

The pattern maker must understand the trade of moulder and founder besides his own highly skilled branch of woodworking. A pattern is a replica in wood of the object to be cast in the foundry. It is used by the moulder to make an exact impression (mould) in a matrix of sand into which molten metal is poured to produce the casting.

The design and making of the pattern is a complex process. So that it can be withdrawn from the sand without damaging the mould, it must be made in separate parts, accurately fitted together, and tapered slightly in one direction. To allow for the shrinkage of the molten metal as it cools, the pattern must be made larger in proportion to the amount of shrinkage expected; for this purpose the pattern maker uses a special rule with a biased scale.

The pattern maker also makes core boxes. These are wooden moulds made in two halves, for producing a core of clay and sand which is hung within a mould. This is done when a hollow space is required inside the metal casting that is to be produced. The clay-sand core is provided with spigots that protrude from its sides or ends for resting in the surrounding sand of the mould. These keep the core in place while the molten metal runs round it.

Woods used in pattern making include straight-grained pine and mahogany. In addition to the usual woodworking tools, the following special tools belong to the trade: *Chisel, Paring; Chisel, Turning* (Scraping Tool); *Gouge, Pattern Maker's; Hammer, Pattern Maker's; Plane, Pattern Maker's* (including the Core Box Plane); *Rule* (Pattern Maker's Construction Rule).

Patterns, Templets and Jigs *Fig. 438*

Patterns (or templets) are like the brown paper patterns used by a dressmaker – thin pieces of material cut to the size or profile of the finished article and used as a guide to their external shape during manufacture.

In this category are the innumerable templets used, for example, by toolmakers as a guide for making Saw handles, Planes, Spoke Shaves, and other tools, and by tradesmen generally for shaping workpieces of many kinds.

Note: The word Pattern is also applied to the wood models used in a foundry for making an impression in the sand in which metal castings are produced – see *Pattern Maker.*

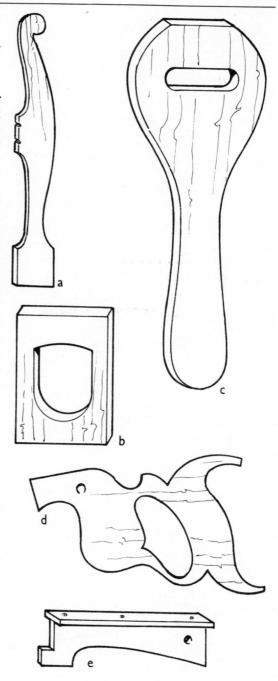

Fig. 438 Toolmaker's templets used when making:
(*a*) the frame of a Bow Saw (*b*) the fence of a Marking Gauge (*c*) an upholsterer's Web-Strainer (*d*) a Saw Handle (*e*) a Spoke Shave

Jigs are devices which are laid over the object to be manufactured to act as a guide to the tool being used to shape it. They are used in order to produce identical dimensions in repetition work.

Every manufacturing trade employs aids of the above kinds which may vary greatly from workshop to workshop. Examples of those used for making Planes will be found under *Plane Maker.* A selection of templets used by toolmakers are illustrated above.

Pawl: see *Ratchet and Pawl.*

Peavey: see *Timber Handling Tools.*

Peeler (Peeling Iron)
A steel blade, shaped like a spade, mounted on a handle about 4 ft long. Used by tree fellers for removing bark, to prevent beetles from breeding behind it. See also *Barking Iron.*

Peg Cutter
A name sometimes given to a *Dowel Plate; Knife, Bench;* Tine Former under *Rake Maker.*

Pellet
A tapered plug or disc of wood with the grain running crosswise. Used by cabinet makers to fill in recesses made in furniture to receive the heads of screws; and by shipwrights to fill in the counter-bore on the deck of a ship in which a bolt-head is sunk. (See *Auger, Deck Dowelling; Trenail.*)

Peltie
Scots shipwright's term for a heavy hammer. See *Hammer, Ship.*

Pencil, Carpenter's *Fig. 439*
The traditional carpenter's pencil is oval rather than round, with a wide lead usually cased in cedar.

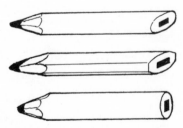

Fig. 439

292

Phillip's Tool: see *Screwdriver, Phillips.*

Pianoforte Maker and Tuner *Figs. 440; 441*
In a piano, the sound is produced by vibrating strings which are stretched over a sound box. The strings of the spinet or harpsichord were plucked; the idea of striking the strings, instead of plucking them, marked the development of the piano.

Piano making involves a number of different processes which are usually carried out in different departments, or indeed by different firms. There are those who make the frame (which must be strong enough to resist the combined tension of the strings amounting to over ten tons); the stringers who also insert the wrest-pins round which the end of each string is wound and by which its tension can be varied for tuning; the piano-action makers who produce the delicate mechanism that connects the key to the hammer which strikes the strings and upon which the quality of the music so much depends; the keymakers who assemble the keyboard and who cut and fix the ivory keys; lastly, the makers of the outer cases which are often outstanding examples of the cabinet maker's art.

The tools used for making the cases are largely those of the cabinet maker. We have not included the many specialised and largely metal-working tools used for making the action and other internal mechanisms.

Piano Action Fig. 440
The example illustrated is an obsolete type which demonstrates the once common use of vellum hinges. Modern actions are made on a similar principle, but the leather hinges have been replaced by yoke joints with centre pins.

Piano and Tuning tools Fig. 441 overleaf
The following tools are from the kit of a piano tuner who also carries out minor repairs:

(*a*) *Tuning Hammer* (Spinet Hammer; Piano Wrester)
A small key-wrench with a square, oblong, or star hole. The steel stem is 5 in long with a hardwood cross handle capped with metal.

Used for tuning by turning the wrest pins on which the ends of the piano wires are wound. The cross handle is occasionally used as a Hammer for tapping home the wrest pins when necessary.

(*b*) *Tuning Lever*
A key-wrench made in the form of a lever.

(*c*) *Key Spacer*
A cranked steel stem with a forked end mounted on a wooden handle, 10 in long overall. Used for

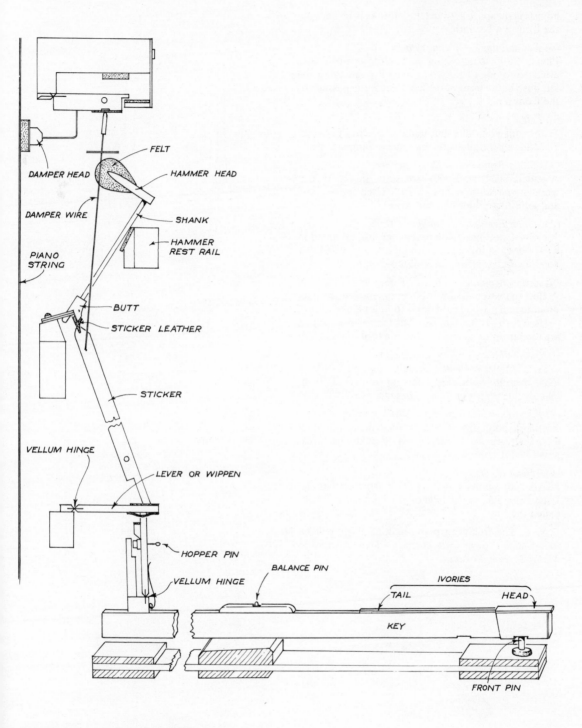

Fig. 440 Diagram of a 'sticker' Piano Action

adjusting the space between the white keys by bending the front pin beneath.

(d) Key Bushing (or Easing) Pliers

Pliers, 7–8 in long, with a loose circular metal disc attached to one of the jaws. Used for squeezing the felt bush in the head of the key to ease the passage of the front pin.

(e) Felt Knife

With cranked handle, but made in various patterns. Used for shaping felts for hammers, dampers, etc.

(f) Action Regulator – Check Bender

A square steel head with two slots, mounted in a wooden handle, 6 in long overall. Used for bending and adjusting under-damper wires.

(g) Action Regulator – Damper Crank

A circular steel head with cruciform slot, mounted in a wooden handle, 11 in long overall. Used for bending and adjusting over-damper wires.

(h) Action Regulator – Hopper Turner or Set-off Tool

A steel rod with a hollow oblong opening at the end, mounted on a wooden handle, 16 in long overall.

Used for twisting Hopper Pins (screw eyes) which are too far inside to reach without a tool of this kind.

(i) Dust bellows

Older examples are made entirely of wood, 20–27 in long overall. Used for blowing out dust from inaccessible places in pianos, organs, etc.

(j) Piano Key Bit (Pianoforte Bit)

A small Spoon Bit with a pointed nose. Made in assorted sizes for boring holes to take the wrest pins (see (a) above).

(k) Toning needle

Needle-like points mounted in a wooden handle. Used for pricking the felts of the hammer heads to soften the tone.

Note: The following piano-making tools will be found under their own entries: *Plane, Pianoforte Maker's; Saw, Vellum.*

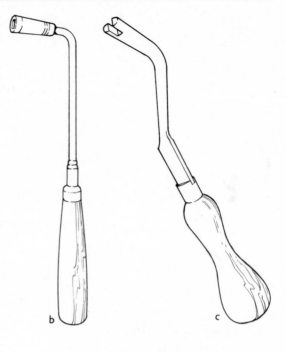

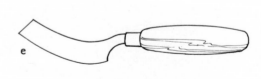

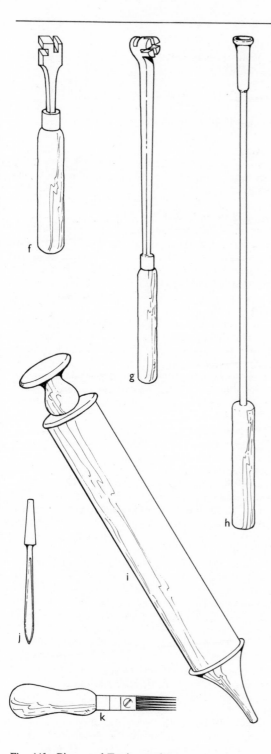

Fig. 441 Piano and Tuning tools

Pickaroon: see *Timber Handling Tools (Sappie)*.

Picture Framing Cramp: see *Cramp, Corner*.

Piercer (Persore; Piercel)
An early term for small boring tools such as the Awl or Gimlet, and later by extension for the Bit of a Brace or Wimble. Moxon (London, 1677) used the word Piercer for the Brace as a whole. Certain Augers are sometimes called Piercers (e.g. see First Pump Bit under *Pipe and Pump Maker*).

Pin Maul: see *Hammer, Ship's*.

Pincer Hammer: see *Hammer, Pincer*.

Pincer Spanner: see *Wrench (1) For Gripping Nuts*.

Pincers, Carpenter's *Fig. 442* See also *Pliers*.
Used mainly for pulling nails, most of these tools have knob-and-claw ends to the grips. The purpose of the knob is doubtful, but it may be to prevent the hand from slipping off the grip. Variants include:

(*a*) *Tower Pincers*. These have rounded jaws and round-shouldered grips with knob-and-claw ends. Sizes from 5 to 12 in long overall. The origin of the term 'tower' is unknown, but it may relate to the name of a maker. In their 1911 catalogue, Ward & Payne of Sheffield apply the term to Shoe Pincers and Gas Pliers as well as to Carpenter's Pincers. Apart from being forgings of good quality, usually left in the black, they do not appear to have any other feature in common.

(*b*) *Shouldered or French Pattern Pincers*. Similar to the Tower but with flatter jaws and square shoulders. Made in sizes from 6 to 8 in long overall. (Shouldered pincers of this type have been found at Roman camp sites in Germany.)

(*c*) *Lancashire Pattern*. Flatter jaws and no shoulder; a conical point is sometimes provided instead of the usual knob. A distinctive feature is the black, hand-forged appearance.

(*d*) *Boxed Pincers*. Rounded jaws and a box joint.

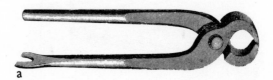

a

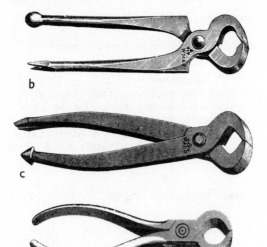

b

c

d

Fig. 442

Pincers, Cooper's: see *Cellarman.*

Pincers, Upholsterer's: see *Upholsterer.*

Pinch Bar: see *Crow Bar.*

Pincher Jack *Fig. 443*
A name given to a travelling smith who travelled over England and Wales stopping at the forges to make pincers out of old rasps for farriers and other tradesmen. Born at Llanbedr, near Crickhowell, about 1860, his real name was Jack Millet. He died in Crickhowell in 1926. In the 1940s he was still remembered by older tradesmen as a man of almost magical skill. His legend illustrates the persistent, though not always justified, belief in the superiority of the hand-made tool over the factory product. The sketch shows a pair of pincers from the forge of Mr. John M. Kaye of Staincliffe, near Dewsbury in Yorkshire. They were made by Pincher Jack *c.* 1905. (See References, Salaman 1960.)

Fig. 443

296

Pinion
A small toothed gear wheel meshing with a larger spurwheel.

Pinker Tool: see *Print and Block Cutter.*

Pinking Tool: see *Coffin Maker* and *Upholsterer.*

Pintle
A pin or bolt, for example one on which some other part turns, as in a hinge; also the spindle of a pulley block.

Pipe and Pump Maker *Fig. 444*
Pipes and pumps for supplying water have been made in wood in this country since the Middle Ages – and most probably before. A well known drawing by Leonardo da Vinci (*c.* 1500) depicts a machine intended for boring logs for this purpose. Wooden lift pumps were still being made near Dublin until 1950.
Materials. Elm was often employed owing to its durability when kept constantly damp, and also because it is easily obtained in long straight lengths in most parts of the country. Elm pipes laid down in London in the seventeenth century have been found to be still sound when unearthed in recent years. Pump barrels were also made of larch and probably other woods.
Process. Holes were bored through the length of the logs with giant-sized Augers. The logs were laid horizontally on a trestle, carefully marked out, and bored halfway through, starting from each end. The Auger was mounted on the end of a 12 ft handle, which was supported on a forked post, as sketched below. According to O'Sullivan (Ireland, 1969) two men turned the Auger, while a third forced the Bit into the log by pulling the shaft of the Auger towards it.
Pumps. Wooden pumps, like the metal lift or yard pump of today, consist of a tube with a valve fixed at the bottom and, above it, a sucker valve (the 'pump bucket') which is moved up and down in the tube by a lever handle. Pump buckets were turned out of wood, usually elmwood, and flanged with leather. The 'clapper' valve was of leather, weighted with a block of wood or lead.
A hooked iron bar was used for inserting or removing the lower valves; and a long pair of 'lazy-tongs' was sometimes used for recovering a pump bucket if it became detached from the pump rod.
Note: Writers who describe Pipe and Pump making include the following (see References): H. L. Edlin (1949); Walter Rose (1937); J. C. O'Sullivan (1969).

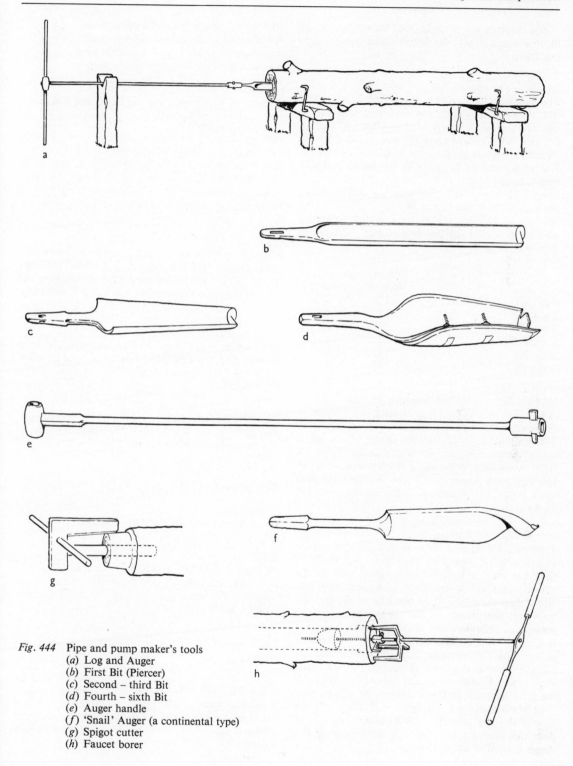

Fig. 444 Pipe and pump maker's tools
    (*a*) Log and Auger
    (*b*) First Bit (Piercer)
    (*c*) Second – third Bit
    (*d*) Fourth – sixth Bit
    (*e*) Auger handle
    (*f*) 'Snail' Auger (a continental type)
    (*g*) Spigot cutter
    (*h*) Faucet borer

The tools used include the following: (some illustrations will also be found under *Auger, Pipe and Pump*.

(a) The boring operation

The hole was bored in two stages, first with a 'nose' type Auger ((b) below), and then reamered out to size with a series of tapered Pump Augers of ascending size (b). These Augers (known as 'Bits') were turned by a cross handle on a long stem or rod (e). The work was done horizontally and the long handle was supported outside the log on a post or rest. Special tools were used for tapering and recessing the ends of the pipes (g and h).

J. C. O'Sullivan (1969) records the following sizes of Bits used for boring out logs for wooden pumps in Ireland.

	Length of blade (in)	Maximum width (in)
First Bit	17½	2¼
Second Bit	15	3
Third Bit	11¾	3½
Fourth Bit	15	4
Fifth Bit	13¾	4⅞
Sixth Bit	14¼	5½

(b) First Pump Bit (Shell Auger or Piercer)

A parallel Shell Auger 16–20 in long and 2–2½ in. in diameter, with in-bent cutter at the nose. This tool bored a pilot hole through the centre of the log, a shallow recess being first cut with a gouge to give it a start. According to W. Rose (Cambridge, 1937), a screw-pointed Auger could not be used for this work because it 'insists on following the direction of the grain'.

(c), (d) Taper or Reaming Bits

Tapered Shell Augers, some with a flat nose, made in sets with a diameter of from about 2–8 in. With these the pilot hole could be reamered out to the size required with a succession of Augers, each one larger than the other. Only one side of the tapered body was sharpened; on the other side there were sometimes provided two or more holes, used for bolting (or tying on with a well-soaped cord) an iron or wood packing piece or 'liner'. This increased the effective cutting diameter of the Bit; it was used when it was desired to enlarge the bore to a diameter less than that made by the next size of Bit in the series.

In Mathieson's list (Glasgow, c. 1910, p. 42) the nose of this tool appears to be a perfectly round flat disc; this presumably acted as a guide to keep the Auger central in the pilot hole.

(e) Turning Handle (Pump Rod)

A rod about 12 ft long with an eye at one end for the cross handle. There is a socket at the foot to take the Bit, with a metal wedge to secure it.

(f) 'Snail' Auger

A type of Auger used on the continent for pump boring (see *Auger, Snail*).

Jointing Tools. The wooden pipes or pump barrels were jointed by tapering one end and driving this into a corresponding recess in the next length of pipe. This work could be done with an Axe and tapered Auger, but F. C. Morgan (Newcomen Transactions, 1942) describes the following tools for this purpose:

(g) Faucet Tool

For cutting the recess. A boring device similar to the wheelwright's Hub Boring Engine with the self-feeding screw supported by a plug driven into the bore of the pipe. Presumably, this was followed by a tapered Auger.

(h) Spigot Cutter

For cutting the tapered end. A cutter or knife is pivoted on a mandrel fitted into the bore of the tube. When revolved it cuts a taper as 'true as if turned in a lathe'.

Pipe Tools: see under *Pipe and Pump Maker; Plumber's Tools; Screw Dies and Taps; Wrench (2) For Gripping Rounds; Yarning Tool.*

Pipe Wrench: see *Wrench (2) For Gripping Rounds.*

Pistol Router: see *Router, Coachbuilder's.*

Pitch

Of tools – the angle at which a cutter or other part is set, e.g. the inclination of the teeth of a Saw or the inclination of the cutting iron of a Plane. Of a gear wheel – see *Millwright*.

Pitch Board: see Patterns and Jigs under *Plane Maker*.

Pitch Pot and Ladle *Fig. 445*

Found in many of the older workshops and where coffin-making is done. See *Coffin Maker;* Pitch Pan under *Brushmakers;* Pitch Ladle etc. under *Caulking Tools.*

Fig. 445

Pivot

The point of support (fulcrum) of a lever; or the pin of a hinge; also the reduced end of a shaft which is held in a bearing.

Plane *Figs. 446; 447*

Note: For other general information on planes see: *Plane, Bench; Plane, Grooving; Plane, Metal; Plane, Moulding; Plane, Rebate; Plane Maker.*

The invention of the Plane was the most important advance in the history of woodworking tools in the last two thousand years and it appears to have taken place during the Roman era since no Planes have been found among Greek or Egyptian remains.

The earliest known Planes, with a stock of wood, or in the form of an iron sole with side plates and a wooden core, were used by Roman joiners at the beginning of the Christian era. The cutting iron was fixed by a wedge driven tight against a bar across the mouth. Tapered grooves for the wedge were introduced early in the sixteenth century. The double cutting iron (i.e. with back or cap iron) is encountered from about 1760.

The essential feature of a Plane is the built-in control provided by the sole which allows the worker to employ his full strength simply as driving force.

The various tasks of which Planes are capable can be performed by simple edge tools, i.e. a Chisel, but so much more skill and sheer physical effort are required to keep the edge of a Chisel on course, and to prevent it from 'digging in', that many woodworking operations which we take for granted would

be impossibly laborious without Planes. A worker with the right Plane for a particular task has only to concentrate on holding it in the correct attitude in contact with the workpiece, and he can then push away freely until the task is completed.

We may consider the operations performed by Planes as being of three kinds: shaping (or sizing), fitting, and finishing. The distinction is to a large extent arbitrary – for example, in many tasks shaping and finishing are done in one operation with a single Plane – but it is sufficiently real to have led to Planes of specialised function being found in the general woodworker's tool kit from an early date.

1. The use of Planes for the preliminary shaping of workpieces – as contrasted with the fitting of the pieces together or the finishing of the completed job – has greatly diminished since timber in a wide variety of dimensions, thicknesses, and profiles has been made available ready-machined from the saw-mill. But when a workman has to prepare his own 'stuff' from the solid or from thick boards sawn over the pit, he keeps Planes specially for removing the waste wood as quickly as possible. For this purpose close accuracy is not necessary and a certain amount of 'tearing out of the grain' can be tolerated; so the Plane is set 'coarse' to take a thick shaving, has a wide open mouth to allow the shavings to pass, and, since the resistance would otherwise be too great, the cutting iron is often narrow and its edge slightly convex. The length of these Planes is immaterial within certain limits; it is, however, necessary to keep the surface of the work reasonably level, and therefore the traditional English Plane for preliminary shaping, the Jack Plane, is 12–17 in long. The German equivalent, the *Schropphobel*, is only about 9 in long. The shortness of the German Plane, together with its upstanding front horn or handle, makes it very handy to use, with the result that examples are often found in English kits and are known as 'Bismarcks'.

2. For the second type of operation – fitting – where the purpose is to true and adjust surfaces so that they exactly match the surface of an adjoining piece, accuracy is all important. The characteristics of Planes for this purpose are therefore fine set irons with only the slightest curvature of the edge, and the longest sole which can conveniently be managed. The obvious example in the English kit is the Trying or Jointing Plane which is used on the edge of boards to make them butt together so closely that the joint is invisible. These Planes are used on the long edge of the board, held clear above the Bench Vice, so they can themselves be long (26–30 in), but other Planes, used for rebates, such as the

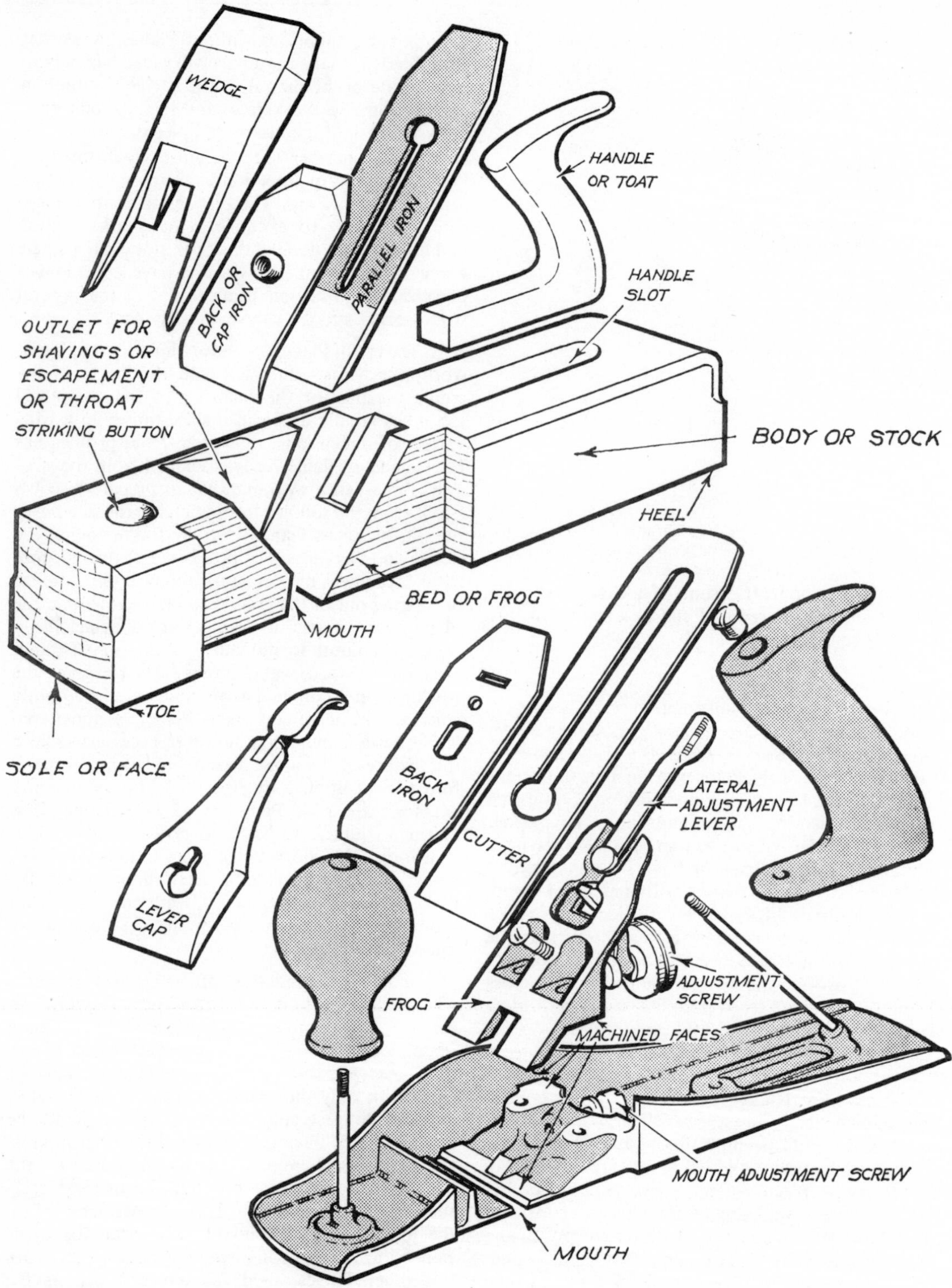

WEDGE

BACK OR CAP IRON

PARALLEL IRON

HANDLE OR TOAT

HANDLE SLOT

OUTLET FOR SHAVINGS OR ESCAPEMENT OR THROAT

STRIKING BUTTON

BODY OR STOCK

HEEL

BED OR FROG

MOUTH

SOLE OR FACE

TOE

BACK IRON

CUTTER

LATERAL ADJUSTMENT LEVER

LEVER CAP

FROG

ADJUSTMENT SCREW

MACHINED FACES

MOUTH ADJUSTMENT SCREW

MOUTH

Fig. 446 Parts of a wood and iron Plane

Shoulder Plane and the Bullnose Plane, although having similar requirements, are shorter, either because the work surface itself is short or because they are designed for use in a restricted space.

3. Finally, Planes for finishing, called Smoothers or Smoothing Planes, are designed to produce a finished surface leaving the minimum of work to be done with Scraper or glass paper. The essential features are a very finely set iron, with the minimum curvature of the edge, or with only the corners rounded so as not to leave score marks, and the narrowest possible mouth to prevent the grain from tearing up. In joinery and cabinet making, the object under construction will normally have been assembled and glued up before the Smoothing Plane is used, so that any remaining broad undulations can be ignored, and the Plane is short to enable it to reach any local depression.

4. Under the individual plane entries will be found many examples of Planes designed for special jobs which perform more than one of the three operations described above. In some applications, e.g. the making of decorative mouldings, there is no element of 'fitting' involved and the Plane which shapes the wood should leave the surface as nearly finished as possible. In other instances, such as 'ploughing' to produce grooves, the result will not be visible, and a certain roughness may even be desirable if it is to form part of a glued joint.

Materials
English Planes are usually made of beech and are rarely carved or decorated; their makers held to a tradition of severe yet graceful lines. The cores of wood-and-metal Planes are often made of rosewood or ebony, and some small Planes were made of boxwood throughout. Boxwood is also widely used for inserted pieces to resist wear. Such Planes are then described as 'boxed'. Metal Planes were made in mild steel or gun metal but are now made in cast iron. On the Continent, many different woods are used, and at least until the mid-nineteenth century Planes were often beautifully carved. See *Plane, Continental.*

NOTE ON SPECIAL PLANES
Throughout this section we have recorded the names of a number of Planes of unknown appearance and purpose, and it may be helpful to quote from a letter about this subject written to the author by Mr. Edward Pinto, a noted authority on woodworking tools (Northwood, 1971).
'There are many ancient planes which make no sense to the modern man, such as the fashion for

putting handles off-centre, which persisted over quite a long period. The plane has always been a challenge to inventiveness, and although there are so many recognised types, there are a number which do not fit completely into any category. There are also a few which turn up from time to time which do not seem to work, and do anything useful. In the Dolling tool chest we had one made of rosewood – a most beautiful tool, but it just would not do anything except choke itself after one or two strokes. I lent it from time to time to older craftsmen than myself, but none of them could do anything with it, and I came to the conclusion that it represented some woodworker's experiment, and that he spent so many weeks making it that he could not bring himself to throw it away.'

Cutting Irons Fig. 447
These may be 'common' with the blade tapering in thickness from the cutting edge, or 'parallel' with the blade the same thickness throughout the length. The latter were used for high-class Bench Planes as the mouth aperture remains constant no matter how often the cutter is ground and sharpened. Special thin irons of parallel thickness are used in metal Planes of the Stanley-Bailey type. (See *Planes, Metal.*)
The following are the present usual angles for the cutting irons (approx.):

Bench Planes:
for softwoods	45°	'Common pitch'
for hardwoods	50°	'York pitch'
Rebating and Grooving Planes	50°	
Moulding Planes:		
for softwoods	55°	'Middle pitch'
for hardwoods	60°	'Half pitch'
Side Rebates, Side Snipers and Toothing Planes	Vertical or nearly so	
Mitre Planes, Shoulder Planes, and others, for fine work across the grain	20° or less, but with the sharpening bevel uppermost	

The following are some of the chief kinds of Plane iron:
(*a*) *Tooth irons*, with the face scored by parallel grooves, are used in Toothing Planes for preparing the surface for veneers. Before the introduction of the double iron they were also used for planing cross-grained material; indeed some Roman Planes had toothed irons for this purpose.
(*b*) *Snecked Irons* (Nibbed Irons). Some irons, such as those on Ploughs, Cooper's Jointers, and some

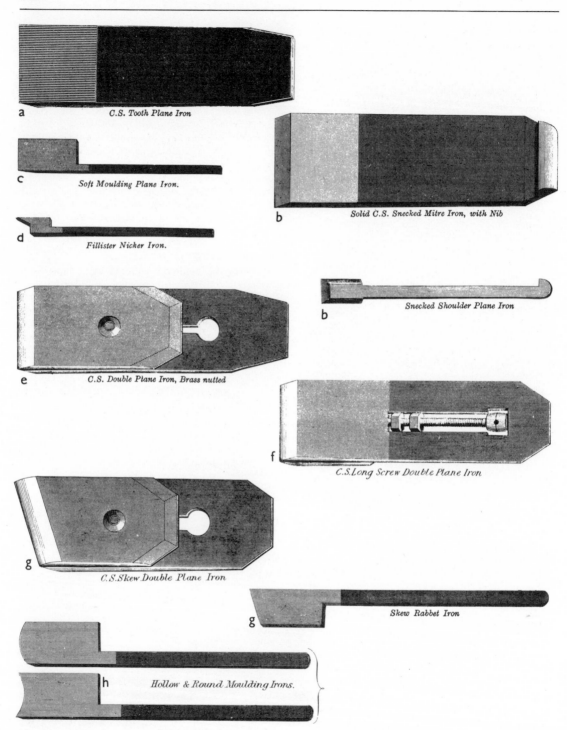

Fig. 447 Plane cutting Irons

metal Mitre Planes, are 'snecked', i.e. they have a projecting piece on the side of the iron near the top to enable them to be adjusted or withdrawn without striking (and damaging) the stock of the Plane itself.

(*c*) *Soft Blank Irons.* A name given to moulding-plane irons supplied in the form of blanks to be ground to the required shape by the customer, and subsequently hardened and tempered.

(*d*) *Spur, Tooth, or Nicker Irons.* These are vertical cutters which cut the fibres at the edge of a groove or rebate when working across the grain. They are used in Fillister, Dado Grooving, and certain other Planes.

(*e*) *Cap Iron* (Also called Break Iron, Top Iron or Back Iron. When a cap iron is fitted, the cutting iron is said to be 'double'.) Up to the eighteenth century all Planes had single irons; such irons are still used in some Jack and general-purpose Planes for rough work. Moulding, Grooving, and most Rebating Planes also have single irons. The cap is the upper member of the 'double iron'. It consists of a steel plate of the same width as the cutter, with one end curving downwards and ground to fit closely across the cutting iron, leaving a small gap behind the cutting edge. The main purpose of the cap iron is to break the shaving immediately it is raised, so that it does not tear out the grain ahead of the cutting edge when planing against the grain. It also tends to curl the shaving, and thus gives an easier discharge through the shaving outlet.

(*f*) *Fixing the cap iron.* The first cap irons may have been loose and held in position by the wedge; this type is still to be found in France. Later, two methods of securing the cap iron were adopted. The 'short screw' irons are secured by a screw passing through a slot in the cutting iron and engaging in an iron or brass nut on the cap iron, as illustrated in (*e*). In the 'long screw' iron, clearance between the end of the cap iron and the cutting edge of the plane-iron is regulated by a screw fitting lengthways in a slot in the cutter itself, as shown in (*f*).

(*g*) *Skew irons.* Those provided for Rebating and Badger Planes have a cutting edge set at an angle. There is a belief that a skewed iron makes a slicing and therefore an easier cut. This is probably a fallacy – at least in the case of any Plane which, guided by a fence, has to move straight forward; but the slanting blade is useful when starting a cut, for it enters like a guillotine. But once a shaving is raised across its edge, it cannot 'slice' unless the Plane is itself moved sideways. The main purpose of the skewed iron on Rebate Planes is that it helps to draw the Plane into the corner of the rebate, and it also throws the shaving to one side which helps to prevent choking.

(*h*) *Moulding irons.* The tails are cut away to leave room for a side outlet for shavings.

Iron removal. To remove the iron in a Plane of medium length e.g. a Jack Plane, the top of the stock is struck near the toe; some Planes have a special knob there for this purpose. With short Planes like Smoothers, there is not sufficient spring in the stock, so the iron is loosened by knocking the back. It would be impossible to hit the top of a Cooper's Jointer because it is too long, and useless to knock the heel; hence the 'sneck' in the iron itself. (See Snecked Irons above.)

Mitre Planes and certain Grooving (Plough) and Shoulder Planes have snecked irons because the top or back cannot be hit without damage to the tool.

Plane, Adjustable Mouth: see *Plane, Reform.*

Plane, Airtight (Case Maker's Plane; Show-case Maker's Plane) *Fig. 448*
A Moulding Plane which produces an interlocking 'airtight' joint between the meeting styles, or part of the frame, of show-cases, museum cabinets, etc. The purpose is to prevent the penetration of dust and damp. There are two varieties:

(*a*) Moulding Planes made in matched pairs with single or double tongue on one, and the corresponding grooves on the other.

(*b*) A Moulding Plane, sometimes called a Hook Joint Case Plane, with an S-profile cutting iron and an adjustable depth stop.

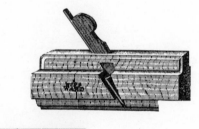

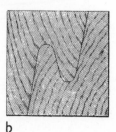

a   b

Fig. 448

Plane, American Wood Pattern *Fig. 449*
A name given to a series of wooden Planes fitted with
the thin Stanley-Bailey type double iron held in place
by a wooden wedge. Melhuish (London, 1912) listed
three types:

(*a*) A coffin-shaped Smoothing Plane 8 in long
with a 2 or 2⅛ in cutter.
(*b*) A Smoothing Plane with handle, 10¼ in long
with 2 or 2⅛ in cutter.
(*c*) A Jack Plane, 16 in long with a 2⅛ in cutter.

Fig. 449

Plane, Astragal
A general term for Moulding Planes which work a
moulding of semi-circular profile like a raised bead,
but usually larger, with a short horizontal flat on
each side of it.

Astragal mouldings are used, like beads, at the
closing joint of doors etc., to separate different
parts of more complex mouldings, and for working
the tracery moulding on the bars holding the glass
on cabinet doors. See *Plane, Moulding.*

Plane, Back Check: see Sash Fillister under *Plane,
Fillister.*

Plane, Backing: see Mother Plane under *Plane
Maker.*

Plane, Backmakers: see *Plane, Cooper's: Backmaker's
Jointer.*

Plane, Badger (Skew Plane; Skew Badger Plane) *Fig.
450*
A wooden Plane of jack or coffin shape, with a
skewed cutting iron exposed at the right-hand
corner. Planes are said to be 'badgered' or 'badger-
mouthed' (in Scotland 'skew' – or 'badger-eyed') if
they have skewed irons. According to the late Mr. A.
Collier (Brixton, 1947), this plane took its name from
a London plane maker named Charles Badger, who
was a member of the firm of Badger & Galpin, of
No. 1 Stargate, Lambeth, in 1863. (See *Plane, Panel*
for a diagram of a Badger and other similar Planes.)

The stock is of jack or smooth size and type,
sometimes with slipped or dovetail boxing to resist
wear. It was occasionally provided with an adjustable
fence slot-screwed to the sole. The cutting iron is
double (or occasionally single), skewed and bedded
at about 50°. The skewed iron has the advantage
that there is no need to rebate the stock; for it
emerges at the right-hand bottom corner of the sole
(where the iron is ground square to the cutting edge
for a short distance), leaving the stock intact to give
a bedding for the iron. It is also less liable to choke
and helps to draw the plane into the corner of the
rebate.

The Badger Plane was used for cutting and cleaning up wide rebates. There is reason to believe that
it gradually superseded the Panel Plane, for it was
still listed after the Panel Plane had disappeared from
the makers' catalogues. Owing to the skewed iron
it was able to dispense with the rebated sole of the
Panel Plane and could therefore cut a deeper rebate.
See also *Plane, Rebate.*

Fig. 450

Plane, Bailey's: see *Plane, Stanley-Bailey.*

Plane, Banding
Mercer (U.S.A., 1929) uses this term for a Plane which cuts wider grooves than the normal Grooving Plane or Plough. (The Planes he illustrates under this heading include a Badger Plane and a Dado Grooving Plane.) Mr. K. H. Basset (U.S.A., 1971) informs us that a Rabbet Plane with side spurs, known as a Dapping Plane, is used for cutting wide grooves in millwright's work. Holtzapffel (London, 1847) used the term Banding Plane for a plane which is 'allied to the gauges, and is intended for cutting out grooves and inlaying strings and bands . . .' He says that it resembled the Plough but was furnished in addition with the double tooth of the Grooving Plane.

Plane, Bannister: see *Plane, Hand Rail.*

Plane, Base Moulding
A term used for a Moulding Plane used for skirting boards and for mouldings immediately above the plinth of a wall or pilaster.

Plane, Bead: see *Plane, Moulding.*

Plane, Bed Mould: see *Plane, Cornice.*

Plane, Bed Rock: see *Plane, Stanley-Bailey.*

Plane, Belt Maker's *Fig. 451*
A small metal Plane consisting of a stock about 6 in long and $2\frac{3}{4}$ in wide, with a flat sole, and with side plates supporting a turned cross-handle. The iron is bedded at a low angle. The front of the sole is adjustable to regulate the size of the mouth.

Used by belt makers and repairers for chamfering down the lap joints of leather transmission belting.

Fig. 451

Plane, Bench (see Table overleaf)
A general term for Planes with flat soles used mainly on the bench for preparing and smoothing the workpiece. The following are some general characteristics of Bench Planes: The stock is a flat sole without a fence, and with the shaving outlet at the top of the stock. The cutting iron is usually double, bedded square at the following angles to the sole: 45° ('Common Pitch') for deal and similar soft woods; 50° ('York Pitch') for mahogany and all hard woods.

Typical examples of Bench Planes are described under the following entries: *Plane, Jack; Plane, Jointer; Plane, Metal; Plane, Smoothing; Plane, Trying.*

Nomenclature
The standard lengths and the corresponding names at various dates and from various sources are given in the table overleaf. An interesting feature of the table is the retention in the U.S.A. of the seventeenth-century English terms 'Jointer' and 'Fore', which have not been much used in Britain for many years. The Scottish term 'Panel' was adopted by the Spiers and Norris firms as a name for a metal Bench Plane of jack size. This Plane should not be confused with the wooden Panel Plane proper (see *Plane, Panel*).

Plane, Bevelling: see *Plane, Chamfer;* Side Chamfer Plane under *Plane, Coachbuilder's.*

Plane, Bismarck *Fig. 452*
A term used for a Plane of Continental type with a front horn or handle, and usually provided with a single iron set coarse for quick removal of waste. See also *Plane, Roughing; Plane, Continental.*

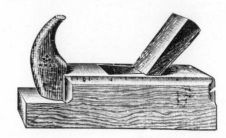

Fig. 452

Plane, Block *Fig. 453*
The term is applied to a range of metal Planes about 3–7 in long with a variety of devices for holding the iron and adjusting the cut or mouth. The single iron is bedded low (12–18°) with the bevel uppermost. (*continued overleaf*)

Table of Bench Planes (See *Plane, Bench*)

| | WOOD PLANES | | | METAL PLANES[3] | | |
| | SOURCES[2] | | | SOURCES[2] | | |
Length[1] (in)	Moxon 1677	Nicholson 1812 Holtzapffel 1847	Sheffield List 1910	Spiers 1845 Norris 1900	Stanley-Bailey, U.S.A. 1870	1960
30	—	Jointer	Trying	Jointing	Jointer	—
28	Joynter	Jointer	Trying	Jointing	Jointer	—
26	—	Long	Trying	Jointing	Jointer	—
24	—	Long	Trying	Jointing	Jointer	—
22	—	Trying	Trying	Jointing	Jointer	Jointer
20	—	Trying	—	Jointing	Fore	—
18	—	—	—	Panel	Fore	Fore
16	Fore or Jack	Jack	Jack	Panel	Jack	Fore
14	—	Jack	Jack	Panel	Jack	Jack
12	Strike Block	Straight[4] Block	—	—	Jenny[5]	—
9	—	—	—	Smoothing	Smooth	Smooth
8	Smoothing	Smooth	Smoothing	Smoothing	Smooth	Smooth
7	—	Smooth	—	Smoothing	Smooth	Smooth

Notes:
  [1] The lengths are approximate, and in the case of Moxon are largely guesswork.
  [2] Particulars of the sources quoted along the top of the table will be found under References.
  [3] Particulars of the makers will be found under *Plane, Metal*.
  [4] This Plane is mentioned only by Nicholson.
  [5] This is a metal Plane with a wooden sole. (See Wood-Bottomed Plane under *Plane, Stanley-Bailey*.)

These Planes are designed for use with one hand for trimming small work. The low-pitch iron is especially suitable for end grain.

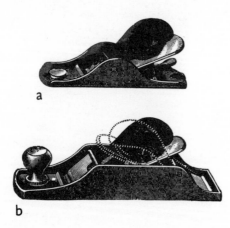

Fig. 453   Block planes
(*a*) Common type
(*b*) Double ended (or 'Duplex') with one end normal and the other bull-nosed
(*c*) Low angle
(*d*) Rabbet type

Plane, Bookcase Shelf *Fig. 454*
Mentioned in the Varvill list (York, *c.* 1870), it may have been used for working the canted grooves for adjustable book shelves. It was probably made right and left hand, and the detachable slip on its side must have been removed to enable the first groove to be worked, but was replaced for subsequent grooves for which it acted as a depth stop. A spur was fitted to cut across the grain at the deep side of the groove.

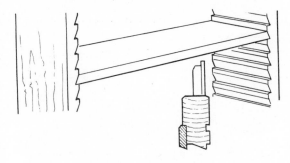

Fig. 454

Plane, Botel: see *Plane, Inbowing.*

Plane, Bowling: see *Plane, Cooper's Smoothing.*

Plane, Box Maker's (Flogger; Box-Maker's Smooth Plane) *Fig. 455*
A coffin-shaped Plane up to 9½ in long, rounded at both ends, with an iron, 2–2⅛ in wide, set at about 35°. Used by box makers for cleaning up rough timber used in box and crate making. See also *Plane, Roughing.*

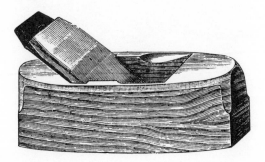

Fig. 455

Plane, Box Scraper *Fig. 456*
A small, metal, plane-like tool pivoted between the prongs of a forked handle. The stock is usually oval and carries a single cutting iron 2 in. in diameter, held in place by a hinged cap and screw. Used for erasing marks and brands on casks, boxes, etc. See also *Scrapers; Drawing Knife, Cooper's Round Shave.*

Fig. 456

Plane, Boxing: see *Plane, Rebate; Plane Maker.*

Plane, Bullnose *Fig. 457*
The planes in this group are made in wood or metal, with the cutting iron set close to the toe end of the stock which measures about 3–6 in long and 1¼ in wide. The iron is single, square, and often bedded at a low angle. Used for fitting, cleaning up small rebates, chamfers, etc. The very short sole in front of the cutting iron enables it to remove shavings close to a corner or stop.

(a) Wooden examples
Early examples are coffin-shaped. In one type the toe end of the stock slopes forward to overhang the sole and is reinforced by a thin wooden or metal plate; in another, the toe slopes backwards and is also fitted with a metal plate.

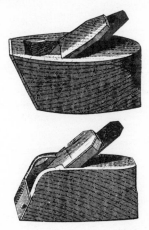

Fig. 457 (a)

307

(*b*) *Metal examples*

Made in iron or gunmetal with the cutting iron exposed at the sides. This, which may be set as low as 15° (but with the sharpened bevel upwards), is bedded on a wooden core and fixed with a wedge of rosewood or ebony. Later models had screw adjustment at the rear.

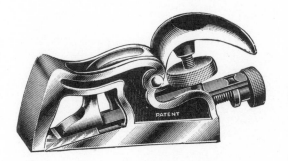

Fig. 457 (b)

(*c*) *Combination Pattern*

Those made by such firms as Stanley and Record, are completely adjustable. Some are combination tools and have two replaceable 'noses'; with the longer one the tool becomes a Shoulder Plane, with the shorter, a Bull Nose Plane, and with the noses removed altogether the tool becomes an Edge Plane.

Fig. 457 (c)

Fig. 457 (d)

(*d*) *Variants in cast iron*

These are japanned or nickel-plated, and rather outlandishly curved and decorated, with the cutter secured by a knurled screw. An adjustable fence is sometimes fitted, and in some cases a depth stop.

Plane, Capping: see *Plane, Hand Rail*.

Plane, Carcase Grooving: see *Plane, Dado Grooving*.

Plane, Casemaker's: see *Plane, Airtight*.

Plane, Casement: see *Plane, Hollow and Round; Plane, Inbowing.*

Plane, Cavetto: see *Plane, Moulding.*

Plane, Centre-Board (Centre-Board or Centre-Bead Moulding Plane; V-Plane; Wagon-Builder's Beading or Side Board Plane) *Fig. 458*
This group includes various narrow Grooving and Moulding Planes, often home made, designed to form a decorative groove, reed, bead, or narrow moulding down the centre of the side and end-boards of wagons and carts, and for decorating boards used in other situations. To guide the Plane, the stocks were often fitted with home-made fences mounted on twin stems, or alternatively a fence stick, consisting of a straight batten, was nailed temporarily to the board to serve the same purpose.

A Moulding Plane of this type with a V-shaped cutting iron, was used for making the grooves in bureaux or desks for holding pigeon-hole dividing pieces.

(*a*) Home-made examples
(*b*) A Centre-Bead Plane from the Stanley range

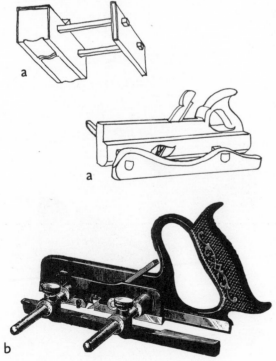

Fig. 458

Plane, Chair-Foot
Plane, Chair Splat
Listed but not illustrated by James Cam (Sheffield *c.* 1800), Varvill (York, *c.* 1870), and the Sheffield Standard List (1862). Appearance and purpose unknown.

Plane, Chair Rail
Listed but not illustrated by Mathieson (Glasgow, *c.* 1900). It is a Moulding Plane with a shifting fence, and the profile produced was an elongated ogee and quarter round. It was presumably used for working a moulding on chair rails.

Plane, Chamfer (Stop Chamfer Plane) *Fig. 459*
(*Note*: a form of Shave called a Chamfer Plane is described under *Drawing Knife, Cooper's Chamfering*.) A chamfer is a bevel formed by removing the sharp corner of a board or beam and leaving a surface, usually at 45° to the workpiece. A stop chamfer is a bevel which does not run all the way along to the end of the workpiece, but instead stops a few inches short and ends with a plain or moulded 'stop'. The name Stop Chamfer Plane does not relate to the chamfer but to the stop fitted to the plane which regulates the depth of the chamfer. The stop chamfer itself is worked with a Draw Knife and not with a Plane. For illustration see under *Chamfer*.

The examples described below all have a stock about 9 in long, 3 in deep, and 2 in wide, with a sole hollowed out in the form of a deep 90° V-shaped groove to fit over the corner of the workpiece to be chamfered. The iron is single, 1½ in wide and set at 45°.

The Plane is used for working chamfers at 45° to any required depth up to the limit of the cutter width. The Plane is held so that one surface of the V-sole bears against the side of the work. It ceases to cut when the other surface of the V reaches the top edge of the work.

The following variants differ mainly in the type of stop provided:

(*a*) A moveable block in front of the cutter and behind the wedge, with its lower end cut to align with the sole of the Plane.

(*b*) A fence which is adjustable laterally by means of two screws by which it is secured through slots to the sole, thus controlling the width and depth of the chamfer. The iron remains fixed and projects slightly below the front of a V-shaped opening in the stock.

309

(*c*) A 'box', which forms the working sole, can be moved up and down inside the stock, its position being fixed by a screw at the side. Its projection from the bottom of the V sets the required chamfer depth.

(*d*) The metal example has a stock about 9 in long with $1\frac{5}{8}$ in cutter and a V-shaped sole similar to the wooden version. There is an adjustable front section, raised or lowered according to the width of chamfer required. The plain cutter can be replaced by profiled cutters for making a moulded chamfer; these are fixed vertically and have a scraping action.

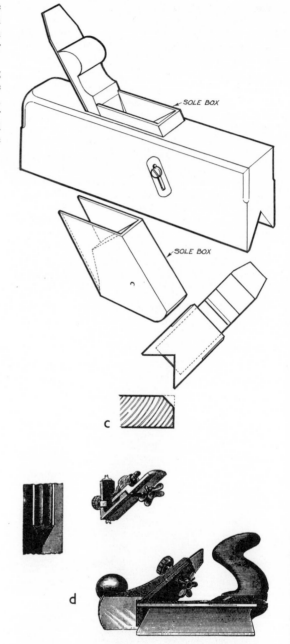

Fig. 459

Plane, Chariot (Scroll Plane) *Fig. 460*
A term applied to small Planes, 3–4 in long, which are similar to Block or Bullnose Planes.

(*a*) The wooden version resembles the eighteenth-century Scandinavian *Skav* or Dutch *Schaaf*, with rounded heel and scroll-shaped front grip curved backwards like the toe of a Persian slipper. The cutting iron is set at a low angle, usually near the front of the stock.

(*b*) The metal version is made in iron or gunmetal, and the wedges are often rosewood or ebony.

(*c*) A so-called Irish pattern, 9 in long, has the sole projecting some distance in front of the cutter. [Not illustrated]

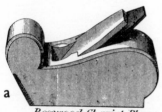

Boxwood Chariot Plane

Iron Chariot Plane

Fig. 460

Plane, Check
A name sometimes given to the *Plane, Rebate*, to the *Plane, Fillister*, and to the Door Check Plane which is described under *Plane, Coachbuilder's*.

Plane, Chinese Types *Fig. 461*
Chinese Planes (which may occasionally be found in British workshops) often have a central ridge running down the top surface of the stock. The cutting irons are sometimes secured in the stock in an archaic manner, like the Roman Planes. Contrary to the general belief, Chinese joiners use a bench for their work and the Planes are pushed away from the body. (cf. *Plane, Japanese.*)

The following are examples:

(*a*) *Jack and Smoothing Planes*
The stocks measure about 14 and 9 in respectively in length, and are about 3 in wide. The cutting iron is secured behind a bar across the escapement, sometimes with an iron wedge. A shaped cross-handle is fitted on the top of the stock, immediately behind the iron.

(*b*) *Moulding Planes*
A small stock about 6 in long. The iron is let into a slot in the side of the stock and held in place with a thin wedge, as in some Roman and medieval Moulding Planes. The most common profiles are various types of bead.

(*c*) *Plough Plane*
The rectangular wooden stock is about 7 in long and $1\frac{1}{2}$ in wide, with a projecting strip $\frac{3}{8}$ in square in the middle of the sole. A narrow fence is let through the stock just in front of the iron and held in place by a wedge. The cutting iron, of the same width as the projecting strip on the sole, is let into a slot cut in the side of the stock, as in (*b*) above.

In use, if this tool was pushed, as is the usual method with Chinese planes, it must have worked from the right-hand face of the material. With the narrow fence considerable skill would be needed to start the cut, but once the groove is partly run, the central strip on the sole, corresponding to the plate of a European plough, would engage in it and keep the tool straight.

(*d*) *Rabbet Plane*
The stock is about 24 in long, $2\frac{1}{2}$ in deep, and 2 in wide, with a wide, thin strip nailed or dowelled to the sole to form a fence. The thin cutter, about $1\frac{1}{4}$ in wide, is let into a slot in the side of the stock and bedded with a slight skew. The upper edge of the slot is shaped to a curve to allow the shavings to emerge. In use, the rebate is worked on the right-hand edge of the material.

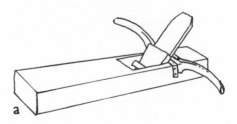

311

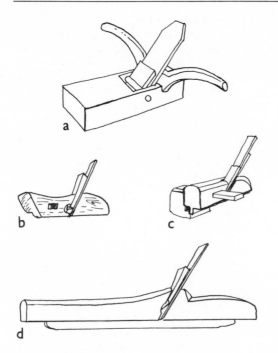

Fig. 461   Chinese planes

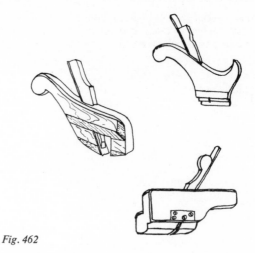

Fig. 462

Plane, Chip: see *Plane, Scaleboard.*

Plane, Chisel: see *Plane, Edge.*

Plane, Circular: see *Circular Cutting Gauge; Plane, Compass; Plane, Rounder.*

Plane, Cleaning-Out: see *Plane, Cooper's Stoup.*

Plane, Coachbuilder's *Fig. 462*
Coachbuilders use many of the ordinary woodworking Planes but those designed specially for coach work are often of smaller size, and have shorter soles adapted for working in confined spaces or on sharp curves. For example, coachbuilder's Smoothing and Compass Planes are often made under 6 in long. The small coachbuilder's Planes illustrated below, which are used for various trimming operations, have a characteristic rear handle, often scroll-shaped, which is probably Continental in origin. Planes of similar shape are illustrated by Diderot under *Menuisier en voitures* (Paris, 1763).

Certain metal Planes of the Smooth and Jack type are sometimes listed as Coachbuilder's Planes, but they are no different from the standard metal patterns sold to joiners and cabinet makers.

In addition to the range of Compass, Rabbet, and Smoothing Planes made in small sizes for coach work, the following special Planes were used:

(a) Door Check Planes
The two sizes made were known as Coach Door Smooth and Coach Door Jack. The sole is flat and is usually rebated on both sides so that the cutter extends right across its width. They were used for cleaning up door checks (i.e. rebates). Unlike the Tee Planes which were also used for cleaning out rebates, this Plane has a double iron, and could therefore deal better with edges where the run of the grain might change direction within a space of a few inches.

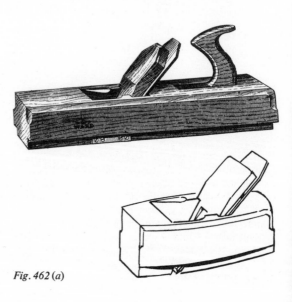

Fig. 462 (a)

(*b*) *Door Rabbet Plane*
A small rebate type stock, $6\frac{1}{2}$ in long, 1 in wide, and $3\frac{1}{2}$ in deep, but the sole is cut away along the right-hand side, front to back, in a coffin-shaped curve. The 1 in iron is single, skewed, and shouldered. It was used, presumably, for cleaning out rebates in curved parts of the frame.

Fig. 462 (d)

Fig. 462 (b)

(*c*) *Carriage Maker's Rabbet Plane* (Railway Carriage Maker's Rabbet Plane)
A metal Rabbet Plane with a stock 9–13 in long. The handle and knob in some models can be tilted clear of adjoining surfaces to avoid hurting the hands. The iron is $2\frac{1}{8}$ in (the full width of the sole) and is sometimes fitted with spur cutters. Used for rebating heavy framing in coach, railway carriage, or wagon building, and for working any wide rebates.

(*e*) *Tee Rabbet Planes*
Rabbet Planes in which the sole is wider than the stock (T-shaped in cross section), thus leaving room for the fingers when working in deep rabbets or in confined spaces. The stock is usually rectangular, measuring about 6 in long. They are made in sets with straight or compass soles varying in width from 1 to $1\frac{3}{4}$ in. In some cases the rear of the stock is shaped to form a handle. The iron is square, single, double-shouldered, with the cutting edge from 1 to $1\frac{3}{4}$ in wide. Used for cleaning up glass frame runs and the rebates on door pillars and elsewhere.

Fig. 462 (c)

(*d*) *Side Chamfer Plane* (Bevelling Plane; Side Bevel Plane)
A Moulding Plane with a shallow V-shaped sole and a cutting iron on one arm of the V, with or without a forward spur, and made in sizes from $\frac{1}{4}$ to $\frac{1}{2}$ in. Probably used for chamfering coach frames and pillars.

Fig. 462 (e)

Note: There are a number of other Planes listed as 'Coachbuilder's Planes' by tool manufacturers. As these are neither described nor illustrated, their appearance and purpose are uncertain. They include:

Coachbuilder's Side Light Plane } (James Cam,
Coachbuilder's Glueing Plane } Sheffield, *c.* 1800)
(possibly a Toothing Plane) }

Coachbuilder's Shaft Ovolo Plane }
Coachbuilder's Sittle Plane } Varvill & Sons,
(described as a 'Double Iron } York, *c.* 1870)
Fence Plane' and priced at }
10/–) }

Plane, Cocked Bead *Fig. 463*

These Planes were listed by Plane makers in the nineteenth century. The only examples known to us are in the Seaton Chest in the Rochester Museum; they were made by Gabriel & Sons of London and bought in 1795. The purpose of these Planes is not certain, but it is probable that they were used for the cocked bead on the front edges of drawers in chests, etc.; (*a*) was used for beading the edge of the planted fillets before fixing, and (*b*) for rebating drawer fronts to take them.

(*a*) *Cocked Bead Plane*

A Moulding Plane with a concave groove along the sole for producing a bead of semi-circular profile standing above the surface of the material, with no quirk on either side.

(*b*) *Cocked Bead Fillister (or Rebating) Plane*

Used in conjunction with (*a*), this has a stock $9\frac{1}{2}$ in long and $3\frac{1}{2}$ in high with a fence 1 in wide, and a square bed $\frac{1}{8}$ in or $\frac{3}{16}$ in wide. It is fitted with a skewed iron sharpened on the cutting edge and on one side, and bedded at about 55°. About $2\frac{1}{2}$ in. in front of this there is another cutter set vertically with one spur on the outside edge.

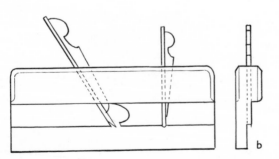

Fig. 463

Plane, Coffin

This term refers to the shape of a Plane and not to any particular use by coffin makers. Smoothing and Compass Planes are usually coffin-shaped.

Plane, Combination *Fig. 464*

Names used by various makers include:

Howkins: See *Plane, Howkins.*
Miller: Combined Plough Fillister and Matching Plane.
Record: Multi-Plane, No. 405.
Sargent: Combination Plane.
Stanley: Universal Plane. (Early term for the 'Fifty-Five'.)

Fifty-Five Plane } So called from the
Forty-Five Plane } number of different
} cutters supplied.

These metal Planes are of somewhat intricate design and have a single main purpose: to combine in the one tool the functions of nearly all Planes other than those performed by conventional flat-soled bench Planes. The most universal of these Planes are capable of ploughing, dadoing across the grain, rabbetting, fillistering, chamfering, beading, tongueing and grooving, moulding, sash moulding, and, when the stock accommodates a slitting cutter, the cutting of thin boards into strips.

The example illustrated is the 'Stanley Patent Universal Plane No. 55'. The following are the chief parts:

(A) is the main stock or bed which carries a depth gauge (F), a cutter, and also the handle of the plane.

Two transverse arms, or stems, branch out from one or both sides of the main stock; on these arms are mounted the sliding sections (B) and the auxiliary soles (C) with cutters and adjustable fences (D) and (E). Spurs are provided for cross-grain work to sever the fibres on the return stroke.

A feature of the Stanley fifty-five Plane is that the soles (B and C) are adjustable vertically so that they can be lowered to align with a moulding cutter which is lower at one side than at the other (as in the second illustration).

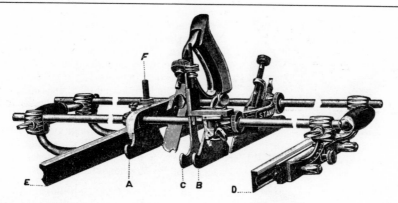

Fig. 464 Combination plane

Plane, Compass (Compass Smoothing Plane; Heel Plane; Scots: Roundsil) *Fig. 465*
Note: The term Compass Plane is sometimes applied to other planes with compass-shaped soles, e.g. Rebating Planes and Coachmaker's T-Planes.)

These Planes have soles shaped to a concave or convex curve in the direction of their length. In some patterns the amount of curve can be adjusted. They are used by wheelwrights, coachbuilders, and other tradesmen for smoothing curved surfaces. Variants include:

Wooden Patterns
(*a*) The ordinary wooden version has a coffin-shaped stock.

(*b*) The stock is made adjustable by a boxwood stop which slides up or down at the toe end where it is secured by a screw – a model which, incidentally, we have found difficult to use effectively.

Metal Patterns (sometimes called Circular Planes)
Metal versions of Compass Planes were developed in the second half of the nineteenth century. The sole is a flexible steel plate adjustable to any required curve, convex or concave. The double cutters and cap irons are usually of the Stanley type, with the usual lateral and vertical adjustments. Variants include:

(*c*) The stock is U-shaped, with the flexible sole fixed to it in the middle, and at each end where it is separately adjustable by thumbscrews.

(*d*) In this improved version the hinged bars at the toe and heel are actuated by a large screw on the front of the stock, working through a system of levers and toothed segments so that both ends are regulated by the movement of one screw.

(*e*) The frog on which the iron is bedded is fixed to the middle of the sole and is free to move up or down inside a boat-shaped iron stock. The front of the

frog has a projecting step, and the amount of curvature of the sole is regulated by a screw bearing on this step.

a

b

b

315

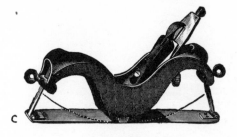

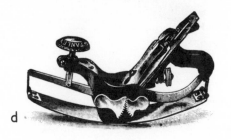

Fig. 465

Plane, Compass Moulding: see *Plane, Sash Moulding*.

Plane, Concave: see *Plane, Compass*.

Plane, Continental Types *Fig. 466*
The following Planes from Continental Europe are occasionally found in British workshops. British Planes were usually made of beech, but many different woods were (and are) used on the Continent, including cormier or service tree (rather like pear wood), evergreen oak, hornbeam, birch, and sometimes lignum vitae for the soles.

(*a*) *Bench Planes* (Not illustrated)
Equivalent to Smoothing and Jack Planes, with parallel stock, rounded at the heel, and with a horn-

shaped front handle. They are listed in British catalogues as German or Continental Planes. They have been in common use on the Continent, particularly in Germany, Scandinavia, and Eastern Europe, since the sixteenth century. The characteristic horn-shaped handle is mounted on a projecting step in the front for gripping with the left hand. In some modern patterns the wedge is tightened against a bar across the mouth as in Roman and early medieval Planes. An improved version has an adjustable mouth – see *Plane, Reform*. The longer Bench planes have closed handles mounted close to the heel of the stock.

(*b*) *Roughing Plane* (not illustrated)
Known in Britain as a *Bismarck*. This is about 7 in long and has the same appearance as the horned Plane (*a*) but is usually fitted with a single iron and used for the quick removal of waste. See *Plane, Roughing*.

(*c*) *Moulding Plane*
The Continental type differs from the British in having the upper part of the stock the same thickness as the sole or lower part. This allows for a wider tongue on the iron and a wider wedge, resulting perhaps in a firmer fixing. They have no 'spring', i.e. they were used vertically. Most have a deep groove along the sides of the plane to provide a grip, often with heavily moulded edges.

(*d*) *Plough Plane*

(*e*) *Moving Fillister Plane*
Unlike the English patterns, the stems are fixed into the stock while the fence is free to move on the stems and can be fixed in position by means of a wooden locking nut.

Carved and Decorated Planes. Though examples are not likely to be seen outside museums, it will be noted that Continental Planes of the seventeenth and eighteenth centuries are often carved with a name and date as a central ornament. The eighteenth-century Smoothing and Trying Planes, often of great beauty, with a carved design incorporating spiral motifs, popular in Holland and Scandinavia, were probably of commercial manufacture from about 1720 to 1820. Planes with overall chip-carving or floral designs are usually of Swiss, Tyrolean, or Austrian origin.

A small Plane, known in Holland as a *Schaaf*, has been much used in the Low Countries and Scandinavia since the sixteenth century. It is gracefully shaped with curved sides, narrowing towards the front. It has a scroll-shaped front grip carved from the solid, sometimes turned backwards like a

Persian slipper, and an opposite-curving spiral at the back of the stock which is made high enough to serve as a bed for the cutting iron.

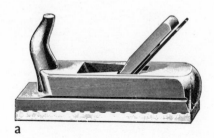

a

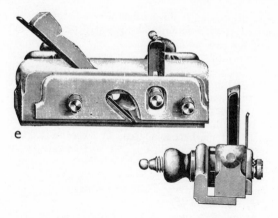

e

Fig. 466

c.

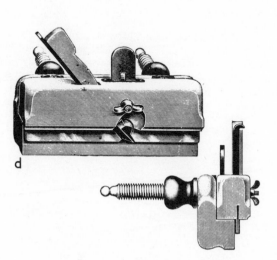

d

Plane, Cooper's: Backmaker's Jointer
A wooden Plane listed in the Plane Maker's List of Prices (Birmingham, 1871) as 3 ft long and costing a penny less to make than the Cooper's Jointer. Its appearance and use are unknown, but it may have been used for trimming the edges of staves used for making 'backs'. This was a term for large shallow wooden vessels, troughs or vats, especially those used by brewers, dyers, and picklers. The trade of backmaker is included in Pigot & Co's Commerical Directory for 1826–7 (London).

Plane, Cooper's Chiv including the Groper, Sloper, and Flincher. (Box Chiv; Cheve; Chive; U.S.A.: Howel – but see also (*e*) below; Ireland: Stock Howel; Scots: Chaif) *Fig. 467*
Note 1. The word chiv is used for both the tool and the surface it is designed to cut. In the U.S.A. the surface is called the howel.
Note 2. For a historical note on the origin of the Cooper's Chiv, and a diagram showing the work done by this tool, see *Cooper* (1) *The Trade.*
Made in many different patterns, this tool could be described as a Plane with a round-both-ways or compass-shaped sole, mounted on a strong fence. This fence is usually circular, thus following the shape of the cask end on which it will bear when in use; and it is sometimes plated to resist the abrasive action of the stave ends. In use, both fence and sole are lubricated with linseed oil.
In many instances the cutting iron is not only convex at the cutting edge (to follow the sole of the Chiv) but it is also forged like a shallow Gouge.
The purpose of the Chiv is to make a smooth surface in the form of a shallow depression near the top of the staves on which the croze groove will subse-

quently be cut. If this surface were left uneven, the croze groove would not be equal in depth all round, and the cask would leak.

In use the Chiv is held with one hand on the circumference of the fence (which is often shaped or grooved to protect the fingers) while the other hand grasps a handle which is usually fitted to the bottom of the stock.

When the cooper does not possess the right size of Chiv for a particular cask, or when he is doing repair or rough work, the chiv surface can be cut with either an Adze (see *Adze, Cooper's Howel*) or a Drawing Knife (see *Drawing Knife, Cooper's Jigger*). The use of a Chiv-type tool to cut the chime bevel is described below (*b*).

Many of the older Chivs seen in cooperages today are home made. They do not appear in the catalogues of nineteenth-century toolmakers; they are not included in Smith's *Key to the Manufacturers of Sheffield* (1816), nor in the 1862 edition of the Sheffield Standard List. The Chiv is not depicted by Diderot (Paris, 1763), though some kind of compass-type Plane is shown, mounted on stems from a fence of the same shape, which may have been used for making a chiv-surface on vats and household utensils. A theory that the Chiv is a development of the Howel Adze is discussed under the entry *Cooper*.

The different types of Chivs may be divided into the following groups:

(*a*) *Chivs for wet casks*
This is the ordinary tool (described above) which is found in British cooperages where casks for beers or spirits are made. Their size corresponds with the size of cask and they bear the same names, e.g. Hogshead, Barrel, Kilderkin, Firkin, or Pin. The cutting irons vary from about $1\frac{1}{4}$ to $4\frac{1}{2}$ in. in width.

In the U.S.A. this type of Chiv is called a 'Beer Howel' and makers such as A. F. Brombacher (New York, 1922) list many varieties, including one with a fence mounted on stems like a Plough Plane, called a 'Shifting Howel'.

(*b*) *Chivs used mainly for dry casks* (Sloper)
Several kinds of Chivs, usually too lightly made to cope with the hard oak of wet casks, are employed to handle the softer wood used in dry work; and they are occasionally used for very small casks, whether wet or dry. These tools can be employed to cut both the chiv surface and also the chime bevel; or they can be used to smooth and taper the inside ends of the staves.

As may be seen in the diagram under *Cooper*, in dry work the head is held in position by various means including the nailing of a wooden hoop around the inside of the cask below and above the head, and

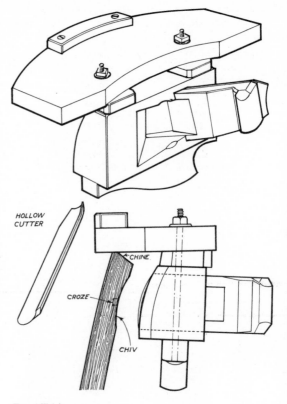

HOLLOW CUTTER

CHINE

CROZE

CHIV

Fig. 467 (a)

consequently the stave ends are tapered to form a long sloping chime so that the head can be dropped in to rest on the lower of these hoops.

One of these smaller Chivs, sometimes called a Sloper, has a compass-shaped sole, slightly rounded or flat across. The outlet for shavings is usually on the side; and the shouldered cutting iron, set at a very low angle, is exposed at the side. The ends of the fence are rounded to be held one in each hand while pushing the tool round the inside edge of the cask. This tool is sometimes called a Jigger but it must not be confused with the Cooper's Drawing Knife of that name.

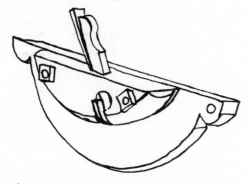

Fig. 467 (*b*)

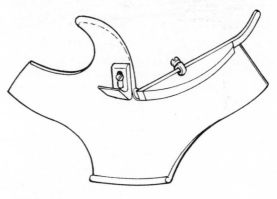

Fig. 467 (*e*)

surface, used particularly for herring casks, allows a head to be 'sprung' into a croze groove below it without removing the topmost hoop. (See sketch under *Cooper*.)

(*e*) In the U.S.A. a Chiv for dry work, often made in iron throughout, is called by Brombacher (New York, 1922) a Close Chamfer Howells Iron. It is used, presumably, for cutting both the chime and howel surfaces. Mr. K. H. Bassett (U.S.A., 1972) tells us that he has heard coopers call this tool a 'go-devil'.

(*c*) Groper

Another similar tool, given this name by a London cooper, is designed to cut a short chime bevel rather than a long sloping one. The compass-shaped sole slopes at about 45° from the fence. The cutting iron is wedged into the back of the fence instead of bedding in the stock.

Fig. 467 (*c*)

(*d*) Flincher (Flencher; Flinch)

In Scotland a tool similar to the Chiv or Groper is known as a Flincher. It is used for working the chime (rather than the chiv surface) and is designed to cut either a shoulder or a quarter curve in the top of the stave. The shoulder is designed to hold the head, with a wooden hoop on top; the concave

Plane, Cooper's Croze (Croze-iron Board; Crowe; Grove; Race) *Fig. 468*

(*Note*: See *Cooper* (*1*) for historical note on the use of the Cooper's Chiv and Croze, and a diagram showing the work done by both these tools.)

The Croze is made in a great variety of patterns and both factory-made and home-made examples may be found in most cooperages. It is a form of Plough Plane often constructed on the lines of a giant Carpenter's Gauge. A narrow cutter (called the Croze-Iron) is mounted on a stem (called the Post) which can be made to shift through a heavy semi-circular fence in which it is secured in any desired position by means of a wedge. Thus the distance between the croze groove and the end of the staves can be varied. In another pattern there are two stems instead of a single post. Croze sizes like the Chiv follow cask sizes.

The purpose of the Croze is to cut a groove (also known as the croze) round the inside of the staves, near each end, to take the heads; or as Cotgrave described it in his Dictionary of 1611, 'the furrow, or hollow (at either end of the pipe-staves) wherein the head peeces be enchased'. This operation is carried out after the ends of the staves have been

319

levelled with the Sun Plane, and after the insides of
the stave ends have been smoothed by the Chiv to
make a level surface on which to cut the croze groove.
In use it is held in the same way as the Chiv. A special
Saw is used occasionally for cutting the croze,
especially on single staves or in repair work (see
Saw, Riddle).

Cutting Irons. The simplest (and probably earliest)
cutting irons are shaped like saw teeth. These are
held in a housing called a box or guard. More
recently, this saw-tooth assembly was improved by
the addition of a hooked router tooth at one end.
Irons which cut a V-shaped groove are used in
Crozes made for certain dry work.

Continental Crozes which may turn up in British
cooperages include a French version with a straight-
sided fence like that illustrated by Diderot (Paris,
1763), and American types, which are very similar
to British ones but are sometimes made in iron
throughout.

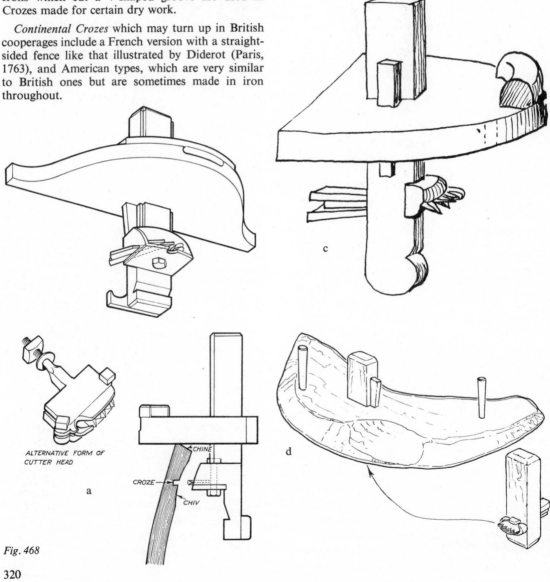

ALTERNATIVE FORM OF
CUTTER HEAD

a

b

c

CHINE

CROZE

CHIV

d

Fig. 468

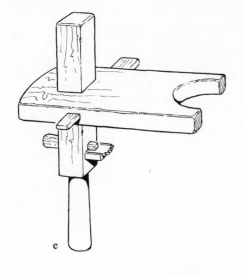

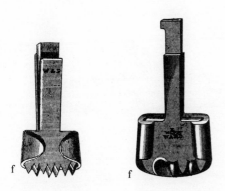

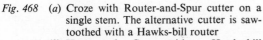

Fig. 468 (*a*) Croze with Router-and-Spur cutter on a single stem. The alternative cutter is saw-toothed with a Hawks-bill router

(*b*) Home-made Croze with a Hawks-bill cutter mounted on a double stem

(*c*) Home-made Croze, with Saw-tooth cutter on a single stem

(*d*) Home-made Croze made in elm, from a Cambridgeshire wheelwright's shop. It was used for crozing 'white work', e.g. buckets etc.

(*e*) A French Croze with Saw-tooth cutter

(*f*) Croze-Iron assemblies: Saw-tooth and Hawks-bill

(*g*) The Croze in use

Plane, Cooper's Jointer (Overshave; Stave Jointer) *Fig. 469*

This largest of all Planes, made as long as 6 ft and more, is designed for shaving the edges of cask staves and head-pieces. Unlike other Planes, it is used when lying stationary with the sole uppermost; the Cooper pushes the stave along the upturned sole of the plane towards the cutter. In the case of a cask stave, he does this with a rocking motion to help produce a slight convexity in length; and he uses only eye and feel to produce the correct angle ('in shot') at the edge. As shown in the sketch, this is done with one end of the plane supported on the Jointer Legs, and for staves of normal size, the other end rests on the ground. But when jointing longer staves for larger butts and vats, the other end, which is stepped for this purpose, rests on a bench.

There are several size styles of Cooper's Jointer, including the following:

(*a*) The Stave or Long Jointer: 4–6 ft long and $4\frac{1}{2}$–6 in wide, with leg stool.

(*b*) Straight Head or Short Jointer: 4–$5\frac{1}{2}$ ft long $4\frac{1}{2}$–6 in wide.

(*c*) The Double-Mouth Jointer: about $6\frac{1}{2}$ ft long with two irons, spaced as illustrated.

(*d*) An all-iron Jointer. A recent development which does not appear to have yet superseded the wooden tool.

(*e*) A Journeyman Cooper's Jointer. The example, with a joint in the middle, was used by the cooper at a Luton brewery when visiting neighbouring breweries to attend to the casks.

Cutting iron. Single or double, 3–5 in wide and 10–12 in long. A special type of iron, 21 in long, with a skewed cutting edge and a small projection called a 'sneck' on the side of the upper end is known as the 'Spanish Sneck Iron'. The sneck eases removal for sharpening.

There is some doubt why Jointers are made in two sizes (*a*) and (*b*), and the reason for providing two mouths (*c*) is not clear. After wear, the part of the sole in front of the iron becomes slightly hollow. Coopers tell us that this is an advantage when jointing staves; but when trimming the edges of the perfectly straight head pieces a perfectly flat sole is required. It seems possible, therefore, that the Head Jointer, and one of the mouths of the Double-Mouthed Jointer, might have been reserved for the heads, while the other, which could be allowed to wear, was kept for stave work.

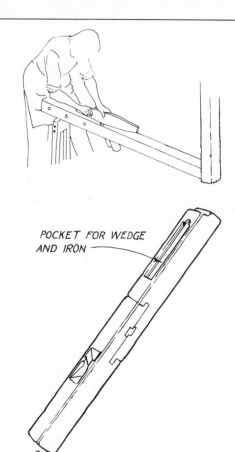

POCKET FOR WEDGE AND IRON

Fig. 469   Cooper's Jointer Planes

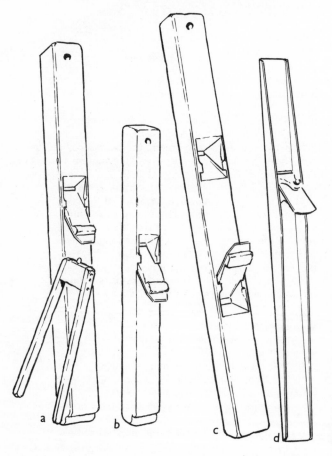

Plane, Cooper's Jointer Machines

Among the many ingenious stave and cask-making machines that were developed in the late nineteenth century, one described by Knight (U.S.A., 1877), can be classed as a hand tool. This is a bench attachment in which the previously bent staves are clamped edge uppermost. A 'double-acting knife' and a 'swing-plane' (like a short Sun Plane), are mounted on swinging arms pivoted from the back of the frame. When pushed, they describe an arc across the stave and shave its edge.

Plane, Cooper's Short Jointer (Pull Plane)

Mercer (U.S.A., 1929) describes this tool. It is a short iron-soled wooden plane, with turned side-handles protruding on either side at the back of the stock. It was used for smoothing the edges of vat staves 14–22 ft in length, which were too long to be worked on the Jointer.

Plane, Cooper's Smoothing
A name given by Mathieson (Glasgow, *c.* 1900) to a set of Planes which have the appearance of Smoothing Planes but with their soles flat, round across, or round lengthways. The latter is called a Bowling Plane. All are listed for 2 in single or double irons.

Plane, Cooper's Stoup (Cleaning-Out Plane; Inside Plane; Round-Both-Ways-Plane.) *Fig. 470*
A coffin-shaped Plane with a 2 in single or double iron and a sole round-both-ways. Like the Inside Shave (to which it is preferred by some coopers because it can be grasped with one hand and with less danger of grazing the fingers), it is made in different sizes to fit the size of the cask. Used for cleaning up and smoothing the inside of a cask. Legros (Liège Museum, Belgium, 1949) illustrates a Plane which has the appearance of a Stoup Plane being used for cutting the chiv surface.

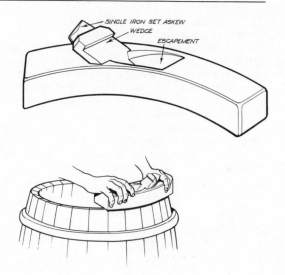

Fig. 471

Fig. 470

Plane, Cooper's Sun (Topping Plane; U.S.A.: Cooper's Leveller) *Fig. 471*
A stock, 10–14 in long and 3 in wide, shaped in plan to an arc of a circle. Made in pairs for right-hand and left-hand working. The single iron, about 2½ in wide, is sometimes bedded on the skew. (A Sun Plane fitted with a fence, which bears inside the cask, can be seen in the Edinburgh National Museum.)

After the chime at the stave ends has been bevelled with the Cooper's Adze, the sharp outer edge may be uneven. The Sun Plane is used for levelling this edge to provide a narrow ledge on which the fence of the Chiv, and later the Croze, will subsequently bear. It can be worked with one hand.

Plane, Coping: see *Plane, Sash Scribing or Coping.*

Plane, Core-Box: see *Plane, Pattern Maker's.*

Plane, Cornice (Bed Mould Plane; Cornish Plane; U.S.A.: Crown or Crown Mould Plane) *Fig. 472*
A general term for a wide Moulding Plane with a simple or complicated profile, e.g. ogee and quirked bead, ovolo and ogee, and other combinations.

The width of the sole varies from about 2 in to as much as 6½ in. Such wide mouldings were in demand for the cornices around the top of wardrobes or book cases, round a room at the junction of wall and ceiling, by church furnishers, and for making heavy picture frames. They were often home made and, unlike most Moulding Planes, are frequently provided with a toat (handle) and, in addition, a bar or hole in the front to which a cord is attached for an assistant to pull while the tradesman guides the Plane with the rear handle. Eric Sloane (U.S.A., 1964) illustrates one of these Planes with a 'pulling stick' attached to the front instead of a cord, and another with a notch in the back for a 'push stick'.

As much as possible of the wood is first removed with a Jack and other Planes, but for the final trueing of the mouldings to shape these large Cornice Planes were useful, if not essential. It was heavy work. Mr. F. H. Wildung, writing in the Chronicle of the Early American Industries Association (U.S.A., 1955), relates that 'a rope would be fixed to this handle to enable the apprentice boy to pull the plane while the master craftsman guided it.

The man guiding the plane would be cursing the other for pulling the blamed thing off the wood, and the panting helper, sore of hand and mind, always wondering why any man wanted to plane such wood, in the first place, and consigning it, the plane, and its user to the Devil! One must have done this sort of thing, to know and appreciate what takes place.'

In many instances the cutter was made in two pieces so that one cutter was about 1¼ in ahead of the other, the two together forming the complete moulding. The reasons for this are not clear. It may have made the job of sharpening easier; more important it avoided weakening the stock by cutting the mouth most of the way across.

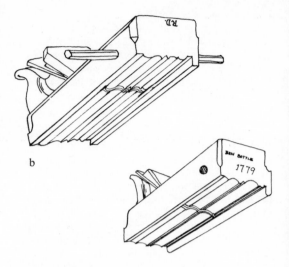

Fig. 472   (*a*) A Cornice Plane with two irons
(*b*) Eighteenth-century examples from a firm of Kentish church carpenters

Plane, Cove: see *Plane, Moulding*.

Plane, Cove Raising
A Panel Fielding Plane which produces a hollow or groove, known as a cove, at the edge of the raised face of the panel, instead of the more usual square list. See *Plane, Panel Fielding*.

Plane, Cow: see *Plane, Roughing*.

Plane, Crown Mould: see *Plane, Cornice*.

Plane, Curl: see *Plane, Spill*.

Plane, Cylinder
Listed by James Cam (Sheffield, *c.* 1800) and by Varvill & Sons (York, *c.* 1870). Appearance and purpose unknown.

Plane, Dado Grooving (Carcase Grooving Plane; Housing Plane; Scots: Raglet or Flooring Raglet; Trenching Plane) *Fig. 473*
The stock is about 9–10 in long, rebated along the left-hand side. There is a depth stop made of wood or metal. The shavings are discharged through an outlet which is often gracefully tapered through the width of the stock. Indeed, the shape of this Plane is an outstanding example of the severe but graceful lines followed by English plane makers. These tools

were later made in metal, with a skewed iron and two separate adjustable spurs.

The cutting iron is a single iron, bedded at about 50°, normally skewed, shouldered, and made in sizes rising in eighths from $\frac{1}{4}$ to 1 in. A double spur is wedged in front of the iron to prevent tearing out when trenching across the grain.

The original use of these Planes was for cutting the grooves in flooring to take the tongue at the bottom edge of the skirting of the dado or half-panelling round the lower part of the wall in a room. It is also used for cutting across the grain in book-case ends for shelving. A straight strip of wood is fixed across the work to act as a fence, and when working across the grain the Plane is drawn backwards in a preliminary stroke to ensure that the sides of the groove are cut by the spur before the cutting iron comes into operation.

Fig. 473 Dado Grooving Planes
(*a*) The traditional pattern in wood
(*b*) A factory-made pattern with an adjustable metal stop
(*c*) An all-metal pattern of *c.* 1910

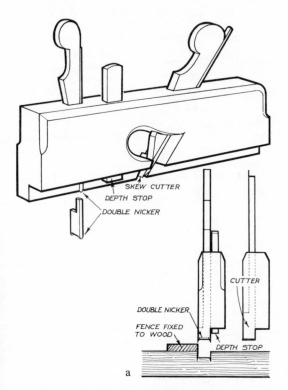

SKEW CUTTER
DEPTH STOP
DOUBLE NICKER
CUTTER
DOUBLE NICKER
FENCE FIXED TO WOOD
DEPTH STOP
a

Plane, Dapping: see *Plane, Banding.*

Plane, Door
A name given by some plane makers, including Mathieson (Glasgow, *c.* 1900), to a Moulding Plane designed to work 'Door Moulds' which are small mouldings planted round the edges of the panels of a door. They consist of a quarter round with fillet, an ogee with quarter round or a small Grecian ogee.

Plane, Door Check and Door Rabbet: see *Plane, Coachbuilder's.*

Plane, Door Trim *Fig. 474*
Listed by Stanley, *c.* 1910, this has an adjustable side fence and interchangeable cutters held lightly in position with a spiral spring, and it is fitted with two handles like a Router. Used for cutting shallow recesses for butt hinges, face plates, strike plates, and other door furniture (U.S.A. 'trim'). With the fence removed and the cutter locked with a screw, the tool becomes an accurate router.

325

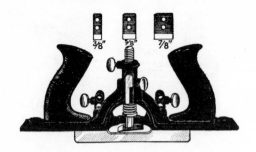

Fig. 474

Plane, Double Chamfering
Illustrated by Mathieson (Glasgow, *c.* 1900) under this name. It appears to be a V-Plane for making an ornamental groove along a board. cf. *Plane, Centre-Board*.

Plane, Double-Ended (Duplex Plane): see under *Plane, Block*.

Plane, Dovetail *Fig. 475*
The Stanley range of metal Planes (U.S.A., *c.* 1910) includes a Dovetail Tongue-and-Groove Plane, which cuts both a slot-dovetail groove and the corresponding dovetail tongue, but it does not appear to have come into very general use. The name is also given to a Plane used for 'dovetail boxing' – see *Plane Maker*. Knight (U.S.A., 1877) uses the same term to describe a 'side-rabbet plane with a very narrow sole, which may be made by inclination to dress the sides of dovetail tenons or mortises. . . .'

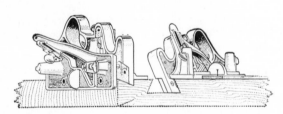

Fig. 475

Plane, Drawer-Bottom Grooving (Drawer Plane) *Fig. 476*
A Grooving Plane with a fixed or adjustable fence projecting below the level of the cutter and handed right and left. Used for cutting the grooves in the sides of drawers to take the bottom.

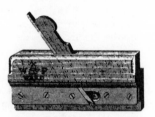

Fig. 476

Plane, Edge (Chisel Plane; Piano Maker's Edge Plane) *Fig. 477*
A term applied to a small Plane, 4–6 in long, with the toe-end open, leaving the iron uncovered and extending to the extreme front edge of the Plane. The iron is single, bedded at a low angle, the bevel uppermost. These Planes are used for trimming mouldings, rebates, etc. by working right into a corner and in places inaccessible to other Planes.

(*a*) A home-made example in wood with the iron wedged behind a pin fitted across the stock.

(*b*) A metal version, which in Stanley's range (U.S.A., *c.* 1910) is called a Piano Maker's Edge Plane. The sole is 10 in long, and the $2\frac{1}{2}$ in iron is secured with a lever cap.

Fig. 477

Plane, Edge Trimming *Fig. 478*
An all-metal stock, 6 in long, with the sole rebated to form a right angle. The horizontal part of the sole bears on the surface of the board while the vertical part of the sole, in which the cutting iron is bedded, bears on the edge of the board which is being trimmed. The cutting iron is skewed and $\frac{7}{8}$ in wide. Its purpose is for trimming the ends of boards to make close butt joints. Wooden blocks with various bevels can be screwed to the fence to enable the tool to make a slanting cut.

Fig. 478

Plane, Facia (Fascia Plane)
Listed under Bench Planes by Varvill & Sons (York, *c.* 1870) and in other early lists including Cam (Sheffield, *c.* 1800). It may have been a Cornice Plane used for working a moulding surrounding a fascia board.

Plane, Falconer's: see *Plane, Plough (Circular).*

Plane, Fielding: see *Plane, Panel Fielding.*

Plane, 'Fifty-Five': see *Plane, Combination.*

Plane, Fillet
A name given by Knight (U.S.A., 1877) to a Plane he describes as 'a molding plane for dressing a fillet or square bead'.

Plane, Fillister (Scots and U.S.A.: Filletster Plane) *Fig. 479;* and *Fig. 519* under *Plane, Rebate.*
A general term for Rebating Planes that are fitted with fences. The stocks measure about $9 \times 4 \times 2$ in. A sectional diagram of the different types can be seen on page 348.

Since these Planes come in for very heavy usage, the bottom right-hand corner of the stock in the better types is strengthened by a boxwood insert. In its simplest form this was let in at an angle and known as 'shoulder-boxed'. In the more expensive tools the insert took the form of a rectangular block with dovetailed tongues on the top and side, and it was described as 'dovetail boxed'. The special Planes for this process are described under *Plane Maker.*

(a) Standing Fillister (Check Planes)
A Rabbet Plane with a fence cut solid in the stock, and used for cutting long runs of rebated stock of fixed width and depth, e.g. for sash or greenhouse work with no moulding.

Fig. 479 (a)

(b) Moving Fillister (Scots: Fore Fillister or Fore Check; Side Fillister). *Overleaf.*
The fence is adjustable laterally to expose as much of the cutting iron as the width of the rebate to be made. There is an adjustable depth-stop in metal or in wood. Shaving discharge is at the side. The cutting iron is single, skewed, bedded at about 50°, shouldered, and occupying only the right-hand two-thirds of the stock. A spur tooth is mounted in the shoulder of the stock, set in advance of the cutting iron to produce a sharp edge. It is used for cutting rebates of different widths on the edge of the work-piece. In use the Plane is drawn backwards in a preliminary stroke, so that the grain is severed by the spur before the cutting iron comes into action.

(c) Sash Fillister (Scots: Back Check). *Overleaf.*
When working sash timbers the rebates may be cut on either the left-hand or right-hand sides of the wood. The ordinary Moving Fillister *(b)* can be used for the left (near) side; for the far side, a plane is used which is similar to the Moving Fillister except that the cutter, stop, and spur are mounted on the left-hand side of the stock, instead of the right. Thus the fence bears on the face of the work and, instead of being fixed to the sole, is carried on two stems like the Plough Plane. The cutting iron is single, skewed, shouldered, bedded at about 50° and extending all the way across the sole.

For other tools used in sash work, see *Window Making.*

(d) Combination Fillisters
1. A wooden Fillister Plane which can be used with either the right-hand or left-hand part of the cutter, thus serving equally as a Moving or Sash Fillister. For this purpose it is provided with spur cutters

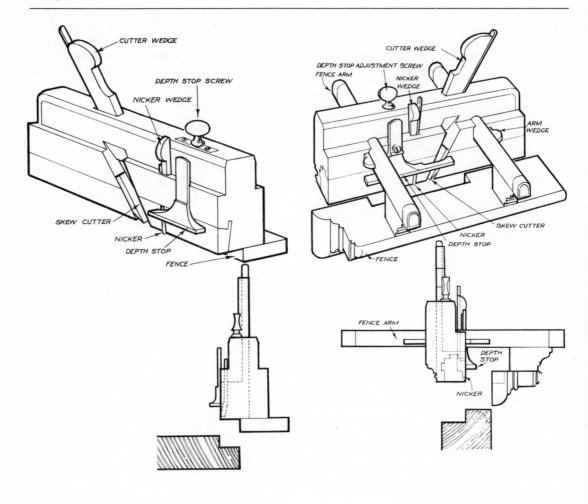

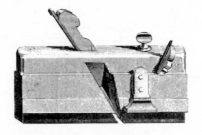

Fig. 479 (b)

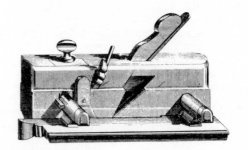

Fig. 479 (c)

and depth stop on both sides of the stock. [Not illustrated]

2. An all-metal plane with two seatings for the cutter, one normal and the other near the front for 'bullnose' work. It is known as a 'Rabbet and Fillister Plane'. It can be used as a left-hand or right-hand Moving Fillister or, when the fence is removed, as a Rabbet Plane.

Fig. 479 (d)2

Plane Fittings: see entry at end of Plane section.

Plane, Flit Plow

A term used by Mathieson (Glasgow, *c.* 1900) for an adjustable Plough Plane with stems. See *Plane, Plough.*

Plane, Flogger: see *Plane, Boxmaker's.*

Plane, Floor *Fig. 481*

A metal Plane, $10\frac{1}{2}$ in long with a $2\frac{5}{8}$ in cutting iron, to which is attached a tubular handle 45 in long. Used for planing floors, bowling alleys, the decks of ships, etc. *Note:* Mercer (U.S.A., 1929) uses the same term for an extra large Trying or Jointing type of Plane, up to 36 in long, with a closed grip and single or double iron, which is used for shooting long joints and for levelling the surface of wooden floors.

Fig. 481

Plane, Fluteing: see *Plane, Moulding.*

Plane, Fore

A term used by Moxon (London, 1677) for a Plane of jack size. The term is also used by the Stanley Works in the U.S.A. for jack-type metal Planes. Mercer (U.S.A., 1929) uses the term as an alternative name for a wooden Jack Plane. (See *Plane, Bench* for table of names etc.)

Plane, Fore Check: see Moving Fillister under *Plane, Fillister.*

Plane, Forkstaff (Forkshaft; Fork Stail Plane; Rounding Plane) *Fig. 482*

A name occasionally given to a Plane of smoothing or jack size with a sole hollow in cross-section, and with a single cutting iron to match. The name is not commonly used; it is mentioned but not illustrated in Mathieson's *List* (*c.* 1900) under Smoothing Planes 'No. 716 – Hollow Sole or Forkstaff'. The term is also used by Mercer (U.S.A., 1929).

This Plane was used for rounding handles and similar work. P. Nicholson (London, 1822) defines its purpose as 'to form a convex cylindrical surface, when the wood to be wrought is bent with the fibres in the direction of the curve; as the convex surfaces of the rims of carriage wheels, or the top rails of camp-bedsteads, and work of a similar nature'. See also *Plane, Rounding.*

(*a*) Examples with square stocks measuring $8 \times 3\frac{1}{2}$ in.

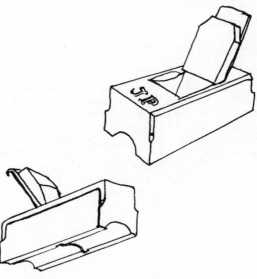

Fig. 482 (a)

(*b*) This home-made little Forkstaff Plane was dug out of a Hertfordshire wheeler's shop about 12 in below the surface of the shop floor. The wheelwright himself, aged almost ninety, had never seen it before.

Fig. 482 (b)

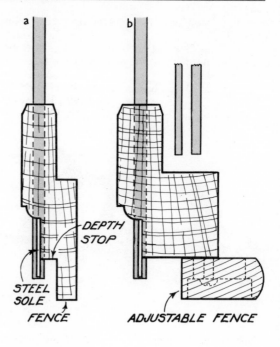

Plane, French Plough
A term for a Plough Plane with screwed stems. (See *Plane, Plough.*)

Plane, Furring: see *Plane, Roughing.*

Plane, Galloping Jack
A name occasionally given to a Plane for smoothing floor boards.

Plane, Gentleman's
A term applied to a light, and sometimes a smaller, version of the Jack or Smoothing Plane. It is intended for amateurs.

Plane, Greenhouse Rabbet: see *Plane, Fillister.*

Plane, Grooving *Fig. 483*
A general term for all Planes which make grooves, but here confined to those which cut a narrow rectangular trench, such as those listed below. The wooden stocks vary in length from 8 to 10 in, with the shaving outlet at the side. The irons are usually single, and bedded at about 50°. Grooving Planes are described under the following entries: *Plane, Combination; Plane, Cooper's Croze; Plane, Dado Grooving; Plane, Matching; Plane, Plough.*

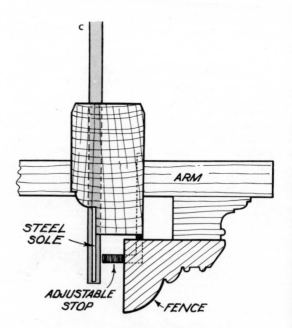

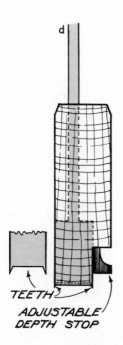

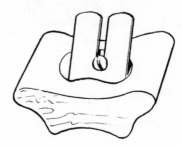

Fig. 483 Planes for Grooving (diagrammatic view)
    (*a*) Fixed Grooving Plane
    (*b*) Adjustable Grooving Plane with fence
    (*c*) The Plough
    (*d*) Dado Plane

Plane, Gunstocker's
At the Birmingham Museum of Science and Industry (1965) there is a set of four round-soled planes which are described as being used by Gunstockers for bedding the barrel into the stock. They are about 8 in long, with a C-shaped outlet for shavings discharge at the side.

Plane, Gutter: see *Plane, Spout.*

Plane, Half Round or Half Hollow: see *Plane, Side Round.*

Plane, Halfin (or Hafflin)
Scots term for a *Trying Plane.*

Plane, Hand Rail (Banister Plane; Capping Plane; Stair Builder's Plane) *Fig. 484*
Planes made in the form of Moulding or Smoothing Planes with ogee profiles, usually made in sets of two for the two sides of a hand rail, and a third sometimes known as a Capping Plane, for the top. Figure 485 shows a 'Stair Builder's Plane' as illustrated in the Shelburne Museum's *Woodworking Tools* (U.S.A., 1957). The square, stubby-looking stock measures 4 by 2 in, has the sole profiled to the contour required, and is sometimes provided with a fence. (See also *Shave, Handrail.*)

Fig. 484

Plane, Handicraft: see Technical Jack Plane under *Plane, Jack.*

Plane, Heel: see *Plane, Compass.*

Plane, Hollow and Round (Scots: Casement Planes; a slang term for these planes – 'Hollows and Bollows') *Fig. 485 overleaf*
Moulding Planes in matching pairs, with concave (hollow) and convex (round) soles with either square or skew cutting irons. A set of these Planes was an important part of a joiner's or cabinet maker's kit, for they can be used for all kinds of shaping and trimming work, as well as for working a moulding.
    The full set of 'Hollows and Rounds' comprises 18 pairs ($\frac{1}{8}$–$1\frac{1}{2}$ in, rising in $\frac{1}{16}$ths) but half sets, including either the odd or the even numbers, were common. It was usual to include pairs of Snipe Bill Planes and some of the Side Planes. With these the tradesmen could work any special moulding required, particularly non-standard cornice moulds.

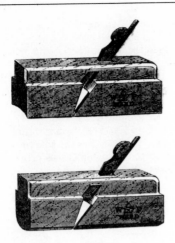

Fig. 485

Plane, Hollowing: see *Plane, Spout*.

Plane, Holtzapffel: see *Plane, Metal Cutting; Plane, Plough, Circular*.

Plane, Hook-Joint: see *Plane, Airtight, Casemaker's*.

Plane, Housing: see *Plane, Dado Grooving*.

Plane, Howkins
A combination metal plane of unconventional design intended for all kinds of ploughing and grooving, and for plain or dovetailed trenching. The cutting iron is held in a tool holder designed like a slide rest on a lathe, and is fed into the work at every stroke by means of a feed screw. (cf. *Plane, Combination*.)

Plane, Hunter: see *Plane, Roughing*.

Plane, Inbowing (Enbowing Plane)
Mr. W. L. Goodman (1972) states that, from the evidence given by L. F. Salzman in his *Building in England down to 1540* (London, 1952) and from his own researches into the Apprentice Enrolments of 1535–1650, this is a carpenter's plane for working mouldings on the underside of joists, beams, and other timbers. The mouldings are usually 'boltel' or 'casement' or a combination of both. A boltel is a concave or ovolo moulding; a casement has a convex scotia or cove-like profile. Both 'Botel Planes' and 'Casement Planes' are referred to in the Bristol Apprentice Book of 1611.
  Note: Hollow and Rounding Planes are called

Casement Planes in the catalogue of Alexander Mathieson & Sons (Glasgow, 1900).

Plane, Inside: see *Plane, Cooper's Stoup*.

Plane Iron
Particulars of cutting-irons used in Planes will be found under *Plane*.

Plane, Jack (U.S.A.: Fore Plane; Moxon (1677) writes that this tool is called a Jack Plane by carpenters, but explains that joiners call it a Fore Plane 'because it is used before you come to work either with the Smooth Plane or with the Joynter'.) *Fig. 486* The stock is 12–18 in long and $2\frac{1}{2}$–3 in wide, with a flat sole and an open or closed handle. The cutting iron is $1\frac{3}{4}$–$2\frac{1}{4}$ in long, slightly convex to prevent the corners of the iron from marking the work when taking a heavy cut, and bedded square at 45° or 50°.
  One of the commonest of all Bench Planes, it is the first to be used for the comparatively rough work of preliminary preparation of the surface before truing up with the Trying Plane. (In exceptional cases a Roughing Plane may have to be used for the grosser preliminary work.)
  Variants include the following:

(*a*) *Wooden Jack Plane* as described above.

Fig. 486 (*a*)

(*b*) *Metal Jack Plane*. Early twentieth-century example made by Stanley.

Fig. 486 (*b*)

(*c*) *German Jack Plane*. (See also *Plane, Continental*.) Two types, both with the characteristic front handle.

Fig. 486 (c)

(*d*) *Technical Jack Plane* (Handicraft Pattern Jack Plane; Technical School Jack Plane; U.S.A.: Razee-Jack). Somewhat shorter than the ordinary Jack Plane, with the stock reduced in depth at the heel to bring the handle down lower. Used in school workshops. The tool is lighter, and this makes it easier for a beginner to get the 'feel' of the planing operation.

Other varieties are described under *Plane, American Wood Pattern; Plane, Spiers and Norris Types; Plane, Stanley-Bailey*. For further information on component parts see *Plane*; on sizes see *Plane, Bench*.

Fig. 486 (d)

Plane, Jack Raising: see *Plane, Panel Fielding*.

Plane, Japanese Types *Fig. 487*
A light, shallow stock of rectangular section without handles. The thick, tapering iron is set well towards the heel and is let into slots in the cheeks of the mouth, thus acting as its own wedge. As Japanese woodworkers usually work on the floor or on a low, sloping bench, the Planes are pulled towards the body. For jointing, the Plane itself is let into the bench with the sole uppermost, with the surface of the bench acting as an extended sole, and it is used like a Cooper's Jointer except that the work is pulled and not pushed over the cutter. (cf. *Plane, Chinese*.)

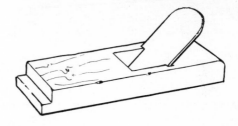

Fig. 487

Plane, Jointer
The name given to Bench Planes over 20 in long. The term has now mostly dropped out in favour of 'Trying', but it is still used to some extent in the U.S.A. and Scotland. See *Plane, Trying; Plane, Cooper's: Backmaker's Jointer; Plane, Cooper's Jointer*.

Plane, Keel or Stem: see *Plane, Shipwright's*.

Plane, Key-Maker's: see *Plane, Pianoforte Maker's*.

Plane, Lagging
A term used for a Plane of jack size with a sole concave lengthways, made for various radii. Used for shaping lags or laggings, which are segments of wood which were bolted to iron drums to form wooden pulley wheels for driving by belt.

Plane, Lamb's Tongue: see *Plane, Sash Moulding*.

Plane, Landing: see *Plane, Shipwright's*.

Plane, Lead: see *Plane Metal Cutting*.

Plane, Long
A term used in the first half of the nineteenth century to denote a Trying or Jointer Plane about 24–26 in long. See *Plane, Bench* for table of names etc., and *Plane, Trying*.

Plane, Low Angle *Fig. 488*
A name given to metal Planes of various sizes up to about 14 in long, with the cutting iron set at the very low angle of 12° or even less. Used for heavy cutting across the grain, trimming mitres, etc. Other low-angle Planes are illustrated under *Plane, Block; Plane, Edge; Plane, Mitre.*

Fig. 488

Plane Maker: see entry on page 370.

Plane, Mast and Spar *Fig. 489*
A Plane of smoother or very occasionally jack size, with a single or double iron and the sole hollowed to a shallow curve to follow the contours of masts varying in diameter from about 5 to 18 in. In the smaller sizes the lower half of the stock is rectangular but the upper half is curved like the coffin-shaped Smoothing Plane. This characteristic shape gives a comfortable hold without reducing the surface area of the sole. Used to smooth masts and spars after they have been trimmed down roughly to the correct taper. Planes of this type are also used for rounding oars. See *Mast and Spar Maker.*

Fig. 489

Plane, Matching (Tongue and Groove Plane; Slit Deal Plane; Match Planes; Match Plows) *Fig. 490*
These Planes are used for cutting a matching tongue

and groove on the edges of boards in order to join them together to form partitions, table tops, floors, etc. Variants include:

(a) Matching Pairs
A pair of Planes, one a Grooving Plane, the other containing a deeply notched iron for making a tongue. Made in matching pairs for boards from ½ to 1¼ in thick, with fixed or adjustable fences. The stock is 8–10 in long with side shaving outlet. The underside is rebated to hold a strip of steel plate which acts both as the sole to the Plane and, at the point of separation, as a bed for the iron, for which purpose it is knife-edged to engage with a groove on the back of the iron, thus keeping it rigid. The iron is single, shouldered, bedded at about 50°, and made in sizes up to ¾ in. There are two kinds of fence: one a rebate in the stock itself, the other movable and screwed through slots to the sole of the stock. See diagram *Fig. 483* under *Plane, Grooving.*

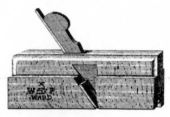

Tongueing.

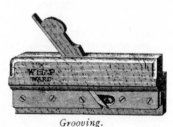

Grooving.

Fig. 490 (a)

(b) Double-ended Match Plane (Twin Grooving Plane)
The pair of Planes (a) are occasionally joined into one stock. For this purpose a central fence with matched grooving and tonguing irons on either side are bedded opposite ways. Sometimes the stock is provided with two opposing handles as here illustrated.

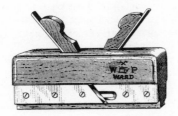

century, a metal form of Mitre Plane with a wooden core was developed in England, and this was later modified to cover the whole range of metal Bench Planes. A famous maker of these Planes was Stewart Spiers of Ayr, and later T. Norris of London. Their handsome products, made from *c.* 1845 onwards, are still sought after by tradesmen for high-class work, and by collectors for their fine design and beautiful finish.

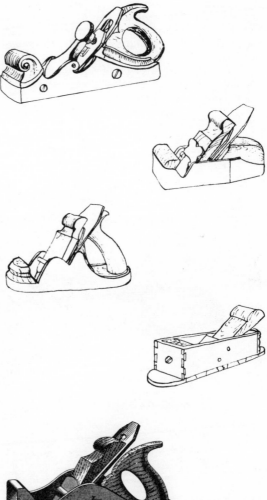

Fig. 490 (b)

(*c*) *All Metal Matching Plane*
The example illustrated comes from the Stanley Works range, U.S.A., *c.* 1910. The fence is pivoted, and when swung round it covers one of the cutters thus converting the plane from a tonguing to a grooving plane. Made in sizes to deal with $\frac{3}{8}$ to $1\frac{1}{4}$ in board.

Fig. 490 (c)

Plane, Metal *Fig. 491*
Although some Roman Planes had metal soles, it appears that the only surviving metal Planes dating from the following centuries, and from the Middle Ages, are small Block Planes. In the early nineteenth

Fig. 491 (a) Metal Planes developed during the nineteenth century

The materials used by Spiers and those who followed his designs were wrought iron, mild steel or gunmetal, and, later, annealed cast iron, with the core usually of wood, often rosewood.

In the U.S.A. Bailey and others developed Bench Planes made of cast iron, dispensing with the wooden core altogether and bedding the iron on a metal frog. The only wood used was in the handles. These planes led to the series of planes made famous by the name of Stanley in the U.S.A., and later by the Record Co. and others in England. This type of Plane, after some hundred years of trial and improvement, is the standard bench plane used in almost every English workshop, both in the trade and at home. It is given a British Standard Specification, No. 3623:1913, where it is called a 'Woodworking metal-bodied plane'. From the middle of the nineteenth century metal versions were made of nearly all the wooden Planes.

For the nomenclature of the types and component parts of modern metal Planes see *Planes* and *Plane, Bench*. Examples of earlier metal Planes will be found under *Plane, Spiers and Norris Types*. The Stanley-Bailey range will be found under many entries, but see *Plane, Stanley-Bailey*, under which Sargent, Preston, and Record Planes are also mentioned.

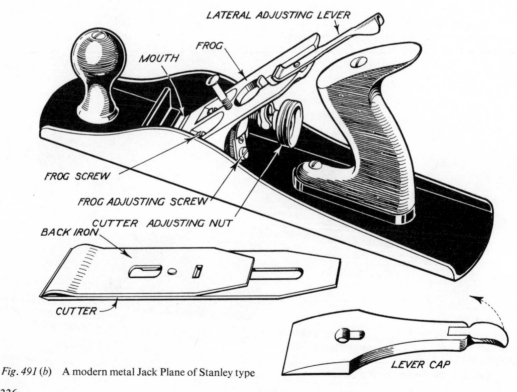

Fig. 491 (b)   A modern metal Jack Plane of Stanley type

Plane, Metal Cutting *Fig. 492*
Planes designed for smoothing and trimming metal include the following:

(a) 'Smith's Plane'
A Plane made by Charles Holtzapffel and described in his book (London, 1846). The stock is of cast iron fitted with a 1 in cutting iron, toothed or flat, set vertically, and secured by a screw acting on a wooden block. It was designed for cutting (or rather scraping) brass, iron, or steel.

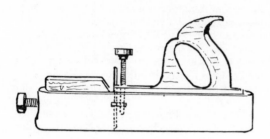

Fig. 492

(b) Lead-cutting Plane
Illustrated by Hellyer (London, 1905), this appears to be a Plane of jack size and type, with a single iron bedded at the usual angle. Planes of this kind were used by plumbers for trimming joints and other surfaces of lead tanks etc., and by shipwrights for trimming lead keels.

(c) Organ Builder's 'Thicknessing' Plane
A jack-size Plane used for trimming the metal sheets used in making organ pipes. See *Organ Builder*.

Plane, Millwright's: see *Millwright*.

Plane, Mitre (Metal) (Pianoforte Maker's Plane) *Fig. 493*
This metal Plane, found in a variety of patterns, has a very different appearance from that of the wooden Plane of the same name. The stock usually consists of a shallow metal box with the sides and sole often dovetailed together, surrounding a hardwood core. The toe is usually square, the heel often rounded, and the sole projects beyond the body of the Plane in front, and often behind also. The overall length ranges from 7 to 12 in. The mouth is very narrow, sometimes as little as $\frac{1}{32}$ in. The iron is single, $1\frac{3}{4}$–$2\frac{1}{2}$ in long, and often nibbed. It is set at a low angle with the bevel uppermost, and usually wedged against a flat bar fitted across the outlet at

an angle to suit the wedge. According to Holtzapffel (London, 1846) the 'iron lies at an angle of about 25°, and is sharpened at about the ordinary angle of 35°, making a total elevation of 60° which, together with the delicate metallic mouth, renders the absence of the top iron unimportant, even when the plane is used lengthways of the fibres. . . .'

Like its wooden counterpart, this Plane is used for shooting mitres of all kinds, but with its rigid bed and narrow mouth it is valued for its ability to perform particularly fine work on end grain and hardwoods. Mercer (U.S.A., 1929) illustrates a similar Plane which he states was probably made in England about 1800, and calls it an 'Iron Plated Joiner's Plane'. He points out that its construction resembles the Roman Plane from Silchester. The Metal Mitre Plane is a descendant of the medieval Block Plane; it was the starting point of the production of metal Planes on a big scale in England, leading to the development of the beautifully made Spiers and Norris pattern metal Planes in the mid-nineteenth century. (See also *Mitre Appliances*.)

Fig. 493

Plane, Mitre (Wood) (Mitre Block Plane; Mitre Shooting Plane; Picture-Frame Maker's Plane) *Fig. 494*
A heavy square stock, with a grip-handle sometimes provided on one side. In some cases the sole, or the sole and one side, is steel-plated to resist wear, which adds greatly to the weight of the Plane.

There are two common sizes:

'Block' – 12 in long with $2\frac{1}{4}$ in iron.
'Shooting' – 20–24 in long with $2\frac{3}{4}$ or 3 in iron.

The cutting iron is single, square or skewed, and bedded at about 50°. When fitted with a skewed iron the Plane is made in handed pairs. A special version has a tapered boxwood block fitted in the front of the shavings outlet. This can be lowered to adjust the mouth opening after wear.

These heavy Planes are chiefly used on a Shooting Board for trimming mitres on wide workpieces or mouldings such as architraves and cornices, and for shooting the mitres of large picture frames. (See also *Mitre Appliances; Plane, Strike Block.*)

Fig. 494

Plane, Modelling: see *Plane, Thumb.*

Plane, Moot: see *Plane, Rounder, Trenail.*

Plane, Mother: see *Plane Maker; Plane, Sash Templet Making.*

Plane, Moulding *Figs. 495–510*

Note. Working a profile with Moulding Planes is known as 'sticking' or in Scotland 'running' the mould. A 'stuck' moulding is worked directly on the framing it is used to ornament; a 'planted' moulding is worked separately, and fixed in position afterwards.

There is an almost infinite number of different moulding shapes and it was customary for the plane makers to produce Planes with any profile to meet the customers' requirements. They are used for decorating architraves, door and window frames, details of cabinet work of many kinds, skirtings, picture frames, etc. Many of the shapes are based on the mouldings which occur in the classic architecture of both the Greeks and Romans. The curves of the typical Greek mouldings were derived from conic sections (the ellipse, hyperbola, and parabola), while the typical Roman mouldings were segments of a circle. Some of the Planes which produced these profiles are listed below.

General description

Since about 1770 the size and shape of British Moulding Planes have been more or less standardised. The stocks are $9\frac{1}{2}$ in long; the width of the upper part varies from $\frac{5}{8}$ to $1\frac{1}{4}$ in, and of the lower part according to the width of the profile. There is therefore a step about half way down the stock, finished square but with chamfers round the upper part which formerly terminated in a handsome stop-chamfer. Later the terminations were reduced to small gouge cuts.

The sole of the Plane is shaped to the reverse of the profile required, usually with a small fence at one side and a rebated step at the other which forms the depth stop. No handles were fitted, except in the case of the wide Crown or Cornice Moulding Planes. These were sometimes provided with a toat type of handle and/or cross-handles, or a hole for a rope, which was pulled by an assistant. (See *Plane, Cornice.*)

The narrower Planes, including the Bead Planes, if made of plain beech, soon lose their shape owing to wear. Consequently, that part of the sole containing the quirk and fence is often reinforced with strips of boxwood (see *Plane Maker*). To allow for wear on the shoulder, a wide beechwood slip is sometimes screwed to the side of the stock.

Cutting irons are bedded at 55–60°. In some wide and intricate profiles the work is divided between two staggered irons, partly in order to avoid weakening the stock, and partly to simplify the process of keeping the correct shape of the cutting edge when sharpening. This is shown under *Plane, Cornice.* Most Moulding Planes from the Continent of Europe differ from English Planes in having the upper part of the stock the same thickness as the lower part. See *Plane, Continental.*

Use

In English practice all Moulding Planes, with the exception of the beads and reeds, are held canted over when in use, the amount of inclination being known as the 'spring'. The purpose of the spring is to avoid as far as possible the 'dead' spots where a profile like an ogee changes direction and where the iron, if the Plane is worked vertically, would scrape rather than cut. It also enables shavings to be thrown clear more easily.

Historical Note

The Romans used Moulding Planes with the iron let into a tapering groove at the side of the stock, and this design persisted throughout the Middle Ages. Moulding Planes of the modern type were being made by specialist plane makers at least as

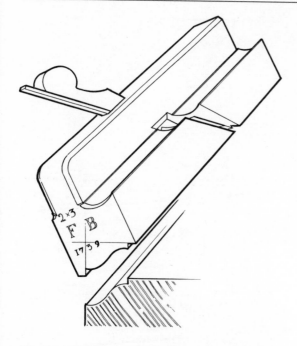

Fig. 495 A typical Moulding Plane of the eighteenth century with ogee profile

far back as the early part of the eighteenth century. One of the earliest makers in this country was Robert Wooding of Queen Street, Cheapside, London, 1704–1728. See *Plane, Panel; Plane Maker.*

Common types of Moulding Planes
Most Moulding Planes are known by the section they produce in the wood; thus an Ogee Plane cuts an ogee moulding. Exceptions are the rounds and hollows; a Round Plane cuts a hollow section, and a Hollow Plane cuts a round one.

Planes for making the most commonly used mouldings are illustrated, but less typical examples are described under the following entries:

Plane, Centre-Board; Plane, Combination; Plane, Cornice or Crown; Plane, Door; Plane, Hollow and Round; Plane, Neck Mould; Plane, Nosing; Plane, Sash; Plane, Side; Plane, Snipe Bill; Plane, Table.

Note: The term 'quirk' when applied to a moulding is a narrow groove at the back of a bead or other moulding. E.g. see *Plane, Moulding, Fig. 505.*

Fig. 496 Astragal moulding
(*a*) A bead with two fillets, as used for traceried bars, etc.
(*b*) A raised bead

Fig. 497 Bead moulding
(*a*) Plain bead
(*b*) Bead with fillet. Often used to hold door panels
(*c*) Cocked bead
(*d*) Cocked bead, made separately and fitting in rebate. Often used on drawer fronts.
(*e*) Quirk or side bead. Often used at the joint between boards as in match boarding
(*f*) Return or staff bead. Used to ornament a corner or to disguise a joint
(*g*) Sunk or centre bead

339

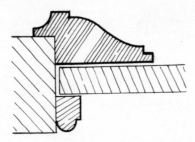

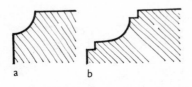

Fig. 498 Bolection moulding. A rebated moulding of varied section, made to fit over the edge of a framework such as a door

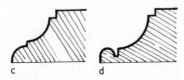

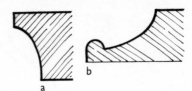

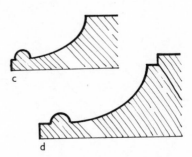

Fig. 501 Cove moulding. A concave profile, one quarter of an ellipse
(*a*) Plain
(*b*) Cove and bead
(*c*) Cove and astragal
(*d*) Cove and astragal, double fillet

Fig. 499 Cavetto moulding (Often called a Scotia, see *Fig. 509*)
(*a*) Plain cavetto
(*b*) With double fillet
(*c*) With quarter round
(*d*) With quirked bead

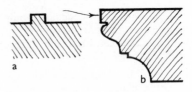

Fig. 502 Fillet or Listel. A rectangular strip:
(*a*) A raised fillet
(*b*) In combination with other curved mouldings

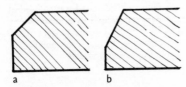

Fig. 500 Chamfer moulding
(*a*) Cut at 45°
(*b*) Other angles

Fig. 503 Flute moulding. A semi-circular, concave moulding; frequently repeated side by side

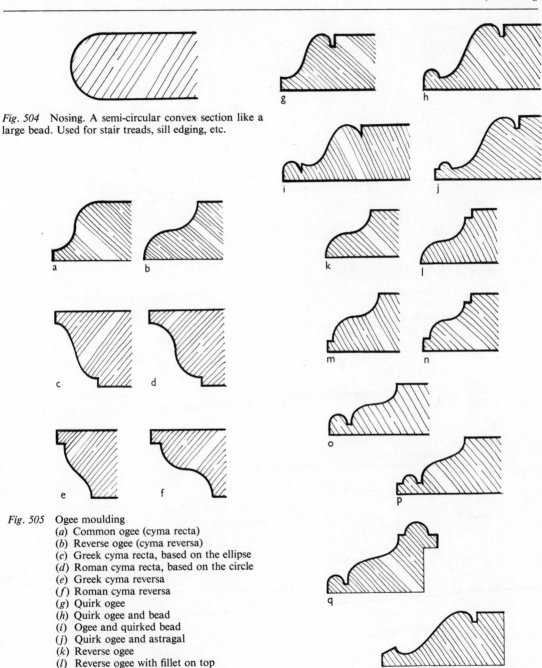

Fig. 504 Nosing. A semi-circular convex section like a large bead. Used for stair treads, sill edging, etc.

Fig. 505 Ogee moulding
    (*a*) Common ogee (cyma recta)
    (*b*) Reverse ogee (cyma reversa)
    (*c*) Greek cyma recta, based on the ellipse
    (*d*) Roman cyma recta, based on the circle
    (*e*) Greek cyma reversa
    (*f*) Roman cyma reversa
    (*g*) Quirk ogee
    (*h*) Quirk ogee and bead
    (*i*) Ogee and quirked bead
    (*j*) Quirk ogee and astragal
    (*k*) Reverse ogee
    (*l*) Reverse ogee with fillet on top
    (*m*) Reverse ogee with fillet on bottom
    (*n*) Reverse ogee with double fillet
    (*o*) Reverse ogee and bead
    (*p*) Reverse ogee and astragal
    (*q*) Reverse ogee and bead with astragal on top
    (*r*) Grecian ogee
    (*s*) Grecian ogee and bead

341

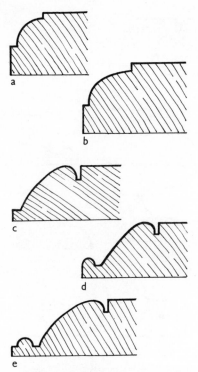

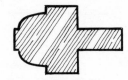

Fig. 506 Ovolo moulding. A convex moulding based on either a circle or an ellipse

    (*a*) Common ovolo
    (*b*) Common ovolo, elliptical
    (*c*) Greek ovolo
    (*d*) Greek ovolo and bead
    (*e*) Greek ovolo and astragal

Fig. 508 Sash Bar moulding. An example of ovolo type. For other patterns; see *Plane, Sash Moulding*

Fig. 509 Scotia. Usually an elliptical hollow. See also the cavetto moulding

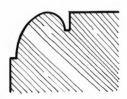

Fig. 510 Torus. A form of bead, usually of elliptical section with quirk and square fillet. Often used on skirtings

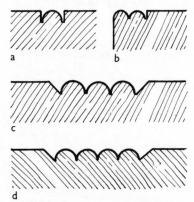

Fig. 507 Reed moulding. A bead usually in groups of two or more.

    (*a*) Single reed or centre bead
    (*b*) Double side reed
    (*c*) Three reeds with bevelled quirks
    (*d*) Four reeds with bevelled quirks

Plane, Multi: see *Plane, Combination.*

Plane, Neck Moulding (Necking-and-Nosing or Neckmould Plane. The term neck is a name given to a moulding just below the capital of a column or around the legs of tables, etc.) *Fig. 511*
A Moulding Plane for working a profile consisting of an astragal and scotia. Made for wood from $\frac{5}{8}$ to 1 in thick.

Fig. 511

Plane, Norris: see *Plane, Spiers and Norris Types.*

Plane, Nosing (Stair Nosing Plane; Step Nosing Plane) *Fig. 512*
A Moulding Plane working a semi-circular convex profile on the edge of the wood, such as a stair tread, window board, etc. Made in sizes up to 1 in. See also under *Plane, Moulding*.

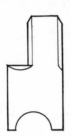

Fig. 512

Plane, Oar: see *Plane, Mast and Spar*.

Plane, Ogee: see *Plane, Moulding*.

Plane, Old Woman's Tooth: see *Plane, Router*.

Plane, Organ Builder's
Two special Planes used in this trade are described and illustrated under *Organ Builder*. One is a Plane of jack size used for trimming the metal sheets used for the pipes; the other is a type of Router Plane used for cutting grooves on the surface of the sound board.

Plane, Ovolo: see *Plane, Moulding*.

Plane, Ovolo Raising: see *Plane, Panel Fielding*.

Plane, Panel *Fig. 513*
The name Panel Plane is given to two different tools: one, a metal Plane of the Spiers or Norris type which is used for fine smoothing and trueing, and the other, a wooden Plane used for working a wide, flat, or sloping rebate (the 'fielding') round the edge of a raised panel. It could also be used for trueing long rebates of all kinds. It was, however, not the ideal tool for panels and was largely superseded by the Badger Plane with a skewed iron and flat sole without a rebate; and by the Panel Fielding Plane which worked a flat tongue that protruded from the margin of the fielding to fit into the groove of the frame. See also *Plane, Rebate*.

(*a*) The diagram demonstrates the differing cross-sections of the wooden Plane of this type. (See *Plane, Bench* for comparative sizes.)

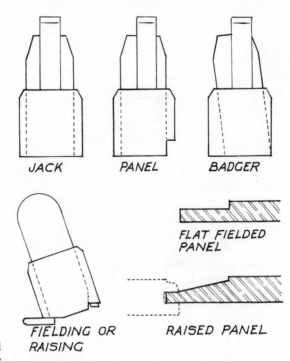

JACK   PANEL   BADGER

FLAT FIELDED PANEL

FIELDING OR RAISING   RAISED PANEL

Fig. 513 (a)

(*b*) The Panel Plane is usually about 14 in long, but occasionally much shorter. The sole is rebated along the right-hand (far) side by about ⅜ in, so that the iron on that side cuts right out to the edge of the sole. This rebate is sometimes fitted with a removable fillet which, when in place, transforms the tool into a Jack Plane. The cutting iron is 2½ in single or double, and usually bedded square in the stock at about 50°.

The Panel Plane illustrated is one of the earliest planes known with an English maker's name on it – ROBERT WOODING. It was found in a wheelwright's shop in Castle Bytham, Lincolnshire. The stock is of beech, 8½ in long and 2½ in square, with a thin, round-topped 2¼ in iron with two almost illegible stamps on it. The right-hand side of the sole has a rebate ⅜ in wide and ¼ in deep, and the skew iron has the corresponding corner bevelled. The sole is pitted with holes along the left-hand half where a

343

fence, now missing, had been screwed or nailed to it. This plane must have been made between 1710 and 1740. Robert himself, however, died before 1736, when the business was being run by his widow Anne Wooding.

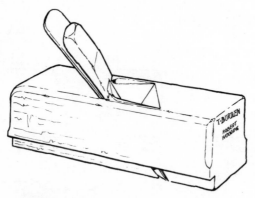

Fig. 513 (b)

Plane, Panel, Fielding (Jack Raising or Smooth Raising Plane; Raising Plane) *Fig. 514*
A Plane used to cut the wide, canted rebate round the edges of a raised panel. It works not only the cant, but also a small list at the edge of the raised face of the panel, and a flat tongue round the edge which fits into the groove of the framework. When the list is replaced by a cove or ovolo, the tool is sometimes called a 'Cove' or 'Ovolo Raising Plane'.

The commonest size is 8–9 in long by about 3 in across the sole. An adjustable, slotted fence is fitted to the part of the sole which works the tongue; and a depth stop, usually of wood, is dovetailed into the other side of the tool at the forward end, often at a slight forward slant. The iron is single or double, skewed, and bedded at about 50°. The double iron gives a better finish, especially across the grain at the top and bottom of the panel. (See *Plane, Rebate* for skewed irons.) The right-hand exposed edge of the iron is ground off at right angles to the slant of the cutting edge, and sharpened here to cut the vertical edge of the list.

In use the Plane is held at an angle, with the fence horizontal. Slight lateral adjustment of the fence enables the width of the tongue to be varied to suit the depth of groove in the frame; the depth stop controls the depth of the list and the thickness of the tongue. It is necessary to mark the rebate width with a cutting gauge before using the Plane, especially along the cross-grain ends which would otherwise be liable to tear out.

One drawback of this tool is that it forms a rebate of one width only. By using a Badger Plane, the rebate can be made of any width, but it does not produce the tongue.

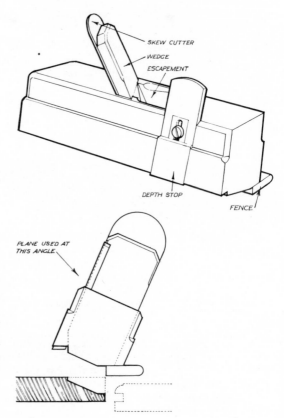

Fig. 514

Plane, Panel (Spiers or Norris Types) *Fig. 515*
The term Panel was used by Messrs Stewart Spiers and T. Norris, and by their imitators, to describe a metal plane measuring about $13\frac{1}{2}$–$17\frac{1}{2}$ in long with a $2\frac{1}{2}$ in cutting iron. They were superbly designed and are used for fine smoothing and trueing work. (See also *Plane, Spiers and Norris Types.*)

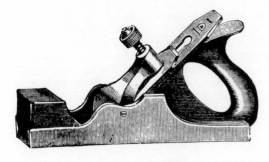

Fig. 515

versions of these rounding (or compass) Planes have been noted. The detachable soles were fixed to the stock by means of pins which fit into key-hole slots, or by dovetailed projections which slide into corresponding sockets in the sole.

The Shelburne Museum Pamphlet No. 3 (U.S.A., 1957) illustrates Pattern Maker's Planes of the late eighteenth century. These are Moulding Planes, about 8 in long but only about 2 in high, 'thus taking up much less space in the patternmaker's chest. They were made in various designs, and often 25 to 30 were found in the older patternmakers' tool chests.'

Fig. 516

Plane, Pattern Maker's *Fig. 516* (Core-Box Plane)
One of the Planes designed specially for this trade is the metal Core-Box Plane listed by Stanley (U.S.A., 1902). It may have been used by other trades, but was recommended particularly for hollowing out accurate semi-circular channels in patterns and core-boxes – the wooden patterns used for moulding sand cores. (See *Pattern Maker*.)

The sole is V-shaped, forming a right-angle in section. A cutter in the apex is secured with a screw lever-cap. Extension pieces can be added to the sides to increase the width of each sole. In use a shallow groove is cut along each edge of the channel and the bulk of the waste removed with other tools. The Plane is then used with its two soles bearing on the edges of the channel, and by slightly turning the plane after each cut a perfect semi-circle is produced.

Another Plane designed, presumably, for the same purpose is listed by Mathieson (Glasgow, 1900) as a Pattern Maker's Jack Plane. This is round across, 'with six removable soles, to suit different circles, and six Single Irons, 2 in to correspond'. Home-made

Plane, Pianoforte Maker's
(See *Pianoforte Maker*.)
The ordinary run of Planes are used in this trade. The metal version of the Mitre Plane is sometimes referred to as a Piano Maker's Plane. Among special Planes the following have been noted:

(*a*) The Sheffield *List* of 1910 illustrates a small 'Piano Maker's Boxwood Badger Plane'. The iron is skewed, its corner appearing on the edge of the sole on the left-hand side. It is made in handed pairs and used presumably for cleaning rebates.

(*b*) A 'Piano Smoothing Plane' is listed but not illustrated in the Plane Makers' List of Prices (Birmingham, 1871), with an alternative in boxwood.

(*c*) The ordinary *Edge or Chisel Plane* is sometimes called a Piano Maker's Edge Plane.

(*d*) Special planes were listed by some makers (e.g. Chas. Nurse, London, 1914) for use by pianoforte key-makers. One such has been seen in the works of Messrs J. F. Pyne, Ltd (London, 1966). This is a metal Rabbet Plane and is used for shooting the heads of ivory keys, i.e. the butt joints where the head (the part pressed by the fingers) joins the rear portion or tail. Only a small part of the $\frac{1}{2}$ in cutting iron protrudes from the sole, and this is ground almost square to cut with a scraping action. The tail of the key is squared with a metal Mitre Plane with a $1\frac{3}{4}$ in cutting iron.

(*e*) Special Planes are used in the piano action-making trade for planing the beechwood cauls used when glueing felt to piano hammers. The width and shape of the caul varies according to the size of the hammers; consequently the Planes have soles round across and are in three sizes known as treble, middle, and bass.

Plane, Picture-Frame Maker's
A name sometimes given to the wooden version of the Mitre Plane and to certain Moulding Planes used for making picture frames.

Plane, Pill Box: see *Plane, Scaleboard.*

Plane, Pipe Lighter: see *Plane, Spill.*

Plane, Plane Maker's: see *Plane Maker.*

Plane, Plough (Plow Plane; Scots: Flit Plow) *Fig. 517*
A Grooving Plane with a stock 7–8 in long, about $2\frac{1}{2}$ in deep and $1\frac{1}{2}$ in wide, rebated along the bottom to take a steel plate (known as a 'skate') which serves as a bed to the cutting iron and as a sole to the plane. This plate is parted where the iron emerges, and is often curved upwards at the front. A slot is cut in the inner side of the fence to make room for an adjustable metal or wooden depth stop, set by a thumbscrew or wedge at the top of the stock. The wedge is cut short at the foot to avoid obstructing the escape of shavings.

The iron is bedded at 50°, and thickened back-to-front at the lower end where there is a V-groove on the underside to engage with a knife-edge on the sole-plate, thus keeping the iron rigid. They are made in sets of eight different sizes from $\frac{1}{8}$ to $\frac{9}{16}$ in wide and snecked at the top for easy removal.

The fence is riveted loosely to two stems, D-shaped in section and often tipped with brass, which slide through the stock and are fixed in the required position by captive wedges, known as 'keys' to distinguish them from the wedge holding the iron. In the so-called 'French' or Screw-Stem Plough, the stems are threaded and held in place by wooden nuts. The characteristic moulding on the underside

and front end of the fence serves to ornament the tool as well as giving a comfortable grip. This fence, with stems sliding through the bottom of the stock, is characteristic of English Ploughs. In most of the Continental tools the fence is deeper than the stock and rebated over it, the stems being fixed to the stock and sliding through the fence. This is exactly the opposite of the usual English practice. (See *Plane, Continental.*)

The Plough Plane is used where a fully adjustable tool is required for grooving all kinds of work to take panels, or for making joints.

Other versions have longer stocks (10–12 in) and sometimes closed handles solid with the stock. There are various methods of fixing the adjustable fence in position, including thumbscrews bearing on each stem or on a bridle joining them; hence the term 'Bridle' or 'Bridge Plough'. As a rule these larger Ploughs also have long, skate-shaped fore plates.

Trade catalogues from *c.* 1837 to 1914 often include a Plough 'to work straight or circular'; and J. Bennett (London, 1837) lists a plough with 'moving plates for sundry sweeps'. We have never seen one of these Ploughs, but N. Lewington (1973) has described one in his possession in which the skate is easily detachable – presumably to permit its replacement by a plate of different profile – and in which the fence could be lowered to facilitate its use on curved work. A Plough capable of working a groove curved in the horizontal plane (the '*Circular Plough*') is described in the next entry.

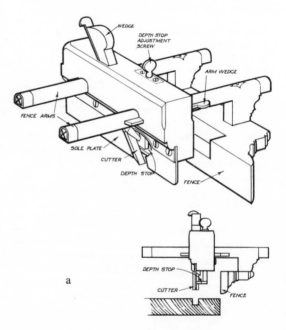

a

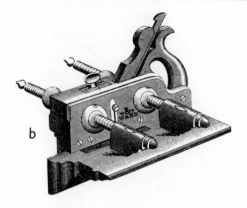

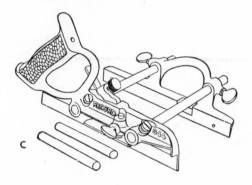

work such as hand-railing and circle-on-circle joinery.

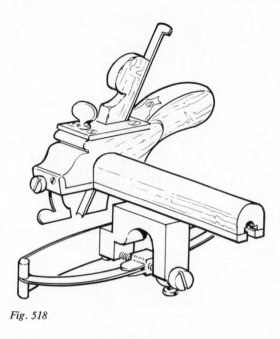

Fig. 518

Fig. 517 (*a*) A wooden Plough Plane of ordinary design
(*b*) A wooden Plough Plane with a handle and screwed stems
(*c*) A metal version

Plane, Plough, Circular (Falconer's Plough) *Fig. 518*
This unusual tool was designed by Mr. Falconer in 1846 and is described by Holtzapffel (London, 1846, Vol. II, p. 979). The illustration is taken from an example in the Science Museum, London. The pistol-shaped handle is solid with the ebony stock and an ebony stem carrying the moveable fence. The plate or skate has a flat sole with the section in front of the iron adjustable in the direction of its length. The plough iron or cutter is hollow-ground on the face, making the action similar to that of a double nicker or spur; it also has the usual V-groove on the back to keep it in position on the plate. The fence is a flexible steel strip and can be adjusted to work straight, concave, or convex. For use on circular

Plane, Plow, Glass Check
A plane illustrated by Mathieson (Glasgow, *c.* 1900), which produces a rebate and tongue. Its purpose is not known.

Plane, Pull: see *Plane, Cooper's Short Jointer; Plane, Japanese.*

Plane, Quarter Round or Hollow: see *Plane, Side Round and Side Hollow.*

Plane, Rabbet: see *Plane, Rebate.*

Plane, Raglet: see *Plane, Dado.*

Plane, Raising: see *Plane, Panel Fielding.*

Plane, Rebate or Rabbet (Boxing Plane; Check Plane; Scots: Geelum) *Figs. 519; 520*
(*Note:* A rebate is a rectangular recess or step along the edge of a piece of wood or other material.)

Rebate Planes are made in wood or metal, and are flat across the sole, in which the cutting iron extends to the extreme edge on one or both sides. The diagram *Fig. 519* illustrates the main types.

347

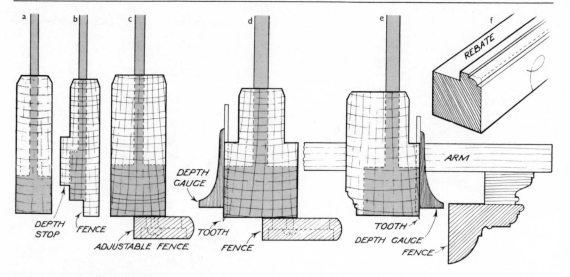

Fig. 519   Diagram of Rebate Planes (seen from the front)
   (*a*) Ordinary Rebate or Rabbet Plane
   (*b*) Standing Fillister
   (*c*) ⎫
   (*d*) ⎬ Moving Fillisters
   (*e*) Sash Fillister
   (*f*) Rebate formed by Sash Fillister

They are used for forming and cleaning up rebates. See also *Window Making*.

Variants include:

(a) Made in wood
The stock is 8–10 in long, about $3\frac{1}{2}$ in deep, and from $\frac{1}{2}$ to $1\frac{1}{2}$ in thick. (Those imported from the Continent may be as long as 11 in.) The sole may be flat, convex, concave, side-round (radiused), or coffin-shaped. Among the home-made examples many minor variations will be found in the size of the stock, angle of iron, direction of shavings discharge, occasional provision of depth stop, etc. The shavings aperture is often gracefully curved. To act as a fence, fillets are sometimes nailed temporarily to the sole, or nailed or cramped to the workpiece itself, though many tradesmen manage without this by tipping the plane for the first few strokes to form a guiding channel. (Where adjustable fences are provided, see *Plane, Fillister*.)

The iron is shouldered, square or skewed, occasionally side-sharpened, and bedded at about 50°. An advantage of the skewed iron is that it helps to draw the Plane into the corner of the rebate, and it throws the shaving to one side which helps to

prevent choking. For extra fine work, capped irons are sometimes used. (See Cutting Irons under *Plane* for the so-called 'slicing cut'.)

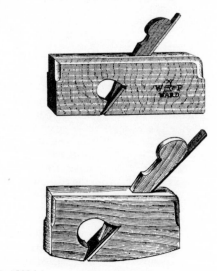

Fig. 520 (a)

(b) Metal – Spiers and Norris type

These were often provided with a rosewood or ebony core and wedge, and were beautifully turned out. They are distinguished from the metal Shoulder Planes by their flat top and steeper iron, which is bedded with the sharpening bevel uppermost. They were sometimes made with two irons, one in the normal position near the middle of the stock, the other near the toe.

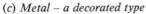

Fig. 520 (b)

Fig. 520 (d)

Note: Other Planes of the rebating type are described under the following entries: *Plane, Badger; Plane, Coachbuilder's* (Door Check, Tee, etc.); *Plane, Fillister; Plane, Panel* (and *Panel Fielding); Plane, Shoulder; Plane, Side Rabbet*.

(c) Metal – a decorated type

These metal examples, which were decorated in a somewhat grotesque manner, were popular during the early years of the twentieth century. Later examples were fitted with a screw adjustment for the cutting iron and a lever-type metal wedge.

Plane, Reed: see *Plane, Moulding*.

Plane, Reform *Fig. 521*
A term applied to a Bench Plane, imported from the Continent, which has an adjustable mouth. A movable block is let into the sole, just in front of the mouth, which is adjustable by means of a screw on the top of the plane. In another version, made in this country, the toe end of a boxwood sole-plate slides in a groove to open or close the mouth.

Fig. 520 (c)

(d) Metal – Stanley types

That illustrated immediately below is the Stanley so-called 'Cabinet Maker's Rebate Plane', which other makers often class as a Shoulder Plane – for which purpose it would serve equally well.

Fig. 521

Plane, Reglet (Not to be confused with the Raglet Plane, a Scots term for the Dado Grooving Plane) *Fig. 522*
A Plane of jack appearance with a fence on each side of the stock, usually fixed but sometimes vertically adjustable. Used for planing printers' reglets (spacing-pieces for type) to standard widths and thicknesses. The wood is sawn slightly full to the required thickness, placed on the bench against the stop, and planed till the fences rest on the bench top. When no more shavings can be taken off, the reglet is of the correct thickness.

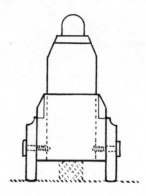

Fig. 522

Plane, Reverse: see Mother Plane under *Plane Maker.*

Plane, Roughing (Cow Plane; Hunter Plane; Scrub Plane; Scud Plane; Scurfing Plane. The term 'cow plane' may be used in the same sense as 'cow cut', which is a country word for a heavy haircut.) *Fig. 523*
Many Roughing Planes are made from old or worn-out Jack or Smoothing Planes, and fitted with a single cutting iron. They are used for rough-shaping or preliminary planing which was necessary before it became customary for timber to be bought machine planed.

Factory-made examples include:

(*a*) Wooden Planes of the Continental type with a front 'horn' handle; hence the term 'Bismarck' often applied to Roughing Planes. (See *Plane, Continental.*)

Fig. 523 (a)

(*b*) In the early years of this century the Stanley Works listed two metal Planes for roughing. One, called a *Scrub Plane*, has a stock $9\frac{1}{2}$ or $10\frac{1}{2}$ in long, with a single $1\frac{1}{4}$–$1\frac{1}{2}$ in iron with rounded corners to the cutting edge, bedded at about 40°.

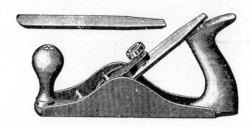

Fig. 523 (b)

(*c*) Another Stanley version, called a *Furring Plane*, is 10 in long and $2\frac{1}{2}$ in wide at the mouth, narrowing at both ends. The underside of this Plane is most unusual, for the sole proper extends only about $\frac{3}{4}$ in on either side of the mouth, leaving a space fore and aft between the underside of the Plane and the timber being worked. Stanley describes it as 'a new tool for preparing the lumber as it comes roughly sawn from the mill, to remove the fur and grit before using a Smooth Plane. . . . The construction of the bottom is such that it will accomplish this very rapidly.'

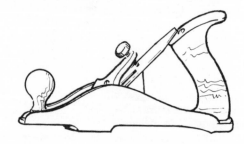

Fig. 523 (c)

350

(*d*) A modern version of the Roughing Plane is made like a rasp – see *Surform Tools*.

Plane, Round-Both-Ways
A term applied to Planes with a sole which is convex both in cross and longitudinal section. See, for example, *Plane, Cooper's Stoup*.

Plane, Rounder *Fig. 524* Alternative names include:

Circular Plane	Stail Engine
Fork Shaft Rounder	Stave Block
Fork Staff Rounder	Tap
Jack Fork Staff	Thole Reamer (Thole is
Jack Fork Stail	the short handle or
Ladder Rounder	nib on a scythe
Moot (see *Plane,*	snaith.)
Rounder, Trenail)	Turning Plane
Nogg	Witchet or Widget
Rung Engine	(mainly in the U.S.A.)

Note: For method of working see *Fig. 734* under *Woodland Trades*.

A rotary Plane, made in various patterns and usually home made, used for rounding components such as rake or fork handles, trenails, and dowels, putting a round taper on ladder rungs, and working spoke tongues. The stick to be rounded is held in a vice or brake; one end is placed in the mouth of the Rounder which cuts circumferentially across the grain when turned hand-over-hand.

Note: The Rounder should not be confused with the various Rounding Planes which have hollow soles, and are worked along the axis of the stick or pole. Rounding Planes are described under the following headings: *Plane, Forkstaff; Plane, Hollow and Round; Plane, Mast and Spar; Plane, Spout.* (See also *Rounding and Hollowing Tools*.)

Rounders are of two kinds:

(*a*) The Solid Rounder
This consists of a block of wood $2\frac{1}{2}$–5 in thick, with two handles, and with a hole bored through the middle, which is usually tapered. The cutter, either a special iron with two slots for screws, or an ordinary plane iron with the cutting edge slightly rounded, is bedded tangentially in a V-shaped mouth.

This type of Rounder is preferred for rounding short sticks such as spoke-tongues, dowels, or ladder rungs, while the adjustable type, (*b*) below, is employed for rounding long handles. When the orifice is much tapered it is suitable for tapering the ends of ladder rungs or the legs of a milk stool. But when such Rounders are worked right through from one end of the stick to the other, the stick is finished parallel.

Factory-made Solid Rounders (which are rarely found in country workshops) often have snail-shaped bodies and are made in sets with a hole rising in eighths from $\frac{5}{8}$ to $1\frac{1}{2}$ in wide.

Some of the home-made Rounders are fitted with an adjustable shoe, controlled from the back of the Rounder by a screw, which protrudes into the mouth to reduce its circumference and allow a slightly smaller stick to be rounded. Another variety has a solid body made entirely of iron or brass, with side bosses in which the wooden handles are mounted. This was the tool apparently favoured by 'Mooters', i.e. the men who made trenails (see *Plane, Rounder, Trenail*).

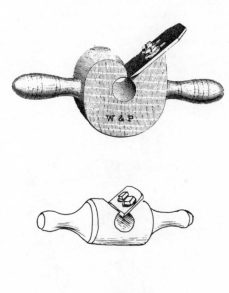

Fig. 524 (a)

(b) The Adjustable Rounder

This consists of two wooden blocks, one (or exceptionally both) fitted with cutting irons, and connected together with wood or metal hand screws. By means of these screws the distance between the blocks can be varied to take sticks of different diameters. The mouth may be slightly tapered, and the leading edge of the cutting iron is well rounded. The handles are fitted in two ways: either as an extension of the adjusting screws, or as separate handles fixed into, or made solid with, the body of the Plane. If the handles are separate, the adjusting screws are usually of metal, with long-eared or rat-tail wing nuts.

This type of Rounder is preferred for rounding longer sticks, such as scythe, rake, fork, or broom handles. To produce a long tapering handle, the screws are adjusted as the work proceeds. For a different method of adjustment see Trap under *Fishing Rod Maker*.

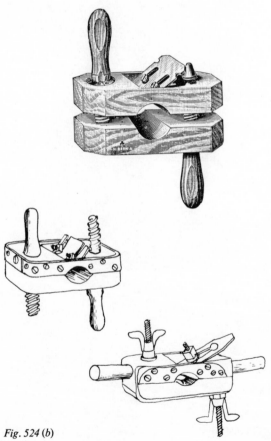

Fig. 524 (b)

Plane, Rounder, Trenail (Also known as a Moot)
Fig. 525

A Rounder used by shipwrights for rounding trenails. Two types have been noted:

(*a*) A heavy wooden Rounder of adjustable type made in two halves. The throat is plated to resist wear.

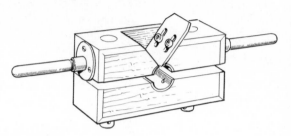

Fig. 525 (a)

(*b*) A solid metal Rounder with wooden handles. The throat is fitted with an adjustable shoe or stop to accommodate different sizes of trenail.

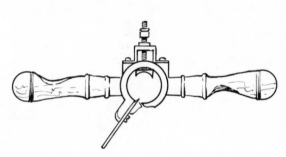

Fig. 525 (b)

Plane, Rounding (In the U.S.A. these planes are sometimes called 'Hollow Planes'.)

A Plane with the sole concave across (i.e. a hollow sole) or otherwise adapted for rounding and smoothing a shaft, stick, or pole.

See: *Plane, Forkstaff; Plane, Hollow and Round; Plane, Mast and Spar; Plane, Pattern Maker's* (Core-Box Plane); *Plane, Rounder; Plane, Rounder, Trenail; Plane, Spout.* (See also *Rounding and Hollowing Tools.*)

Plane, Roundsil: see *Plane, Compass*.

Plane, Router (Granny's Tooth; Depthing Router; Old Woman's Tooth) *Fig. 526*
(*Note:* The term is also applied to a group of tools used by coachbuilders and others for working curved parts, see *Router.*)

This tool is used for routing out 'housing waste', i.e. cleaning the bottom of wide grooves below the surface of the wood, and for 'depthing' a flat recess in a carving design.

The following are typical examples of the tool:

(*a*) A narrow iron is bedded at about 45°, or at a steeper angle, in an ordinary plane stock with the front hollowed out as far as the mouth.

Fig. 526 (a)

(*b*) A wooden block with an iron bedded at about 45° in the middle. The block is often notched out in front to the edge of the mouth.

Fig. 526 (b)

(*c*) The metal version consists of a flat casting forming the stock, often cut away for lightness, and so that the cutter can be seen while at work.

The narrow cutter, which is mounted vertically in the centre of the stock, is bent at right angles beneath the sole. Consequently it works almost horizontally and therefore cuts more readily than the wooden version with its higher angle of cutter.

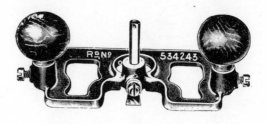

Fig. 526 (c)

(*d*) Smaller patterns of the Router Plane are used by carvers. These may have a plain stock about 4 in long and a narrow cutter fixed with a wedge.

Fig. 526 (d) (*continued overleaf*)

(*e*) Two home-made examples in wood.

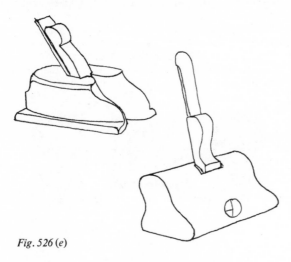

Fig. 526 (e)

Plane, Rule Joint: see *Plane, Table.*

Plane, Sash Fillister: see *Plane, Fillister.*

Plane, Sash Moulding *Figs. 527; 528*
The Sash Moulding Planes are designed to produce the mouldings on glazed windows, doors, etc. The standard profiles are shown in *Fig. 527.*

The size of Plane depends on the thickness of the sashes and the width of the glazing bars which vary from about $\frac{1}{2}$ to 1 in. Each size is made in pairs, No. 1 for removing the main bulk of the waste, and No. 2 for planing down a shade further and for giving the finishing cut to the profile. The bars are held in a Sticking Board while being planed. The rebates are cut with a Sash Fillister Plane.

Each pair is accompanied by the appropriate Saddle or Side Templets for cutting the mitre on the moulding and as a guide to correct scribing. For the simpler profiles, special Scribing or Coping Planes are available, as well as the corresponding Scribing Gouges. See *Window Making.*

Other Planes used for the above purpose include (*Fig. 528*):

(*a*) *Circular Sash Plane* (Double Compass Moulding Plane)
Sash Moulding Planes compassed both ways and fitted with a fence. Used for sticking mouldings on curved sash stuff, known as 'circle-on-circle-work'.

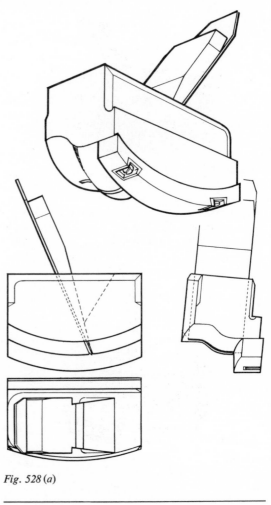

Fig. 528 (a)

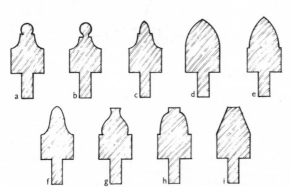

Fig. 527 Sash Moulding Profiles (*a*) Astragal and Hollow (*b*) Astragal and Quirk Hollow (*c*) Astragal and Scotia (*d*) Gothic (*e*) Gothic and Fillet (*f*) Lamb's Tongue (*g*) Ogee Sash (*h*) Ovolo (*i*) Rustic.

(b) Stick-and-Rebate Sash Plane

A Sash Moulding and Sash Fillister Plane combined to work sash stuff in one operation. Each plane has two irons with a regulating screw for different widths of bar. These combined planes are rarely seen in this country but are more common in the U.S.A.

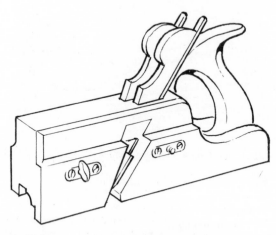

Fig. 528 (b)

one with the iron bedded at the usual angle of 50°, with the fence on the left-hand side of the stock in the ordinary way; the other with the iron nearly upright, and the fence on the right. The first would be used for roughing out the profile and the second as a kind of scraper to give a clean, accurate finish. See *Templet, Sash; Window Making.*

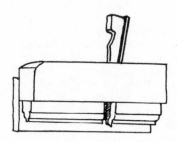

Fig. 529

Plane, Sash Scribing or Coping *Fig. 727* under *Window Making.*

A special Moulding Plane with a reverse or counter-profile corresponding to the simpler sash mouldings and used for scribing the shoulders of sash bars in order to fit one moulded profile over the other. The bar was laid in the Saddle Templet and it is thought that the material was probably cut with the Plane laid flat, using the side of the tenon as a guide. The Saddle Templet, being an exact reverse of the moulding section, supported the wood of the sash bar and so helped to avoid splintering out the end grain of the wood.

According to the late Mr. A. Collier, the use of Planes for scribing was discontinued after about 1890. At present only one actual example is known, in the Collier Bequest in the Science Museum Reserve Collection. The modern method is to use a Sash Gouge or the Coping Saw. (See *Window Making.*)

Plane, Sash Templet Making *Fig. 529*

A Moulding Plane for making the Saddle or Double Templets used for scribing sash bars. It consists of a stock with the sole made to the exact profile of the moulded part of the bar with a deep fence screwed to one side. There appear to have been two types:

Plane, Scaleboard (Chip Plane; Pill Box Plane; Scabbard Plane; Spelk Plane; Slitting Plane; Splint Plane) *Fig. 530*

Planes mostly home made, designed to produce a long thin strip or shaving of wood. These wooden strips, called 'scaleboard' or 'spelks', are used for making chip or spale baskets, hat boxes, and pill boxes, and for tying up goods such as tobacco. (They were not used to make the tough oak strips from which the traditional spale or spelk baskets were woven – see *Basket Maker, Spale.*)

Varieties of this plane include:

(a) *Chip Plane.* This Plane is described by the late Charles Freeman (1953). It was used in the eighteenth century and later for making wooden shavings ('chips') usually of willow. These were plaited for making hats. The Plane had a broad sole and an iron bedded at about 40°. It was preceded by a tool known as a *Width Maker* – an appliance set with a number of steel blades, which could be regulated according to the width of the strips required. It was drawn along the flat surface of the wood to make a number of parallel cuts, after which the Chip Plane raised a broad shaving already divided into strips.

(b) *Pill Box Plane.* Listed by H. Griffiths (Norwich, *c.* 1850). It had a $3\frac{1}{2}$ in iron and was priced at 9/–, when the best Smoothing Plane with a double iron cost 3/10d. Its appearance is unknown, but it was probably used for cutting thin wood strips for making pill and ointment boxes.

(c) *Spill Plane.* A Plane for producing flat or tightly curled shavings for lighting pipes etc. See *Plane, Spill.*

(d) *Scaleboard Plane.* Described but not illustrated by Knight (U.S.A., 1876). For 'planing off wide chips for fruit, hat and bonnet boxes . . . a plane loaded with weights and dragged or driven over the surface of the board'.

(e) *Splint Plane.* Illustrated by Knight (U.S.A., 1876) and intended for making blind-slats. It has a fairly long rebated sole with a cutting-iron set very low. The slat, after being detached from the board by the Plane, travels through the length of the stock, 'each one pushing the preceding one out'.

(f) *A Scaleboard Plane from abroad. Fig. 530.* A heavily-made Plane, often fitted with four or more handles. Though seen occasionally in Britain, they appear to have been imported from Europe. Some have conventional plane irons, sometimes skewed, while others have a very low-angle iron of the spokeshave type. In some, the long shaving produced by the Plane is made to travel through an orifice in the length of the stock and out at the back. One such Plane in the Edward Pinto collection is dated 1776 and probably comes from Holland. He describes it as follows: 'The sole is grooved out to a depth of $\frac{1}{2}$ in between solid fences, $2\frac{1}{2}$ in apart. It is believed to have been used for making hazel, poplar or willow slats for continental woven spelk baskets. I do not think that these planes could have been used for the tough oak, from which the traditional English spelk baskets are made. Even for comparatively soft timbers, such a plane must have needed tremendous force to drive it as the three side handles and the one upright handle at the rear, indicate. The heavy iron can be set to various depths, and the mouth at the rear, through which the spelk emerges, measures $2\frac{1}{2}$ by $\frac{5}{8}$ in.'

Plane, Scoopmaker's *Fig. 531*
A special Plane or Shave illustrated in the Shelburne Museum pamphlet No. 3 (U.S.A., 1957). Made with a sole round-both-ways, it has a long handle projecting from the toe by which it is pulled towards the user. Used for making wooden scoops for taking flour or sugar from a sack.

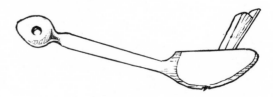

Fig. 531

Plane, Scoring: see *Organ Builder.*

Plane, Scotia: see *Plane, Moulding.*

Plane, Scraper *Fig. 532*
Not really a Plane, but a trade name for various handles or stocks to hold a Scraper. Those made by Stanley Works in metal have a blade which leans toward the toe of the stock at about 25° from the vertical. Used by cabinet makers and others for scraping large veneers and hardwood surfaces, and also for scraping off old paint.

There are advantages to be gained by using this tool instead of holding the scraper in the bare hands. It avoids burnt fingers which are apt to result from the use of an ordinary scraper by all but those with the horniest hands; and a flat sole prevents the scraper from digging into parts of the timber with softer grain. By changing the blade the tool can also be used as a Toothing Plane for roughing up the surface of wood before glueing veneers.

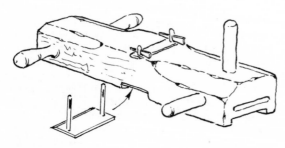

Fig. 530

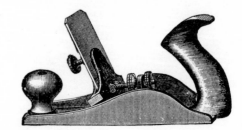

Fig. 532

Plane, Scratch: see *Scratch Stock.*

Plane, Scribing: see *Plane, Sash Scribing.*

Plane, Scriff-Scraff
The term 'Scriff-Scraff Planes (German Jacks)' occurs in the index of Mathieson's catalogue (Glasgow,
c. 1900). This refers to a Jack Plane with a front
'horn' which is presumably intended as a Roughing
Plane.

Plane, Scud, Scrub, or Scurfing: see *Plane, Roughing.*

Plane, Seam: see *Plane, Shipwright's.*

Plane, Shaft
A name sometimes given to hollow-soled or Rounding Planes. See list under *Plane, Rounding.*

Plane, Shave
The name Shave Plane is sometimes given to Shaves
of the type used by coopers or wheelwrights. These
tools have the form of a Plane, though they look
more like very large Spokeshaves. We have grouped
them under their more usual name – *Shave.*

Plane, Shipwright's
We have not seen any Planes made specially for
shipbuilding – apart from the Planes described under
Plane, Mast and Spar and *Plane, Rounder, Trenail.*
But Messrs A. Copley (London, *c.* 1925) list (but do
not illustrate) the following:

Boat-Builder's Plane, Stem or Keel	11/0 pair
,, ,, ,, Mast or Pole	6/0 pair
,, ,, ,, Ogee	6/9 pair
,, ,, ,, Ogee with Fence	8/6 pair
,, ,, ,, Oar and Scull	4/9 each

(In the same list a Smoothing Plane with a 2 in
iron costs 3/0.)
The 'Mast or Pole' and the 'Oar and Scull' Planes
are presumably Mast and Spar Planes, but we do
not have any information on the 'Stem or Keel'
Planes. When we asked some shipwrights if they had
any knowledge of them, we received the following
replies:

Mr. R. Paterson (U.S.A., 1968) told us that he
had seen the following special Planes in use:
A round-soled Plane for forming hollow grooves
on the inside of carvel-type planking to make them
fit snugly against the curved ribs or frames. (On
carvel jobs, with very heavy planking, a Hollowing
Adze with a 3 in wide blade and a short handle was
used for the same purpose.)

A Seam Plane: a home-made Side Rabbet Plane
used by caulkers for 'fairing up damaged seams after
reaping out' (i.e. taking out old oakum) and before
re-caulking.
A special Rebating Plane with a 'skewed rabbeting
iron projecting through the right hand side', which
American shipwrights made for themselves and used
for finishing the 'heavy planking rebate on the stem
or keel'.

Shipwrights working in the Victory Workshop of
H.M. Dockyard (Portsmouth, 1969) told us of the
following special Planes they had seen being used
some years ago:
A Keel Plane: used for smoothing the lead keels
of yachts, an operation known as scruffing or scrunt-
ing. (This may be the same type of Plane as that
described under *Plane, Metal Cutting.*)
Landing Plane: a Rebating Plane 4–5 in long used
for levelling the exposed upper edge of boards in
clinker-built boats.

Plane, Shooting
A name given to a long Plane such as a Trying Plane,
used for 'shooting' the edge of a board. See *Shooting
Board; Mitre Appliances.*

Plane, Shoulder *Fig. 533 overleaf*
These specialised, low-angled Planes of rebate type
seem to have first appeared in the nineteenth century
and, so far as we know, were made only in metal.
The stock is usually about 8 in long, 3 in deep,
and 1½ in wide, with a wooden core and wedge, often
of rosewood or ebony. The top is arched with shaped
ends, often very handsomely designed, and turned
out with a high finish. Smaller examples are also
found. The modern versions made by such firms
as Stanley and Record have a screw to alter the
mouth opening for coarse or fine work. The iron is
usually square and shouldered, but sometimes
skewed and exposed at the sides. It is set at a low
angle of about 20°, but since the sharpening bevel
is uppermost, the effective angle remains at about
50°. The object of this is to support the iron right
up to the cutting edge, and so avoid 'chattering'
when working on end grain. Their purpose is to
clean rebates, shoulders of tenons, etc. across the
grain.

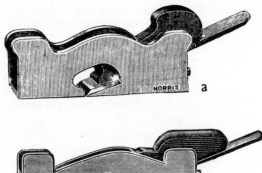

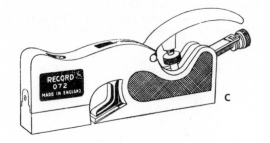

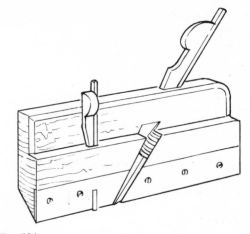

Fig. 534

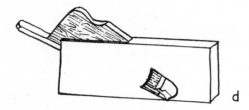

Fig. 533   Shoulder Planes
(*a*) Spiers or Norris type: 'Yorkshire Pattern'
(*b*) Spiers or Norris type: 'London Pattern'
(*c*) Modern adjustable type, all metal
(*d*) A home-made example in gunmetal

Plane, Shouldering *Fig. 534*
A Plane of moulding type with the iron bedded at
50°, the left-hand working edge plated with metal
and provided with a spur cutter. The V-shaped sole
slopes upwards sharply at an angle of about 60°.

The purpose of this plane is doubtful. It may have
been used for cutting and cleaning shoulders of
tenons etc. across the grain.

Plane, Show-Case Maker's: see *Plane, Airtight.*

Plane, Shutter
Included in Bennett's *Artificier's Lexicon* (London,
1837). Appearance and purpose unknown, but see
also *Bit, Pin.*

Plane, Side Bevel: see Side Chamfer Plane under
Plane, Coachbuilder's.

Plane, Side Rabbet *Fig. 535*
These Planes are intended for trimming or widening
the sides of rebates or grooves, and also for trimming
acute corners as in a dovetail groove. They can also
be used when shaping non-standard moulding profiles. Variants include:

(*a*) A wooden stock of moulding-plane type made
in right-handed and left-handed pairs, with its sole
tapering almost to a point. The iron is skewed and set
vertically, with a cutting edge on the side.

(*b*) A wooden stock, L-shaped in cross-section,
about 6 in long and 1¾ in wide. The iron, also L-
shaped, is about two-thirds the width of the sole and
bedded at an angle of about 60°. In use the Plane
would be held on its side. A Plane of this type,
illustrated in the Shelburne Museum Booklet (U.S.A.,
1957), has both right-hand and left-hand cutters in
one stock, which is consequently T-shaped in cross-
section.

(*c*) All-metal versions are made in handed pairs,
or they are combined with the cutters either at opposite ends or crossing each other.

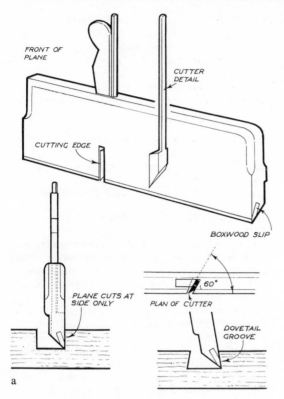

FRONT OF PLANE

CUTTER DETAIL

CUTTING EDGE

BOXWOOD SLIP

PLANE CUTS AT SIDE ONLY

PLAN OF CUTTER

60°

DOVETAIL GROOVE

a

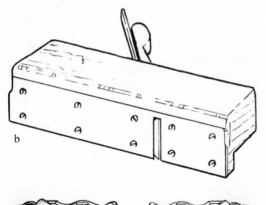

b

c

Fig. 535

Plane, Side Round and Side Hollow (Half or Quarter Round Plane; Half or Quarter Round Hollow Plane) *Fig. 536*

Moulding Planes with a profile of a quarter or half round, or of a hollow. They are handed right and left in pairs. Used with the Hollow and Round Planes for shaping non-standard profiles, or for trimming existing mouldings.

Fig. 536

Plane, Side Snipe and Snipe Bill *Fig. 537*

Both are Moulding Planes made in handed pairs with an ogee-shaped sole about $\frac{3}{4}$ in wide.

(a) Snipe Bill

The cutting iron is bedded square at about 50°, and only the lower edge of the cutter is sharpened for cutting. Used mainly for shaping non-standard mouldings after working the main profile with the Hollow and Round Plane. It is useful for cleaning the rounded edges of quirked mouldings.

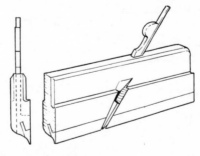

Fig. 537 (a)

359

(b) Side Snipe
The cutting iron is skewed and bedded vertically or even sloping slightly forward. Only one side (vertical) edge is sharpened; the curved profile at the sole takes no part in the cut. A variant which avoids the necessity of having handed pairs is fitted with two opposing irons bedded at an angle of about 80° on the same stock. Used mainly for trimming and cleaning the vertical edges of quirked mouldings.

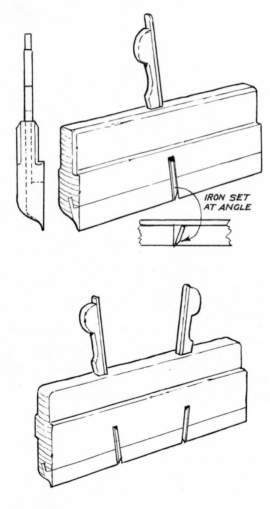

IRON SET AT ANGLE

Fig. 537 (b)

Plane, Sittle: see under *Plane, Coachbuilder's.*

Plane, Skew: see *Plane, Badger* (For skew cutting iron see Cutting Iron under *Plane.*)

Plane, Slit Deal: see *Plane, Matching* ('Slit deal' is a term for tongue and groove boarding, also called 'match-boarding'.)

Plane, Slitting: see *Plane, Scaleboard; Gauge, Cutting.*

Plane, Smith's: see *Plane, Metal Cutting.*

Plane, Smooth Raising: see *Plane, Panel Fielding.*

Plane, Smoothing (Coffin Plane; Smooth Plane) *Fig. 538*
The shortest, and perhaps the most generally used, of all Bench Planes. Though increasingly replaced by its metal counterpart, it is still to be found in most woodworking shops, often so worn on the sole that its original thickness is considerably reduced.

The stock is about $6\frac{1}{2}$–9 in long and $2\frac{1}{4}$–$3\frac{1}{4}$ in wide, its upper edge rounded and chamfered for easy handling. Since about 1700, the stock has usually been coffin-shaped. The flat sole is sometimes plated with lignum vitae or metal, either in front of the iron or all over. The cutting iron is double, $1\frac{1}{2}$–$2\frac{1}{2}$ in long, very slightly convex at the cutting edge, and bedded at 45° or 50°. Though intended primarily for smoothing and producing a finished surface, it is often found in general use as an all-round bench tool and is sometimes called upon to perform rougher service.

Typical Smoothing Planes and their variants are illustrated. Other types are described under *Plane, American Wood Pattern; Plane, Spiers and Norris Types; Plane, Stanley-Bailey.*

For information on component parts see *Plane;* on sizes, see *Plane, Bench.*

Plane, Snipe Bill: see *Plane, Side Snipe.*

Plane, Spelk: see *Plane, Scaleboard.*

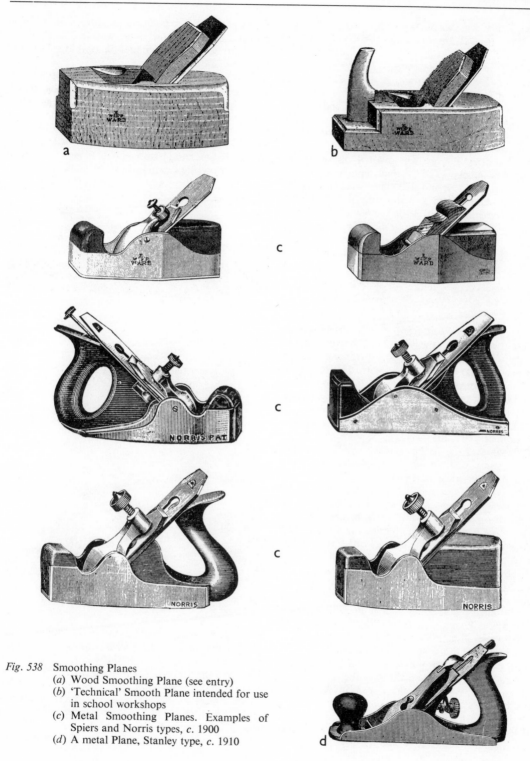

Fig. 538   Smoothing Planes
    (*a*) Wood Smoothing Plane (see entry)
    (*b*) 'Technical' Smooth Plane intended for use
        in school workshops
    (*c*) Metal Smoothing Planes. Examples of
        Spiers and Norris types, *c.* 1900
    (*d*) A metal Plane, Stanley type, *c.* 1910

Plane, Spiers and Norris Types (For range of sizes, see *Plane, Bench.*) *Fig. 538 (c)* and *Fig. 539*

Stewart Spiers, Ayr, Scotland (*c.* 1840–1920)
These handsome and still sought-after planes were developed from the early nineteenth-century Mitre Plane by Stewart Spiers of Ayr (Scotland) from about 1845 onwards. The stock took the form of a steel box or channel with a rosewood core. The cutting iron was secured with a wedge, or with a screwed gunmetal lever cap, pivoted on two pins engaging in the cheeks of the mouth opening.

The mild steel sole and side plates were frequently dovetailed together in a highly ingenious manner which is sometimes invisible on the plane itself. They were thus described as 'dovetailed' to distinguish them from those made from iron or gunmetal castings. The planes were made in the following sizes:

Smoothing – $7\frac{1}{2}$–$8\frac{1}{2}$ in long with 2–$2\frac{1}{4}$ in iron
Panel – $13\frac{1}{2}$–$17\frac{1}{2}$ in long with $2\frac{1}{2}$ in iron
Jointing – $20\frac{1}{2}$–$27\frac{1}{2}$ in long with $2\frac{1}{2}$ in iron

The Jointing and Panel Planes had a characteristic square, cushion-shaped front grip made of wood, and a closed handle at the back. The Jointing patterns had the sole projecting for a short distance at the toe and heel, similar to the earlier Mitre Planes. There were three patterns of Smoothing Plane, one with round bulging sides, one with parallel sides, and one with a square heel but tapering towards the toe.

These planes were intended for high-class joinery and cabinet work, and were so popular that other makers adopted the designs extensively. The genuine Spiers planes were, in fact, too good for anything but a narrow range of first-class work; ordinary joinery could be dealt with quite adequately by the more popular makes, which were roughly a half or a third the price. Some few years before the firm went out of business, in the period 1920–1930, they put a cheaper plane on the market which they called the Spiers 'Empire' Plane. This appears to be a tool of the Stanley type, with a cast-iron stock.

T. Norris & Son, London (*c.* 1860–1940)
A wood-lined metal plane similar in most respects to the Spiers Planes. The Bench Planes were in two main forms: with mild-steel sole and sides and a wood lining; and a cast-iron (or occasionally gunmetal) channel, with the wood lining held with countersunk screws driven through the metal sides. Later another type was introduced in which the metal body was made from rolled steel channel.

In 1913 Norris patented a device for the lateral and longitudinal adjustment of the iron, a feature which had not hitherto been provided for in the Spiers pattern. This assembly of cutter, screwed lever cap and single control lever, was frequently fitted to otherwise standard wooden Smoothing, Jack and Tryplane stocks.

Illustrations of Norris Planes are included under a number of entries, including those for Jack, Panel, Smoothing, and Mitre Planes. It may be assumed that Spiers Planes are the same in general appearance though sometimes of superior workmanship and finish. *Fig. 539* shows a typical plane of the Spiers or Norris type, and also wooden Planes fitted with Norris adjustable cutting irons.

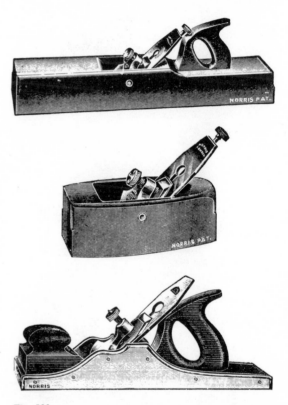

Fig. 539

Richard Melhuish & Sons, London (*c.* 1828)
A Plane marketed in three sizes: Smooth, $7\frac{1}{2}$ in long with 2–$2\frac{1}{4}$ in iron; Panel, $13\frac{1}{2}$–$17\frac{1}{2}$ in long with $2\frac{1}{2}$ in iron; and the Jointer, $20\frac{1}{2}$ in long with a $2\frac{1}{2}$ in iron.

This range of Planes was, in fact, a cheaper version of the Spiers or Norris type, the main difference being the omission of the wood core, the handles being fixed directly to the top of the metal sole. The side-plates of the stock are, however, similar in shape to those of the prototype. The turned wooden front knob borrowed from the Stanley-Bailey range makes this a curious hybrid.

Plane, Spill (Curl Shaving Plane; Pipe-lighter Plane)
Fig. 540

Although this is not strictly speaking a woodworking tool, and sometimes not even a Plane in the strict sense, some patterns, of which the variety is almost endless, look and work like Planes, and so merit inclusion.

Almost any cutting iron, if set at an angle in a suitable sole, will produce a tightly curled spiral shaving which can be used as a spill. If a flat spill is required, the cutting iron is set square and the tool then resembles a Scaleboard Plane – see (*d*) below. The material used was usually a soft wood of about $\frac{1}{2}$–$\frac{3}{4}$ in thick, and some skill is needed to keep the tool moving over the wood at a fairly fast and constant speed. Philip Walker writes as follows concerning their use (London, 1968):

'Judging by the various old tool chests I have explored, most joiners had one of these spill planes, usually home-made. I suppose he used it to use up any odd scraps of softwood (it was 'waste not, want not' in those days) provided they gave about 9 inches at least along the grain. Set right, and used with vigour, the plane will produce a tight-rolled, stiff spill which is ideal for taking a flame from the fire to light your pipe. I have no doubt that, tied up in bundles, they made an acceptable present for the 'Missus' to use in lighting a candle or lamp from the kitchen grate . . . when matches were a really expensive item, there was a great need for something to carry fire (I think housewives frequently borrowed a hot coal from each other) and some people felt a match spoiled the taste of tobacco.'

Variants include:

(*a*) *V-soled*. A cutter with a skewed cutting edge about $1\frac{1}{2}$ in wide is bedded vertically but set at an angle to the length of the plane across a deep V-shaped groove in the sole.

(*b*) *Rebated sole*. A rabbet type of Plane, with a fence down one edge, and a skewed cutter fitted horizontally across the sole instead of penetrating the stock from above.

(*c*) *Channelled sole for flat spills*. This has a square channel along the sole with the iron held horizontally and set square to produce a flat spill. Like the Scaleboard Plane (for which it may well have been intended though now found serving a different purpose), the shaving travels through the stock itself and emerges at the back. The tool is used like a

Plane, or it can be held in a vice and the board pushed over it.

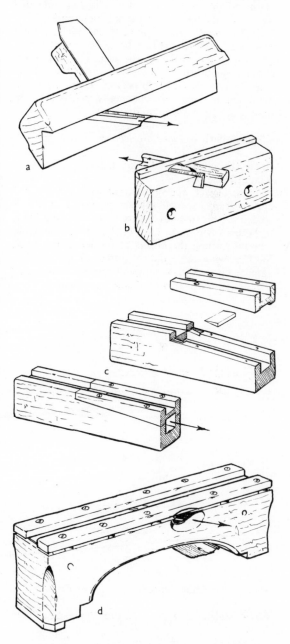

Fig. 540 Spill Planes. *Note*: The arrow indicates the point where the spill-shaving leaves the plane. The plane is moved along the wood in the opposite direction.

363

(*d*) *Channelled sole for curled spills.* This type has a square channel along the sole formed by fencing the edges. The iron is skewed, and is bedded at about 50° to produce a curled spill. It is intended for use on the bench with the sole uppermost.

Factory made. One example is the 'Presto Patent Spill Machine'. A cast-iron base, 7 in long overall, has a $\frac{1}{2}$ in groove along the top. The adjustable cutter is mounted on the skew across the groove near the end. A short nib projects from the right-hand side to deflect the shaving, which issues in a tight spiral from $\frac{1}{8}-\frac{1}{4}$ in. in diameter.

Plane, Splint: see *Plane, Scaleboard.*

Plane, Spout (Gutter Plane;. U.S.A.: Convex or Round Plane) *Fig. 541*
A Plane of jack appearance, but with a single iron and a deep, rounded sole. A horn-like handle is sometimes provided at the front to allow the Plane to be used below the level of the top of the spout without hurting the fingers of the left hand. Used for hollowing out wooden gutters or spouts. Triangular pieces known as 'spout screeds' were sometimes sawn out before the Plane was applied.

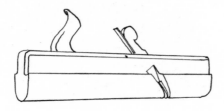

Fig. 541

Plane, Stair Builder's: see *Plane, Hand Rail.*

Plane, Stair Nosing (Step Nosing Plane) See *Plane, Nosing.*

Plane, Stanley-Bailey
Formerly the Stanley Rule & Level Co. of New Britain, Connecticut, U.S.A., established *c.* 1843.

Now manufacturing under the name of The Stanley Works both in the U.S.A. and in Sheffield, England. *Figs. 542; 543.*

See *Plane* for diagram of component parts; and *Plane, Bench* for table of sizes. Illustrations of Stanley models will be found under other entries, including: *Plane, Combination; Plane, Jack; Plane, Jointer; Plane, Metal; Plane, Rabbet.*

These Planes were based on Leonard Bailey's patents for all-metal Planes, taken out in 1867 and acquired by the Stanley Rule & Level Co. in 1896. The lateral adjustment for the cutter was patented by J. A. Traut, an employee of the firm, in 1888. (It is pleasing to note that even today Stanley planes still bear the name Bailey on the toe of the stock.)

The present Stanley model (*Fig. 542*) is the end product of some hundred years of experiment and gradual improvement both in efficiency and appearance. The stock is made of cast iron with a sole sometimes corrugated (fluted) to reduce friction. The frog (or bed) for the iron is adjustable to regulate the mouth aperture. Special devices are incorporated for adjusting the cutter both laterally by means of a lateral adjustment lever which engages with a slot in the cutting iron, and vertically, by means of a knurled-headed screw operating a forked lever which feeds the cutter up or down. The cutter is held in place by a wedging device known as a lever-cap, with a spring cam. The standard model of *c.* 1910 is illustrated and has remained virtually unchanged except for certain refinements until the present time.

No attempt is made here to present a complete record of these famous Planes. The following are some of the chief variants (see *Fig. 543*).

Fig. 542   A Stanley Plane of *c.* 1910

(*a*) *The Wood-Bottomed Plane* – 'Bailey's Adjustable Wood Plane', *c.* 1867. This had a boat-shaped iron casting to take the adjustable cutter, fitted on a wooden base or stock. Patented by Leonard Bailey in 1867 and made by Stanley, the double iron was secured with a spring-lever cap and adjusted with the knurled screw behind the frog, both of which were standard fittings on the contemporaneous all-metal Bailey Planes. The Traut patent lateral adjustment was added in 1888.

These Planes were made in the range of Stanley Bench Planes but with the addition of a 'Jenny Smooth', 13 in long with a $2\frac{5}{8}$ in cutter. That these rather unusual wood-bottomed Planes survived until recent years in competition with the same makers' all-iron Planes may have been due to their sweeter movement over wood.

(*b*) '*Stanley's Adjustable Plane*' was made with a mild steel or wood-bottomed sole (*c.* 1876). It was made in two sizes, Smooth, 9 in long, and Jack, 14 in long, both with a $2\frac{1}{8}$ in cutting iron. It included a reversion to an earlier type of gunmetal lever cap working against a bar across the mouth, with a knurled screw instead of the spring cam. (The device of a bell appears on this lever cap. This was to commemorate the hundredth anniversary of the ringing of the Liberty Bell on 4th July 1776, announcing the independence of the United States.)

When made in metal, the sole, instead of being cast, was made of mild steel, $\frac{1}{16}$ in thick, bent up to form the side plates. The handle and front knob were screwed to a separate casting riveted inside the sole. This also carried the frog, and by raising or depressing a two-prong lever at its side, moved the cutter up and down. This device was quicker than the screw of the original Bailey-Stanley pattern, but not capable of such fine adjustment. Although these Planes were 10–15 per cent cheaper, the design was withdrawn about 1910.

Fig. 543 (*b*)

Fig. 543 (*a*)

(*c*) *The 'Bed Rock Plane'* was introduced by Stanley towards the end of the century and gave two advantages not then available in the Bailey range: a frog with a more rigid seating, giving support for the cutter right to the heel of the bevel; and screw adjustment for width of mouth. Later, these improvements were effectively incorporated into the standard models.

Fig. 543 (c)

(*d*) *All-Metal Combination Planes.* By the end of the nineteenth century, Stanley had developed metal versions of almost every wooden Plane. In addition they produced the well known 'Stanley Fifty-Five', a Combination Plane which combined the function of grooving, rebating, moulding, and other operations (see *Plane, Combination*). The earlier iron castings used for the stocks and fences of these Planes were often decorated with designs of leaves, scrolls, and flowers, and were also nickel-plated. Figure 543 (*d*) shows Stanley's 'Traut Adjustable Dado and Plow Plane', which is typical of this type.

Fig. 543 (d)

Other Makers of Stanley Type Planes. The original Stanley-Bailey patterns were extensively copied by makers in the U.S.A., in Britain and on the Continent, with variations in minor details. One of the early imitators, who also contributed certain improvements, was the U.S. firm of Sargent & Company. This firm made a wide range of metal Planes closely resembling the Stanley-Bailey types, including the wood-bottomed pattern. Among early British imitators was Edward Preston & Son Ltd (Birmingham, *c.* 1825–1931) who described their version as a 'Patent Malleable Iron Adjustable

Plane'. (This firm also made metal Planes of the Spiers type as well as a range of wooden Planes.) At the present time, a well known British Plane of the Stanley type, known as Record, is made by C. & J. Hampton Ltd, of Sheffield.

Plane, Steel Dovetailed
A general term applied to British metal Planes of hollow-box construction in which the sides and sole were dovetailed into each other. See, for example, *Plane, Spiers and Norris Types.*

Plane, Step: see *Plane, Nosing.*

Plane, Stop Chamfer: see *Plane, Chamfer.*

Plane, Stower
Listed in Varvill's Catalogue (York, *c.* 1870) as a Bench Plane: 'Stower Planes to Round, 3 in with Handle'. Also listed in James Cam's Price List (Sheffield, *c.* 1800). Since stower is a dialect word for a bar or pole, this Plane may well be one of the type used for rounding poles, e.g. *Plane, Forkstaff* or *Plane, Rounder.*

Plane, Strike Block (Straight Block)
A name given to a short Jointer Plane with a square stock, 11–12 in long, with the iron bedded at a low angle with its sharpening bevel uppermost. It is used for shooting mitres or the butt ends of boards across the grain.
  Peter Nicholson (London, 1823) calls it a 'Strike Block', although in his earlier book *Mechanical Exercises* (1812) he calls it a 'Straight Block', which is the term used in a list issued by the plane makers H. Griffiths of Norwich (*c.* 1850). It is described by Moxon (London, 1677) as a '*Plane* shorter than the *Joynter* having its sole made exactly flat, and straight, and is used for the *shooting* of a short *Joint*; because it is more handy than the long *Jointer*'. He later describes how it can be held upside down for smoothing a mitred end of framing 'and so thrust it (the workpiece) hard and upright forwards, till it pass over the edge of the *Iron*, with several of these thrusts continued, cut, or plane off your *stuff* the roughness that the *Teeth* of your Saw made'. (See also *Plane, Mitre* (*Wood*) and, for table of names etc., *Plane, Bench*.)

Plane, Stringing: see *Stringing Router* under *Router, Metal Types.*

Plane, Stryking

This term occurs in conjunction with the set of 'six graven (carving) Tooles' included in a joiner's kit in the Bristol Apprentice Book, 1594. According to Mr. W. L. Goodman (Bristol, 1972), this tool is different from the Strike Block Plane. He thinks it was probably a Carver's Router of the Old Woman's Tooth type (illustrated under *Plane, Router*), and that it was used for levelling the background of a carved panel like the 'strike' used for levelling the corn when filling a corn measure.

Plane, Sun: see *Plane, Cooper's Sun.*

Plane, Surbase

Listed by John J. Bowles, Hartford, Conn., U.S.A. (*c.* 1840), this is probably a Moulding Plane intended for shaping a surbase – the moulding immediately above the base or lower panelling of a wainscoted room.

Plane, T-Rabbet: see *Plane, Coach builder's.*

Plane, Table (Rule Joint Planes) *Fig. 544*

Moulding Planes in matched pairs, one to work a profile of a quarter circle with a small square fillet, the other hollow to match. Gauges were supplied to mark the depth of the square on the quadrant profile. They were used for making the so-called rule joint on the flaps of folding tables, desks, etc. One purpose of the joint is to avoid a gap across a table through which bread crumbs or other small objects can fall; another is to improve the appearance of the joint when the table flaps are down. The term rule joint is borrowed, presumably, from the makers of folding rules. (See *Rule*.)

Both Planes were worked on the edge of the wood, the fence bearing against the underside. That working the hollow had a depth stop, but that planing the rounded member could not have a stop because it worked across the entire thickness of the wood. It had to be worked down to a gauge line level with the top fillet, and this may be the reason for the gauge with the single marker. Possibly the double marker gauge may have been for marking the recess for the knuckle of the hinge.

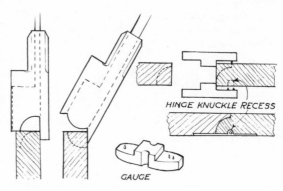

HINGE KNUCKLE RECESS

GAUGE

Fig. 544

Plane, Tambour

Listed in the Plane Maker's List of Prices (Birmingham, July 1871), it is probably a Moulding Plane with a hollow sole for shaping the strips forming the tambour or roll-top desk lid. The term 'tambour' refers to the drum-like shape; Sheraton called it a 'Cylinder desk'.

Plane, Technical: see *Plane, Jack.*

Plane, Tee: see under *Plane, Coachbuilder's.*

Plane, Thicknessing: see *Organ Builder.*

Plane, Throating

A Grooving Plane with a narrow round sole for working the throating or 'drip' under the outside edges of window sills. Its purpose is to prevent rain water from running back to the wall. See *Window Making.*

Plane, Thumb (Modelling Plane) *Fig. 545*

A general term for miniature Planes of many types, including smoothing, rabbet, hollows and rounds, side rabbet and side rounds, straight-soled and compassed, etc., more or less identical with the full-size tools. They are 1–5 in long, usually of beech or boxwood, and often beautifully made by the owner himself.

Used by first-class joiners, shipfitters, and other tradesmen for making and cleaning up a run of moulding too short or too sharply curved to be stuck by a normal plane. Examples of such mouldings may be seen in the wardroom panelling of battleships of *c.* 1910, the panelling of mid-nineteenth-century public buildings, or on the cases of pianos.

A metal version, together with the more usual

wooden examples, is illustrated. Other Planes, which are often equally small, include the *Block*, *Bullnose*, *Chariot*, and *Violin Planes*.

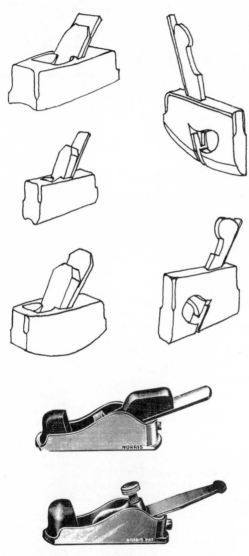

Fig. 545

Plane, Tongue and Groove: see *Plane, Matching*.

Plane, Toothing *Fig. 546*
Like a Smoothing Plane in appearance, but with a single iron set almost vertically in the stock. The iron has a series of vertical serrations which, when

ground, give the edge a saw-like appearance. Used for roughening (known as scoring or toothing) the surface of the groundwork to provide a key for the glue in preparation for glueing veneers upon it. It is also used for such work as the fixing of ivory facings to pianoforte keys, or when glueing wide laminated joints, or for removing saw marks from the back of saw-cut veneers before laying.

Fig. 546

Plane, Topping: see *Plane, Cooper's Sun*.

Plane, Torus: see *Plane, Moulding*.

Plane, Trenching
A name sometimes given to Planes that work grooves See list under *Plane, Grooving*.

Plane, Trying (Jointer; Long Plane; Scots: Halfin or Hafflin) *Fig. 547*
The names given to these long Bench Planes require some explanation. As may be gathered from the table of Bench Planes under *Plane, Bench*, the term Jointer was used in England in the seventeenth century for Bench Planes over 20 in long. In the nineteenth century the terms Jointer, Long, and Trying were used for this range in descending order of length. Holtzapffel (London, 1846) lists them as follows:

	Length (*in*)	Width (*in*)	Width of cutting iron (*in*)*
Trying	20–22	$3\frac{1}{4}$–$3\frac{3}{8}$	$2\frac{3}{8}$–$2\frac{1}{2}$
Long	24–26	$3\frac{5}{8}$	$2\frac{5}{8}$
Jointer	28–30	$3\frac{3}{4}$	$2\frac{3}{4}$

* The cutting irons are double and bedded square at 45–50°.

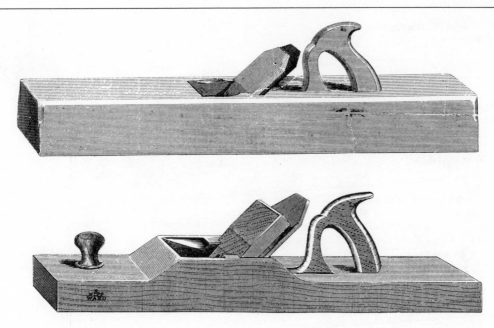

Fig. 547

But after about 1870, the terms Jointer and Long began to drop out, and although at the present time the range of Bench Planes from 20 to 30 in are usually called Trying, in the U.S.A., and to some extent in Scotland, the older term Jointer has been retained, sometimes for the whole range.

The handle of Trying Planes is usually closed (hollow clasp) and, until about 1820 and occasionally later, it was often placed off centre towards the right or left of the user. This curious feature is of unknown purpose, but it is almost universal on large Planes of the eighteenth century, British, Dutch, and Scandinavian alike.

These Planes are used for truing up the surface and edges of long boards. '. . . for shooting the edges of boards perfectly straight, so that their juncture may be scarcely discernible when their surfaces are joined together' (P. Nicholson, London, 1822).

Plane, Turning: see *Plane, Rounder.*

Plane, Universal: see *Plane, Combination.*

Plane, V: see *Plane, Centre-Board.*

Plane, Veneer Scraper: see *Plane, Scraper; Plane, Toothing.*

Plane, Violin (Oval Planes) *Fig. 548*
A very small Plane usually made in metal, but sometimes home made in wood, rounded or oval in plan, and up to 2½ in long. The sole is flat or round-both-ways, and the cutting iron plain or serrated (toothed) and from ½ to ¾ in wide. These irons are now secured with screwed lever caps, but earlier types had a wooden wedge against a cross-bar. Violin Planes, now in the Rochester Museum of Arts and Science (U.S.A.), had been fitted with awl handles extending from the heel of the Plane.

Used by violin makers for shaping the back and front of the instrument. This is done after roughing out with a Gouge, and is followed by scraping. See *Violin Maker.*

Fig. 548

369

Plane, Wagon Builder's, Side Board: see *Plane, Centre-Board.*

Plane, Washboard
Listed by J. B. Shannon, Philadelphia, U.S.A. in their catalogue of 1873 'with or without handle', but no sizes or illustration are given. Purpose unknown, but possibly for reeding a washboard, i.e. cutting the grooves in it.

Plane, Wedge Making: see *Plane Maker.*

Plane, Whip *Fig. 549*
A short wooden Plane with a steel sole, and a single iron set at 45°. The stock is about $3\frac{1}{4} \times 2\frac{1}{8}$ in, and sometimes plated. Used for paring down the tapering canes which, when glued and bound together, form the handle of a whip.

Fig. 549

Plane, Wood-Bottomed: see *Plane, Stanley-Bailey.*

Plane Fittings *Fig. 550*
Plane Fittings were provided by the makers for home repairs. Several examples are illustrated.

Moving Fillister Stop.

Solid Plough Stop.

Iron Front for Smoothing Plane.

Iron Shoe for Smoothing Plane.

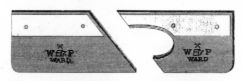

Plough Plate.　　*Skate Front Plough Plate.*

Jack Plane Handle.　　*Try Plane Handle.*

Brass Plane Lever.

Fig. 550

Plane Maker *Fig. 551* indicates the terms used by plane makers; *Figs. 552–560* illustrate the tools used. (The following account is based on investigations made by W. L. Goodman in Sheffield and elsewhere in 1966. For further information see his *British Plane Makers from 1700* (Bell, London, 1968.)

Up to the middle of the seventeenth century at least, most joiner's Planes were made by the craftsman himself. As they were largely confined to the ordinary Bench Planes and only one of each type would be required, this was a fairly simple matter. From about this period the increasing use of mouldings in joinery and cabinet making called for a wide range of special Moulding Planes which could only be made economically by men devoting themselves exclusively to this trade and able to build up the large range of special tools and equipment necessary to engage in it.

The earliest English plane maker known by name is Robert Wooding. He was apprenticed in 1693 and elected to the livery of the Joiners' Company of London in 1710. (One of his Planes is described and illustrated under *Plane, Panel.*)

From about 1800 the number of makers in Great Britain rose to a peak of around 180 in 1855, after which there was a rapid decline, owing to the introduction of machinery for making mouldings and the change in the organisation of the trade from small family workshops to larger units. The continued existence of large numbers of Moulding Planes, which had an average working life of several generations of tradesmen, was a contributory factor in this decline.

In the later stages of plane making by hand the trade was divided into two distinct sections: one group specialising in Bench Planes and the other in Moulding Planes. The Bench-Plane specialists also made coopers' tools such as Shaves, Sun Planes, Chivs, and Crozes, and also wheelwright's and coachbuilder's Routers and Shaves.

Materials

In England beech is the wood preferred for plane making, mainly because it is the only timber large enough and common enough, possessing the required close grain, stability, and freedom from defects, as well as being hard, cheap, and comparatively heavy. In France, however, other woods are used occasionally, such as service tree (cormier), evergreen oak (chêne vert), or orchard trees such as apple or pear. In Scandinavia birch is sometimes used. In England small planes, and tools composed of comparatively small parts, such as ploughs, fillisters, and so on, were sometimes made of boxwood, which was also widely used for inserted 'boxing' provided to resist wear (see below).

Process

Before the plane maker started his work the stocks were 'cultivated', that is, the timber, which had been seasoning for five years or more, was taken from the stack, sawn to the length and size required, and in the case of some Moulding Planes rabbeted if necessary.

(a) Bench Planes

The first step in the process of making Bench Planes is to mark the position of the shavings outlet (and the handle, if provided) on the cultivated stock, using the patterns and jigs described below. Holes are bored into the position of these recesses and then cut out with special chisels (see Plane Maker's Chisels below). The iron and wedge are then fitted and the sole trued up with them in place. The chamfers round the top are cut with a special Moulding Plane and the returns chiselled out and finished with small gouge cuts. Lastly, the stock is given a final clean up with Floats, Files, and glasspaper.

(b) Moulding Planes

When making an ordinary Moulding Plane, the mortice in the top and the 'spring' or angle of the profile is scribed on the toe end with the Spring Guide (see Patterns and Jigs below) and the fence and depth-stop squares worked with the appropriate Rabbet Planes. If any boxing is required it would be done at this stage (see Boxing below). The profile of the moulding is then worked with the appropriate 'Mother' or Reverse Plane. The bed and mouth are sawn out on the side of the lower (thicker) part of the stock with a Tenon Saw and chopped out. The mortice in the top is started by boring two holes and chopped out to meet the previously-sawn mouth and cleaned out with narrow Floats. The wedge blanks sawn roughly to size are prepared in batches of 16 to 20 at a time in a screw clamp, the notch being shaped out across the grain with a special Wedge Notching Plane (*Fig. 556*). The step between the upper and lower part of the stock is hollowed out with a special Plane; the chamfers round the top and ends are worked with another, and finished off with Chisel and File, with the customary small gouge cuts on each side. The stock is finally cleaned up with Files and glasspaper.

As there are about 50 different standard profiles of Moulding Planes, most of which can be made in up to ten different sizes, with Sash-Moulding Planes in pairs, a very large number of Mother Planes, one for each size and shape, are essential. Similar Mother Planes are also necessary for Sash Templets, single and double. (When, in 1949, the Author visited the derelict workshop of the Norwich plane makers H. Griffiths & Co, a pile of some hundred Mother Planes were being used to support the roof). In addition to these, special Planes are also required for shaping the hollow of the rebate on the stock, shaping the fence and depth stop, working the slits for 'boxing', etc. In other words, when anything could be done with a Plane, the plane maker made a special Plane for the purpose.

As the moulding-plane specialists also made Ploughs, Fillisters, Dado Grooving Planes, etc., they also required further sets of special Planes for dovetailed boxing, which were in handed pairs according to the side of the sole on which the boxing was required (see Boxing the Fillister Plane below).

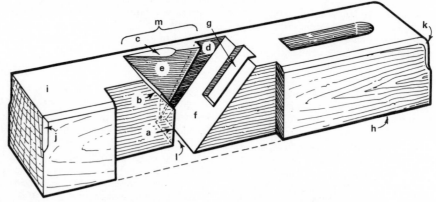

Fig. 551
Parts of a Plane: terms used by Bench-Plane Makers
(*a*) Throat or Underfront
(*b*) Top Front
(*c*) Turn out (U.S.A.: Eye)
(*d*) Buttlements or Abutments (U.S.A.: Wedge Grip)
(*e*) Cheek
(*f*) Bed
(*g*) Notch Hole
(*h*) Sole or Face
(*i*) Top
(*j*) Trim up
(*k*) Back End Horseshoe (on Smoothing Planes)
(*l*) Mouth
(*m*) Outlet for shavings (or Escapement)

Tools used by plane makers are described and illustrated in the following paragraphs:

Patterns and Jigs: *Fig. 552*
Holding Devices: *Fig. 553*
Plane Maker's Chisels: *Fig. 554*
Miscellaneous Plane Maker's Tools: *Fig. 555*
Plane Maker's Planes: *Fig. 556*
Boxing the Fillister Plane: *Fig. 557*
Working the Plane stock for boxing: *Fig. 558*
Working the boxwood insert: *Fig. 559*
Boxing Moulding Planes: *Fig. 560*

PATTERNS AND JIGS *Fig. 552*

(*a*) *Bedding Guide.* Shaped like a bench-plane wedge 5 in long, with one leg longer than the other. Used for checking the angle of the bed on a Bench Plane. The longer leg is held on the bed, the shorter overhanging one side and checked against the Pitch Board.

(*b*) *Pitch Board.* A triangular piece of thin wood with a fence along the lower edge. Used for checking the angle of the bed on a Bench Plane with the Bedding Guide.

(*c*) *Skew Iron Mortice Guide.* The ends of the slot are cut at an angle of 75° to the sides. Two small pins, about $\frac{3}{8}$ in apart, are used for marking the sides and ends of the mortice for a skewed iron and wedge fitted in Planes such as Sash Fillisters, Rabbet Planes, and Hollows and Rounds.

(*d*) *Sinking Guide for Technical Jack Plane.* A strip of wood with one end shaped to a quarter circle, screwed on a flat fence. Used for marking the sinking on the stock of a Technical Jack Plane.

(*e*) *Stock Templet for Smooth Plane.* A strip of hardwood, 9 in long, with an iron plate $\frac{1}{16}$ in thick let into a groove along one side. The plate is curved to the shape of the smoother stock and is used for marking and testing its shape.

(*f*) *'Spring' Guide or Adjustable Square for Moulding Planes.* Two wooden strips $14 \times 1 \times \frac{3}{8}$ in, with a gap of $\frac{1}{16}$ in between them, are held together by brass tips riveted at each end. About $4\frac{1}{2}$ in from one end a steel square is pivoted on the centre-line of the strips. Used for marking the 'spring' of the profile on the toe end of a Moulding Plane, both as a guide for working the fence and depth stop, and when sticking the moulding.

(*g*) *Bed and Mouth Guide for Moulding Plane*. A piece of wood the size of the cultivated moulding-plane stock, $9\frac{1}{2} \times 3\frac{1}{2}$ in, with another piece screwed along the bottom edge at right angles. Slots are cut out to the shape of the bed and mouth on the sole or face. The bevelled strip on the top edge marks the position of the mortice for the iron and wedge on the top of the stock – at right angles if the iron is bedded square, and at the appropriate angle if the iron is skewed.

(*h*) *Bed and Mouth Guide for Fillister Planes*. A jig similar to (*g*) but provided with two sets of slots, one for marking the skewed iron and the other for the spur cutter. These are in handed pairs for either Moving or Sash Fillisters.

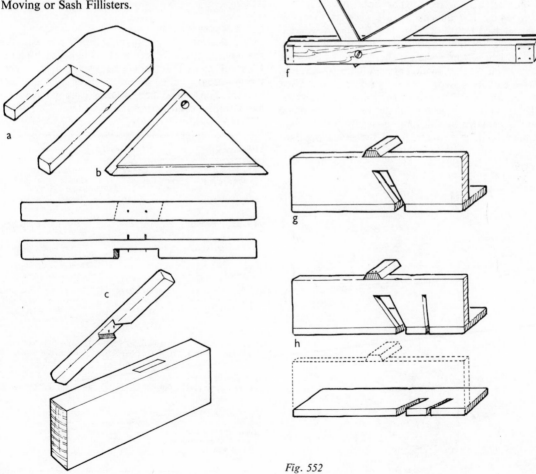

Fig. 552

HOLDING DEVICES *Fig. 553*

(*a*) *Holder for Moulding Plane Stocks.* A block of hardwood 15½ in long, 5 in wide, and 4 in deep, with a fillet worked on the lower face for holding in the vice. A wedge-shaped block running in a dovetailed groove across the block tightens up to the standard length (9½ in) of the Moulding Plane stocks. Used for holding the stock when working the bed and mouth from the side of the stock.

(*b*) *Vice Chops for Smoothing Planes.* Two wooden blocks L-shaped in section and 9 in long, with the opposing faces shaped to the curve of a coffin-shaped smoother stock. Strips of leather or canvas are glued to the faces to prevent damage to the stock, and the blocks are held together at the correct distance by a U-shaped iron hoop. Used for the final trueing of the smoother sole.

(*c*) *Holder for Shooting Ends of Plane Stocks.* A box-like frame about 12 in high, designed to be held in a vice, with a rectangular hole in the middle 4¾ in × 4 in. The dimensions are such as to admit the standard length of Moulding Planes. A wooden hand screw is fitted about half way down one leg, and this bears on a sliding cheek which holds the stock in place. Two notches are cut in the lower cross pieces to take a wedge-shaped strip of wood. Used for shooting the end grain of Moulding Plane stocks. The amount of projection of the stock of the plane can be regulated by adjusting the wedge up or down. Similar holders were also used for other standard plane stocks.

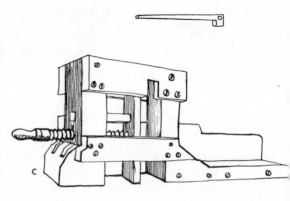

Fig. 553

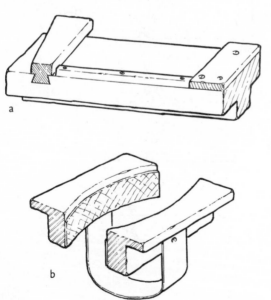

PLANE MAKER'S CHISELS *Fig. 554*

The following Chisels (and Gouges) are used for shaping the outlet for shavings cut in a Bench Plane. Most of them are Chisels of strong firmer type, though usually called by other names by the plane-makers. Those described as 'long' usually have long handles for working from the shoulder.

Operation	*Name of Chisel*	*Description*
(*a*) Preliminary chopping out	Sinking-down Chisel and Gouge	Short, stout Firmer Chisels and Gouges about 1 in wide. Usually socketed. Driven by mallet.
(*b*) Trueing the bed	Bedding Chisel	Long, socketed chisels 1¼ in wide.
(*c*) Paring the cheeks or sides	Cheeking Chisel	Firmer type ¾–1½ in wide, ground on the skew in right- and left-handed pairs.
(*d*) Final trueing of the bed by scraping	Scraping Chisel	Firmer, 1 in wide, with the cutting edge ground off leaving a flat of $\frac{1}{32}$ in to use as a scraper.
(*e*) Cleaning out awkward corners	Cleaning-Out Chisel (U.S.A.: Chamfering Chisel)	A long Chisel with a knife-shaped blade.
(*f*) Cutting the turn-out (U.S.A.: eye)	U.S.A.: Eyeing Chisel	A short Firmer Chisel with square cutting edge.
(*g*) Chiselling out the throat and for roughing out the wedge slots before finishing with a Float	Throat Chisel [not illustrated]	Narrow Chisels of various kinds.

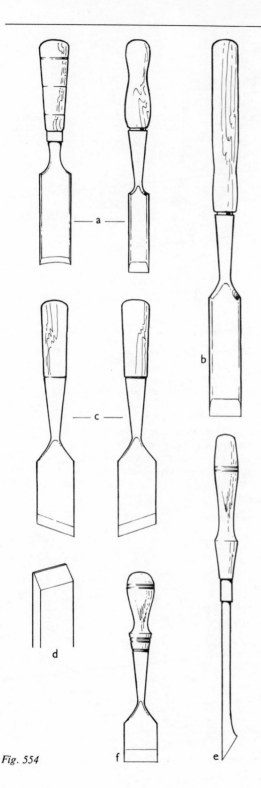

MISCELLANEOUS PLANE MAKER'S TOOLS *Fig. 555*

(*a*) *Plane Maker's Saw.* A straight blade, 9 in to 16 in long, about $\frac{1}{8}$ in thick and $1\frac{1}{2}$ in wide at the handle, tapering to a point and with a straight or slightly-curved back. The teeth have no set and are filed alternately from one side or the other at an angle. For sawing the mouth, bed, and wedging slots on a plane.

(*b*) *Plane Maker's Floats.* These tools are usually made from old Files and fitted into any suitable handle. The teeth are cut square with from 4 to 6 points to the inch. Several types are used for cleaning out the mouth, cheeks, bed, and wedge slots. Tapered Floats are used for Bench Planes, and narrow, parallel-sided Floats for Moulding Planes. Narrow, bevelled-edge Floats are used for cleaning the mortice to take the tongue of a skewed iron.

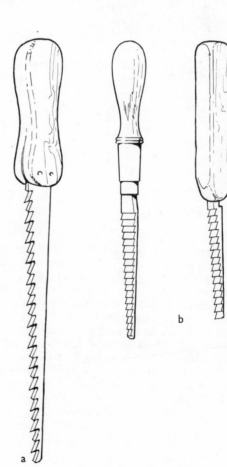

Fig. 554

Fig. 555 (a)–(b)

(*c*) *Wrest*. An iron plate about 16 in long with a slot 4 in long in the middle. Used for straightening twisted Bench Plane irons.

(*d*) *Router*. Plane makers use a small router with a $\frac{3}{8}$ in cutting iron for levelling the recess to take the metal plates often fitted to the sole of such tools as Coopers' Shaves.

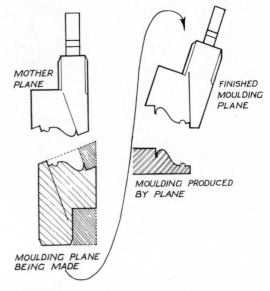

Fig. 556 (*a*)

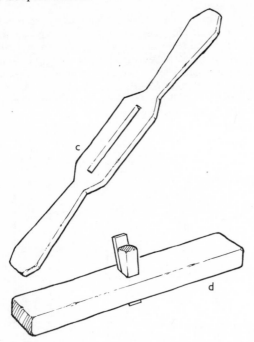

Fig. 555 (c)–(d)

PLANE MAKER'S PLANES *Fig. 556*

(*a*) *Mother Plane* (Backing Plane; Reverse Plane). For making each standard moulding and the simpler combinations, and for each size of each type, a Plane known as a Mother Plane was used, with the profile of the sole exactly the reverse of the Moulding Plane required. The profile of the Mother Plane was that of the moulding itself. Judging from surviving examples, these Planes appear to have been used for finishing only, as there is very little in the way of a fence to work from. The profile of the Moulding Plane was first shaped out with the hollows and rounds etc. and the Mother Plane run down a few times to give the true, standard profile.

(*b*) *Wedge-Notching Plane*. A rectangular stock $11\frac{3}{4}$ in long and $1\frac{5}{8}$ in wide. The sole is shaped to cut the notch on a moulding-plane wedge. The cutting iron is shaped like a letter 'L' with a curved corner, and is bedded on the skew. A vertical spur cutter in front of the iron cuts the other sharp corner of the notch. A wooden fence is screwed to the left-hand side of the stock. The roughly sawn wedge blanks are fixed in a screw clamp in batches of up to 20 and the grooves cut across the grain, the rounded end being finished off with Jack and Hollow Plane and File.

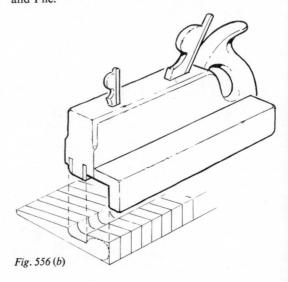

Fig. 556 (b)

BOXING (The making and inserting of Boxwood inserts)

The sole of certain Planes is subject to considerable wear, and to enable the plane to resist this, a strip of boxwood was sometimes inserted in a groove or rebate. This process was known as Boxing, and it called for the use of several special Planes to form the necessary grooves which could be either straight-sided or dovetailed. These special Planes were similar in size and type to ordinary Moulding and Rebate Planes and were often made in handed pairs. The particulars given below are based on a set of five such planes made by John Moseley & Son (London, *c.* 1900) lent to us by Mr. K. Hawley of Sheffield and used until recent years in the Hibernia Works of Messrs William Marples & Sons of Sheffield.

BOXING THE FILLISTER PLANE *Fig. 557*

In a Standing or Moving Fillister Plane, the bottom left-hand corner (as seen from the front) is the most subject to wear; in a Sash Fillister it is the right-hand corner. The boxing may be of the following kinds: (*a*) corner, (*b*) shoulder, (*c*) shoulder dove-tailed. The tools and methods for working the sloping grooves for these boxings are not known. The so-called double-dovetail boxing is shown in (*d*) and (*e*). The sizes vary with different makers, but an average arrangement, measured from a processed blank from the former plane makers Messrs H. Griffiths of Norwich, is shown at (*f*).

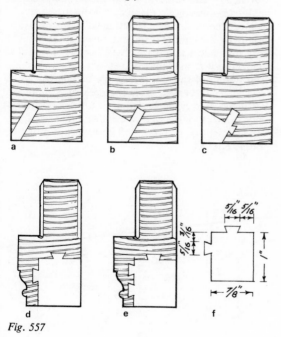

Fig. 557

Taking the dovetail-boxed Plane (*d*) of *Fig. 557* as an example, the sequence of working is as follows:

Working the Plane stock for boxing Fig. 558. The rebate is first cut out $\frac{7}{8}$ in wide and 1 in deep (*g*). A groove is then cut in the top and side of the rebate, working in each case from the corner of the rebate, with the Grooving Plane (*h*). The sides of these grooves are then shaped with the Dovetail Plane, each side being opened out, working in opposite directions, as at (*i*).

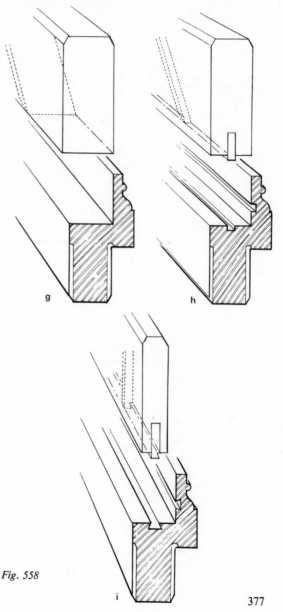

Fig. 558

377

Working the boxwood insert Fig. 559. The insert (*j*) is prepared in boxwood ⅛ in wider and deeper than the rebate. Square-sided tongues are worked on the top and side with the Tonguing Plane (*k*), which leaves a small projection at the top and bottom outer corners (*kx*). This is removed with a grooved-sole Rebate Plane producing the section (*l*). The sides of the tongues are planed to the dovetail shape with the Bevelled Shoulder Plane (*m*).

Lastly the boxwood block is marked for the sloping and slightly skewed bedding for the iron, and sawn in two, each half presumably driven in tight from opposite ends.

At first sight this may seem a rather roundabout way of working the insert. The point seems to have been that if the Tonguing Plane (*k*) were designed to take off the whole of the rebate, or if the projecting piece (*kx*) were removed (as it could have been with an ordinary Shoulder Plane), the slightest error in taking a shaving or so too many would mean that the tongue would have to be re-worked to bring it square again. This would put the tongue on the other face out of register with the corresponding groove. By leaving the piece (*kx*) projecting, Plane (*k*) acted as its own depth stop. Similarly Plane (*l*), which just fitted over the tongue, with about $\frac{1}{16}$ in of the sole between the iron and the edge of the groove, also acted as an automatic stop.

Boxing Moulding Planes Fig. 560. The larger Moulding Planes are not usually boxed, but those with narrow projections, which are liable to more rapid wear, are often boxed over the whole of their profile. Narrow Bead Planes in particular are boxed, as are those for working reeds. The smaller sizes are also provided with a detachable strip of beech to form the stop. When worn with continuous use, it can be replaced. Such a plane is said to be 'slipped' (*b*).

A Bead Plane without a slip usually had vertical boxing (*a*), but since a Plane with a slip is necessarily rebated, the boxing is inserted at an angle (see (*b*)). Otherwise it would fit merely in the rebate and might drop out. There appears to have been no invariable rule, however, as some Planes have angle boxing even when not slipped. Where the boxing has to be inserted at the edge, as in the case of the Snipe's Bill Plane, it is either sloped as at (*d*) or fitted vertically as at (*c*). In the case of a rather wider bead, the boxing might be as at (*e*). Still larger beads might be double boxed (*f*).

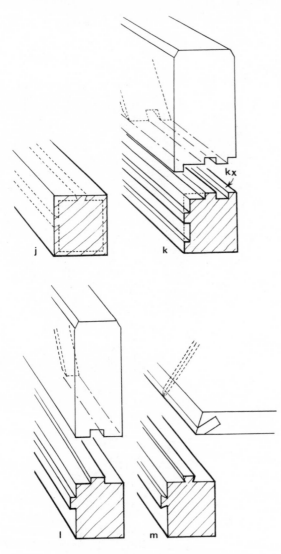

Fig. 559

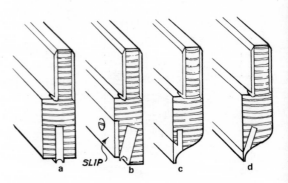

378

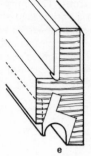

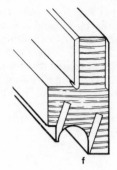

Fig. 560

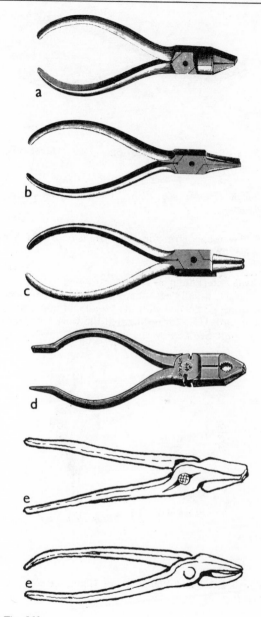

a

b

c

d

e

e

Fig. 561

Planted Moulding

A separate moulding attached to a workpiece and not formed in the solid.

Plate

A flat piece of stone, metal or timber, on which the end of a beam rests. Of tools: a metal insert on the sole of a Plane or Router to take the wear.

Pliers *Fig. 561*

The following examples are often found in woodworking shops where they are used for holding small objects, bending wire, turning nuts, and similar purposes. Common variants include:

(*a*) *Bell Pliers*. These usually have a flat nose with side cutters. 'Gates' are sometimes provided in the side of the joint for cutting wires. The serrated hole between the jaws is for gripping round objects; it is called a 'burner grip' owing to its use for removing the jets from gas lamps. Bell pliers were originally used by bell hangers for bending and cutting the wires connecting household (non-electric) bells; later they became popular for general use in the woodworking and other trades.

(*b*) *Flat Nose Pliers*.

(*c*) *Round Nose Pliers*. Used mainly for bending wire or thin metal.

(*d*) *Bell and Burner Pliers*. See (*a*) above.

(*e*) *Home-made Pliers*. These usually have flat noses and are without the means of wire-cutting; those with round noses may have been made originally for pig-ringing and are therefore often known as 'Hog Pliers'.

(*f*) *Combination Tools*. These are Pliers combined with a hammer and even an axe-blade. See *Hammer Pincers* and *Fencing Tool*.

Saw Setting Pliers. See *Saw Setting Tools*.

Pliers, Glazier's *Fig. 562*

A flat-nosed Plier, 6–8 in long, with jaws meeting only at their extreme ends. Used after a Glass Cutter on sheet glass, for breaking off narrow strips, or for trimming uneven edges.

Fig. 562

Pliers, Key Bushing: see *Pianoforte Maker and Tuner.*

Plough (Plow): see *Plane, Plough; Bookbinder's Plough.*

Plucker (Scots term for *Downwright* or *Swift*): see *Shave, Cooper's.*

Plumb Bob (Plummet) *Fig. 563*
A weight, usually pear-shaped, of lead, brass, steel, or stone, suspended with the broad end uppermost, on a cord. Used for testing that the work is upright ('plumb'). See *Levels.*

Iron Plumb Bob   *Brass Plumb Bob*

Fig. 563

Plumb Level and Plumb Board (Plumb Rule): see *Levels.*

Plumber's Tools *Fig. 564*
Tools of this trade are often found in woodworking shops. Common examples are illustrated:

   (*a*) *Melting Pot and Ladle*
   (*b*) *Pipe Grips*
   (*c*) *Pipe Cutter* for cutting iron pipes
   (*d*) *Putty Knife*
   (*e*) *Chase Wedge* for drawing lead sheet into angles
   (*f*) *Dressing Stick* for shaping lead sheet

   (*g*) *Plumber's Iron* (Grozing Iron). This ancestor of the blow lamp was used for jointing lead sheet and pipes. After heating, it is held against the joint to keep the solder fluid while being wiped smooth.
   (*h*) *Soldering Iron* for joining tinned iron sheet etc.
   (*i*) *Gas Fitter's Pliers.* Often found in general service owing to its handy shape for gripping nuts, bolts, etc.

   See separate entries for *Knife, Putty; Saw, Plumber's.*

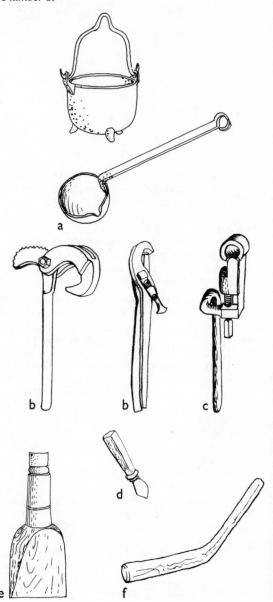

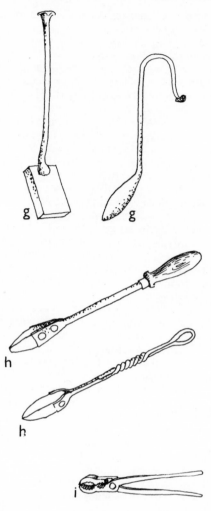

Fig. 564

involved the burning of incised patterns or pictures on a wooden surface. This was done with metal rods known as 'points' with flat, claw- or horn-shaped ends, fitted to cork-covered handles to protect the hands. The points were heated in a flame and then applied to the wood surface.

Poll Prytch: see Shepherd's Bar under *Hurdle Maker.*

Pompeyed
A cooper's term for burning the inside of a cask to give it a charred surface.

Post Bender: see Tyre Bender under *Wheelwright's Equipment (3) Tyring Tools.*

Post Vice: see *Vice, Woodworker's.*

Pot Hanger *Fig. 565*
An adjustable hanger for suspending a pot over a fire; e.g. a glue or pitch pot, or a kettle.

Fig. 565

Poll (Butt)
Part of the head of an Axe or Adze opposite to the cutting edge. See diagrams under *Adze; Axe.*

Poker, Mastmaker's
A heavy iron rod about 8–10 ft long and 1½ in. in diameter, with one end spade shaped. Used for driving iron bands on a mast. Two men, each holding a Poker, stand on each side of the mast and simultaneously drive the bands on. (See *Mast and Spar Maker.*)

Poker Work Tools (Relief Burning Tools)
Poker work was a popular indoor occupation which

Presser: see Rubber under *Sailmaker.*

Pricker
'Is vulgarly called an Awl' – Moxon (London, 1677). See also *Sailmaker.*

Pricker Pad: see *Awl, Pad.*

Print and Block Cutter *Fig. 566*
The carving of wooden blocks for printing designs on textiles and wallpaper derives from the art of making woodcuts (see *Wood Engraver*). The blocks were carved in wood, sometimes laminated to prevent warping. Towards the end of the eighteenth

century, while the background design was still cut in wood, the more intricate patterns (e.g. flowers and leaves) were 'coppered'. This process involved the insertion of countless brass pins, and strips and rods of various sections, into the working surface of the block. According to Christine M. Vialls (London, 1969), a rough count of the number of separate pieces of metal in a block of simple design amounted to 1450, and it might take a man a month to complete. (To reduce the enormous labour involved, a system of casting the surface pattern in metal was evolved in the mid-nineteenth century.)

The tools used for inserting the metal pieces into the block are described below. The Pinker tool and Curfing iron are used for cutting the channels into which the metal pieces are inserted. These are left standing proud by about $\frac{3}{16}$ in above the surface of the block.

At the time of writing Morris wallpapers are still hand-printed from the original nineteenth-century pear-wood blocks, some with metal inserts, at the factory of Sanderson & Sons of Perivale. The illustrations show some of the tools in use at that factory.

Divisions within the trade

1. The Block Cutter carves out the pattern on a wood block, using wood-carver's Chisels.

2. The Print Cutter makes the 'coppered' type of block. He cuts and shapes the metal parts and inserts them into the surface of the wooden block.

3. The Stamp Cutter makes small blocks with delicate metal inserts for printing monograms and trade-marks on textile materials. This trade, which involves exceedingly tedious work, is now almost extinct.

4. The Roller Cutter. In recent times, wooden, metal and composition rollers have superseded blocks for continuous production. In some instances, the Block and Print Cutters continue to perform the same type of work on the rollers as they did on the blocks, except that most of the carving operations are now done by Machine Router.

Materials

The woods used for the blocks are mainly pear and, more recently, sycamore. The wood is soaked in water before being cut, in order to soften it. The metal inserts are mainly of brass and copper.

Illustrations (Fig. 566)

(*a*) *Block*. An example with brass inserts.

Print Cutter's Chisels and Gouges. Besides using ordinary Carving Chisels, special Chisels are made for this trade which have straight parallel blades,

from $\frac{1}{8}$ to $\frac{5}{8}$ in wide, and taper down to the cutting edges on the face side.

(*b*) *Print Cutter's Chisel*. (Print Cutter's Chisels are also illustrated on page 142.)

(*c*) *Print Cutter's Gouge*, made in varying curves, described as 'flat, middle, quick, or fluted'.

(*d*) *Print Cutter's Dog Leg Chisel*.

(*e*) *Pinker Tool*. This is the most constantly used of the Print Cutter's Chisels. About $4\frac{1}{2}$ in overall, it has a narrow blade tapered down on both sides by slightly hollow grinding to a fine edge. It is designed for making fine vertical cuts in the block into which the copper strips are inserted.

Note: Most of the Chisels are left unshouldered. This may be done because unshouldered Chisels are cheaper (they are quickly worn down by sharpening). But a more likely explanation may be that, when loosened in the handle by backward tapping, an unshouldered Chisel can be driven further in. Their handles are mushroom-shaped to facilitate their extraction from the block by hammering backwards (see (*f*) below).

(*f*) *Curfing Iron* (Striking Iron). A bat-shaped iron tool, about 9 in long, fitted with a wooden handle. Used as an alternative to a light hammer for driving the Print Cutter's Chisels in or out. Its square edges are well adapted for extracting the Chisels by tapping under their mushroom handles.

(*g*) *Knocker-Up*. A small metal tool with end bent downwards which is used for lifting or straightening any part of the metal insertions if they become bent or injured.

Metal working tools used in this trade include: Shears; end cutting Pliers; flat and round nosed Pliers; light-weight Hammers; special hollow-mouthed Pliers for making a V-shaped bend in the strip; fine Files, including a flat 'screw-head' File which, cutting only on its edges, is used for making a kerf in the strip in order to produce a sharp angular bend.

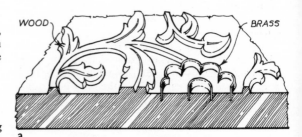

WOOD   BRASS

a

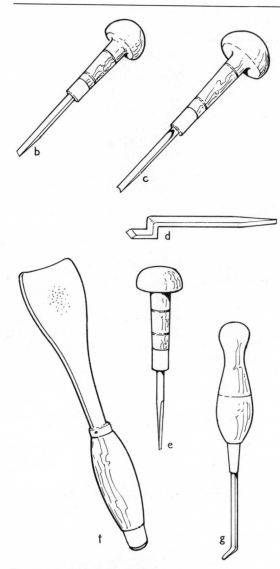

Prying Rod: see *Flagging Iron.*

Pulling-Up Hook: see *Knife, Hooked.*

Pulphook: see Woodhook under *Timber Handling Tools.*

Pump Augers: see *Pipe and Pump Maker.*

Pump Drill: see *Drill, Pump.*

Punch
A steel tool used for driving, piercing, and penetrating. Examples used in woodworking and allied trades are described in the following entries.

Punch, Astragal: see *Chisel, Astragal.*

Punch, Carver's (Buhl Punch; Ground Work Punch; Frosting Punch) *Fig. 567*
Steel punches with points of various shapes including floral and other designs. Those that give texture to the background of low-relief woodcarving by producing groups of dots, stars, etc. are called Frosting Punches.

Fig. 567

Punch, Cooper's: see *Cooper (3) Hooping Tools.*

Punch, Coppering, Shipwright's *Fig. 568*
A round steel Punch with a pointed end, often mushroom headed. Used for piercing the copper sheeting fitted on the ship's bottom. See *Hammer, Ship's Coppering and Sheathing.*

Fig. 568

Punch, Eyelet *Fig. 569*
A metal Bed and Punch used for inserting metal eyelets in sails, tarpaulins, upholstery, etc. A pliers-type of tool is used for very small eyelets.

Fig. 566   Print and Block cutter's tools

Pritchet
An iron-pointed tool made like a Cold Chisel and used by farriers for piercing holes in horse-shoes. The term is also applied to other tools of similar appearance including that used for cutting down high spots on a millstone before using the Mill Bill. (See *Millstone Dresser.*)

Prover and Proof Staff: see *Millstone Dresser.*

Fig. 569

Fig. 571

Punch, Frosting: see *Punch, Carvers.*

Punch, Grommet: see *Sailmaker.*

Punch, Handrail *Fig. 570*
A short steel bar, square in section, with one end tapered to a chisel point and bent slightly backwards. Used for tightening the nuts of handrail bolts. These bolts, which are double ended, are used for joining wooden handrails end to end, the bolt being sunk half way in each. Holes are cut through the side of the handrail at right angles to the bolt to hold the nuts. One nut is wedged fast while the other is tightened by striking the nicks in its circumference with this Punch.

Fig. 570

Punch, Letter or Figure (Letter or Figure Stamp; Marking Punch) *Fig. 571*
Sets of punches carrying a single letter or figure, sometimes packed in a metal box. For marking workpieces etc. of wood or metal.

Punch, Nail (Driving Punch; Nail Set; Pin Punch) *Fig. 572*
A short steel rod, round or square in section, tapered at one end. The round pattern is sometimes knurled at the centre for easy holding. Used for driving the head of a nail below the surface of the wood. The point is sometimes cupped to avoid slipping off the nail.

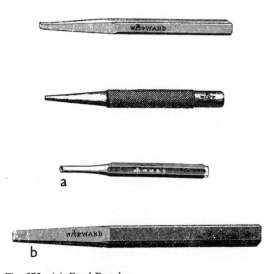

Fig. 572 (*a*) Brad Punches
(*b*) Flooring Punch

Punch, Sash: see *Chisel, Astragal.*

384

Punch, Shackle
A Steel Punch with a parallel stem and hollowed face, used by shipwrights and others for driving the pin from the shackle. See also Shackle Maul under *Hammer, Ship Maul.*

Punch, Starting
A term used for a plain steel rod, about 8–10 in long, usually tapered towards one end. Used by wheelwrights, coachbuilders, and other tradesmen for driving a bolt out of its hole.

Punch, Upholsterer's *Fig. 573*
These tools include serrated Punches, used for cutting out leather tufts for mattresses and for ornamental work. See *Upholsterer.*

Pinking Irons

Fig. 573

Punch, Veneer: see *Veneering and Marquetry.*

Purfling Tool: see *Violin Maker.*

Putter: see *Cellarman.*

385

Quadrant
Literally a quarter of a circle; but also a slotted segmental guide through which a lever such as a brake or reversing handle works, or on which the swinging arm of a Bench Drill can be fixed in any desired position.

Quannet: see *Combmaker*.

Quirk
The narrow groove at the back of a bead or other moulding. See *Plane, Moulding*.

Quirk Cutter: see *Router, Metal Type*.

R

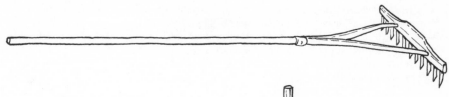

Rabat
A medieval term for a carpenter's Plane.

Rabbet: see *Rebate.*

Rabetstock
A tool mentioned by Thomas Tusser (London, 1573), probably a Plane.

Race
A name sometimes given to the *Plane, Cooper's Croze.*

Race Knife: see *Timber Scribe.*

Rack
A toothed (or cogged) bar as used, e.g., for the bar of certain Joiner's Cramps; or when made in the form of a rack-and-pinion (in transforming rotary into linear motion), for certain Lifting Jacks. (See also *Rachet-and-Pawl.*)

Raglet: see *Plane, Dado Grooving.*

Railway Tools
Woodworking tools used by trackmen, platelayers, and builders of railway wagons are described under the following entries: *Adze, Platelayer's; Adze, Trackmen's Spiking; Auger, Skewnose; Bit, Railway Carriage; Bit, Wagon Builder's; Brace, Nut; Brace, Platelayer's; Brace, Wagon Builder's; Chisel, Trenail; Chisel, Wagon Builder's.*

Rainette: see *Saw-Setting Tools.*

Raising Iron: see *Jumper; Vice, Cooper's.*

Rake: see Hoe under *Caulking Tools.*

Rake Maker *Fig. 574*
The following notes apply to the making of small Hand Rakes from coppice material. The large Rakes ('Drags') with five-foot heads and chamfered timber frame were usually made by wheelwrights.

Rake handles or 'stails' are generally made of coppice-grown ash, but hazel, alder, birch, or willow

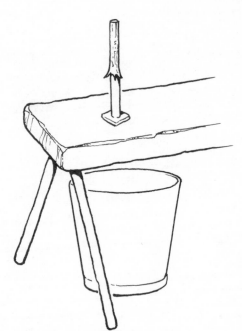

Fig. 574   A wooden Hand Rake
Tine Former on a Tining Horse

are also used. The sticks, about 6 ft long, are put in a Shaving Brake to be trimmed. The rounding and tapering is done with the 'Stail Engine' (Plane, Rounder). One end of the handle is sawn or split down the middle for about 20 in and a piece of tin or other material nailed round to prevent further splitting. The rake heads are usually made of cleft willow and holes are bored for the tines to marks made with a pattern. The tines are made from short lengths of cleft ash or willow, about 6 in long.

Most of the tools used in this trade are illustrated under *Woodland Trades*, but see also *Knife, Bench*. A more specialised piece of equipment is the:

Tining Horse (Driving Stool). This is a low bench fitted with an upright hollow cutter known as a Tine Former or Peg Cutter. This is like an inverted leather punch; it consists of an iron tube about 6 in long and $\frac{1}{2}$ in internal diameter, with the top end ground on the outside to leave a sharp cutting edge. It is fitted upright on the stool and the wooden sticks intended for tines are driven through it, each one driving the previous one through. The ends of the tines are then tapered with a Drawing Knife or Bench Knife and finally driven into the head of the rake. Another type of Tine Former resembles a *Dowel Plate*.

Ramhead: see *Froe*.

Rampin *Fig. 575*
A trade name given to a narrow tube fitted into a handle about 9 in long overall, with a spring-loaded magnetic ram (or piston) inside. A nail is put into the tube, which is then held against the work; when the handle is depressed the nail is driven home.

MAGNETIC RAM

Fig. 575

Rank Set
Of a Plane – when the cutting iron is set to take a coarse cut.

Rasp *Fig. 576* (see also *File, Float; Rifler*.)
An abrading tool of hardened steel, the teeth of which are formed separately with a pointed punch – as distinct from the straight grooves cut with a Chisel on the File. Most Rasps designed for use on wood are either half-round or flat in section, taper towards one or both ends (fusiform), and are tanged for fitting into a wooden handle. The range of cut varies from smooth, with about 24 teeth to the inch, through second, bastard, middle, and rough, with 7–8 teeth to the inch. (The farrier's Horse Rasp is coarser still, with 6–7 teeth to the inch.)

Rasps are used for the preliminary shaping of curved surfaces. Variants used for woodworking include:

(a) Cabinet Rasp
Flat or half-round, from 4 to 16 in long. Since about 1900 the flat pattern seems to have died out. The distinction between the Cabinet and the ordinary Wood Rasp (*h*) below seems to be that the Cabinet Rasp has a thinner blade, a flatter curve, and a finer cut.

(b) Gunstocker's Rasp
Flat or half-round from 4 to 10 in long. Used for smoothing the stocks after the initial shaping of the blank.

(c) Last Maker's Rasp
A heavy half-round Rasp, 14–18 in long, with the teeth punched in parallel lines sloping at an angle of about 20°. Used by boot-last makers for shaping the lasts.

(d) Rifler Rasp: see *Rifler.*

(e) Round Rasp (Rat's Tail)
4–24 in long, circular in section. Used for removing heavy waste and enlarging holes.

(f) Saddle Tree Maker's Rasp [Not illustrated]
A heavy Rasp, similar to the Last Maker's Rasp, slightly rounded on both faces and the teeth punched in parallel curves. Used for shaping wooden saddle trees.

(g) Saw-Handle Rasp
Intermediate in width and thickness between the Last Maker's and Gunstocker's Rasp. Used for shaping Saw handles etc.

(h) Wood Rasp
Flat, square, or half-round, in a wide range from 4 to 24 in long. It may be distinguished from the Cabinet Rasp by the thicker blade and the fact that it is made up to 24 in long. Used by various tradesmen working in wood for speedy removal of waste, particularly on curved work.

Rasp, Baker's (Bread Rasp) *Fig. 577*
A flat plate, about $2\frac{1}{4} \times 6$ in, rasp-cut on the face, and fitted with an offset handle like that of a flat-iron. Sometimes mistaken for a woodworking tool, it is used by bakers and housewives for removing the burnt crust from the base of the loaf. A similar tool called a Quannet is used when working horn – see *Comb Maker*.

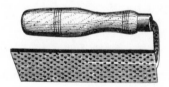

Fig. 577

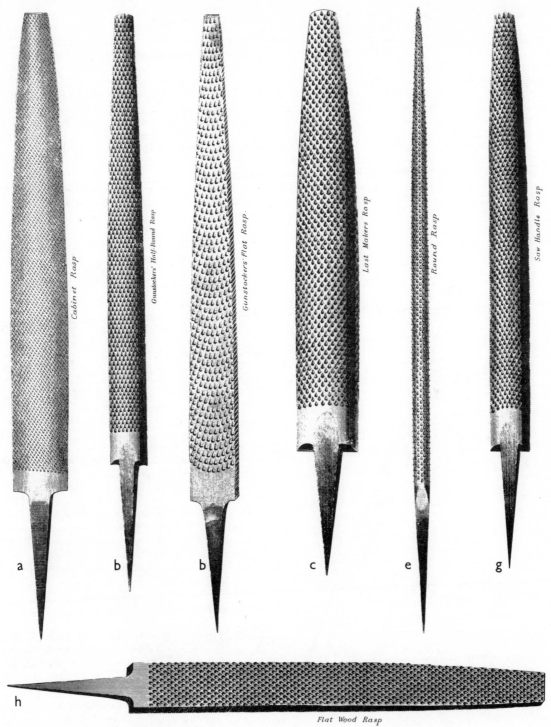

Cabinet Rasp

Gunstockers' Half-Round Rasp

Gunstockers' Flat Rasp

Last Makers Rasp

Round Rasp

Saw Handle Rasp

a   b   b   c   e   g

h

Flat Wood Rasp

Fig. 576 Rasps for different trades

Rasp, Two Handed (Long Rasp; Hot Rasp) *Fig. 578*
This remarkable tool is occasionally found in wheel-wright's and coachsmith's workshops. It was made in iron, about 4 ft long overall, with or without wooden handles at each end, for use by two men together for rasping off hot metal. We have seen only home-made examples. The centre rasping portion was about 1–2 in wide, and sometimes made from a plate of steel, coarsely jagged to a float cut and then riveted to the iron face of the tool.

Axles were brought in from outside in two halves and had to be joined to the length required. The ends were upset, scarfed, welded together, and finally smoothed down with this Rasp. It was also used for reducing the axle arm itself after it had been refaced by welding.

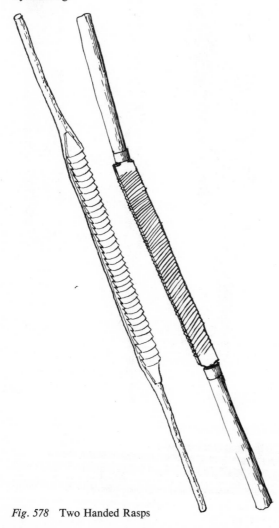

Fig. 578 Two Handed Rasps

Rasp Tool: see *Surform Tools.*

Rat's Tail
A name sometimes given to a round *File* or *Rasp.*

Ratchet-and-Pawl
A set of slanting teeth on the rim of a wheel into which a pawl (or finger) may catch for the purpose of preventing reversed motion; or, in the case of a ratchet Brace, in order to convert reciprocating motion into an intermittent rotary motion.

Ratchet Tools
Tools containing some form of ratchet mechanism include: *Brace, Corner; Brace, Ratchet; Drill, Archimedean, Double Spiral; Screwdriver, Ratchet.*

Rave Hook (Rake; Rove Hook; Reeve Hook): see Hoe under *Caulking Tools.*

Ream Iron: see Reaming Iron under *Caulking Tools.*

Reamer
A tool used for enlarging a previously bored hole. For examples, see under *Auger, Cooper's Bung Borer; Auger, Taper; Bit, Rimer; Bit, Taper; Broach; Fishing rod maker* (Spoon); *Pipe and Pump maker* (Augers); *Rimer, Hand; Violin Maker; Woodwind Instrument Maker.*

Reaping Iron: see Jerry Iron under *Caulking Tools.*

Rebate (Boxing; Check; Rabbet)
A rectangular recess or step along the edge of a piece of wood or other material.
   Noun – Rabbet or Rebate, e.g. a $\frac{5}{8}$ in rabbet or rebate.
   Verb – Rebate, e.g. to rebate the wood.
   Adjective – Rabbet, e.g. Rabbet Plane.
   Tools used for rebating include *Plane, Fillister; Plane, Rebate;* Boxing Router under *Router, Coachbuilder's;* Rabbeting Router under *Router, Metal Types.*

Recessing Tool: see *Chisel, Hinge.*

Reed and **Reeder:** see *Plane, Moulding;* Hand Reeder etc. under *Router, Coachbuilder's* and *Router, Metal Types.* (For reeds used in cooper's work see *Flagging Iron.*)

Reed Maker: see *Woodwind Instrument Maker.*

Reeve Hook: see Hoe under *Caulking Tools.*

Reglet: see *Plane, Reglet.*

Regulator: see *Upholsterer.*

Riffler *Fig. 579*
A double-ended File or Rasp, provided with two curved heads, usually tapered, about 2 in long, connected by a shank which is often swollen for better handling. The heads are made with various cross-sections: round, rectangular, oval, or with sharp edges. Sometimes the heads are rasp-cut one end and file-cut the other. Used by carvers, pattern makers, masons, and other tradesmen, mainly for smoothing concave parts of carvings etc. made in wood, stone, or metal.

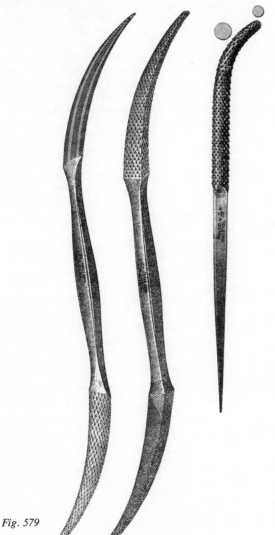

Fig. 579

Rigger and Rigger's Screw: see *Sailmaker.*

Rimer or **Rinder, Hand** (Square Rimer or Rinder) *Fig. 580*
A square shank tapering from about $\frac{3}{8}$ in to a point, fitted to a cross-handle. For enlarging holes. See also *Reamer.*

Fig. 580

Rinding Iron: see *Barking Iron.*

Ring Dog *Fig. 581:* see *Timber-handling Tools.*

Fig. 581

Ring Punch: see Taphole Auger under *Auger; Cooper's Bung Borers.*

Ringer: see Carpet Ringer under *Upholsterer.*

Ritter
Scots term for *Drawknife.*

Rive or **River** (Riving Axe)
A term sometimes used for a *Froe* or for a *Cleaving Knife.* See *Lath Maker.*

Rod, Prying: see *Flagging Iron.*

Rod, Surveyor's: see Folding Lath under *Rule.*

Round Square: see *Square, Radial.*

Rounder: see *Plane, Rounder; Plane, Rounding.*

Rounding Cradle *Fig. 582*
A wooden block with the top formed to make a V-shaped trough. A stop is made near one end. It is used to hold the workpiece when planing to an octagonal section, when rounding, or when working chamfers etc. (See also *Bevel Block.*)

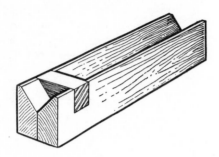

Fig. 582

Rounding and Hollowing Tools include: *Adze, Hollowing; Bit, Dowel Trimmer; Dowel Plate; Drawing Knife* (various hollowing patterns)*; Gouges; Plane, Cooper's Stoup; Plane, Core Box; Plane, Forkstaff; Plane, Hollow and Round; Plane, Mast and Spar; Plane, Rounder; Plane, Rounding; Plane, Spout; Shaves* (various); Tine Former under *Rake Maker.*

Roundsil
Scots term for a *Compass Plane.*

Rouser: see *Jumper.*

Router
The term Router has come to be applied to a number of tools which differ from one another in purpose, operation, and appearance. The name is applied most appropriately to the Router Plane which is in fact used for 'routing out' a depression in the surface of the work (see *Plane, Router*). But there is another group of tools known as Routers which resemble Spokeshaves in outward appearance though they differ in most other respects, including their narrow, frequently profiled cutters. Many of these are used by coachbuilders, but a larger version, the Sash Router, has large profiled cutters bedded like a plane and is used for making bow windows. Other tools occasionally referred to as Routers will be found under *Shaves* and *Scratch Stocks.*

392

Router, Carving: see *Plane, Router.*

Router, Circular: see under *Circular Cutting Gauge; Router, Sash; Shave, Handrail.*

Router, Coachbuilder's *Fig. 583*
A wooden stock, about 14–17 in long, with the ends shaped to serve as handles. The cutters are either bedded in the stock itself, like a plane iron, or in a special metal fitting screwed to it. The factory-made examples are usually plated on the sole to resist wear.

They were used to work grooves, rebates, and mouldings on the curved parts of carriage frames, and occasionally for decorating parts of carts and vans. The cutting irons are often mounted almost vertically, some are hooked, and the user could cut or scrape by varying the pressure and angle at which the tool was held.

The reason why many Routers were made in pairs (or provided with two irons and a central fence) is explained as follows by a Manchester coachbuilder, Mr. Philip Clarke (1966). 'If we look at almost any old carriage, say a Brougham or a Landau, it will be noticed that the shape is made up of either curves or compound curves. Therefore the grain must run sometimes one way and sometimes in the opposite direction, so it is evident that any tools, especially those with a fence, must be made in pairs right and left so that the work may be done with the grain and not against it, and also so that the Router could be made to run round the shape of a piece of wood either internally or externally.'

Variants illustrated include (for *Beading Router* see *Moulding Router* below):

(*a*) *Boxing Router* ('Boxing' or 'check' are coach-builders' terms for a rebate.)

This has a single iron ¼–¾ in; no fence, and is similar in construction and working to the Router Plane. Used for finishing rebates to the depth required, and for cleaning out grooves already made and testing them for depth. They were used when one could not employ a coachbuilder's Tee Plane owing to sharp curves in the timber.

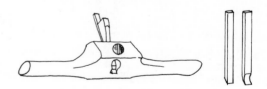

Fig. 583 (a)

(b) Boxing Draw Knife
Another tool for the above purpose, which is designed and operated like a Drawing Knife, is illustrated in the Shelburne Museum pamphlet No. 3 (U.S.A., 1957). A short blade, up to about 2 in wide, is fitted to the back-bar of the tool.

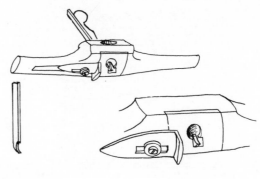

Fig. 583 (d)

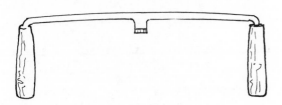

Fig. 583 (b)

(c) Corner Router
A V-shaped notch in the lower part of the stock, in which is wedged a straight or profiled cutting iron. Used for cutting chamfers or beads on the corners of a frame etc.

(e) Jigger (Side Router)
The cutters are parallel to the sole, and carried in a metal housing. The single-iron type has two hooked cutting edges fixed with two screws; the double iron (London pattern) has two separate plain cutters set at 45°, secured with thin metal wedges. Used for cutting glazing or panel grooves in frames and pillars. It began to replace the Pistol Router in the mid-nineteenth century.

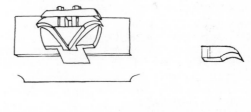

Fig. 583 (c)

(d) Grooving Router (Fence Router)
The iron, $\frac{1}{8}$, $\frac{3}{16}$ or $\frac{1}{4}$ in wide, has a hooked cutting edge, and is wedged in the stock sideways. Made in pairs for working on either hand, with a metal fence adjustable within 5 in limits and fixed by various means including a screw engaging a nut which runs in a slide within the stock. The round outlet for shavings is known as the 'eye hole'. It is used for working grooves for taking a panel or glass, and occasionally for cutting away waste timber from a rebate at some distance from the edge of the work-piece. For this purpose a groove was cut to the required depth as close as possible to the finished outline; then with either a Chisel or a Gouge the intervening wood was chopped away and finished with a Boxing Router.

Fig. 583 (e)

(f) Continental type Jigger
A Jigger made in France and Sweden (known also as a Coachmaker's Plow or Grooving Router) has an S-shaped stock with the cutter mounted at one end, leaving the lower curve to be used as a handle. The cutters were about $\frac{1}{4}$ in square and held by a closely fitted steel wedge which followed the same curve as the cutter.

393

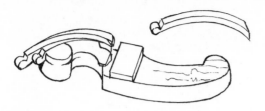

Fig. 583 (f)

(g) Listing Router
Made like a Moulding Router but with single or separate square cutters to cut a small square rebate to decorate the outside face of the frame.

Fig. 583 (g)

(h) Moulding Router (including Beading Router, Ovolo Router, Rounding Tool, etc.)
The two irons, $\frac{1}{8}$–$\frac{5}{8}$ in wide, fitted on either side of a centre fence (which is part of the metal sole plate), are secured by a single wedge or sometimes wedged separately. They are shaped to a bead or other profile in opposite hands. These irons are interchangeable with those of square ('listing') or rounding irons of quarter-circle or ovolo profile.

Used for working mouldings and other decoration on the edges of coach parts; the two cutters enable the tool to be used in the direction that suits the grain.

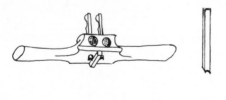

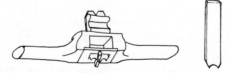

Fig. 583 (h)

(i) Home-made Moulding Routers
Another type, usually home-made, is used by wheelwrights on carts and wagon framing and undercarriage. It is made like the Grooving Router with an adjustable fence, but the iron is about $\frac{3}{4}$ in wide and cut to various decorative profiles. (See *Plane, Centre Board* for a tool with a similar purpose.)

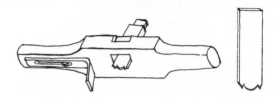

Fig. 583 (i)

(j) Pistol Router
An L-shaped stock with one handle straight and the other pistol-shaped. The iron, $\frac{3}{16}$ or $\frac{1}{4}$ in wide, with hooked cutting edge, lies vertically in a dove-tailed groove flush with the step of the stock, and is wedged on its side. An adjustable fence is fitted. Made in pairs for right and left-hand working.

Pistol Routers are used for cutting a groove in the pillars or frame to carry the panels. The specially shaped stock allows the cutter to operate at the lower face edge of the frame. This Router was replaced by the Jigger Router *c.* 1850.

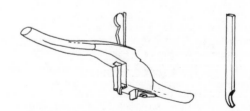

Fig. 583 (j)

(k) Double Pistol Router
A variant is designed to be worked left-hand or right-hand with the same tool, and consequently the stock is yoke-shaped and has two irons.

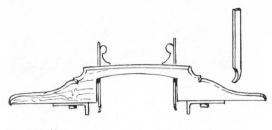

Fig. 583 (k)

(l) Reeding Router
With a fence similar to the Grooving Router, the single cutting iron is bedded and wedged like that of a Plane and provided with reeding teeth up to eight in number, in which case the iron is 2 in wide. It is said to have been used for ornamenting axle-blocks.

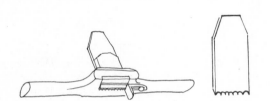

Fig. 583 (l)

(m) Side Cutting Router
Illustrated by Preston (Birmingham, 1914), this has a flat sole, with a cutter shaped like the spur of a Cutting Gauge let into a small step on the stock and secured parallel to the sole. Purpose unknown.

Note. The wheeler's Jarvis and Nelson are sometimes called Routers. See under *Shaves.*

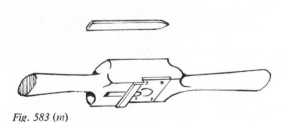

Fig. 583 (m)

Router, Metal Types *Fig. 584*
Towards the end of the nineteenth century a number of metal Routers (usually cast iron) were developed which aimed at replacing the older wooden Routers. These were used by joiners, cabinet makers, shop fitters, and others for circular work on sashes, doors, hand rails, and furniture. Metal Routers were seldom used by coachbuilders: they clung to the wooden version until the end – see *Router, Coachbuilder's.* And under that entry will be found an explanation of the fact that many Routers are made in handled pairs or, for the same reason, are provided with twin cutters bedded on either side of a central fence.

Varying in length from 6 to 12 in (and often wrongly called Spokeshaves which they resembled), they were made in many different profiles, including the following:

(a) Hand Beader
A flat sole, slotted to take an adjustable fence (two are supplied, right or left-hand) and a set of double-ended blades. Used for sticking a bead of the required size on straight or curved workpieces.

Fig. 584 (a)

(b) Double Bead Router (Circular Bead Router)
A sole with two side-by-side circular grooves with a fence between them. The two cutters of opposing hands are in twelve sizes from $\frac{1}{8}$ to 1 in wide. Used for working quirked beads on circular members.

Fig. 584 (b)

(c) Quirk Router (Quirk Cutter)
In its simplest form the cutter is set upright in the middle of the stock and fixed at the required height by means of a thumbscrew. The cutters take the form of two hooked teeth facing each other. One

395

tooth is of V-section so that its sharp corners cut the sides of the quirk. The other is square and scoops out the waste. An improved version mounts a similar cutter assembly on the upright of an L-shaped casting, the base of which forms the sole of the tool, and the upright one of the handles. Provided with two fences, one for straight and the other for circular work, they are used for cutting narrow grooves or the quirks of mouldings on stopped, straight, or circular work.

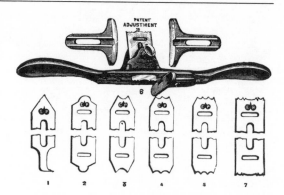

Fig. 584 (*d*)

(*e*) *Rabbeting Router:* see *Sash Router* below.

(*f*) *Sash Routers* (Circular Sash Router)
Formerly made in wood on the lines of the Coachbuilder's Moulding Router, but often in bigger sizes. The metal version has two cutting irons, with fences made solid with the stock. The sole is shaped to the profiles of the standard sash ovolo or lamb's-tongue mouldings for either right-hand or left-hand working, and the irons fitted in pairs accordingly. A wooden version, illustrated under *Router, Sash*, is used for working mouldings on circular members of sash frames. (A similar tool designed to work an ovolo moulding is sometimes called an Ovolo Router.)

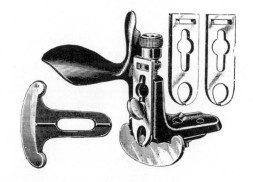

Fig. 584 (*c*)

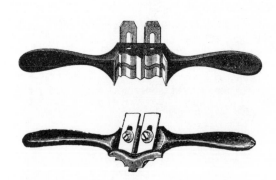

Fig. 584 (*f*)

(*d*) *Hand Reeder and Moulder*
A flat sole, with interchangeable cutters bedded in the middle of the stock, and with fences for straight or circular work on either hand. The cutters are shaped to give various profiles and various combinations of reed mouldings. Used for working reeds and other mouldings on straight or circular parts.

(*ff*) A *Rabbeting or Fillister Router* is used as a counterpart to the Sash Router for working the rabbet for the glass on circular sash stuff and bars, and used like a Sash Fillister. It has a single square iron bedded in the middle of the stock, and two pairs of adjustable fences with concave and convex faces for straight or circular work.

Fig. 584 (ff)

Fig. 585

(g) *Stringing Router* (Lining Router)
A comparatively long narrow sole with interchangeable irons of various widths and straight cutting edges, and an adjustable fence. Used by cabinet makers and others for cutting grooves for inlaid strings, bands, or lines. It takes the place of the home-made Scratch Stocks for this purpose.

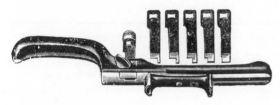

Fig. 584 (g)

Note See also:
Tectool – a modern Router which combines the function of the Quirk and Stringing Router; and *Plane, Door Trim* – a router-like tool used for recessing door furniture.

Router, Old Woman's Tooth: see *Plane, Router.*

Router Plane: see *Plane, Router.*

Router, Sash *Fig. 585*
These are often more heavily built than other wooden Routers. They are about 11 in long and have the cutters bedded on both sides of the fence. The sole and cutters are profiled to the sash ovolo or other moulding for either right-hand or left-hand working. Used for moulding circular members of sash frames (e.g. bow windows) and other curved parts. (A metal version is illustrated under *Router, Metal Types;* see also *Window Making.*)

Rove Hook: see Hoe under *Caulking Tools.*

Roving Iron
An iron Punch used by shipwrights. One end is hollowed for driving the roves (or washers) upon the ends of the rivets on the inside of small boats.

Rubber: see *Sailmaker.*

Rubber Tapping Tools: see *Knife, Rubber Tapping.*

Rubbing Stick
A strip of wood about 14 in long, faced with leather. Used for the final finishing of highly polished, rounded or shaped objects such as Gunstocks.

Rule *Figs. 586; 587*
Measurements based on the length of the human foot and the outstretched hand or arm were used in many of the ancient civilisations. They were standardised by the Egyptians who, as early as *c.*2500 B.C., used calibrated stone or wooden rods. Bronze folding foot-rules were used by the Romans but more commonly they used a plain strip of wood, graduated in various ways, and this type was common throughout the Middle Ages. In the seventeenth century, Rules were graduated in inches and quarters, but it was not until the eighteenth century that the inch was commonly divided into eight and, later, sixteen parts.

The mass production of the modern brass-mounted, boxwood folding Rule dates from about 1800. This commonest type of woodworking Rule (also used by other trades) consists of four 6 in strips in boxwood, hinged together in pairs, each pair connected by a brass knuckle-hinge, so that all four strips can be folded flat. Similar rules are made in different lengths and folds; a common variant was a 2 ft two-fold Rule, often containing a brass slide. Graduated slides were fitted to measure depth or extend the effective length of the rule.

One of the best looking of these rules was made in ivory with German silver mounts. The white background adds legibility to the markings. A modern version is made in white nylon which has the added advantage of being almost unbreakable.

The Joint. Folding Rules are hinged at the centre by what is known as a rule joint. This consists of a thin disc on the head of one strip and a forked disc at the head of the other. The thinner disc is held in this fork and is riveted to move smoothly but with sufficient friction to stay where set.

The term 'arched' or 'square' joint etc. relates to the shape of the brass mount of which the rule joint is a part. Variations in these shapes are illustrated (*Fig. 586*).

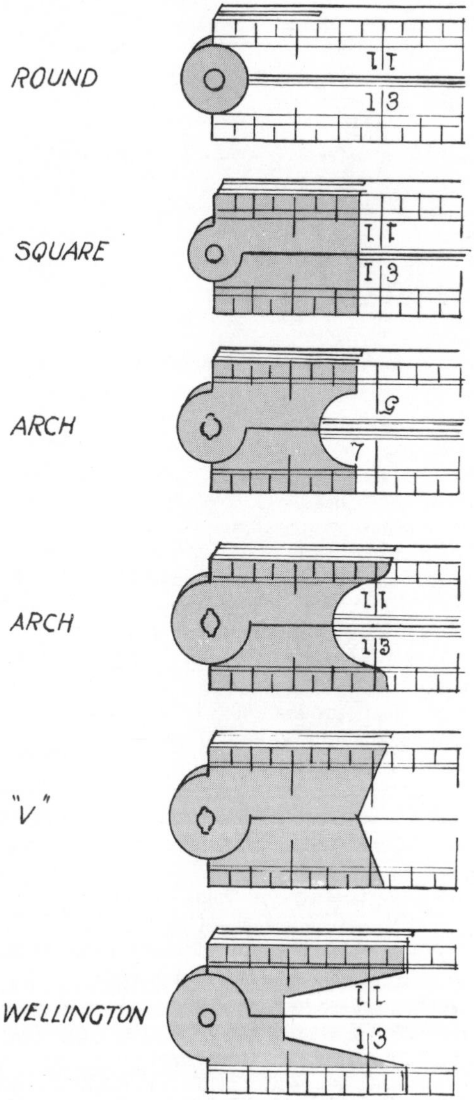

ROUND

SQUARE

ARCH

ARCH

"V"

WELLINGTON

Fig. 586　Folding Rule: variations in mounts

Rules used in woodworking shops include the following (*Fig. 587*):

(a) Folding Rules
A selection from innumerable variants are illustrated.

(b) Folding Lath
(Surveyor's Lath; Multi-Folding Lath; Surveyor's Rods; Zig-Zag Rule)
A long Rule, made up to 6 ft or more in length, with up to six or more folds. The strips are fitted with spring self-locking joints so that the rule remains rigid when extended. Used by surveyors, but also by tradesmen (particularly in the U.S.A.) as a substitute for the traditional boxwood Rule.

(c) Bench Rule [Not illustrated]
A name given to straight, graduated Rules without folds, about 1–1½ in wide, usually brass-tipped, and from 12 to 36 in long. Used by joiners and others on the bench.

(d) Calliper Rule
A Rule with a graduated brass slide which terminates in a brass jaw, and acts as a Calliper Gauge.

(e) Coachbuilder's Rule [Not illustrated]
A 4 ft four-fold rule, 1½ in wide. Used by coach and motor-body builders.

(f) Glazier's Rule
Up to 6 ft long and 2½–3 in wide. Used for measuring sheet glass before cutting. Also made in the form of a T-square. See *Square, Glazier's*.

(g) Log Stick (Board Stick; Lumber Stick; Measuring Rod) [Not illustrated]
A graduated rod of varying length used by timber merchants and others to measure length and diameter.

(h) Pattern Maker's Contraction Rule (Contraction Lath; Shrinkage Rule)
A straight lath, 24 in long, graduated on one side with standard inches, and with 'contraction scales' for iron, brass, etc. on the other. Used by pattern makers so that the wooden patterns they make will come out slightly larger than if an ordinary rule were used, and consequently produce a slightly larger mould in the sand. This is done to allow for the shrinkage of the metal when it cools after being poured into the mould.

(i) Protractor Rule
A Folding Rule marked with angles on the joint so that it can be used for marking mitres and other angles. A spirit level is sometimes added.

(j) Steel Rules

Many of the wooden Rules have their metal equivalents, but the commonest steel Rule used by woodworkers is probably the so-called Twelve Inch Rule. This is an accurately made steel strip with one end square, and the other rounded off with a small hole for hanging up. In addition to measuring it is used as a straight edge for testing workpieces.

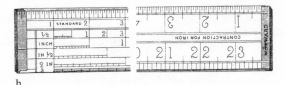

h

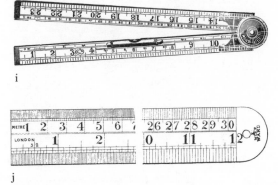

i

j

Fig. 587   Rules used in woodworking trades

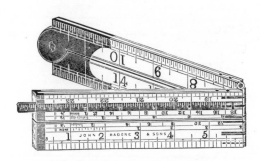

a

b

d

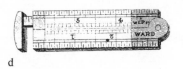

f

Mast and Spar Maker's Rules

Two-foot two-fold rules with a brass slide in one half. The scales marked '8-Square Lines' or 'E and M' are used for calculating how much to cut off the corners of squared timber to make it eight-sided. Provision is made to measure from the middle ('M') of the timber instead of the edge ('E') if the timber has defective corners. The slide can be used for quick calculation of the octagon-side length, given the size of any square figure.

Rule, Flexible *Fig. 588*

A graduated ribbon of linen or steel contained in a leather, metal, or plastic case. Variants include:

(*a*) A long linen or steel tape contained in a leather case with a handle for re-winding.

(*b*) A linen or steel tape contained in a metal case. After pulling out the length required against the pressure of a coiled spring, the tape is held by a ratchet mechanism until released by pressing a button, after which it returns to its case.

(*c*) Known as Concave Rules or Spring Rules, the tape is slightly concave in cross-section and so remains straight when extended. It is returned to its case either by an internal spring, or by pushing with the hand. See also *Timber Girthing Tape and Sword*.

399

a

b

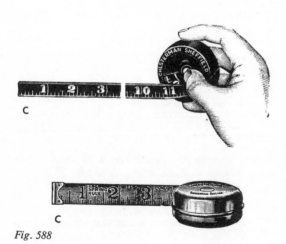

c

c

Fig. 588

Rule, Zig-Zag: see Folding Lath under *Rule*.

Rule Joints
A joint commonly used for Folding Rules, Compasses, etc. Described under *Rule*.

Rung Engine
A laddermakers' term for *Plane, Rounder*.

Rush Iron (U.S.A.: Rushing Lever): see *Flagging Iron*.

Rush Knife: see Chincing Iron under *Flagging Iron*.

S

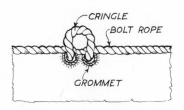

Fig. 589   Parts of a typical mainsail and detail of roping

Sabot Tools: see *Clog-sole Maker.*

Saddlebacked
Term used to describe the shape of a bar or beam which is sunken in the middle.

Sage
Dialect term for *Saw* (Wright, London, 1904).

Sailmaker *Figs. 589; 590*
It is not known when man first used sails but they are represented on tomb carvings in Egypt *c.* 2400 B.C. The trade is described by Robert Kipping in his *Elementary Treatise on Sails and Sailmaking* (London, 1862) as follows: 'Sails are an assemblage of several breadths of canvas, or other texture, sewed together, and extended on or between the masts, to receive the wind and impel the vessel through the water. The edges of the cloths, or pieces, of which the sail is composed, are generally sewed together with a double seam, and the whole is skirted round at the edges with a cord called the bolt-rope.'

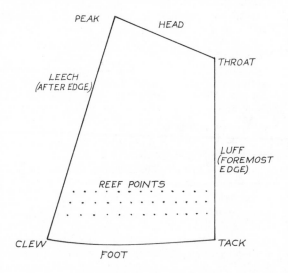

Until the nineteenth century almost all sailcloth was made of flax. Later cotton canvas was introduced, and today both flax and cotton are being replaced by synthetic fibres. The same applies to the ropes, which were at one time made only from hemp. (The tools of the ropemaker have not been included here.)

The tools of the rigger are similar to some of those employed by the sailmaker and are included below. Daniel Steel (London, 1794) describes the trade as a 'Mode of applying ropes to the several other parts and combining the whole, so as to produce the means of navigating a vessel'. The Admiralty's *Manual of Seamanship* (1951) contains an illustrated description of recent practice in rigging.

The sailmaker's workshop is usually called a Sail Loft probably because a large uninterrupted floor space needed for laying out the spread of a ship's sail was easier to find on an upper floor.

Tools used for sailmaking and rigging are illustrated as follows:

(a) Sailmaker's Bench
About 6 ft 6 in long and 14 in high, and used as both seat and work bench. Tools are kept in the holes provided on the worker's right hand.

(b) Sail Hook (Bench Hook)
A metal hook about 4 in long with a swivel-eye. Two hooks are attached by cords to the front of the bench and used as a 'third hand' for holding the cloth while being stitched.

(c) Grease horn
Often made from a cow's horn, filled with tallow, and kept hanging from the Bench. Used for holding the Sewing Needles to keep them free of rust.

(d) and *(e) Palm*
This is the sailmaker's 'thimble' and consists of a ring of leather worn round the hand, in which a dented metal plate is mounted to hold the back of the needle steady when pushing it through tough canvas or rope. The small indentations on a household

401

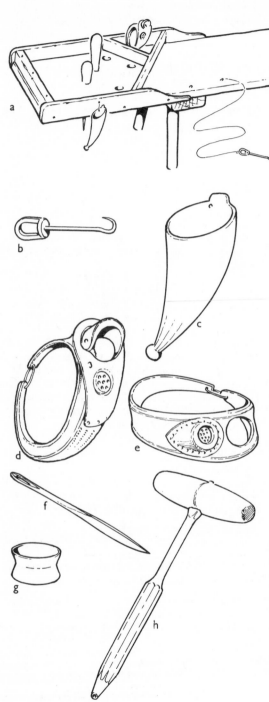

Fig. 590 (a)–(h)

thimble are designed for the same purpose, i.e. to hold the back of the needle. The plate on the Roping Palm (*d*) has larger indentations than that of the Seaming Palm (*e*) because roping needles are coarser. The Roping Palm is also provided with a guard at the thumb hole around which the twine can be wound for pulling it tight.

(*f*) *Needle*
Three-cornered and varying in length from $2\frac{1}{2}$ to 4 in, or longer for very heavy rope work. A lump of bees-wax is kept for waxing the sewing twine.

(*g*) *Thimble*
A metal ring 1 in wide and outwardly concave in cross-section. Used as an alternative to the thumb guard (*d*) to protect the finger after winding on the end of the thread and pulling tight.

(*h*) *Stitch Mallet* (Heaving Mallet or Heaver)
A metal rod, about $7\frac{1}{2}$ in long, with an octagonal shank and a wooden cross handle, but occasionally made in wood altogether. The indentation in the end of the shank is for pushing a needle; the shank of the tool is for pulling the thread tight, particularly when sewing rope. This is done by winding the thread round the octagonal part of the shank and using the tool as a lever.

(*i*), (*j*), and (*k*) *Prickers*
An iron spike, round or three-cornered, mounted in a wood or horn handle, 5–6 in long overall. Used for opening small holes in canvas and for opening the lay of small ropes when splicing.

(*l*) *Hand Fid*
A smooth cone, 10–24 in long, and made in lignum vitae, hard wood, or occasionally bone. It is kept hanging in a rack, or in one of the holes of the bench, to avoid becoming scratched, for a smooth surface is essential. It is used for stretching holes in canvas, and for stretching and sizing rope grommets and cringles, sometimes in order to admit a metal thimble.

Grommets are rope rings to which the edge of a purpose-made hole in the sail is sewn. Cringles are eyelets formed of rope loops fitted over the bolt ropes (the ropes sewn round the edges of the sail). Hand Fids are also used for opening the lay in ropes when splicing.

(m) *Set Fid* (Standing Fid)
A much larger version of the Hand Fid, from 25 to 42 in. in height, and designed to stand on the ground. It serves the same purpose, but for larger work; grommets and cringles are hammered over the Fid with a mallet known as a Wheeler (see below).

flattening seams and also for rubbing down the stitching to make it lie flat. Robert Kipping (London, 1862) writes: 'The seams of sails are generally sewed twice from the foot to the head – that is, the selvage of one cloth is sewed to the edge of the other, turned in to the required breadth and, when finished, it must be well pressed down with a "rubber" and turned over to sew the second side, and again rubbed down.'

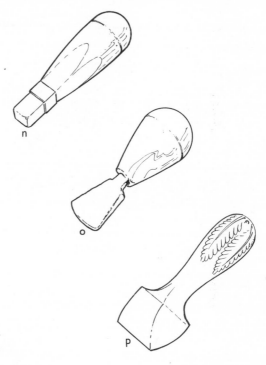

Fig. 590 (n)–(p)

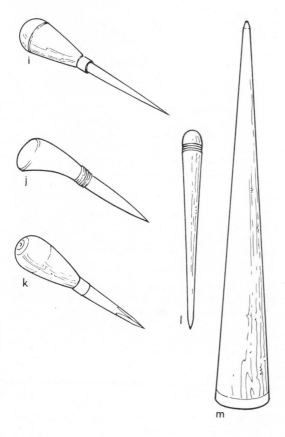

Fig. 590 (i)–(m)

(n), (o), and (p) *Rubber* (Liner; Presser)
From 4 to 6 in long, these small tools have metal heads if factory made, but if home made they are often found in hard wood, ivory, or bone, and are sometimes pleasantly decorated. They are used for

(q), (r), and (s) *Serving Mallet* (Serving Board)
One of the commonest types (q) bears a superficial resemblance to a round-headed mallet. It has a groove (called a score) along the top, used for 'riding' the rope to be served. Another type (r) has a T-shaped head which is hollowed on one side to bear on the rope. Grooves worn by straining the yarn are usually apparent on the back.

A modern version (s) is made in gunmetal with a hard-wood roller to hold the yarn. The claw on the top bears on the rope.

These tools are used for serving (i.e. binding) a splice, or protecting a length of rope with close turns of spun yarn. The score is held against the rope and the yarn passed over the head and round the handle

403

under the neck of the tool. As the tool is turned round and round the rope, each turn is held taut. Before serving, the rope may be 'wormed' (to fill the space between strands with yarn, to round it) and 'parcelled' (wrapping the rope with tarred canvas to prevent chafing). There is a rhyming formula for this work:

> 'Worm and parcel with the lay
> and serve the rope the other way'

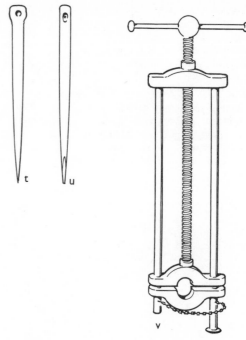

Fig. 590 (*t*)–(*v*)

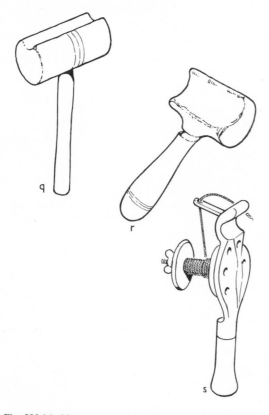

Fig. 590 (*q*)–(*s*)

(*t*) and (*u*) *Marline Spike*
A pointed iron spike, 6–18 in long, usually pierced at the head. Used for opening the strands of ropes when splicing. The example with ground faces at the point (*u*) is used for parting the strands of a wire rope. A very small version, about 4½ in long, is called a Shackle Spile, and is used as a tommy-bar for unscrewing pins of small shackles.

(*v*) *Rigger's Screw*
A strong iron cramp with a central screw. Used to hold and compress the wire strands when splicing wire ropes.

Mallets for use with the Fid (Wheeler)
These are used for beating a grommet or cringle over a Standing Fid. One is a mallet with a square head known as a Wheeler; another is club-shaped and called a Bottle Wheeler.

Punch and Block for metal Grommets
Smaller holes in the sail are today protected with brass eyes instead of rope grommets and these are inserted with punches similar to those used by shoemakers for inserting eyelets for laces. The work is done on a wooden block.

Sampling Tools: see Sampling and Tasting Tube under *Cellarman*.

Sampson: see Strake Tools under *Wheelwright's Equipment* (*Tyring Tools*).

Sand Paper: see *Glass Paper*.

Sand Tray: see *Veneering and Marquetry*.

Sappie: see *Timber Handling Tools*.

Sash Maker and Sash Tools: see *Window Making*.

Sash Punch: see *Chisel, Astragal.*

Saw *Figs. 591; 592* (See also *Sawyer; Tree Feller.*)
A Saw is a strip of thin metal with teeth cut along one edge. When the strip is comparatively narrow, it is called a 'web' and is held taut in a wooden or metal frame; wider strips have a handle or handles attached to the blade itself. Saws are used for cutting wood, metal, stone, and other materials.

The following is a general classification based on function; nomenclature and general information of each type will be found under the entries concerned.

1. Large Saws for converting trees into planks and smaller sections and normally worked by two men, see *Saw, Cross-Cut; Saw, Felling; Saw, Pit.*

2. Saws for preliminary cutting up of timber in the workshop or on the site: mainly tapering hand saws, see *Saw, Hand.*

3. Saws for making joints and for general bench work with parallel blades strengthened on the back; see *Saw, Back.*

4. Saws for cutting curved work with blades sufficiently narrow to follow a curved path; e.g. *Saw, Frame and Bow.*

Saw Blades

The blades of good Hand Saws are ground so that the back of the saw is thinner than the toothed edge, a feature known as 'taper ground', thereby reducing friction in the kerf. The blades are 'tensioned' through careful hammering during manufacture, whereby the blade is stiffened while remaining flexible. The process of tapering and tensioning is described by Moxon (London, 1677) as follows:

'But if it bend into a regular bow all the way, and be stiff, the Blade is good: It cannot be too stiff, because they are but Hammer-hardned, and therefore often bow when they fall under unskilful Hands, but never break, unless they have been often bowed in that place. The Edge whereon the Teeth are, is always made thicker than the Back, because the Back follows the Edge, and if the Edge should not make a pretty wide Kerf, if the Back do not strike in the Kerf, yet by never so little irregular bearing, or twisting of the Hand awry, it might so stop, as to bow the *Saw*; and (as I said before) with often bowing it will break at last.'

Saw Teeth (Fig. 592) *overleaf*

The following are the chief differences in the cutting edges of teeth:

(*a*) 'Rip'. For cutting *along the grain*. The teeth are filed straight across at right angles, and so present a series of chisel-like edges which cut away the wood. The sawdust is held in the gullets of the teeth and is pushed forward with each down stroke; for this purpose the teeth of the Pit Saws were particularly deep and were known as Gullet or Briar teeth (see *Saw, Pit*).

(*b*) 'Cross-cut.' For cutting *across the grain*. The teeth are filed at an angle in the region of 60° in alternate directions and thus have knife-like corners. These points sever the wood fibres, the waste between crumbling away as sawdust. A bewildering variety of teeth have been designed for large Cross-Cut Saws to meet varying tastes (see *Saw, Cross-Cut*).

Set. Most saws have their teeth bent sideways, alternately right and left, so that they cut a kerf slightly wider than the thickness of the blade to allow the blade to move freely without binding in the wood. The amount of set required depends on the nature of the work to be sawn. As Moxon puts it (London, 1677):

'This *Setting* of the Teeth of the *Saw* (as Workmen call it) is to make the Kerf wide enough for the Back to follow the Edge: And is set *Ranker* for soft, course, cheap Stuff, than for hard, fine and costly Stuff: For the *Ranker* the Tooth is set, the more Stuff is wasted in the Kerf. . . .'

Note: Some makers perforate the Saws with a series of holes beneath the teeth so that, as the cutting face of the Saw is worn down, the perforation acts as a gullet for a new tooth when the time comes for it to be cut. Some Hunting and Pruning Saws, and also saw blades in pen-knives, are provided with so-called double teeth. For further details see *Saw, Hunter's.*

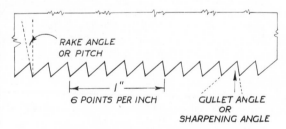

Fig. 591 Saw teeth: nomenclature

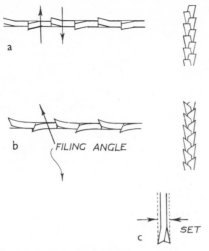

Fig. 592 Saw teeth: cutting edges

Use of Woodworking Saws

Tree fellers and sawyers work in pairs in the forest and at the saw pit; carpenters and joiners usually saw while holding the work on a trestle. The cabinet maker's traditional method of sawing is different: he does it with the work clamped to the bench and he holds the saw almost vertical, with the cutting edge pointing forward. (Illustrated under *Sawing Horse*, page 442.)

Historical Note

The first known metal Saws were used about 2500 B.C. by the carpenters and stone-masons of the Middle East. Owing to the soft nature of the metal used, and to the lack of set, the teeth had to be shaped to cut on the pull stroke. The earliest iron Saws of about the eighth century B.C. were also pull Saws, but by the beginning of the Christian era the Romans began to set the saw teeth, which made the Saws less liable to choke and buckle, and consequently they could be used with the more powerful and accurate push stroke. The Romans used small Frame Saws, narrow-bladed Hand Saws, large Cross-Cut Saws with a handle at each end, and large Framed Saws for ripping down logs and planks.

The Pull Saw may still be seen in use in various parts of the world, for example in China and Japan and in some countries of Europe, such as Greece. Pruning Saws designed to cut on the pull stroke are still made in western countries.

Saws of the Roman type continued in use almost unchanged during the Middle Ages, but towards the

end of the seventeenth century improvements in steel technology made wider strips available for 'open' hand and pit-saw blades, thus making it possible to dispense with the cumbersome wooden frames. By the middle of the eighteenth century English Hand Saws, made usually of Sheffield steel, were approximating to their modern forms, and Backed Saws were developed with thinner blades such as Tenon and Dovetail Saws. Framed Saws have survived in Britain for curved work.

Saw, Amputation: see *Saw, Surgical.*

Saw, Angle (Corner Saw; Dovetail Saw; Square Hole Saw) *Fig. 593*

There are several variants of this unusual Saw which is designed for cutting out surplus wood between multiple tenons or dovetails, and for cutting out square holes.

(*a*) A narrow, tapering Saw of the Keyhole type, with a shorter blade set at right angles to it extending about half way down the blade from the heel. The operation would appear to be as follows: The toe end of the long blade is used to cut a 'vertical' kerf, after which the Saw is pushed further into the work, when the shorter blade comes into operation and saws the 'horizontal' kerf at right angles to it. In the case of dovetails or tenons this will release the waste.

(*b*) A Bow-type Saw, probably made only on the Continent of Europe, has a narrow tapering blade, with a thin plate set at right angles to it at one end. The plate is introduced into a kerf already cut with another Saw, after which the 'horizontal' saw blade cuts out the bottom of the notch.

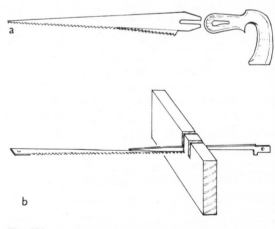

Fig. 593

Saw, Armchair Maker's *Fig. 594*

A French Saw known as the *Scie à araser de Menuisier en Sièges* (Chairmaker's Shoulder or Tenon Saw). This curious tool consists of a narrow saw blade set in the edge of a wooden stock about 10 in long and provided with an open handle at one end and an upright 'toat' at the other (*a*).

It is used for cutting the shoulders of tenons on the ends of large curved members of easy chairs, settees, and so on. The workpiece is held in a cramp in the form of an open box (*Boîte à Tenons*) (*b*) and the tenon marked out with a special templet known as the '*Bilboquet*' (*c*). This is laid on the upper surface of the box and the cheeks marked with the plates, and the shoulders from the upper edge of the base. The shoulders are then cut by drawing the saw across the flat upper surface of the box.

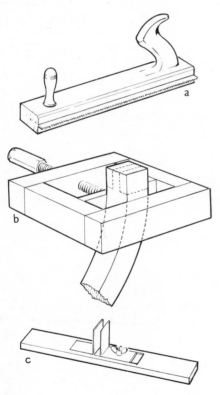

Fig. 594

Saw, Back

A general term for Saws with thin, rectangular blades, stiffened on the back with a strip of iron or brass folded over the top edge of the blade. The teeth are small, usually of the cross-cut type, and only slightly set. Before *c.* 1800 Back Saws were frequently tapered slightly from heel to toe. The modern Saws are parallel, with square toe and heel, but many are found reduced in width by repeated sharpenings.

They are used for cutting joints, and also for fine bench work – hence the thin blades and small teeth.

Little is known about the origin of the Back Saw, but it appears to have been developed by English saw makers about the middle of the eighteenth century or perhaps a little earlier.

The sizes of nineteenth-century Back Saws are listed by Holtzapffel (London, 1846) as follows:

Type of Back Saw	Length of blade (*in*)	Width of blade (*in*)	Points per inch
Tenon	16–20	3¼–4	10
Sash	14–16	2½–3¼	11
Carcase	10–14	2–2½	12
Dovetail	6–10	1½–2	14–18

Back Saws are described under the following entries: *Saw, Bead* (inc. *Jeweller's Saw*); *Saw, Carcase*; *Saw, Dovetail*; *Saw, Kerfing*; *Saw, Mitre Box*; *Saw, Sash*; *Saw, Screw-Head*; *Saw, Tenon*; *Saw, Toy*; *Saw, Vellum*.

Saw, Barrel: see *Saw, Crown*.

Saw, Bead (Beading Saw; Fancy Back Saw; Gent's Dovetail Saw; Jeweller's Saw) *Fig. 595*

A small Back Saw with a blade from about 3 to 6 in long, with a turned handle. The teeth are often very fine. One variety, sometimes called a Jeweller's Saw, has 24–30 points per inch, and is intended for very small, fine work; and being often sharpened without set, it leaves the work so smooth as to need little or no trimming.

Fig. 595

Saw, Beet Root

Listed by Spear & Jackson (Sheffield, 1880), this was a Web Saw, toothed on both sides, 12–14 in long × 1–1¼ in wide. Its purpose is unknown.

Saw Bench: see entries following *Saw, Woodcutter's*

Saw, Bettye (Betty Saw; Chairmaker's Saw; Donkey
Saw; Felloe Saw; Jesus Saw; Up-and-Down-Saw)
Fig. 596
Mr. L. J. Mayes (High Wycombe, 1960) relates that
he has heard chairmakers call the saw a 'Dancing
Betty' – 'I seems to be a-dancing with 'er when I'm
a-using of 'er', and on another occasion a 'Jesus
Christ Saw', with the explanation 'I keeps on a-
bowing to 'er all the time I'm a-using of 'er'. It may
be noted that the Italian *Sega di San Guiseppe*, (St.
Joseph's Saw), is a Framed type of Saw, frequently
shown in pictures of the Holy Family.

These are Framed Saws which correspond with the
three types described under *Saw, Frame*. The specimens illustrated below were all found in wheelwrights' shops in southern England. With the exception of the iron example (*c*) they are all home made.

(*a*) A wooden four-sided frame with a centre
blade, 30 in long, tensioned by a wing nut.

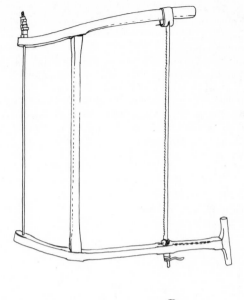

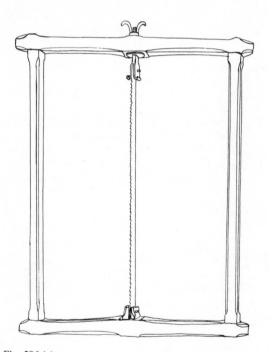

Fig. 596 (*a*)

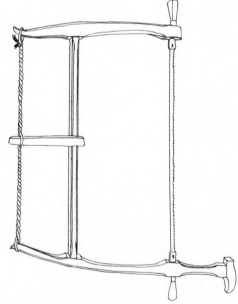

Fig. 596 (*b*)

(*b*) Wooden cheeks and centre stretcher, with a
blade about 27 in long, tensioned by a twisted cord
or metal rod. This tool is made like a large Bow Saw,
but one cheek is extended below the level of the blade,
with a cross-handle at the end.

(*c*) An iron bow, with the blade about 20 in long,
tensioned by a wing nut, and with two spade-type
handles on the upper arm of the bow.

These saws are operated with a characteristic up-
and-down movement, and are used for cutting all

kinds of curved work. Wheelwrights use them for cutting felloes. Chairmakers often use type (*a*) and (*b*) for sawing out chair arms and other curved parts. It is not uncommon to find this Saw with the side bars almost worn through at the point where the hands grasp them.

Fig. 596 (c)

Saw, Bilge: see *Saw, Crown.*

Saw, Billet: see *Saw, Woodcutter's.*

Saw, Blitz *Fig. 597*
A name given to a small Back Saw with interchangeable blades for wood or metal. A small hook is provided at the toe for the thumb of the left hand. Intended for the use of watchmakers and mechanics.

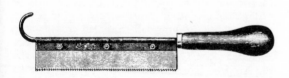

Fig. 597

Saw, Block
A small Saw, probably home made (we have not found one in a tool catalogue), consisting of a short strip of saw blade screwed to a wooden grip. Used for working on the floor near a wall or with a Mitre Shooting Block. The teeth, having no set, do not injure the face of the block. See *Veneering and Marquetry* for illustration.

Saw, Board
A name sometimes given to a *Pit Saw.*

Saw, Bow (Sweep Saw; Turning Saw) *Figs. 598; 617*
A name given to small Frame Saws, and in particular to those in which the blade is strained by means of a twisted cord and toggle stick. A description and diagram of this Saw, together with some historical notes, are given under *Saw, Frame.*

Bow Saws usually consist of two side-pieces (known as cheeks) with a central bar or stretcher between them. The blade, which in British examples is usually very narrow, is mounted between their lower ends, and can be turned to any angle in relation to the frame; the twisted cord joins the top ends of the cheeks. Its construction is well described by Moxon (London, 1677):

'The Office of the Cheeks made to the *Frame Saw* is, by the twisted Cord and Tongue in the middle, to draw the upper ends of the Cheeks closer together, that the lower end of the Cheeks may be drawn the wider asunder, and strain the Blade of the *Saw* the straighter. The *Tennant-Saw*, being thin, hath a Back to keep it from bending.'

Though Bow Saws are still widely used on the Continent of Europe for all types of sawing, in this country they are confined to the cutting of curved work for which the very narrow saw blade is specially suitable.

(*a*) Home-made examples. These are often gracefully curved, sometimes with the top ends of the cheeks turned to take the cord, and with the edges of the frame stop-chamfered.

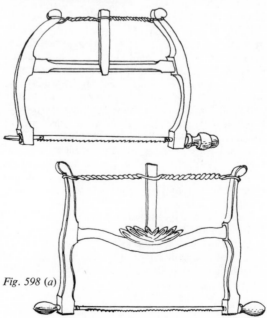

Fig. 598 (a)

(*b*) The factory-made Bow Saws of today have wisely retained some of the good lines of the earlier home-made tools.

Fig. 598 (b)

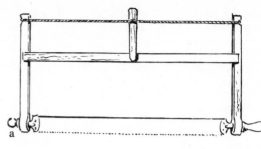

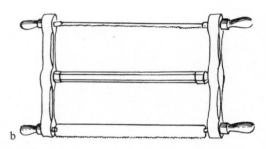

Fig. 599

Saw, Bow, Continental (Continental Frame Saw) *Fig. 599*

A Frame Saw, usually with straight sides, and with a blade from 15 to 40 in long and 1–3½ in wide, tensioned by a cord and toggle, or by a metal rod with wing nut. The larger sizes have a fixed blade and are usually without handles. The medium-size frames with narrow blades and finer teeth are usually fitted with handles which can be turned when sawing curved work. The stretcher is carried across the side arms, usually with a bridle joint – unlike the British Bow Saw, in which the stretcher is usually stub-tenoned.

It is a constant source of surprise to British tradesmen that on the Continent of Europe generally the British type of wide-bladed Hand Saws and Backed Saws are rarely found in the workshops. There is no particular reason for this except that framed saws are usually cheaper, especially when the tradesmen makes his own frames and only has to buy the narrow blade. It is largely a matter of use and custom, like the survival of the Pull-Saw for fine bench work in Eastern Europe. A man learns his trade with a particular type of tool and there is no reason why he should adopt another, unless its advantages are so overwhelming that it pays him to acquire a new skill in using it.

Two types are illustrated:

(*a*) A typical Continental Frame or Bow Saw. The stretcher overlaps the cheeks with a bridle joint.

(*b*) Saw of Continental origin with two blades, and an unusual central stretcher that pushes the cheeks of the Saw apart when twisted.

Saw, Bow, Tubular (Bushman Saw; Forester's Saw; Log Saw; Swedish Saw) *Fig. 600*

A metal Frame or Bow Saw consisting of a saw blade, 24–36 in long, held under tension in a tubular steel bow. The blade is ¾–1 in wide, with plain, raker, or grouped peg teeth, sharpened for cross-cut work. The tubular bow of modern examples is often oval in cross-section, and the blade is tensioned by lever action.

Used for general farm work, forestry, and for cutting up firewood. It is now widely used in British woodlands for felling small trees and cross-cutting small timber. This Saw, which appears to have been developed since 1914, is a modern reversion to the Iron Age Bow Saw which was held in a bent wooden frame.

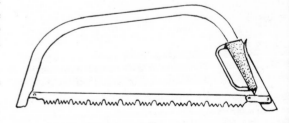

Fig. 600

Saw, Bracket: see *Saw, Fret*.

Saw, Brashing: see *Saw, Pruning (f)*.

Saw, Brickmaker's
A square-shaped Framed Saw, 18–24 in wide. Easily mistaken for a woodworking tool, but for the blade which consists of a twisted wire. Used by brickmakers for cutting clay.

Saw, Bridecake
Listed by Howarth (1884) as being 10 in long, without a back, and with a turned handle. Used presumably for cutting the icing of a wedding cake.

Saw, Broken Space (Fine Hand Saw)
A name given by Holtzapffel, (London, 1847) to a Hand Saw 22–26 in long: see Table under *Saw, Hand* for dimensions. We have found no mention of this Saw in trade catalogues, nor any explanation of the words 'broken space'. It is described by Knight (U.S.A., 1877) as a 'Fine Hand Saw'.

Saw, Buck: see *Saw, Woodcutter's* (also *Saw Buck* under *Saw Horse*).

Saw, Buhl: see *Saw, Fret*.

Saw, Bushman's: see *Saw, Bow, Tubular*.

Saw, Butcher *Fig. 601*
Made on the lines of a Hack Saw, and used by butchers for cutting through bone.

Fig. 601

Saw, Button: see Cylinder Saw under *Saw, Crown*.

Saw, Carcase *Fig. 602*
A Back Saw illustrated in Smith's *Key* (Sheffield, 1816) from which the sketch below is taken. Its dimensions will be found under *Saw, Back*. It is known to be a woodworking and not a butcher's Saw, but its particular purpose is not clear. The term 'carcase' may imply cutting joints in the frames of wardrobes and other furniture, but it would seem

that the normal Tenon Saw would have served as well for this purpose.

Fig. 602

Saw, Chain (Flexible Saw) *Fig. 603*
A Saw in the form of a flexible chain, about 4 ft long. Cutting teeth are mounted in each link, and there is an eye at each end for attaching a rope. Used for pruning high forest trees or for cutting through timber which cannot be reached with a conventional saw.

The same name is given to the modern machine-driven Saw in which the Saw chain is driven round the edge of a steel blade known as a guide bar.

Fig. 603

Saw, Chairmaker's
See *Saw, Armchair; Saw Bettye; Saw, Cross-Cut; Saw, Tenon*.

Saw, Chest
A name sometimes given to a Saw made short enough to fit into a carpenter's tool chest.

Saw, Chinese *Fig. 604*
Many of the Saws used in China and the Far East are of the Bow type and are similar to ours except that the blade is held in the slotted ends of wooden pins, which can be twisted to turn the blade. The teeth of their large two-man Framed Saws are often cut to point in opposite directions in each half of their length, thus giving equal work to both men on the cutting stroke. See *Saw, Cross-Cut* (Tooth form). Many Chinese Hand Saws are shaped like large knives with long, rounded handles. The teeth are sloped to cut on the pull stroke.

411

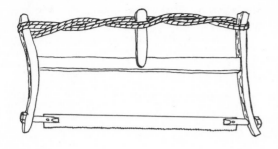

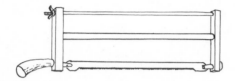

Fig. 604

Moxon (London, 1677), describes the Compass Saw blade as follows: 'The *Compass-Saw* should not have its Teeth *Set* as other *Saws* have; but the edge of it should be made so broad, and the back so thin, that it may easily follow the broad edge, without having its Teeth *Set*; for if the Teeth be *Set*, the Blade must be thin, or else the Teeth will not bow over the Blade, and if it be thin, (considering the Blade is so narrow) it will not be strong enough to abide tough Work, but at never so little an irregular thrust, will bow, and at last break; yet for cheapness, they are many times made so thin that the Teeth require a setting. Its Office is to cut a round, or any other Compass kerf; and therefore the edge must be made broad, and the back thin, that the Back may have a wide kerf to turn in.'

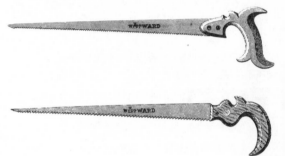

Saw, Cock

Illustrated by Peter Stubs (Warrington, *c.* 1845), this has the appearance of a small Hack Saw. Its purpose is unknown. Messrs Peter Stubs (1970) informed us that they have reason to believe that it was used in connection with cock spurs in the days of cock fighting.

Saw, Comb Cutter's: see *Saw, Stadda.*

Saw, Compass (Lock Saw; Scots: Port Saw) *Fig. 605*

A Hand Saw with a narrow blade, about 10–18 in long, tapered almost to a point, with teeth cut to 10 points to the inch, and fitted to a pistol-shaped handle. In another pattern, a compass Saw blade is included in a 'Nest of Saws' combining Keyhole, Compass, and the so-called Table Saw blades in a set with a handle and tightener. (See *Saw(s), Nest of*)

Used for cutting curved shapes in wood, particularly interior curves where it would be difficult to work the Bow Saw, e.g. in cutting a large hole in the centre of a board. The teeth are very little set, but the blade is tapered from the tooth edge to the back to give clearance, and to enable the user to keep the Saw cutting in the right direction by a slight twist of the handle. The blades are toothed to the end of the point, so that the cut can be started by inserting the Saw into a small hole previously bored in the waste wood. As sketched below, the blades are often found much shortened by wear.

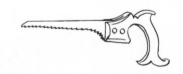

Fig. 605

Saw, Compass (Two-Handed) *Fig. 606*

A narrow Saw with blade 24–30 in long, tapering from $1\frac{1}{4}$ in to $\frac{3}{4}$ in, with a turned handle at each end. Its purpose is not clear, but it may have been used for pierced work. A similar Saw called a Turning or Whip Saw is used by masons for cutting soft stone.

Fig. 606

Saw, Concave: see *Saw, Crown.*

Saw, Continental Frame: see *Saw, Bow, Continental.*

Saw, Cooper's Bilge: see *Saw, Crown.*

Saw, Cooper's Head (Cooper's Frame Saw) *Fig. 607*
A bow-type Saw with a blade about 22 in long and
$\frac{1}{2}$–$\frac{3}{4}$ in wide. The blade is fixed, i.e. it cannot be
turned. The same Saw, but with a blade $1\frac{1}{2}$ in wide, is
called a Side Saw in Mathieson's List (Glasgow, *c.*
1900). Used by coopers for sawing the jointed pieces
of the cask heads into a circular shape. The specimen with a chamfered frame comes from a Scottish
cooperage. A Machine Saw for cutting heads is
illustrated under *Saw, Crown.*

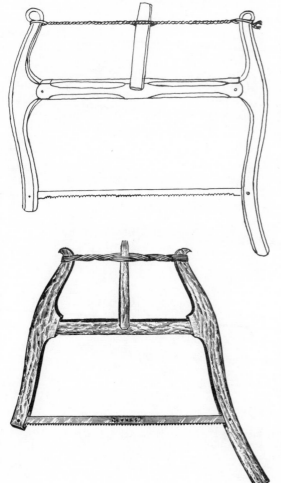

Fig. 607

Saw, Cooper's Riddle *Fig. 608*
A Cooper's Saw described by Coleman (Ireland,
1944). It is like a Compass Saw, but the blade is
bowed to expose a convex cutting edge. Used for cutting a croze groove in a single stave, when doing
repair work. Cf. *Plane, Cooper's Croze.*

Fig. 608

Saw, Coping (Scribing Saw) *Fig. 609*
A small Bow Saw with a very narrow blade, about 6
in long, held in tension within a small metal bow to
which is attached a turned handle. This Saw, which
does not appear in the makers' catalogues until *c.*
1920, was intended for coping, i.e. the fitting of one
moulded workpiece over the moulded surface of
another (also known as 'scribing'), for all kinds of
fine curved work, and for the removal of waste
between dovetails or short tenons etc. (Scribing by
Plane is described under *Window Making.*)

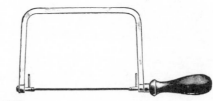

Fig. 609

Saw, Corner: see *Saw, Angle.*

Saw, Cranked (Dowel Saw) *Fig. 610*
A small Back Saw with the tang of the handle cranked
away from the blade which is itself slightly offset
from the back. It is used for cutting off dowels etc.
flush to the surface of the work.

Fig. 610

413

Saw, Cross-Cut (Over-thwart Saw; Thwart Saw) *Fig. 611*

Any Saw can be sharpened for cross cutting; in many workshops it is usual for one of the Hand Saws to be so sharpened and kept for cross-cutting work. This particular Saw is sometimes referred to as a 'cross-cut', but the term is more generally applied to the large one or two-man Saws for cutting logs or baulks across the grain.

(a) One-man Cross-Cut Saw

A blade up to 5 ft long and 6–9 in wide, fitted with a large handsaw-type handle at one end. A supplementary turned handle is often provided which can be fixed on the toe; when so fitted it enables the Saw to be worked by two men.

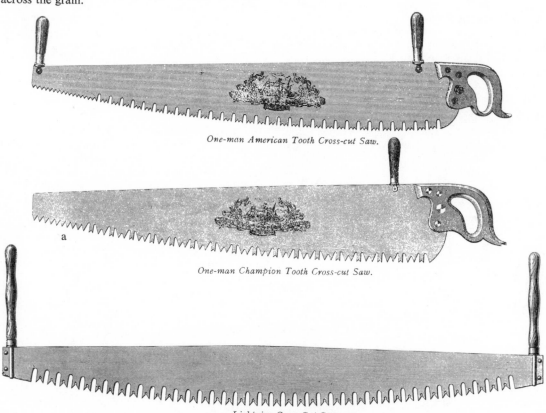

One-man American Tooth Cross-cut Saw.

One-man Champion Tooth Cross-cut Saw.

Lightning Cross Cut Saw

Tuttle's C.S. Patent Cross Cut Saw

C.S. Russian Cross Cut Saw, with M Tooth

Fig. 611

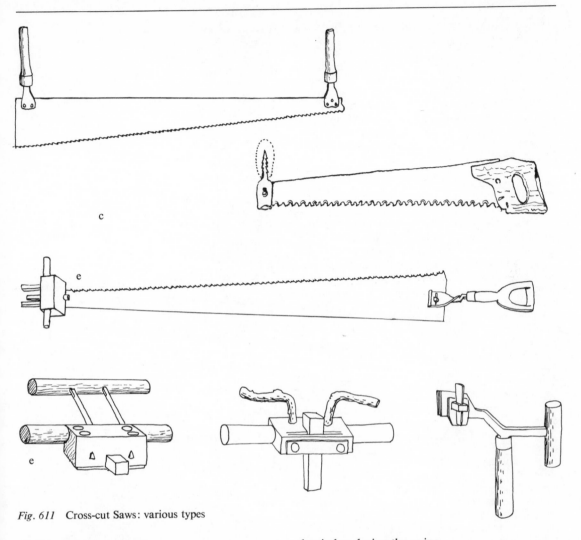

Fig. 611   Cross-cut Saws: various types

(b) Two-man Cross-Cut Saw

The usual purpose-made tool has a blade from 4 to 10 ft long and about 9 in wide in the middle, and is provided with upright handles at both ends. Holtzapffel (London, 1846) writes that some Cross-Cutting Saws used in the colonies for very large logs were as long as 16 ft 11 in wide in the centre, and 6–7 in wide at the ends. Many English patterns have a convex ('bellied' or 'breasted') cutting edge and a slightly convex back (known as 'fish-backed'). The reason for the convex cutting edge is not clear. Spear & Jackson Ltd., the saw makers (Sheffield, 1972), inform us that they believe the shape is associated with the action of cross cutting in which the arms and the shoulders swing through an arc. The curve on the cutting edge keeps the teeth in contact with the timber during the swing.

It is not unusual to find that repeated sharpenings have reduced the blade width to something under 2 in, e.g. that illustrated under *Chairmaker* (2) *The Bodger*, a specimen which comes from the Windsor chair-making trade in High Wycombe.

(c) Pit-type Cross-Cut Saw

It should be noted that until the early years of this century Pit Saws were often used for cross cutting and felling, even though their tapered blades only cut in one direction. Perhaps it was cheaper to adapt a spare saw from the pit for the occasional heavy cross cutting, rather than acquire the proper tool for the job. The tiller and box handles were removed, and short upright handles were riveted to the back or end

415

of the Saw instead. The gulleted teeth were sharpened for cross cutting. (See also *Saw, Pit.*)

(d) Cross-Cut Saws for Tree Felling [Not illustrated]
Purpose-made Felling Saws are similar to the normal two-man Cross-Cut Saws described above, except that the back of the blade is straight or even concave. This may help to prevent the blade from being trapped by the weight of the tree, and allows more room for wedges that are driven behind the saw to keep the cut open. For method see *Tree Feller.*

(e) Pit Saws for Tree Felling
Tapered Pit Saws were also commonly used for felling. Holtzapffel (London, 1846) illustrates these Saws with their home-made handles exactly as they are to be found today in derelict nineteenth-century saw pits. The cross-handle was removed from the tiller and replaced by a D-shaped spade handle; the box handle at the toe end of the Saw was replaced by a special 'felling box' of which home-made examples are here illustrated. One is provided with four handles, two of which are natural branches with a crook in them. The saw could be guided by one man at the tiller and pulled by two or even three men at the box end, sometimes by rope.

(f) Tooth form
The basic form is illustrated in Fig. 591 under *Saw,* but many variants have been designed for Cross-Cut Saws to meet varying tastes both here and abroad. Each is claimed to cut faster or more sweetly than others, or to clear sawdust more efficiently. These range from the common 'peg' or 'fleam' teeth (1) to such elaborate types as the 'Lightning' (3), 'Tasmanian', and 'Great American'. A popular form, which has survived from the fifteenth century, is the so-called M-tooth (2) with spaces between the groups of teeth to hold the sawdust and carry it clear of the cut. A further development, probably American in origin but now widespread, is the placing of 'rakers' (3) between the groups of small teeth. This raker tooth, as its name implies, rakes the sawdust out of the cut. Such Saws are especially suited to the cross-cutting and felling of softwoods, since the fibres of these are longer and more flexible than those of hardwoods.

A type of two-man Cross-Cut Saw found occasionally in Britain, but common in the Far East countries, is that with 'opposed teeth' (4). The teeth are slanted, half in one direction, half in the other, the purpose being to give both men the same amount of work on each stroke. It is called a Horizontal Saw in the 1910 edition of the Sheffield *List.* The so-called Gullet tooth is illustrated under *Saw, Pit.*

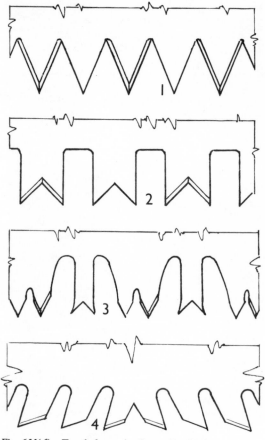

Fig. 611(f)   Tooth forms in Cross-Cut Saws

Saw, Crown (Cylinder Saw; Round Saw; Sheave Saw; Trepan Saw) *Fig. 612*
A metal cylinder with teeth cut on one end. They are designed for cutting cylindrical objects or discs in wood, horn, metal, and other materials. The Saw may be rotated by hand, brace, or machine. Varieties include:

(a) Cylinder Saws
These are called Button, Plug, Sheave, or Shive Saws, according to their purpose. They are used for cutting wood, bone, or horn for buttons, or for forming wooden discs for bung shives or pulley sheaves etc., and for making pegs or plugs.

(b) Crown Saws for Cutting Holes: See *Auger, Crown Saw; Bit, Annular.*

(c) Concave Saw

This is a machine-driven dish-shaped Circular Saw which is able to cut on the arc of a circle, e.g. cask heads and chair parts. The workpiece is placed on a turntable, which, as it revolves, carries the timber into the side of the saw. According to Disston's *Saw in History* (U.S.A., 1925), when making cask heads, the timber is rotated obliquely against the Saw, which 'cuts it round, and makes the long bevel at the same time.'

(d) Bilge Saw (Stave Saw)

A machine-driven cylindrical Crown Saw with bulging sides like a cask. It is used for shaping staves for small dry casks, such as nail casks. The raw billet is secured on a carriage which is moved by hand in the direction of the arrow in *Fig. 612 (d)*. This produces a stave that is 'ready bent' and needs no subsequent steaming and setting.

(e) Surgeon's Trepan

A small Crown Saw, worked by hand or Brace, and used for removing a circular disc of bone from the skull. When provided with a centre point it is known as a Trephine.

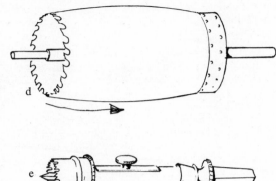

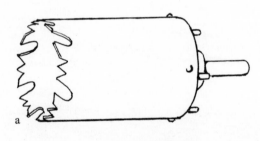

Fig. 612

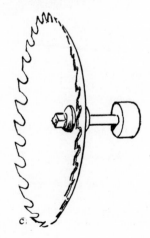

Saw, Cylindrical: see *Saw, Crown.*

Saw, Disston Henry Disston & Sons Ltd., Philadelphia, U.S.A.

Henry Disston was born at Tewkesbury, England, in the year 1819, and emigrated with his parents to America at the age of fourteen. He commenced a small saw-making business in Philadelphia *c.* 1840, which was to become the largest firm of its kind in the world. Most of their products went out of production during the 1920's except for the Skewback Saw which is still made by the H. K. Porter Company of Pittsburgh who absorbed the original firm *c.* 1955. Among the special Saws developed by Disston are the following:

(a) The Skewback Handsaw *Fig. 621* under *Saw, Hand*

This was one of Disston's most successful designs, introduced in 1874. The slight hollowing of the back of the Saw gave the tool a more graceful appearance, lightened it, and improved the balance.

(b) Saws without 'Set'

A Hand Saw of high temper ('Acme'), especially ground thinner than usual towards the back to permit its use without set in dry, seasoned timber. Also a Back Saw ('Mechanics Own'), similarly ground to work without set, and claimed to give a sufficiently smooth cut to eliminate planing before glueing.

(c) 'Square Hole Saw' *Fig. 593* under *Saw, Angle*

This unusual blade was a Keyhole or Compass Saw with an additional toothed blade set at right angles to the main blade, thus permitting both sides of a

417

square hole to be sawn from the usual previously bored hole.

(d) 'Ship' Handsaws

Disston gave this name to a series of strong Hand Saws of reduced width, possibly for their advantage to shipwrights in cutting a slow curve.

Saw, Docking

A name given by Henry Disston (U.S.A.) and Spear & Jackson (Sheffield) to a strong hand saw with an iron handle. It was intended for rough work on buildings and the like, where there was 'a risk of the Saw suffering frequent falls from upper storeys'.

Saw Doctor: see entries following *Saw, Woodcutter's.*

Saw, Donkey

A term sometimes applied to various Saws, including the *Saw, Bettye* and *Saw, Gang.*

Saw, Double: see *Saw, Stadda.*

Saw, Dovetail *Fig. 613* (Note: the Saws described under *Saw, Angle* and *Saw, Bead* are sometimes called Dovetail Saws.)

A small Back Saw with a 6–10 in blade and the teeth cut to 15–22 points to the inch. Used for cutting dovetails, small mitres, and other fine bench work.

(a) An older example with an open handle.
(b) A more recent example with a turned handle.

a

b

Fig. 613

Saw, Dowel: see *Saw, Cranked.*

Saw, Drum: see *Saw, Crown.*

Saw, Drunken (Wobble Saw)

A name given to a small Circular Saw fitted between

two collars with inclined faces, so that the Saw is set at an angle to the axis of the spindle and consequently its cutting edge 'wobbles' from side to side as it revolves. This enables the Saw to cut grooves wider than the thickness of the saw-blade; the more 'wobble' is given to the blade, the wider is the groove cut.

Saw, Duplex: see *Saw, Pruning (d).*

Saw, Electrician's *Fig. 614*

A name given to a small Back Saw with a $3\frac{1}{2}$ in blade, with the toe end sloping towards the cutting edge. Made to cut either wood or metal, and used before c. 1918 for cutting the wooden 'capping and casing' in which the wires were run.

Fig. 614

Saw Fancy Back: see *Saw, Bead.*

Saw, Farmer's *Fig. 615*

A name given to a Hand Saw, 18–28 in long, with coarse teeth, designed for cutting green timber.

Fig. 615

Saw, Felling: see *Saw, Cross-Cut.*

Saw, Felloe: see *Saw, Bettye; Saw, Pit.*

Saw, Firewood: see *Saw, Woodcutter's.*

Saw, Flexible: see *Saw, Chain.*

Saw, Flooring *Fig. 616*

A special Hand Saw with a blade 14–18 in long. The lower edge is often convex, and the teeth are sometimes carried round the curved toe of the Saw and along the back. Used by electricians, gas fitters, plumbers, and other tradesmen for cutting out a section of floor board or partition. The curved end of the Saw enables a particular board to be sawn across without damaging its neighbour. The convex edge and toe of the Saw are used to make a concave kerf almost penetrating the board. The pointed end of the Saw is then pushed through, and after penetration, the Saw, which cuts with both edges, completes the cut. (The handle is made reversible in American patterns.)

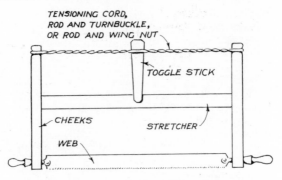

Fig. 617　Parts of a typical framed Bow Saw

Fig. 616

Saw, Folding: see *Saw, Hunter's; Saw, Pruning.*

Saw, Forester's: see *Saw, Bow, Tubular; Saw, Felling; Saw, Pruning; Saw, Woodcutter's.*

Saw, Frame (Bow Saw; Sweep Saw; Turning Saw) *Fig. 617*
The name Frame Saw is given to certain power-driven Saws (see *Saw, Gang*). But in the field of hand-driven Saws the term is applied in a general way to those Saws, including the so-called Bow Saws, in which a narrow blade, known as a web, is stretched between two points of attachment within a wood or metal frame. These are the main types:

(*a*) Framed Saws with blades of relatively large size which are kept in tension by screws or wedges acting directly on the blade itself. For examples see *Saw, Bettye; Saw, Pit (f); Saw, Veneer.*

(*b*) Framed Saws of the Bow type in which the blade is held between the lower ends of wooden side-pieces and is kept in tension by the cantilever action exerted when the top ends of the side pieces are drawn together by a cord. For this purpose the cord is shortened by twisting it with a toggle-stick; or if a metal rod or wire is used instead of a cord, by a thumb-screw or turn-buckle. For examples see Fig. 617, and also the following variants: *Saw, Bow; Saw, Bow, Continental; Saw, Cooper's Head; Saw, Woodcutter's.*

(*c*) Small framed Saws, usually of the Bow type, and often constructed in metal. For examples see: *Saw, Butcher's; Saw, Coping; Saw, Fret; Saw, Hack; Saw, Piercing.*

Historical note
The framed handsaw of antiquity, of which an example of *c.* A.D. 100 has survived, had the tension provided by a piece of springy wood. It was made like a bow, with the saw blade in the position of a bow string. It survives today in the form of certain metal Bow Saws, e.g. *Saw, Bow, Tubular.* A Bow Saw with a twisted cord (but without toggle-stick) is depicted in a stone relief of Roman chairmakers (Goodman, 1964 p. 121).

The original purpose of the Framed Saw must have been to keep a narrow and relatively weak blade in tension and thus avoid buckling when in use. Later, the frame was used to hold the very thin blades which are needed for cutting veneers in valuable wood without undue waste; and also for holding the long narrow blades needed for sawing out curvilinear work. Though framed Saws are still used on the Continent for all ordinary hand-sawing operations (see *Saw, Bow, Continental*), in Britain they survive only for dealing with curved work (see *Saw, Bow*).

Sizes of Framed Saws
The sizes of nineteenth-century Frame Saws other than Pit Saws are listed by Holtzapffel (London, 1846) as follows.

Type of Saw	Length of blade	Width of blade (in)	Points per inch
Veneer Saw	4–5 ft	4–5	2–4
Chairmaker's Saw (see *Saw, Bettye*)	20–30 in	1½–2½	3–4
Woodcutter's Saw	24–36 in	2–3½	3–4

419

Continental Frame Saw (See *Saw, Bow, Continental*)	15–36 in	1–3	4–12
Turning or Sweep Saw (See *Saw, Bow*)	6–22 in	$\frac{1}{10}$–$\frac{5}{8}$	10–20
Piercing Saw	3–5 in	$\frac{1}{30}$–$\frac{1}{40}$	40–60
Inlaying or Buhl Saw (See *Saw, Fret*)	3–5 in	$\frac{1}{25}$–$\frac{1}{30}$	15–40
Ivory Saw	15–30 in	$1\frac{1}{2}$–3	4–6
Smith's Frame Saw (See *Saw, Hack*)	3–12 in	$\frac{1}{4}$–$\frac{7}{8}$	10–14

Saw, Fret (Bracket Saw; Buhl Saw; Jig Saw; Scroll Saw.) The word 'Buhl' is the German form of the name of the French craftsman A. C. Boulle, Paris, 1642–1732, who specialised in the decoration of furniture with brass and tortoise-shell inlay or marquetry work. *Fig. 618*

In nineteenth-century catalogues, and until *c*. 1914, the name Fret Saw was sometimes given to the narrow, tapering blade of the Keyhole or Pad Saw, while the Fret Saw, as we know it, was called a Buhl Saw. Today, the term Buhl Saw has dropped out, and the name Fret Saw is confined to the familiar deep, U-shaped bow-type Saw. The frame, sometimes of wood, but usually of metal, measures 12–20 in. in depth. The blades, some of which are so fine as to resemble a serrated wire, are 5–6 in long and are secured in various ways, but usually by a small clamp and thumbscrew. A handle is provided in line with the blade.

Used for cutting curved shapes in thin wood, plywood, and veneer. The unusually deep frame enables shapes to be cut out at a considerable distance from the edge of the workpiece.

Fret Saws were used in the sixteenth century for cutting veneers for intarsia work. By the midnineteenth century they were sometimes replaced by treadle-operated Saws. These are illustrated under *Veneering and Marquetry* and *Saw, Jig*.

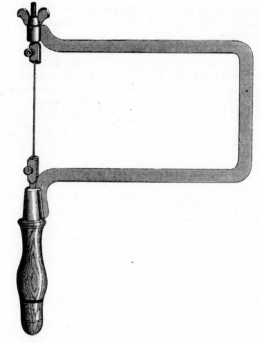

Fig. 618

Saw, Futtock: see *Saw, Shipwright's*.

Saw, Gang (Donkey Saw)
A Saw blade similar to the Mulay and Mill Saws, but shorter and thinner. A number of these Saws were mounted vertically in a frame and reciprocated by machine. By making as many as 20–30 cuts simultaneously a log could be ripped up into planks in one operation.

Saw, Gent's
A name sometimes given by the maker's to the *Saw, Panel* and *Saw, Bead*.

Saw, Grafting (Grafter Saw) *Fig. 619*
A Hand Saw with a blade 10–30 in long. It is illustrated with a closed handle in Smith's *Key* (Sheffield, 1816), and with an open handle in the Sheffield *List* (1888).

Though Saws are used in grafting operations to clear intervening branches and to prepare the stock for the scion, we have no evidence that this Saw, which appears to be an ordinary Hand Saw of medium size, is intended for this purpose. On the other hand Goldenberg's catalogue (Alsace, 1875) illustrates a 'Grafting Saw' which is evidently designed for grafting, made like a heavy bread-knife with coarse teeth, (though in the same volume a Grafting Saw of the English handsaw type is also shown); and Knight (U.S.A., 1884, Supplement) defines a 'Grafter' as a 'fine-toothed, pointed, narrow bladed hand-saw, used in sawing off limbs and stocks for the insertion of grafts'.

Tradesmen whom we have consulted tell us a short Hand Saw is used for splicing (also known as grafting): for example, the splicing of new material to the foot of a rotten door post, the repair of a cart shaft, or the making of a scarfed joint. There does not seem to be a need for a special Saw for these operations, though a fairly short one might be an advantage for working in confined spaces, e.g. when sawing off the foot of an existing door post on site.

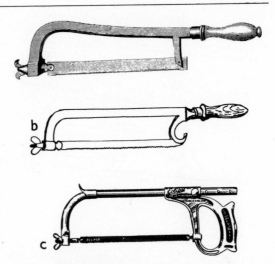

Fig. 620   Hack Saws
(a) Smith-made
(b) Nineteenth-century factory-made
(c) A modern Hack Saw with adjustable frame

Fig. 619

Saw, Grecian: see *Saw, Pruning.*

Saw, Grooving: see *Saw, Drunken; Saw, Stairbuilder's.*

Saw, Hack (Smith's Frame Saw) *Fig. 620*
A metal-cutting Bow Saw with a shallow metal frame, taking specially tempered saw blades from 8 to 12 in long. The blades have a hole at each end which fit over pins in the ends of the frame, and are tensioned by a thumbscrew. In later versions the frame is extendable to take different sizes of blade. Home-made examples sometimes have blades made from an old piece of Scythe blade, jagged along the edge. Used primarily by metal workers, but also found in most woodworking shops for cutting off bolts and similar work.

Saw, Half Rip
A lighter version of the Rip Saw, 26–28 in long. See Table under *Saw, Hand.*

Saw, Ham: see *Saw, Kitchen.*

Saw, Hand *Fig. 621*
The name given to the group of Saws which have wide blades, varying in length from about 10 to 30 in, tapering in width from the handle to the toe. They are often taper-ground, i.e. tapering in thickness from the cutting edge towards the back in order to reduce friction in the kerf, and to reduce the amount of set required in the teeth, thus saving energy and timber. The name Hand Saw is also commonly given to the ordinary general-purpose cross-cut Hand Saw to be found in most woodworking shops. (See Table below.)

Hand Saw teeth are sharpened for ripping along the grain or for cross-cutting. (See diagram under *Saw.*) Some are provided with increment teeth, i.e. small teeth at the toe increasing in size towards the handle. This gives a quicker cut where the operator can exert most force, and, by using the toe end of the saw when cross-cutting narrow pieces, he can avoid some tearing of the grain.

Before the eighteenth century many Hand Saws were fitted with a pistol-shaped handle, but by the beginning of the nineteenth century most of them

421

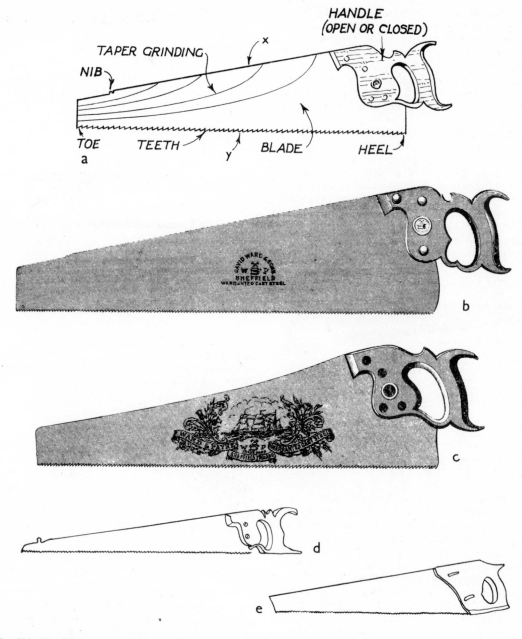

Fig. 621   Hand Saw
  (*a*) Parts of a Hand Saw
      X – Known as 'skew-backed' if hollowed
      Y – Known as 'breasted' or 'bellied' if convex
  (*b*) Typical Hand Saw
  (*c*) Hand Saw with 'skew' (or hollow) back
  (*d*) Hand Saw narrowed by long wear
  (*e*) A broken saw blade fitted to a home-made handle

have the characteristic flat, moulded handle we use today, slotted to take the heel of the saw and secured with rivets or screws. This handle is either 'open' at the bottom of the grip, or 'closed', when the hand-hole is entirely enclosed. The grip is shaped to fit the hand comfortably and to give the user control over the pressure to be applied. Until recently the handle was rather elaborately carved, with 'fish-tail' extremities and decorative curves; today these have been mostly eliminated, and wood is giving way to plastic.

Until recently a small notch was cut in the back of most Hand Saws a short distance from the toe, leaving a projecting tooth or 'nib'. Its purpose is not known, but it may be a surviving vestige of decorative features to be found on seventeenth and eighteenth-century Saws, especially in Scandinavian or Dutch tools of this period.

The terms Skew-back, Hollow-back, or Sweep-back relate to a popular form introduced in 1874 by Henry Disston, the famous American Saw maker, whose designs may have been influenced by those of the older and equally famous firm of Spear & Jackson in Sheffield. This form gives the back a slightly hollowed and, incidentally, a more graceful appearance. See *Saw, Disston.*

Varieties of Hand Saws

The following table includes the Hand Saws listed by Holtzapffel (London, 1846), and will also serve to direct the reader to examples of Hand Saws to be found on other pages.

	Length of Blade (in)	Width at heel end (in)	Width at toe end (in)	Points per in
Rip Saw	28–30	7–9	3–4	3½
Half Rip Saw	26–28	6–8	3–3½	4
Hand Saw	22–26	5–7½	2½–3	5
Broken Space or Fine hand	22–26	5–7½	2½–3	6
Panel Saw[1]	20–24	4½–7½	2–2½	7
Fine Panel Saw	20–24	4–6	2–2½	8
Chest Saw (for tool chests)	10–20	2½–3½	1¼–2	6–8
Table Saw	18–26	1¾–2¼	1–1½	7–8
Compass Saw	8–18	1–1½	½–¾	8–9
Key Hole Saw	6–12	½–¾	⅛–¼	9–10
Pruning Saw	10–24	2–3½	½–1¼	4–7
Grafting Saw[2]	10–30	?	?	?

Notes:

[1] Panel Saws today may have 7–12 points per inch.

[2] This Saw is not included in Holtzapffel's table but is added here because it appears in lists of the time, including Smith's *Key* (1816) and the Sheffield *List* (1888).

Saw, Heading: see *Saw, Cooper's Head.*

Saw, Hollow-Back: see *Saw, Hand.*

Saw, Horizontal *Fig. 622*

A name given by the Sheffield *List* (1910) to a Saw web with 'opposed' teeth, i.e. with gulleted teeth cut to point in opposite directions in each half of the length, presumably for the purpose of giving the same amount of work to men holding each end of the saw. Such blades are more commonly used in the Far East. (See *Saw, Chinese.*)

Fig. 622

Saw, Horn: see *Saw, Ivory; Saw, Stadda.*

Saw Horse (Sawing Horse): see page 441

Saw, Hunter's *Fig. 623*

A name given to long, thick, sword-shaped Saws carried at one time by hunters. A later version is a small tapering Saw with an 8–12 in blade folding into a handle, designed for carrying in the pocket; or a similar Saw which folds into the handle of a Hatchet (see *Hatchet, Hunter's*). The teeth of these Saws are often of a type known as 'double'; they are bevel-filed at 45°, and thus appear to overlap each other. The teeth are not set, but the blade is relieved towards the back to avoid binding. The blade is thicker than normal, and a tooth cut in the ordinary way across its full width might offer too great a resistance. This may be the purpose of the so-called double teeth, for they provide a larger number of smaller teeth in the same space.

Fig. 623

Fig. 624

Saw, Ice
Similar to a woodworker's Hand Saw in appearance, 24–30 in long, with very coarse triangular teeth. Another type is called a Pond Ice Saw: this is made like an open Pit Saw with a tiller on the heel but no lower handle. Used for harvesting ice.

Saw, Ivory
This is described by Holtzapffel (London, 1847) as a Frame Saw with a blade 15–30 in long, 1½–3 in wide, with 4–6 points per inch. Saw webs made for the same purpose (and also for bone and horn) were listed by Spear & Jackson (Sheffield, 1880). These webs are held in a metal frame like a Butcher's Saw, for to avoid wasting valuable ivory the blades are made very thin.

Saw, Jesus: see *Saw, Bettye.*

Saw, Jeweller's: see *Saw, Bead.*

Saw Jig (Fretsaw machine) *Fig. 624*
A very narrow blade held in a U-shaped wooden frame. It is given a reciprocating motion by a treadle and used for all kinds of fretwork, particularly 'jigsaw' puzzles. These machines were developed towards the end of the eighteenth century. The return throw of the saw-blade was sometimes effected by means of a bow string or a spiral spring. (See also *Saw, Fret.*)

Saw, Kerfing *Fig. 226* under *Coffin Maker*
A Back Saw with a blade 24 in long, with teeth cut to 6½ points per inch. Used by Coffin Makers to kerf the inside faces of coffin sides for bending at the shoulder. Six kerfs (saw cuts) were made at a slight angle to the square to within a bare ⅛ in of the face at the top, and less at the bottom. This was usually ample for elm; for oak the kerfs were filled with boiling water to ease the bending.

Saw, Keyhole (Pad Saw; often referred to in nineteenth-century catalogues as a Fret Saw) *Fig. 625*
A narrow tapering blade, about 6–14 in long (see Table under *Saw, Hand*). The blades, which are relatively thick and soft-tempered to prevent breaking when bent, are designed to be held in a special 'pad', from which the tool takes its alternative name. This wooden handle (later made in metal) is perforated down the centre, and fitted with a ferrule in which there are set-screws to hold the blade in position within the handle with no more projection than is

necessary for a reasonable stroke. Used for cutting out small shapes, curved work, and apertures such as key holes, especially when the work is too far from the edge to be reached by a Bow Saw.

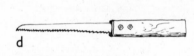

Fig. 625   Key-hole Saws
(*a*) With a wooden handle
(*b*) A wooden pad for holding a keyhole Saw
(*c*) A metal pad for holding a keyhole Saw
(*d*) With a home-made wooden handle

Saw, Kitchen (Ham Saw) *Fig. 626*
Made either as a lighter version of the Butcher's Saw or as a Back Saw, 10–16 in long. Used for cutting through bone in the kitchen.

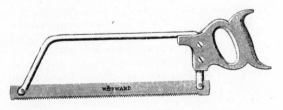

Fig. 626

Saw, Lock: see *Saw, Compass.*

Saw, Log
A name sometimes given to Saws used in the woodlands, e.g. *Saw, Cross-Cut*, or *Saw, Bow, Tubular.*

Saw, Mill *Fig. 627*
A machine-driven reciprocating Saw, consisting of a blade 3–10 ft long × 3–10 in wide, held in a frame.

Fig. 627

Saw, Mitre-Block
A name given by Coventon (London, 1953) to a larger version of the Veneer Saw, which, having no set, can be used on a Mitre Block for cutting thin sections of moulding.

Saw, Mitre-Box *Fig. 628*
A Back Saw with blade 22–30 in long and 4–6 in wide, designed for use with the Mitre Box for cutting mitres. The wide blade enables the thickened back to clear the top of the box. (See also *Mitre Appliances.*)

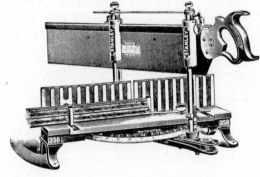

Fig. 628

Saw, Monkey: see *Saw, Turkish.*

425

Saw, Mulay (Muley Saw) *Fig. 629*
According to Corkhill (London, 1948) this is a Pit Saw blade operated by one sawyer, the upstroke being operated by a spring device. But the 7-ft Muley Saw webs illustrated in nineteenth-century tool Lists were used in a mechanical Saw reciprocated by a crank which drove the blade up and down in a guide.

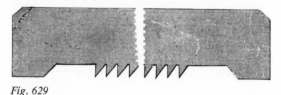

Fig. 629

Saw(s), Nest of *Fig. 630*
A set of three Hand Saws of varying width and length, made to fit one pistol-shaped handle. The largest Saw is often known as a Table or Pruning Saw, the middle a Compass Saw, and the smallest a Keyhole Saw.

Used by woodworkers for curved work, but also by electricians, plumbers, and others for cutting curved holes in floors, skirting boards, etc.

Fig. 630

Saw, Nock
A name given to a small Saw of unknown appearance used for cutting nocks in a bow, i.e. the notches at the tips of the bow for holding the string. The notch in the tail of the arrow is also called a nock.

Saw, Notching: see *Saw, Stairbuilder's.*

Saw, Overthwart: see *Saw, Cross-Cut.*

Saw, Pad: see *Saw, Keyhole.*

Saw, Panel (Gentleman's Saw)
Illustrated in Smith's *Key* (Sheffield, 1816) as a light Hand Saw about 10–24 in long, with fine teeth. (See Table under *Saw, Hand*.) Used for cross-cutting softwoods where a cleaner finish is required than can be obtained with a normal Cross-Cut Hand Saw. Also used for cutting plywoods, larger tenons, etc.

Saw, Patternmaker's
A name given by Disston (U.S.A., *c.* 1915) to a small Hand Saw with a thin parallel blade, $7\frac{1}{2}$ in long and $1\frac{1}{2}$ in wide, with teeth cut to 15 points to the inch, and provided with an open handle.

Saw, Piano Maker's: see *Saw, Vellum; Pianoforte Maker and Tuner.*

Saw, Piercing (Jeweller's Piercing Saw) *Fig. 631*
A small framed Saw intended for cutting curved shapes in thin metal and other materials. A narrow blade, 3–5 in long, is held in a light metal frame which is sometimes adjustable to take different lengths of blade. The ·frame is often beautifully finished and chamfered in the same tradition as the Lancashire Calliper. Used by jewellers and others for cutting curved shapes in thin metal and other materials.

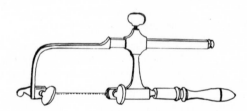

Fig. 631

Saw, Pit (Board Saw; Plank Saw) *Fig. 632*
Note: For general information on the use of Pit Saws, see *Sawyer*. For Saws made from old Pit Saws see *Saw, Cross-Cut Fig. 611*

The word 'pit' to describe these long Saws is mainly confined to English and U.S.A. usage. The use of a pit with Saws of this type is authenticated for England from the late seventeenth century onwards, but in most countries on the Continent of Europe the timber was supported above ground on trestles, as it still is in parts of Africa and Asia.

It is fairly certain that before the mid-eighteenth century all Pit Saws consisted of relatively narrow saw webs held taut within a wooden frame. The broad, open Pit Saw, without a supporting frame, came afterwards, made possible by improvements in steel making. And thereafter, this longest and most graceful of Saws became a standard form in England and was exported all over the world.

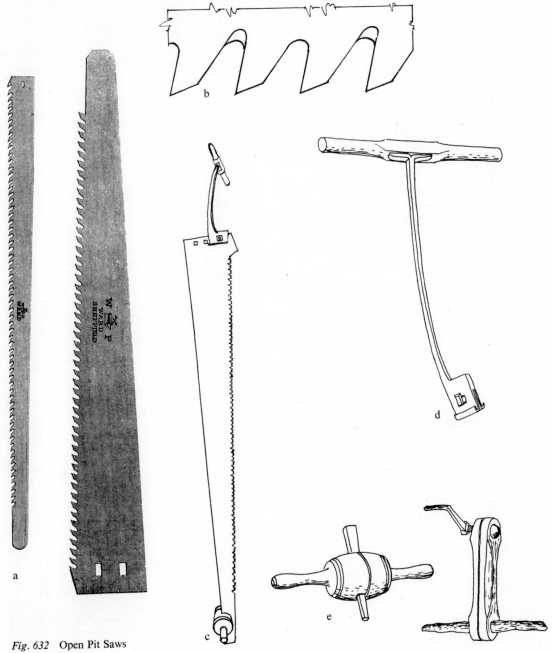

Fig. 632   Open Pit Saws

(a) 'Open' Pit Saw Blades

These are of two kinds, broad and narrow. The broad blades, which are used mainly for ripping down logs and baulks into planks, taper from about 10 in at the heel to 4 in at the toe. The length of blade most commonly seen is 5–7 ft, but blades up to 10 ft in length were also made, particularly for export.

Narrow blades, which are sometimes called Felloe Saws or Turning Saws, are used for curved sawing, e.g. ship timbers, large wheel felloes, and plough and cart shafts. They taper from about $3\frac{1}{2}$ in at the heel to $1\frac{1}{2}$ in at the toe and are usually 5 ft in length.

(b) Pit Saw Teeth

Pit Saw teeth are raked one way to cut only on the down stroke, and are spaced at $\frac{5}{8}$–1 in. They are called gullet or briar teeth owing to the hollow cut away in front of each tooth; this, it is said, allows more room for the sawdust.

(c) Pit Saw fitted with Tiller and Box handles

The handles of the open saw blades are removable to allow them to be taken out of the kerf, and for the convenience of travelling sawyers who carried their Saws from pit to pit.

(d) Tiller

The top cross handle (called the 'tiller') is mounted on a 2 ft stem which curves backward towards the user, so bringing his hands in line with the cutting edge of the saw. The foot of the stem is forked to receive the saw blade, with its lower edges folded outwards to prevent it from entering the kerf.

(e) The Box

The 'box' handle at the lower end of the Saw is made in various forms, the commonest being a turned barrel-shaped stock of about 4–5 in diameter, with peg handles extending on both sides which are often worn thin from long use. A saw cut is made at the centre of the stock to admit the back edge of the blade, with space at its side for a wooden wedge, by which it is secured. By knocking out the wedge the box can be removed. (Box handles for Felling Saws are illustrated under *Saw, Cross-Cut Fig. 611.*)

(f) Framed Pit Saws

Thin or very narrow saw blades are stretched within a frame to prevent distortion of the blade when in use. The blades are often attached to the top and bottom of the frame by means of an iron ring which is split half way to receive them. They are tensioned by folding wedges below the lower ring, or by means of a thumbscrew at the top.

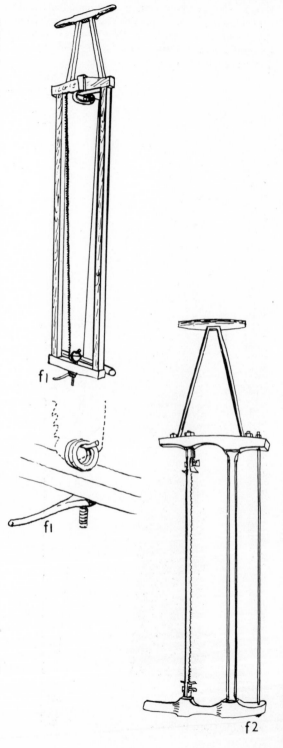

f1

f1

f2

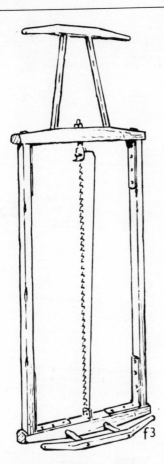

Fig. 632(f) Framed Pit Saws

1. *Broad Bladed.* These have blades similar to the Open Saw (*a*) above, but are thinner and are thus held in a frame. Saws of this type are often listed in the nineteenth-century catalogues as 'Russian Pit Saws'. The example illustrated, found in Hertfordshire, was used for sawing thin boards. The teeth are only slightly set.

2. *Narrow Bladed.* These narrow blades are either tapered or parallel webs and could be used for cutting out work with gentle curves. There are three main forms of this Saw all with frames, and handles at the top and bottom. In (*f*)2 the blade is on one side, a tension rod on the other, and a strut near the middle. This particular Saw was mainly used for sawing out plough shafts. In (*f*)3 the blade is in the middle. In a third form (not illustrated) the tension rod of (*f*)2 is replaced by a wooden strut, and the saw blade is tensioned by a wing nut at its upper end.

(*g*) *The Dimensions of Pit Saws* according to Holtzapffel (London, 1846) are shown in the Table:

	Length of blade (ft)	Width at heel (in)	Width at toe (in)	Spacing of teeth (in)
'Long, pit, or whip saw', (i.e. Open Pit Saws (Broad))	6–8	9–12	$3\frac{1}{2}$–5	$\frac{5}{8}$–1
'Felloe, or pit turning saw' (i.e. Open Pit Saws (Narrow))	4–6	3–4	2–3	$\frac{1}{2}$–$\frac{5}{8}$
Framed Pit Saw (Broad)	4–6	7–11	3–$4\frac{1}{2}$	$\frac{1}{2}$–$\frac{3}{4}$

Saw, Plane Maker's: see *Fig. 555* under *Plane Maker*.

Saw, Plug: see Cylinder Saw under *Saw, Crown*.

Saw, Plumber's
A Hand Saw, 10–18 in long, shaped like a Compass Saw. The blade is specially tempered to cut wood or metal. It sometimes has teeth cut on both edges, one with coarser teeth for wood, the other with finer teeth for lead, or to cut through the occasional nail. (cf. *Saw, Flooring*.)

Saw, Pocket Knife
Short Saws included in pocket knives often have relatively thick blades with 'double-teeth' which are suitable for cutting greenwood and are probably less liable to damage by nails etc. than ordinary saw teeth. (See *Fig. 623 Saw, Hunters* for view of teeth.)

Saw, Port
Scots term for Compass Saw.

Saw, Pruning *Fig. 633*
There are many different patterns but most show some or all of the following characteristics:

Blade. About 10–24 in long, tapering in width toward the toe, and often crescent shaped, i.e. with a concave cutting edge.

Teeth. Relatively coarse, and often formed to cut on the pull stroke. Sometimes they are provided with so-called 'double' teeth. (See *Saw, Hunter's* for the shape of these double teeth.)

Handle. Open, closed, or pistol grip. Some are mounted like carving knives, or made to fold into a handle; others are socketed for mounting on poles.

Pruning Saws are used in the garden and orchard for pruning fruit trees and for shaping individual trees for ornament. Foresters use them for brashing,

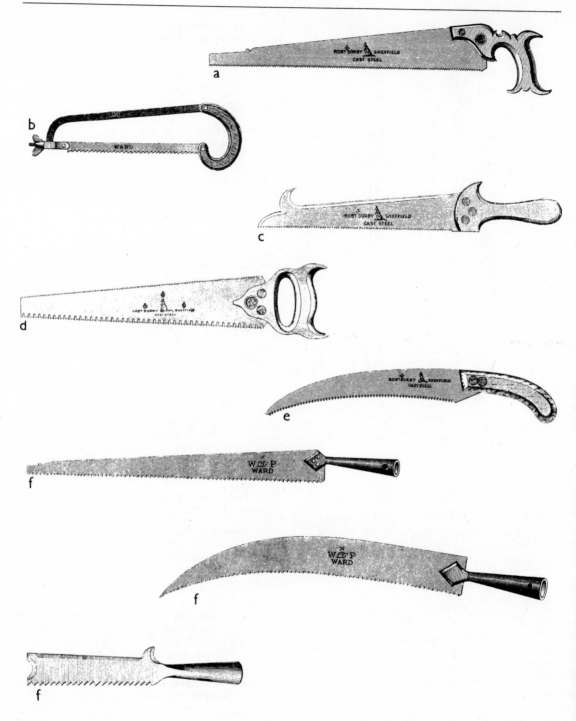

Fig. 633 Pruning Saws

i.e. the removal of dead lower branches from coniferous trees, to permit access and to lessen the risk of fire.

Variants are illustrated:

(*a*) Hand-Saw shaped. Like a small Hand or Compass Saw.

(*b*) Bow-type. An imported Saw, made like a Hack Saw.

(*c*) Bill-hook shaped. A straight blade with a hook on the back.

(*d*) Double Edged (Duplex Pruning Saw). A tapering blade with teeth cut on both edges, one with peg, and the other with coarser M-teeth.

(*e*) Grecian Saw. A tapering, crescent-shaped blade, with a concave cutting edge.

(*f*) Brashing Saw. The blade, straight or crescent-shaped, is socketed and mounted on a pole for cutting branches normally out of reach. Sometimes provided with a sharpened end or hook for cutting.

(*g*) Turkish. See *Saw, Turkish.*

Saw, Pull: see *Saw, Turkish.*

Saw, Rabbet
A variant of the Stairbuilder's Saw in which a Saw blade is secured to the side of a plane stock. There is an adjustable fence on the sole, so that kerfs can be cut at different distances from the edge of the workpiece in order to form a rebate.

Saw, Rack
A name sometimes given to a Saw with wide-set teeth.

Saw, Rib: see *Saw, Shipwright's.*

Saw, Riddle: see *Saw, Cooper's Riddle.*

Saw, Rip
The longest of the Hand Saws (28–30 in) and with the coarsest teeth. The teeth are sometimes smaller at the toe and increase in size towards the handle (known as 'increment'). The gullet angle is 60° and the rake angle 0–5°. Used for ripping soft wood down the grain. See *Saw* and *Saw, Hand* for further information.

Saw, Round: see *Saw, Crown.*

Saw, Russian Frame
A term applied in the Sheffield *List* (1910) to a broad-bladed Pit Saw, 5–7 ft long, 9½–12 in wide at the heel, and about 4 in at the toe. It is drilled at both ends to be held in a frame.

Saw, Salt *Fig. 634*
A small Hand Saw with a zinc or copper blade, 14–16 in long, with coarse teeth. Used for cutting blocks of salt.

Fig. 634

Saw, Sash
A name given by Smith's *Key* (Sheffield, 1816) and Holtzapffel (London, 1847) to a Saw somewhat smaller than a Tenon Saw. (See *Saw, Back* for dimensions.) It was presumably intended for cutting the tenons on sash stuff, but why a special saw was made for this particular purpose is not clear.

Saw, Screw-Head (Smith's Screw Head Saw)
Included here because it looks like a Bead Saw. As described by Holtzapffel (London, 1847) it is a small metal-cutting Back Saw with a turned wooden handle, with a blade 3–8 in long and ½–1 in wide, and with 12–16 points per inch. Used for cutting the slot in the heads of screws.

Saw, Scribing: see *Saw, Coping.*

Saw, Scroll: see *Saw, Fret.*

Saw Set: }see entries following *Saw, Wood-*
Saw Sharpening: } *cutter's.*

Saw, Sheave: see *Saw, Crown.*

Saw, Sheephorn
Listed by Robert Sorby (Sheffield, 1907) as a small Hand Saw with an open handle, 10–12 in long. This firm says that it was used for sawing off ram's horns.

Saw, Shipwright's (For the term 'Ship' Handsaw see *Saw, Disston.*
Large Saws used in the Shipyards include the following:

(*a*) *Pit Saw*
Before the introduction of power-driven Saws, the larger ship's timbers were sawn out over the pit. The

narrow Pit Saw used for curved work was known in the shipyard as a Ribsaw. The late Mr. Skinner, a Sussex sawyer (1949), related that 'in shipyards it was necessary very often for the bottom sawyer to work almost on his knees, whilst the top sawyer was half way down the pit, so curved was the timber as it lay on the rollers'.

(b) Futtock Saws
A name given to Saws used for sawing the futtocks (foot-hooks) which are the curved timbers of the ship's framing below deck. In the saw makers' catalogues (e.g. Henry Disston, U.S.A., 1915) these appear as parallel webs up to 5 ft long and $1\frac{1}{2}$–$4\frac{1}{2}$ in wide, with coarse teeth shaped for ripping, and perforated at the ends to be held in a frame of some kind.

Saw, Shive: see Cylinder Saw under *Saw, Crown*.

Saw, Side: see *Saw, Cooper's Head*.

Saw, Siding
A name given by Spon (London, 1893) to a small Hand Saw that appears to be identical to the Chest Saw, a tool of specially small size to fit into a carpenter's tool chest. See Table under *Saw, Hand* for dimensions.

Saw, Skew-Back: see *Saw, Disston*.

Saw, Smith's: see *Saw, Hack*.

Saw, Span: see *Saw, Web*.

Saw, Square Hole: see *Saw, Angle*.

Saw, Stadda (Comb Maker's Double Saw) *Fig. 635*
A double-bladed Saw used for sawing horn by comb makers. It is composed of two short blades tapering towards the cutting edge, fixed in grooves in the wooden back by means of wedges, and kept apart at a distance equal to the width of the teeth of the comb by a thin slip of metal known as a languid. One blade projects slightly more than the other, and in the process of cutting the teeth of the comb the first slot is taken as a guide for the second and so on.

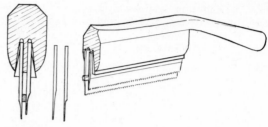

Fig. 635

Saw, Stairbuilder's (Grooving Saw; Trenching Saw; Notching Saw) *Fig. 636*
A short length of saw blade with a shaped wooden grip fitted along its upper edge, and sometimes made adjustable for depth of cut. Though common on the Continent of Europe and in the U.S.A., this type of Saw is made in Britain mainly for export. Used with or without a fence to cut the sides of grooves across or at an angle with the grain, e.g. in staircase strings. See *Saw, Wood-Backed* for a list of Saws mounted in a similar way.

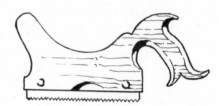

Fig. 636

Saw, Stave
A term applied in the Sheffield *List* (1910) to a broad-bladed Pit Saw held in a frame, $4\frac{1}{2}$–6 ft long, 9 in wide at the heel, and about 2 in at the toe. The name Stave Saw is also given to a *Bilge Saw* (see *Saw, Crown*).

Saw, Stone
Various saws used for cutting soft stone may be mistaken for woodworking saws, e.g. an 8 ft two-man Saw which looks like a Cross-Cut Saw. There are eye sockets at each end to take vertical handles, and the cutting edge is usually bellied and provided with simple triangular teeth.

Saw, Surgical (Amputation Saw) *Fig. 637*

It is curious that Amputation Saws are occasionally found in the workshops. They were probably bought at local sales after being discarded by doctors or veterinary surgeons, and were valued for their high quality.

There are two types: a Back Saw with an 8–10 in blade, usually with the back removable; and various bow-type Saws with metal frames. The handles of nineteenth-century examples are often made in ebony incised with characteristic cross-hatching to give a firm grip. (See also *Saw, Crown*.)

This connection with woodworking is touched on by Bernard Shaw in *The Doctor's Dilemma*, Act I (Constable, 1911, p. 15):

Walpole. . . . Sir Patrick: how are you? I sent you a paper lately about a little thing I invented: a new saw. For shoulder blades.

Sir Patrick (meditatively) Yes: I got it. It's a good saw: a useful, handy instrument.

Walpole (confidently) I knew you'd see its points.

Sir Patrick. Yes, I remember that saw sixty-five years ago.

Walpole. What!

Sir Patrick. It was called a cabinetmaker's jimmy then.

Walpole. Get out! Nonsense! Cabinet maker be —

Ridgeon. Never mind him, Walpole. He's jealous.

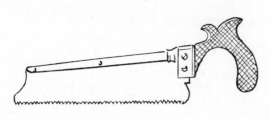

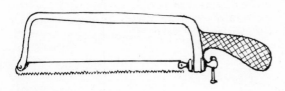

Fig. 637

Saw, Swedish: see *Saw, Bow, Tubular*.

Saw, Sweep

A name sometimes given to Saws designed for curved work, e.g. *Saw, Bow; Saw, Frame*.

Saw, Table *Fig. 638*

A relatively light Hand Saw with a tapering blade and open handle. The length varies from 12 to 26 in tapering from $1\frac{3}{4}$–$2\frac{1}{4}$ in at the heel to 1–$1\frac{1}{2}$ in at the toe. There are 7–8 teeth per inch, and according to illustrations in some lists, e.g. Atkin & Son (Sheffield, 1894) and James Howarth (Sheffield, 1884), the teeth slope backwards to cut on the pull stroke.

There is some confusion in both nomenclature and purpose. Owing, perhaps, to its possible use as a pull saw, Atkin and Henry Disston (U.S.A., 1915) list it under the heading 'Table and Pruning Saws'. On the other hand it is listed by several makers as the largest blade of a nest of three Saws, the others being Compass and Keyhole. This occurs in a recent list of Marples (Sheffield, 1965).

As to the purpose of the Saw, Holtzapffel (London, 1847) writes that it is for 'Circular and Curvilinear Works'; and Spon (London, 1893, p. 211) states 'the "table" or "ship carpenter's" saw . . . has a long, narrow blade intended for cutting sweeps of long radius'.

We have not come across the Table Saw in shipwrights' work, and indeed no convincing explanation of the term 'Table' has emerged so far, in spite of extensive inquiries. Yet they are listed in nearly all the catalogues from the middle to the end of the nineteenth century and were evidently in fairly common use at that time. One conjecture is that they were used for cutting out round table tops, but there seems to be no particular reason why a special Saw should be devoted to this, with Bow Saws available in most workshops, and it affords no clue to explain why some are clearly pull saws.

A rather more plausible theory is that they were used for trueing the flats of tabled scarfing joints in beams and girders. The joint would be assembled temporarily and the Saw run between the two pieces to give a good fit, a technique commonly used by carpenters. A narrow, tapering Saw cutting on the pull stroke would be an advantage for this, as an ordinary Panel Saw might buckle if it stuck and was pushed too hard. This may also explain why the tool seems to have dropped out of use after about 1920, when this form of construction was largely replaced by steelwork and reinforced concrete.

Fig. 638

Saw Teeth: see notes under *Saw; Saw, Cross-Cut; Saw, Hunter's; Saw, Pit.*

Saw, Tenon *Fig. 639*
Mr. L. J. Mayes (High Wycombe, 1960) relates that the word 'tenanting' was quite generally used in the Windsor chair-making trade instead of tenoning, one craftsman pointing out that the tongue of wood was the 'tenant of the mortise'. The same term is used by Moxon (1677) for what is presumed to be a Tenon Saw (it is not illustrated): 'The Tennant Saw, being thin, hath a Back to keep it from bending'. The same term is also used in the Apprentice Indenture Records (Bristol, 1532–1646).

A Back Saw with a parallel blade, normally about 10–16 in long but sometimes as long as 24 in, with teeth cut from 12 to 14 points per inch. The blade is stiffened on the back, and the handle usually closed like that of a Hand Saw. (Variants are listed under *Saw, Back*.)

Used for sawing tenons and for general work on the bench. The teeth are normally of cross-cut type, like the other Back Saws. But some workshops keep a special Tenon Saw with its teeth filed at right angles for cutting along the grain.

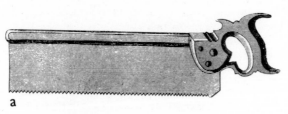

a

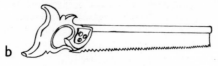

b

Fig. 639 (*a*) A modern example
(*b*) An example made *c.* 1810

Saw, Thwart: see *Saw, Cross-Cut.*

Saw, Toy *Fig. 640*
Very small Hand Saws (4–10 in long) or Back Saws (4–8 in long). Though probably intended for children, some were well enough made to be used in the workshop.

Toy Back Saw

Toy Hand Saw

Fig. 640

Saw, Tree Felling: see *Saw, Cross-Cut.*

Saw, Trenching: see *Saw, Stairbuilder's.*

Saw, Trepan (Trephine): see *Saw, Crown.*

Saw, Tub: see *Saw, Crown* (The cooper's Saw Tub is described under *Cooper* (2) *Furniture.*)

Saw, Tubular: see *Saw, Bow, Tubular.*

Saw, Turkish (Monkey Saw; Pull Saw) *Fig. 641*
A parallel blade, 10–20 in long and 2–3 in wide, with a pistol-type handle. This saw is usually grouped in the English lists with the Pruning Saws, and in the sizes 10–14 in long is said to be used for sugar cane. The smaller sizes, however, with a blade up to 10 in long, and fine teeth cutting on the pull stroke, are still widely used in Greece, Crete, and Turkey for the type of work for which we normally use the Dovetail or small Tenon Saw.

Fig. 641

Saw, Turning (or Turning Web)
A name sometimes given to Saws used for curved work (see, for example, *Saw, Frame; Saw, Bow*) or to narrow Pit Saws.

Saw, Two-Man: see *Saw, Compass (Two-handed); Saw, Cross-Cut.*

Saw, Up-and-Down: see *Saw, Bettye.*

Saw, Vellum
Made like a Bead Saw, with a thin blade 3½ in long, with about 24–30 teeth per inch which are not set. Used by piano-action makers and piano repairers for cutting a kerf to hold vellum hinges. (See *Pianoforte Maker and Tuner.*)

Saw, Veneer *Figs. 642; 713*
The following Saws are known as Veneer Saws:

(*a*) See *Fig. 713* under *Veneering.* A small saw blade 4–6 in long and 2–3 in wide, with a convex cutting edge to avoid 'digging in'. A wooden grip is screwed to the side of the blade at the top. The teeth are not set. Used with a straight edge for cutting veneers.

(*b*) See *Fig. 642.* A saw web, 4–8 ft long and 4–5 in wide, is stretched in a four-sided wooden frame. Used for producing 'sawn' veneers. The hardwood log is usually held vertically in a Vice, but sometimes the logs are cut on the Saw Pit.

Fig. 642

Saw, Web *Fig. 643*
A general term for narrow saw blades, usually parallel, which are intended for use in Frame or Bow Saws. There are many variations in size, tooth form, and the method of fixing at the ends, including:

(*a*) *Billet, Canada,* or *Woodcutter's Web,* 28–60 in long and 2–3½ in wide. Used in large Framed Saws, Woodcutters' Saws, etc.

(*b*) *Cabinet, Chair, Cooper's,* or *Joiner's Web,* 8–42 in long and ⅜–¾ in wide. The lower end of the range covers the common Bow Saws, the upper the Bettye (or Chair Maker's) Frame Saws.

(*c*) *Mill Web:* see *Saw, Mill.*

(*d*) *Muley Web:* see *Saw, Muley.*

(*e*) *Span Web:* (Span Saw; Tab Web), 12–40 in long and ¾–2½ in wide, with necked and shouldered tangs. Used in large framed Saws and also bought by tradesmen to put in their home-made Bow and Frame Saws.

(*f*) *Veneer Web:* see *Saw, Veneer.*

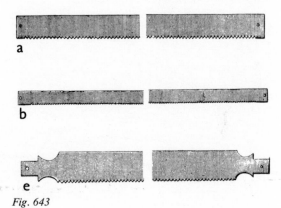

Fig. 643

Saw, Wheelwright's and Coachbuilder's
For special Saws used in these trades see *Saw, Bettye; Saw, Pit* (Felloe).

Saw, Whip
The term 'Whip' when applied to saws is probably derived from the German 'wippen' – to rock or move up and down. Moxon (London, 1677) calls a two-man Cross-Cut Saw a Whip Saw. 'Whip' is used as a synonym for Pit Saws by Holtzapffel (London, 1847) and by Mathieson (*c.* 1900).

Saw, Wobble: see *Saw, Drunken.*

Saw, Wood-Backed
Several Saws are backed with wood to act both as a stiffener and as a form of handle. These include: *Saw, Armchair; Saw, Block; Saw, Mitre-Block; Saw, Rabbet; Saw, Stadda; Saw, Stairbuilder's; Saw, Veneer.*

Saw, Woodcutter's (Buck Saw; Billet Saw; Firewood Saw) *Fig. 644*
A bow-type Saw with a saw blade 24–30 in long, held in a wooden frame with one side-piece extended to form the handle. It differs from the usual Bow Saw in

having a rigid blade which cannot be turned, for it is intended for cross-cutting firewood, etc. Mercer (U.S.A., 1929) writes that 'it is still used among farmers etc. who grease the blade with a piece of hog fat, kept hung on a nail in the wood shed'. (See also *Saw, Bow, Tubular*.) Variants include:

(*a*) *The British pattern* has a single stretcher with toggle stick and cord like the ordinary Bow Saw. The cheeks are often sharply curved. Both factory- and home-made examples are illustrated.

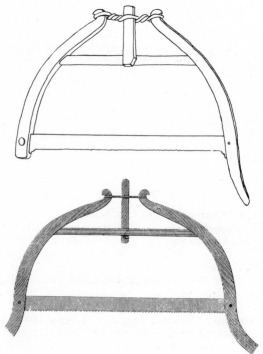

Fig. 644 (b) Woodcutter's Saws

Fig. 644 (a)

(*b*) *The American pattern* has two curved stretchers, or two diagonal struts, sometimes with the one sliding through a slot cut in the other at the inter-section, and the blade is tensioned by a metal rod and turn-buckle.

Saw Bench
A name given to a bench or table on which is mounted a Circular Saw driven by hand-wheel, treadle, or mechanical power.

Saw Box
A removable handle. See *Saw, Pit; Saw, Cross Cut.*

Saw Buck: see *Sawing Horse.*

Saw Chops: see *Saw Sharpening; Holding Devices.*

Saw Doctor and Sharpener *Fig. 645*
Pit sawyers, and country tradesmen generally, sharpened their own Saws. Although some joiners and cabinet makers do the same, the work is often given to specialist saw sharpeners who, through

doing this work every day, acquire a high degree of skill. Often known as 'saw doctors', they undertake to sharpen, set, and repair Hand Saws. (The same term is applied to the mechanics who maintain Machine Saws in working order.)

For a description of the tools used, see under *Saw Setting Tools; Saw Sharpening*. Straightening, levelling, or tensioning is done with a Saw Maker's Hammer on a steel Anvil, or on a hardwood post, the end grain at the top serving as the anvil face.

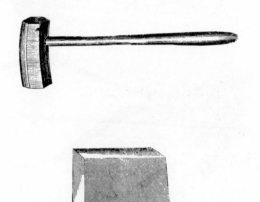

Fig. 645

Saw Flatter: see *Saw Sharpening: Files*.

Saw Gauge: see *Gauge, Saw*.

Saw Guard
A narrow strip of wood (or plastic) made to fit over the cutting edge of a saw, and sometimes held on with a leather strap. Used to protect the teeth when the saw is not in use, or when it is being carried, e.g. by travelling sawyers.

Saw Horse: see *Sawing Horse* on page 441.

Saw-Setting Tools *Fig. 647*
These tools are designed to set saw teeth, i.e. to bend them sideways to left and right alternately, so that they cut a groove a little wider than the blade itself and so reduce friction. (See *Fig. 592* under *Saw*.)

The Romans were probably the first to adopt the practice of setting the teeth of saws; their Saw Sets, which are of the gate type, are sometimes combined with a file. The tools used include:

(a) Saw-Setting Hammer
A symmetrical head, tapering to a flat cross pane on both sides of the eye. Weight 4, 6, or 8 oz. Used mainly by professional saw sharpeners for striking the teeth in order to bend them. Ordinary riveting-type Hammers are also used for setting the teeth of larger Saws.

(b) Saw-Setting Plate
The operation described in *(a)* above was done on a steel block or on a rounded bar known as a Stake.

(c) Gate-type Saw Set (Saw Wrest)
These are the commonest type and are made in a great variety of shapes, some with wooden handles and some in metal throughout. Most consist of a steel bar, 3–12 in long, with tapered edges which contain the 'gates'. These are narrow slots corresponding to the thickness of saw blades to be dealt with. In operation the appropriate gate is fitted over the tooth and the handle used as a lever to bend the tooth over in the desired direction. Variants include:

1. Hand Saw Set with a plain or slotted turnscrew on the end, used for tightening the screws in the saw handle.

2. A Hand Saw Set with a sliding Guard (or 'Gauge') which can be fixed at any point on the blade of the tool by means of a thumbscrew. This acts as a rough regulator for limiting the amount by which the tooth is bent over.

3. Heavy Saw Sets for Pit and Circular Saws have a small hole at the bottom of each gate, which is probably provided to ease the removal of metal during manufacture. The home-made example illustrated has a hammer head at one end which was probably intended for removing the tiller and box (handles) from a Pit Saw.

4. A gate-type Saw Set used on the Continent (called in French a 'Rainette') has a 'tracing end' at the foot. This acts as a Timber Scribe.

(d) Plier-type Saw Set
Various types have been developed since about 1880. They contain an adjustable 'anvil' and a triangular-faced jaw made in the form of a pair of Pliers. In a later model the handle operates a plunger which forces the tooth over.

(e) Punch or Hammer-type Saw Sets
These mechanical devices provided an easier and quicker method of setting the teeth of larger Saws, including Pit Saws. The Saw is laid on a shelf con-

taining an anvil, bevelled to allow for the required degree of set. Above is a small plunger or lever which, when struck with a hammer, bends the tooth. There is usually a rubber cushion or spring to return the plunger to its original position.

c 4

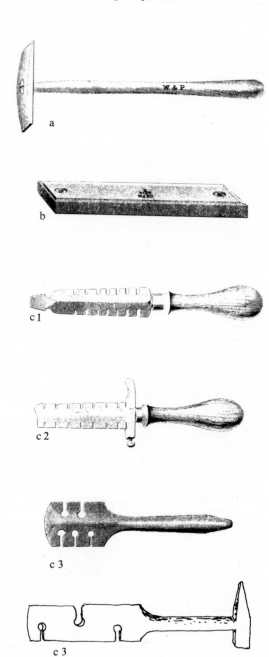

a

b

c 1

c 2

c 3

c 3

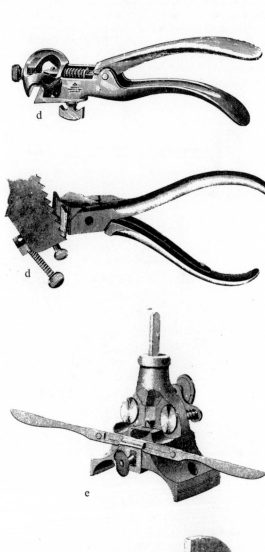

d

d

e

e

Fig. 647 Saw-Setting Tools and Appliances

Saw Sharpening: Files *Figs. 648; 649; 650*

(1) *Typical shapes Fig. 648*
The commonest shape is three sided, but other shapes were used including a round section for gulleting the bottom of the teeth, e.g. for Pit Saws. Another pattern, known as a Double-Ended Taper Saw File, tapers from the middle to both ends ('fusiform') with

(*a*) Half round. 'An old fashioned type; we used the half round side for the gullet and the flat side for the top of the teeth'.

(*b*) Round. 'Towards the end of our time with the pit saws we used a round file for the gullets.'

(*c*) Flat. Used for the tops of the teeth.

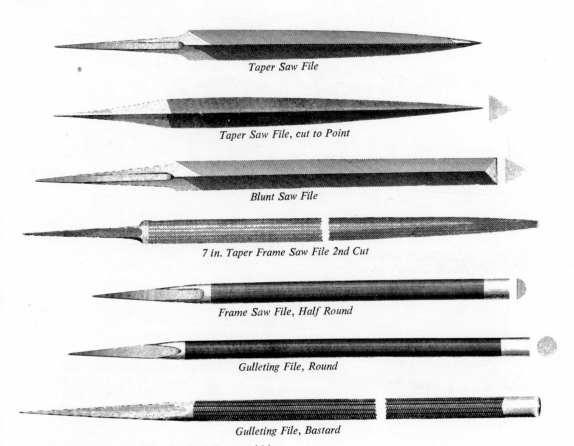

Taper Saw File

Taper Saw File, cut to Point

Blunt Saw File

7 in. Taper Frame Saw File 2nd Cut

Frame Saw File, Half Round

Gulleting File, Round

Gulleting File, Bastard

Fig. 648　Saw-Sharpening Files (Sheffield Illustrated List, 1888)

a short length in the middle and at each end left uncut. It is usually supplied with a wooden handle which has two saw kerfs at right angles into which the end of the file not in use can be inserted.

(2) *For Sharpening Pit Saws Fig. 649 overleaf*
The following Files were used by a Sussex sawyer for sharpening Pit Saws (Mr. J. Skinner, Sompting, 1949):

(*d*) Flatters. Used for dressing ('flatting') the teeth before resharpening until they are of equal height. Two home-made types are sketched – one an iron plate about 7×2 in, screwed to a wooden block. Two flat Files are held between the plate and the block, one in each angle, and the block is chamfered on each side to allow for the set of the Saw. The other consists of a flat or half-round File let into the bottom of a slot in a wooden block.

(Commercially made American tools of this type consist of a casting in two parts, connected by a hinge forming the handle. The File is held in the bottom of the groove formed when the two flanges are closed.)

(3) *For sharpening large Cross-cut and machine-driven Circular Saws etc. Fig. 650*

Files with the following names and cross sections are used for this purpose: (*a*) *Mill* (*b*) *Cant Fleam Tooth* (*c*) *Feather Edge* (*d*) *American pattern Cross Cut* (*e*) *Knife* (*f*) *Fish Back*.

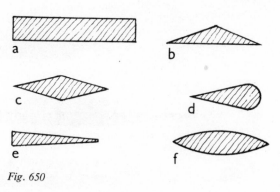

Fig. 650

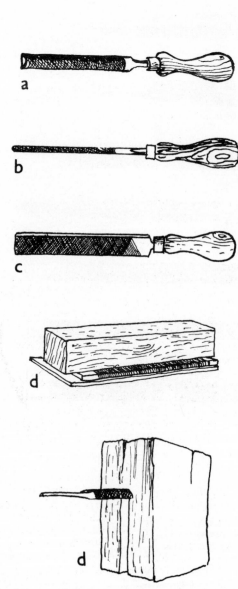

Fig. 649   Saw-Sharpening Files for Pit Saws

Saw Sharpening: Holding Devices *Fig. 651*

Devices for holding a Saw while being sharpened and set include:

(*a*) *Saw Sharpening Vice* (Saw Cramp)

An iron Vice with about 9 in jaws, often provided with a grip and release device operated by a cammed lever. It is fixed to the bench top at a convenient height for holding the Saw.

(*b*) *Saw Chops* (Saw Clamp)

A home-made wooden frame with two uprights about 3 × 2 in, designed to be held in the bench vice. The wide jaws or chops which hold the Saw are wedged into tapering notches in the ends of the uprights.

(*c*) *Saw Sharpening Horse*

A wooden bed, up to about 5 ft long, supported on splayed legs, in which are a number of mortice holes to take four or more upright blocks slotted to hold a Saw blade.

Used for holding Pit and other large saws when sharpening and setting.

Whetting Block

A term used by Moxon (London, 1677) for a block of wood, held flat on the bench, in which the web of a Bow Saw can be wedged for sharpening. The web is held in a kerf made in the block for this purpose. The term Whetting Block is also used by Hummel (U.S.A., 1965) for a Saw chop.

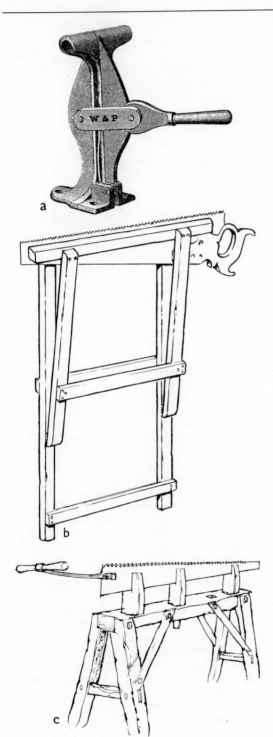

Fig. 651

Saw Teeth: see notes under *Saw; Saw, Cross-Cut; Saw, Hunter's; Saw, Pit.*

Saw Tiller: see *Saw, Pit; Saw, Cross-Cut.*

Saw Tub: see *Cooper* (2) *Furniture.*

Saw Wrest: see *Saw-Setting Tools.*

Sawing Board: see *Bench Hook.*

Sawing Horse *Fig. 652*
The following are among the most common appliances used for supporting timber while being sawn:

(a) Sawing Stool or Trestle
A pair of these trestles are to be found in almost every workshop. They consist of a stout bed or bearer, about 4 in square and 30 in or more long, supported on four splayed legs about 2 ft high. In older workshops the bed, often worn thin by saw and chisel cuts, is supported on legs made from oak or ash stakes, sometimes crossed over at the top. When sawing small pieces, the timber is held steady on the bed with one knee; larger boards etc. are laid across a pair of trestles.

(b) Saw Buck (Wood Horse)
A frame consisting of two crossed timbers, about 3 ft long, held apart by 2 ft stretchers, to form a V-shaped bed, in which the log lies while being sawn. Much used for sawing firewood.

(c) Sawing Dogs (Saw Props)
A pair of wooden bars about 7 ft long, supported at one end on splayed legs, the other resting on the ground. A series of holes in the upper part of the bars provide adjustable housing for wooden pegs. A log intended for sawing is rolled up the inclined bars to the required height and the pegs are dropped in behind it; its own weight holds it firmly enough for sawing. When the Dogs are used singly they are called Saw Props and are used for sawing lighter timber. This device is used in many woodland trades and also by chair bodgers (see *Chairmaker*).

(d) Cabinet Makers
They usually clamp their work on the bench and saw overarm as shown.

(e) Cooper's Sawing Stool
See *Saw Tub* under *Cooper's* (2) *Furniture.*

(f) Saw Box
In some workshops, including cabinet makers' shops, it is customary for each man to make a square box,

441

about 20 in high, which he uses as a chair for his tea and as a horse for sawing.

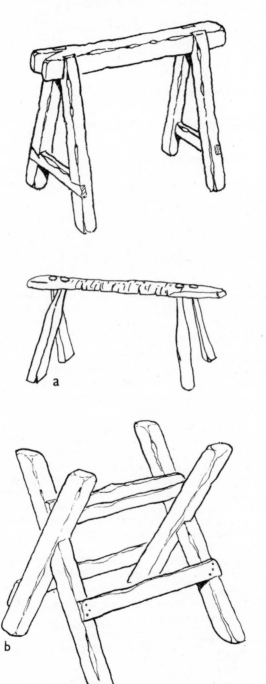

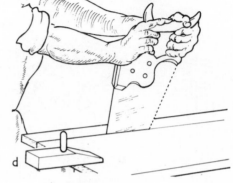

Fig. 652 Sawing Horses and Stools

Sawyer *Fig. 653*

The sawyer's chief work consists of converting logs of timber into beams and planks. Until about 50 years ago this work was done on the saw pit. The log was laid over the pit with one man standing on top of the log and the other standing in the pit below, with the saw held between them. According to W. L. Goodman, the saw pit is a relatively modern device. From evidence so far available, it seems that pit-sawing was common in England and Denmark, at least from the seventeenth century onwards, but although pits were used in other parts of Europe, for example in north-west Germany and Belgium, the normal practice on the Continent, from Roman times until the advent of machinery, was to support the log on trestles above ground. This method survives in Africa and Asia.

The pits were about 6 ft deep, 12–14 ft long, and 3–4 ft wide, a short ladder being used for getting in and out of the pit. A niche was formed in the side wall of the pit in which to place an oil pot and brush, and sometimes a candle-holder. The top of the pit was surrounded by a stout timber sill of oak or elm (called side strakes or head sill). The log was laid over the pit, one end resting on a cross bar (the transom) and the other on an octagonal bar or roller; it was then made fast by means of spiked Timber Dogs driven into the log and adjacent timbers.

Though mostly used for converting large timber, lighter work, such as the cutting of wheel felloes, cart shafts, and plough handles, was also done on the pit.

The Sawyer's Work

After felling, the log was hewn on two sides so that it would lie firmly. The knuckle-end would be hewn off the butt, unless the turn of the grain was useful for curved pieces, e.g. for wagon or ship building.

The planks were set out by 'snapping' a chalked line along the top and then plumbed down each end; the log was then turned over and lined out on the other side in the same way. To avoid waste it was important for each sawyer to follow this guide line very closely. One of their rhymes defends the sawyer who fails to do so:

'A sawyer's no robber –
What he takes from one side he gives to the other.'

The top sawyer guides the cut with the upper handle (the Tiller) while standing on the trunk. The bottom sawyer holds the lower handle (the Box) which can be removed when it is necessary to take the Saw out of the kerf.

Many sawyers travelled about the country from job to job, carrying their Saws with them. The top sawyer was the senior; he was responsible for marking out the log and sharpening the Saws. The bottom sawyer prepared the logs for the pit. Their work was arduous, as one of their proverbs suggests:

'Strip when you're cold
And live to grow old.'

As the sawing proceeded, the log was moved along the pit. First the Timber Dogs were temporarily removed and then the log could be moved along by turning the octagonal roller by means of a tommy-bar inserted into a hole at the end of the roller. (The rhymes quoted above were told to the author in 1948 by the late George Casbon, wheelwright, of Barley, Hertfordshire.)

Tools used by sawyers will be found under the following entries: *Adze, Carpenter's; Axe, Side (Hewing); Chalk Line and Reel; Dog, Timber; Lighting, Workshop; Plumb Bob (Plummet); Saw, Pit; Saw Sharpening* (various entries); *Timber-Handling Tools*.

Fig. 653

Scale

A term given to the flat pieces of wood or other material which cover the metal heel of a Knife or Saw to form a handle. See diagram under *Knife*. Also the name of a jig – see *Brush Maker*.

Scantillon (Scanteloun)

Obsolete term for Measuring Stick.

Scarfing

A method of straight jointing in which the two ends are tapered and overlapped.

Scillop

Scots term for certain half-funnel shaped Shell and Taper Bits and Augers, e.g. *Auger, Blockmaker's; Auger, Cooper's Bung Boring; Bit, Bobbin; Bit Taper*.

Scorer: see *Timber Scribe*.

443

Scorper (Scamper; Scooper; Scorp): see under *Draw Knife, Cooper's Round Shave; Wood Engraver.*

Scotch

A wedge or block to prevent an object, such as a cask or wheel, from moving.

Scotching Hatchet: see *Hatchet, Scotching.*

Scotia

A form of moulding, see *Plane, Moulding.*

Scraper (Devil; Dumb Scraper) *Fig. 654*

The term Scraper is given to a number of different tools which consist of a sharpened steel plate which is held about 30° from the vertical and is pushed along the work. This plate may be held in the bare hands or may be mounted in a wooden or metal handle or stock. It is used for the final smoothing of hardwoods, but also for scraping off paint etc. The Scrapers used by painters (*Knife, Chisel*) or by shipwrights (*Scraper, Ship's*) are ground with plain, square or bevelled edges and may be termed true Scrapers.

Scrapers used by cabinet makers and other tradesmen may more truly be considered cutting tools, for the edges are 'turned up' with a hard steel rod which produces a burr which bites into the wood and consequently takes off a shaving like a finely set Plane or Spokeshave. Much has been written on ways of sharpening Scrapers of this kind. Essentially the process consists (as shown in the diagram) of (*a*) filing the edges square and honing on the oil stone, and (*b*) subsequently turning one or both of the edges by rubbing with a rod of hard steel (or the back of a Gouge). The steel rod used for this purpose by chairmakers is called by them a 'Ticketer'. Other tools which operate by scraping (and are sometimes called by that name) are described under: *Drawing Knife, Cooper's Round Shave; Knife, Chisel; Plane, Box Scraper; Plane, Scraper; Plane, Toothing; Scratch Stock.*

Fig. 654

Scraper, Box

A name often given to Scrapers of varying design, including:

(*a*) A tool similar to a Ship Scraper with a triangular blade $2\frac{1}{2}$–$4\frac{1}{2}$ in long, used for erasing marks and brands from boxes or casks. (See *Scraper, Ship.*)

(*b*) A small metal Plane pivoted on the end of a long handle. (See *Plane, Box Scraper.*)

Scraper, Cabinet *Fig. 655*

Steel plates of different shapes, 4–6 in long and 2–3 in wide, with the cutting edge formed by 'turning' the edge of the blade with a steel rod. To avoid burnt fingers, the blades are often fitted to a wooden holder or even mounted in a form of Spokeshave or Plane. The blade was often made by the user from a piece of saw blade; a nice piece of steel which took and held a good cutting edge was much prized. Scrapers are used for the final cleaning up of most hardwoods, and especially cross-grained or curly-grained wood.

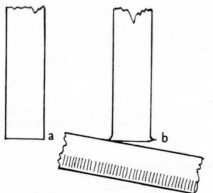

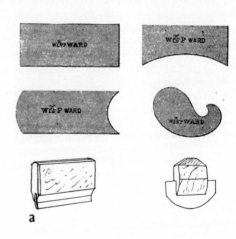

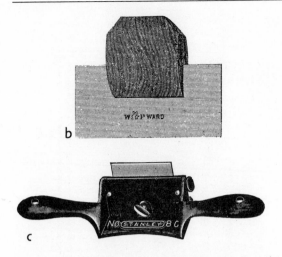

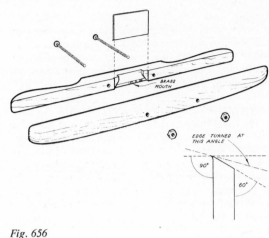

Fig. 656

Fig. 655 (*a*) Scraper plates, and some home-made holders
(*b*) A typical Cabinet Scraper
(*c*) All-metal Cabinet Scraper

Scraper, Chairmaker's (Devil) *Fig. 656*
As well as using ordinary Cabinet Scrapers, Windsor chair makers made their own Scraper, which is a lighter, more refined version of the Coopers 'Buzz' Scraper. This is a wooden stock about 11–13 in long and divided in half lengthwise with the blade mounted between. Mr. Charles Hayward writes as follows concerning this tool:

'To one who has never before handled it this tool comes as somewhat of a revelation. It takes off shavings like a fine-set plane, and with complete freedom from tearing out. Yet it is as simple a tool as one could wish for. The chairmaker may have anything up to a dozen or more of them of various shapes, some with varying degrees of roundness in their section, others round or hollow in length. It will be realised that an elaborate chair with its curves, simple and compound, presents many grain problems, and to the chairmaker it is essential to have cleaning-up tools which will not tear out the grain. The chairmaker's scraper is the perfect answer, but do not expect to buy one in a shop. The chairmaker always makes his own to suit the work he has to do. . . . The tool cuts by virtue of a turned-up edge, but whereas the ordinary cabinet scraper is filed and oil-stoned square and has all four edges turned, the chairmaker hones a bevel of about 60 degrees and turns just the one edge.'

Scraper, Cooper's Buzz (Cooper's Scraper Shave) *Fig. 657*
A steel scraper blade, 3–5 in wide, set upright in a stock about 14 in long. An old Cooper's Downright (Shave) is often adapted to hold the blade. This tool is a larger and rougher version of the Chairmaker's Scraper and the edge of the blade is 'turned' with a steel in the same way. It is used for the final cleaning down on the outside of the cask, or at any time when the grain is difficult. (A scraper blade with a convex edge, held in the hands, is sometimes used for finishing the inside of the cask.) A Continental version of a Cooper's Scraper consists of a steel bar, sharpened along the edges, with a handle at each end.

Fig. 657

Scraper, Dumb
A shipwright's term for an ordinary Cabinet Scraper.

445

Scraper, Glue
A name given to a Scraper, home made from an old plane iron, which is fixed at right angles to the end of a wooden handle. Used for scraping off superfluous glue before planing, or for scraping floors and other rough work.

Scraper, Hook (Skarsten Scraper) *Fig. 658*
Originally known as the Hook Scraper and probably first made in America, the tool has been popularised in recent times by the firm of Skarsten in Sweden. This firm's scraper has an interchangeable cutting edge, fitted to a plain or pistol-grip handle. The cutting edges may be straight, concave, or convex; or serrated for rough work, such as the removal of paint or varnish.

Fig. 658

Scraper, Millstone: see *Millstone Dresser.*

Scraper, Paint: see *Knife, Chisel.*

Scraper Sharpener: see under *Scraper.*

Scraper, Ship *Fig. 659*
Scraping tools are used by shipwrights for cleaning the deck and for removing surplus pitch after caulking. Some shipwrights make their own Scrapers from pieces of old saw blades; the edges are ground and fitted to the end of an iron handle. Others are forged out of old files. The following are some of the types in common use:

(*a*) The usual factory-made version has a triangular steel blade 4–5 in across. It is fitted to the end of an iron socket into which a wooden handle is fitted. The overall length varies from 10 to 20 in but a much longer handle is sometimes fitted.

(*b*) The Yacht or Boat Scraper resembles a Plumber's Shave Hook. It is 6–9 in long and the

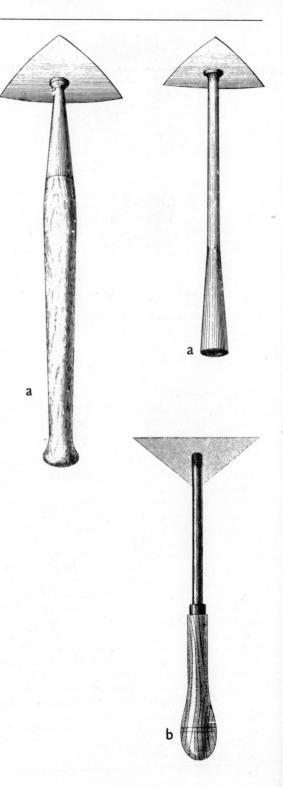

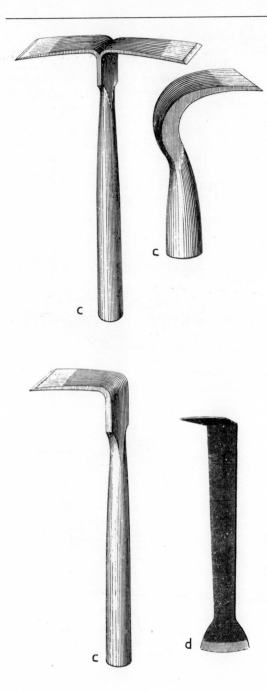

blades are heart-shaped or triangular. Another pattern has a rectangular blade with one serrated edge.

(*c*) A Hoe-shaped Scraper.

(*d*) A type of Scraper often made from an old file. One end is splayed and sharpened.

Scraper, Spokeshave

A name given to a scraper blade mounted in a spokeshave type of stock. See *Scraper, Cooper's Buzz; Scraper, Chairmaker's.*

Scraper, Veneer

An ordinary Cabinet Scraper. See *Scraper, Cabinet.*

Scratch Stock (Scratch Plane; Scratch Router)
Fig. 660

A scraping tool, designed to work simple mouldings, up to about 1 in wide, on the edge of straight or curved workpieces, or to form grooves for inlays. It is made in various forms, including:

(*a*) Two pieces of wood, screwed together, to provide both handle and fence. The cutter, made of a piece of old saw or other suitable steel plate, is filed or ground to the reverse of the profile required, and held upright between the two pieces. The edge of the cutter is filed square, and works in both directions with a scraping action. When used with a Moulding or Turning Box, the stock is shaped to provide two fences which bear on each side of the box. This ensures that the cutter is prevented from 'drifting', i.e. it is kept firmly in line with the moulding being worked. The examples illustrated were made by a London cabinet maker for his own use.

Fig. 659 Ship Scrapers

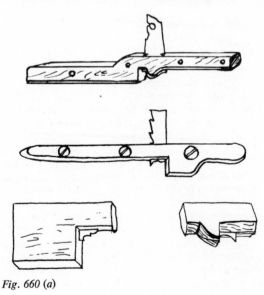

Fig. 660 (a)

447

(*b*) A Scratch Stock made from an old Marking or Cutting Gauge. A saw cut is made in the end of the stem to receive the cutter, and bolts are inserted to hold it in position.

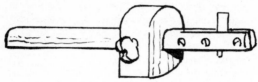

Fig. 660 (b)

(*c*) Francis Young (London, 1893) describes how to make a 'Scratch Router' for working one or two beads along the edge of a piece of wood. This is merely a block of wood in which a wood-screw is screwed down until the head stands a little proud of the surface to act as a cutter. He continues: "Ah", says a sharp and clever reader, "but how are you going to get rid of the rectangular arris left on the other side of the second bead?" Well, this might be a poser to some people, but I should manage it by clearing it off with the cutter of a little toy plane $\frac{7}{8}$ inch by $\frac{1}{4}$ inch that I wear at the end of my watch-chain, after the manner of a charm, and which you can buy for the small sum of 1s. of Mr. E. Walker, 20, Legge Street, Birmingham.'

Scratcher or Scratch Awl: see *Timber Scribe; Marking Awl and Knife*.

Screw Box and Tap for wooden screws. *Fig. 661*
A wooden block made in two pieces held together with screws, and with a cutting iron held in place under a hooked bolt. A hole is bored through both parts of the block, smooth at its entrance to act as a guide, but threaded for the rest of the way through. A V-shaped cutter is mounted at the beginning of the threaded part, ready to start cutting the previously rounded peg as soon as it enters the threaded part in the block. In the larger sizes the block is provided with two solid handles. In operation the peg to be screwed is held in a Vice (or Lathe) and the Screw Box is put over it and twisted. The cutter can be released for sharpening by parting the two halves of the block in which it is held. Screw Boxes are made to take pegs from $\frac{5}{8}$ to 3 in. in diameter. The corresponding Taps are made in iron with a square

shank for turning with a key, or they may be handled like an Auger. Like their metal-working counterpart, flats or flutes are cut in the thread to impart a cutting edge. Screw Taps imported from the Continent have a continuous thread without flats, but a hole is cut through the crown of the thread leaving a sharp profile which serves as a cutter.

Though metal screws were made at least as far back as A.D. 100, it is not known when wooden screws were first made. Tools for making wooden screws are not mentioned by Moxon (London, 1677) nor by Diderot (Paris, 1763), though wooden screws were being made at the time.

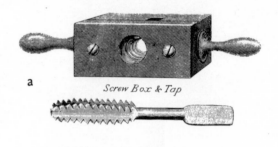

a

Screw Box & Tap

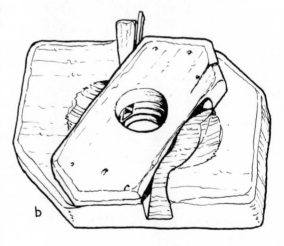

b

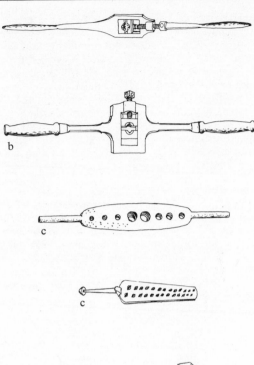

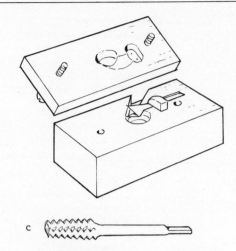

c

Fig. 661 Screw Boxes
    (*a*) Factory-made (for larger screws)
    (*b*) Home-made
    (*c*) Factory-made (for smaller screws)

Screw Dies and Taps for metal screws *Fig. 662*
These metal tools for forming threads on nuts and
bolts are used in many trades including wagon
building.

    (*a*) Screw Dies made by the local smith – a common
practice at least until the end of the nineteenth
century. Other examples are illustrated under *Metal
Working Tools*. They form threads by a pressing or
rolling action, rather than by cutting. Length c.
36 in.

    (*b*) Factory-made Die Stocks of the nineteenth and
early twentieth centuries.

    (*c*) 'Screw Plates' used for cutting threads on
smaller bolts and rods.

    (*d*) Examples of smith-made Screw Taps.

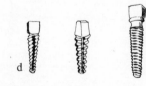

Fig. 662

Screw Remover: see *Bit, Annular.*

Screwdriver: (*1*) *General* (Turnscrew) *Figs. 663* and
664
Note: Although nowadays the generally accepted
name is Screwdriver, it appears from the trade catalogues and other literature that, at least in the Mid-
lands and the North of England, the usual name was
Turnscrew. The earliest reference we know of to the
term Screwdriver comes from Nicholson's *Mechanical Exercises* (London, 1812).

    The tool consists of a steel blade ground at one
end to a flat edge for fitting into the slot on the head
of a screw in order to turn it. The length of the blade
varies from 2 to 24 in, the width of the edge from
about $\frac{1}{16}$ to 1 in, depending on the use of the tool.
In all but the smaller sizes, the flat heel of the blade,

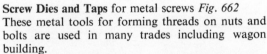

a

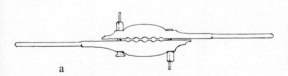

just above the tang, is fitted into a slot in the ferrule. As Young remarks in *Every Man His Own Mechanic* (London, 1893) '. . . additional firmness is thus imparted to the tool, and the blade is prevented from turning in the handle, as bradawls will often turn, much to the vexation of the operator.'

Though plates of sporting guns were fastened with screws early in the seventeenth century, wood screws were not extensively used by carpenters until the mid-eighteenth century, and consequently the Screwdriver does not appear to have been commonly employed until after that time.

Smith-Made Screwdrivers Fig. 663
Those made in the early nineteenth century often have long flat blades (16–24 in) with two or more 'waists'; and occasionally have a hole in the base of the blade for a tommy-bar to help in turning the tool. The very short examples illustrated may have been used for gun adjustments rather than for turning wood screws. The longest is dated 1868 and was used by a Manchester millwright; it measures 28 in overall.

Factory-Made Screwdrivers Fig. 664
The following variations in the type of blade are illustrated:

(*a*) *Cabinet Pattern.* A round blade, 3–12 in long, squared at the tang end. According to Young (London, 1893) this type was introduced *c.* 1880.

(*b*) Another *Cabinet* type.

(*c*) *London Pattern.* A flat blade, 3–18 in long, and waisted.

(*d*) *Scotch Pattern.* A flat, straight-sided tapering blade 3–15 in long.

(*e*) *Spindle-bladed Pattern.* Round, stepped blade, 6–18 in long.

(*f*) *'Gentleman's' or 'Ladies'' Pattern.* A name given to all kinds of small screwdrivers, often made to a high finish with finely polished handles.

(*g*) *Scale-handled.* A recent type with a strong blade made solid with the handle which is covered by hardwood scales on each side. Intended for rough work, it can be used like a cold chisel.

'Stubby' types. See 'short' and 'gun' Screwdrivers under *Screwdriver (2) Special Types.*

Brace-driven Screwdrivers. See *Bit, Screwdriver.*

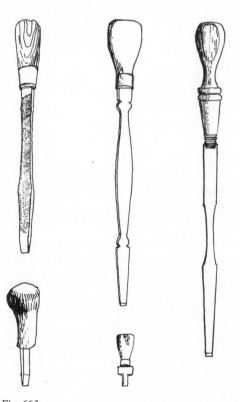

Fig. 663

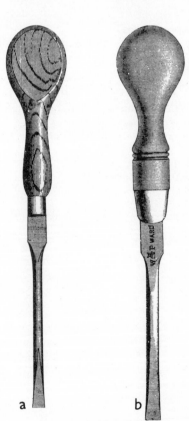

a   b

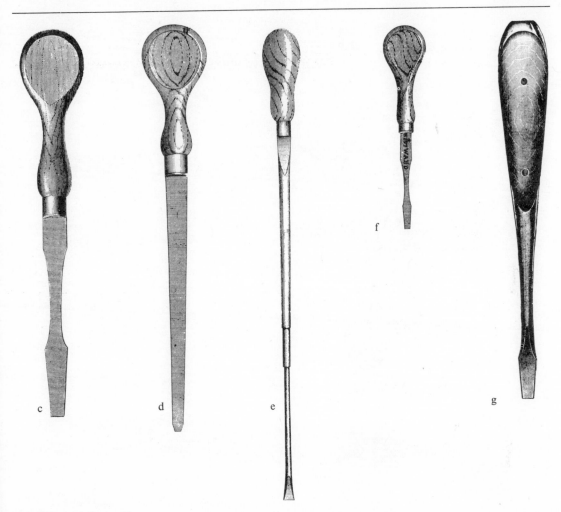

Fig. 664 Screwdrivers (1)

Screwdriver: (2) *Special Fig. 665 overleaf*
Screwdrivers of unusual design are often found in woodworking shops. It is quite common for a tradesman to use one which according to the manufacturer's list belongs to a quite different trade. We have included a number of these special types to facilitate identification.

(*a*) *Cranked Screwdrivers* (Offset, Round-the-Corner, or Right-Angular Screwdriver). A steel bar, about 5 in long, with the ends bent at right angles and ground to the shape of a screwdriver blade. Used to turn screws which are inaccessible to an ordinary Screwdriver, e.g. for the catch-plate of drawer locks.

(*b*) *Forked Screwdrivers*. A forked blade for screwing up the slotted screws in saw handles, skates, firearms, etc.

(*c*) *Short Screwdrivers*. Short, stubby screwdrivers, 1–3 in long and of London or Cabinet pattern. They are sold under the following names: Short Bench Screwdrivers; Plane-Iron Screwdrivers (for removing cap irons); Motor Turnscrews.

(*d*) *Magazine Screwdriver* (Pocket Screwdriver). From innumerable patterns we illustrate a relatively simple model with four interchangeable blades. When closed, the tool is 3 in long.

(*e*) *Pianoforte Maker's Screwdriver* (also called an Electrician's Screwdriver). A name given to a slender Screwdriver with a round blade up to 10 in long.

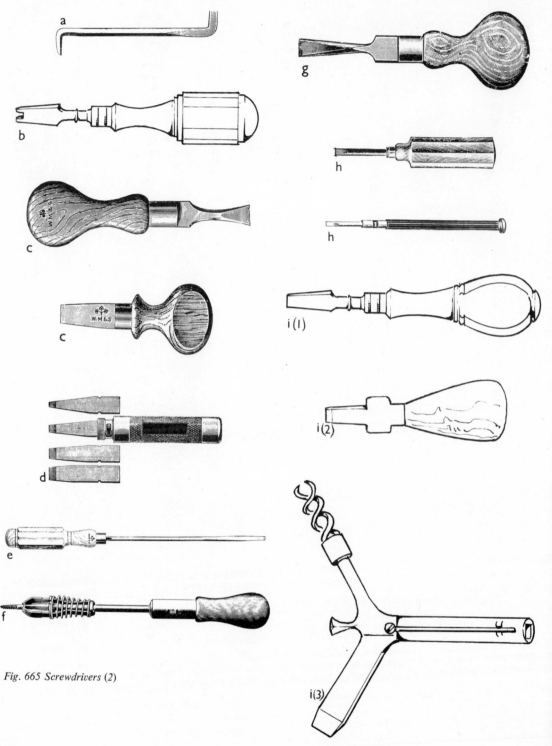

Fig. 665 Screwdrivers (2)

(*f*) *Screw-holder Screwdrivers.* The specimen illustrated is one of many attempts to provide some means of holding a screw on the end of the screwdriver without using one's hands. (The same operation is sometimes performed with the help of a little butter.)

(*g*) *Undertaker's Screwdriver* (Coffin Screwdriver). A short blade, 2–2½ in long, usually with a flat oval handle. (See *Coffin Maker.*)

(*h*) *Model-maker's Screwdrivers.* Small Screwdrivers 2–3 in long, often with a fluted metal handle, and with interchangeable blades. Also used by watch makers, jewellers, opticians, and instrument makers.

(*i*) *Gun, Sportsman's, and Military Screwdrivers* (or Turnscrews). Names given to various small Screwdrivers, including those listed under (*c*) above. The examples shown are occasionally seen in the workshops.

    1. Gunmaker's: in the nineteenth century these were rather elaborately made – as that sketched here from the catalogue of Wynn Timmins (Birmingham, 1892).

    2. Sportsman's: a short blade of cruciform shape with two blades at right angles to the main blade. Used, presumably, for adjusting firearms.

    3. Military: with three arms; two with turnscrew blades and the third with a worm. Another pattern has a small hammer head or a square box spanner in place of the turnscrew blades. The worm was used for extracting wadding etc. after a misfire.

Billiard Table Screwdriver: see *Bit, Screwdriver.*

Screwdriver, Phillips *Fig. 666*

A recently developed Screwdriver with a round shank and a cross-point. It is designed to fit a wood screw with a new type; the usual slot in the head is replaced by a cruciform depression which gives a more positive grip.

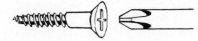

Fig. 666

Screwdriver, Ratchet *Fig. 667*

Introduced from the U.S.A. during the later years of the nineteenth century, the Ratchet Screwdriver gives the operator the enormous advantage of driving

a screw with one hand. Instead of having to shift the position of the hand on the tool as each turn is made, one merely twists the handle back and forth with a rocking motion. It is usually made with a round blade, 2–12 in long, with the ratchet mechanism concealed in the ferrule or handle. By moving a small slide, ring, or knob, the tool can be made to drive clockwise or anti-clockwise or remain fixed. A knurled ring is sometimes added for making the first few turns of the blade with the finger and thumb, the handle remaining still.

Typical variants include:

(*a*) *Gay's Ratchet Screwdriver* (Double-Action Ratchet Screwdriver)

An early form patented in the U.S.A. in December 1878. The blade is 4–12 in long, with the ratchet mechanism concealed in the wooden handle.

(*b*) *'Yankee' Ratchet Screwdriver*

A blade 2–12 in long, with a ratchet mechanism concealed in the ferrule. Smaller sizes may have a knurled ring for turning with finger and thumb to start the screw.

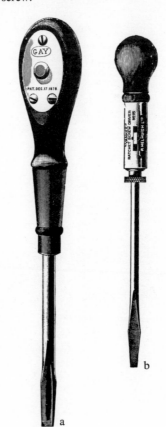

Fig. 667   a   b

Screwdriver, Spiral Ratchet (Double Spiral Screwdriver; In-and-Out Screwdriver; Pump Screwdriver; Spiral Screwdriver; Yankee Screwdriver: see note under *Yankee*.) *Fig. 668*

This modern tool is made like the Double Spiral Drill (see under *Drill, Archimedean*). Working on this principle, screws can be driven in or out by simply pushing the handle downwards. Developed in the U.S.A. during the later years of the last century, the same system has been applied to a whole group of tools for drilling, driving screws, driving nuts, and even for tapping screw threads.

The commonly used pattern has a double spiral cut on the stem and a reversing device in the handle. Some are fitted with a spring which causes the handle to come back of itself on its idle return stroke, ready for the next push. The mechanism can be locked and the tool used as an ordinary Screwdriver. They are made in various sizes, up to 26 in long, and are provided with interchangeable blades of different widths. An earlier pattern made by Goodall-Pratt in the U.S.A. had a separate spiral for right-hand and left-hand cut on different parts of the same stem.

Fig. 668

454

Screw Nail
Scots term for a wood screw; also applied to a spiral roofing nail.

Screw Plate: see *Screw Dies and Taps.*

Scriber or Scribe: see *Marking Awl; Timber Scribe.*

Scribing or Coping
The fitting of one piece with an irregular surface on to another; for instance, in the case of moulded workpieces, where one piece is cut out to fit over the profiled shape of the other. Also known as 'coping', hence the name 'Coping Saw' for the special Saw used for this purpose.

Skirting boards, picture rails, sash rails, etc. are scribed at internal angles (rather than mitred) to prevent the joint from showing a gap as the wood shrinks. (See *Window Making*.)

Scrieve, Scrive, or Scriving Knife: see *Timber Scribe.*

Scutch (Bricklayer's Scutch)
A steel strip, or a hammer head with chisel-shaped panes, mounted on a wooden handle. Used for trimming bricks. (cf. *Hatchet, Scotching*.)

Serving Board (Server): see Serving Mallet under *Sailmaker;* Truss Hoop under *Cooper (3) Hooping Tools.*

Set
The bending sideways of Saw teeth to provide clearance. See diagram under *Saw;* also *Saw-Setting Tools.*

Set (Nail Set): see *Punch, Nail.*

Set Square: see *Square, Set.*

Set Stick: see Spoke Set under *Wheelwright's Equipment (Spoke Tools).*

Setting Pin (or **Brake**): see *Handle Maker.*

Setting Plate: see *Saw-Setting Tools.*

Shackle Punch Spike, or Hammer: see *Punch, Shackle; Hammer, Ship's Maul;* Marline Spike under *Sailmaker.*

Shaft Rounder: see *Plane, Rounder.*

Shake
A term used in woodworking trades for a crack or split.

Shaper: see *Surform Tools.*

Sharpening Tools: see under *Grinding Appliances; Grindstone; Oilstone; Saw Sharpening: Files.*

Shave
This term is often applied to tools as different as a Drawing Knife ('Draw Shave') and a Basket Maker's Willow Shave. But, for convenience, we have confined this entry to tools with a cutting iron mounted in a wooden stock which is usually shaped at the ends to form handles, with the exception of the Scrapers made in the form of Shaves – see *Scrapers.*

To distinguish between the various methods of mounting the cutting iron, we have used the following terms:

(*a*) 'Spokeshave type'. These have the iron bedded horizontally like a Spokeshave, e.g. *Shave, Chairmaker's; Shave, Cooper's* (*Pail*).

(*b*) 'Plane type'. These are a cross between Plane and Spokeshave and have the cutting iron wedged and bedded like a Plane, e.g. *Shave, Cooper's* (Swift and Downright); *Shave, Jarvis; Shave, Nelson.*

(*c*) 'Router type', e.g. *Router, Coachbuilder's.* These have a superficial resemblance to Spokeshaves. The difference is in the size and shape of the iron. The cutting edge of the Spokeshave iron is wide and it is held horizontally; Router irons are bedded at a steep angle and are usually very narrow and profiled for cutting beads or narrow mouldings.

Note on Nomenclature
The following are some of the names of tools belonging or related to the Shaves:

Under *Shave, Cooper's:* Barrel; Bent; Bottle; Bucket; Cross; Heading; Outside; Pail; Straight; Tub Shave; also Downwright; Plucker; Swift.

Under *Shave, Chairmaker's:* Bottoming Iron; Smoker Back Hollow Knife; Travisher; Devil Shave; Double Iron Shave.

See also *Scraper, Cooper's Buzz; Scraper, Chairmaker's* for Scrapers sometimes classed as Shaves.

Shave, Basket: see *Basketmaker.*

Shave, Bottle: see Inside Shave under *Shave, Cooper's.*

Shave, Bucket: see Pail Shave under *Shave, Cooper's.*

Shave, Buzz: see *Scraper, Cooper's Buzz.*

Shave, Chairmaker's (See below for alternative names.) *Fig. 669*
A wide variety of Shaves of the 'spokeshave' type are used by Windsor chair makers. The stocks were usually made by the chair maker himself, and the irons by a local smith. Those with straight irons have a cutting edge up to 4 in long with the usual tapered, square tangs, and they are sometimes provided with a double or cap iron (like a Plane), held in place by two bolts. This cap iron is provided to give the Shave the properties of a Plane – a tool rarely used by Windsor chair makers. The stock is sometimes made with an open back, like the ordinary wooden Spokeshave, or with a closed back ('boxed in') like some of the larger Spokeshaves used by coopers. The examples illustrated came from a Windsor chair maker's shop near High Wycombe. There are two main types:

Straight Shaves, including:
(*a*) An example with a cap iron – sometimes called a 'Double Iron Shave'. A shave of this type is called a 'Travisher' by Hennel (London, 1947).

(*a*1) A typical example of the Chairmaker's straight shave.

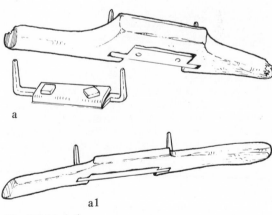

a

a1

Fig. 669 (a)–(a1)

Curved Shaves
Shaves with curved cutting irons were made in a great variety of shapes; a framer might have from thirty to forty of them in his kit. The term 'Bottoming

Iron' was used for a curved Shave for taking out the Adze marks on Windsor seats.

(*b*1) Smoker-Back Hollow Knife. A name used by Hennel (London, 1947) for a sharply curved Shave of spokeshave type, used presumably for hollowing the back splats of the heavy Windsor armchairs known as 'smoker backs'.

(*b*2) An ingenious and good-looking version of the curved Shave is illustrated in the Shelburne Museum pamphlet (U.S.A., 1957). It resembles the Shovel-Maker's Shave, except that it is grasped by a handle mounted above the stock, instead of at both sides.

(b1)

(b2)

Fig. 669 (b1)–(b2)

Shave, Chinese *Fig. 670*

The example illustrated, which is typical of a type found in the Far East, is made in a hardwood and measures $14\frac{1}{2}$ in overall. The iron, $\frac{3}{4}$ in across, is tanged like Spokeshaves in the West; but instead of lying in a recess made in the sole, it rests flat upon it.

Fig. 670

Shave, Coachbuilder's: see list under *Shave, Wheelwright's;* also *Router, Coachbuilder's.*

Shave, Cooper's *Fig. 671*

The term Shave is used by coopers to describe several different cutting and scraping tools used for the final smoothing and cleaning down of the inside and outside of the casks and also the heads. They are mostly of the 'plane type' though the Pail Shaves are made like large Spokeshaves. (See general note under *Shaves*.) Many of the 'plane-type' shaves are plated to resist the tendency for the wedge to split the stock apart. More recently some have been made in metal throughout.

Coopers' Shaves measure about 10–14 in overall. They are frequently home made by the cooper for his own use. The following descriptions apply to typical examples which may be either factory or home made.

(*a*) *Downright* (Barrel Jarvis; Barrel Shave: Scots: Plucker; Outside Shave)

A 'Plane type' Shave with a 2 in straight iron. The side handles are bent forward and the sole is sometimes concave lengthways. Metal Downrights have a slotted cutting iron fixed to the bed with a wing nut or lever cap.

Used for smoothing the outside of the staves after assembly. In good-class work, and also when the grain is difficult, the Downright is followed by the Buzz (see *Scraper, Cooper's Buzz*).

When Downrights were provided with an iron ring on the front of the stock, they were used hanging on a chain from the ceiling in order to shave the outside of casks being revolved on a Lathe.

1. A factory-made example.
2. A home-made example from Ireland with a concave sole.
3. Downright with iron ring for suspension.

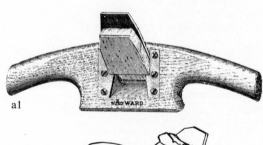

a1

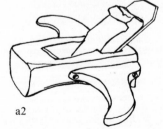

a2

b2

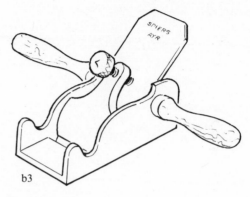

b3

Fig. 671 (b1)–(b3)

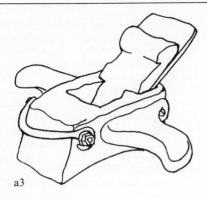

a3

Fig. 671 (a1)–(a3)

(b) *Heading Swift* (Heading Shave; Scots: Plucker or Heading Plucker)
A 'Plane type' Shave, resembling the Downright but larger, and often having a heavy square-shaped stock. The side handles are sometimes turned slightly upwards to prevent the hands from being grazed. The iron is 2½–5 in across, usually straight but slightly convex for cross-grain use. Used for smoothing the heads which, for this purpose, are held on the Heading Board. Planing across the grain is quicker, but in most cooperages this is only permitted for the undersides of the head because of the rougher finish.
  1. A factory-made example in wood.
  2. A home-made example in wood.
  3. A factory-made example in metal, made by Spiers of Ayr (1909) and called a 'Cooper's Plane or Plucker'.

(c) *Inside Shave* (Tub Shave; a very small version is known as a Bottle Shave.) Illustrated overleaf.
A 'Plane type' Shave similar to the Downright, but the sole is convex and the short handles are upturned to avoid hurting the fingers when shaving the inside of the cask. The cutting iron is convex and made with different curvatures according to the size of the cask.
  Used for cleaning the inside of a cask, for instance, if it becomes foul. (Such casks are known as 'stinkers'.) In operation the Shave is pushed forward with the thumbs placed at the back; this eventually produces two depressions which are often visible on the back of the Shave after it has been used for any length of time. An alternative to this tool, preferred by some coopers, is the Stoup Plane; and, when the grain is awkward, a Round Shave is preferred (described under *Drawknife, Cooper's Round Shave*).
  1. A factory-made example.
  2. A home-made example.
  3. A continental example from Italy known as a *Bodda*. Made in the form of a small curved Drawing Knife, it is strapped to the palm of the hand. (cf. *Shave, Chairmaker's. Fig. 669 (b2)*.

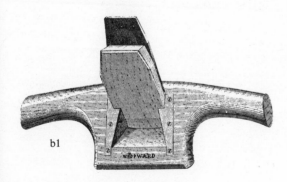

b1

457

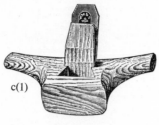

c(1)

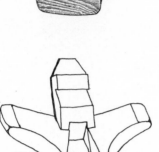

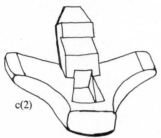

c(2)

from 4 to 5 in long and sometimes double, i.e. 'capped'. The stock is made with either an open or closed back; in the 'closed back', the stock behind the iron is not cut away, but solid for the whole depth, the shaving emerging through the mouth, as in a Plane. Pail Shaves are used for finishing pails and other 'white' coopered vessels. When smoothing the inside of a pail, the bottom is removed and the Shave introduced lengthwise to cut across the grain; it is held by placing one hand through each end of the pail.

1. Straight Shave. With 4–5 in iron for smoothing the outside of pails, etc. normally along the grain.

2. Cross Shave. With 1½–6 in straight iron. Similar to (1) but used for smoothing the inside of the pail which is done across the grain. The raised face of the stock gives added clearance for the hands.

3. Bent Shave. As above, but with a curved stock, upturned handles, and a 4–5 in convex iron. Used for smoothing the inside of pails when working along the grain.

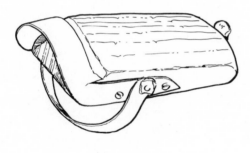

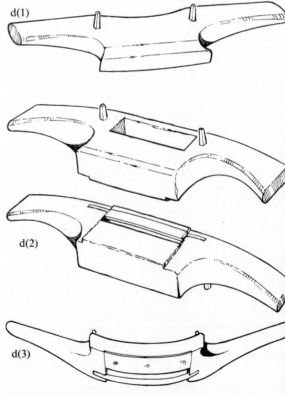

d(1)

d(2)

d(3)

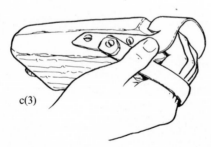

c(3)

Fig. 671 (c1)–(c3)

Fig. 671 (d1)–(d3)

(*d*) *Pail Shave* (Bucket Shave; Cooper's Spokeshave). These are all 'spokeshave type' tools, of heavy make, up to 18 in overall, and with strong cutting irons

(*e*) *Scrapers*. Some of these are referred to as Shaves. See under *Scraper, Cooper's Buzz; Drawknife, Cooper's Round*.

Shave, Handrail (Circular Router) *Fig. 672*
A 'spokeshave type' of shave with an adjustable fence and an unusual cutting iron. This is set horizontally, as in a Spokeshave, but is curved to the profile of the moulding required – often a gentle ogee – and is screwed into a recess on the face of the stock. The Sheffield Illustrated *List* of 1910 shows a tool to perform a similar task; it is called an 'Improved Sash Router'. It is believed that this tool is used to work a mould on circular work, e.g. for bow-windows or curved handrails.

Fig. 672

Shave Hook *Fig. 673*
This is a plumber's tool for cleaning the surface of lead pipes, but the name is sometimes given to a curved knife used by chair makers and others (see *Knife, Hooked*).

Fig. 673

Shave, Hop: a name given by the Curtis Museum (Alton, 1946) to Drawknives used for smoothing hop poles.

Shave, Jarvis *Fig. 674*
A heavy Shave with a concave sole about 12 in long overall, with two rounded handles on either side. The double iron, 2–2¼ in wide, is bedded and wedged like that of a Plane. The top of the stock is sometimes strapped to prevent the short grain of the shoulders from splitting, and the sole is usually plated to resist wear. Used by wheelwrights and others for rounding spokes, poles, etc.

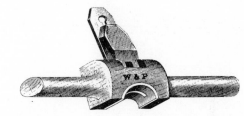

Fig. 674

Shave, Mast: see *Drawing Knife, Mast Maker's*.

Shave, Nelson *Fig. 675*
A Shave similar to the Jarvis Shave but with a flat sole. Used by wheelwrights and others for trimming and chamfering.

Fig. 675

Shave, Pail: see Pail Shave under *Shave, Cooper's*.

Shave Peg: see *Wheelwright's Equipment* (2) *Spoke Tools*.

Shave, Rounding: see *Drawing Knife, Cooper's Round Shave;* also list under *Rounding and Hollowing Tools*.

Shave, Shovel Maker's (Spade Maker's Shave) *Fig. 676*
A heavy wooden stock of spokeshave type, but only about 8 in across with very short stub handles. The

459

sole is convex, and the curved iron is fixed by the tangs like the iron of a Spokeshave.

This tool is used for hollowing wooden shovel blades, sometimes after previous excavation by an adze. A similar tool is also used for hollowing 'dairymaid' yokes. See *Spade and Shovel Maker; Shave, Chairmaker's.*

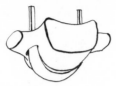

Fig. 676

Shave – Spokeshave (Wood) *Fig. 677* (See also note under *Shave.*) *Opposite.*

The wooden version consists of a straight stock, usually of beech or boxwood, from 4 to 16 in overall. The ends are shaped and cut away underneath which gives the oval handles an upturned or 'winged' appearance. The middle section has a triangular notch cut out at the back to about half the width of the sole of the stock to take the characteristic iron, a wedge-shaped steel cutter, $1\frac{1}{2}$–5 in long and $\frac{1}{4}$–1 in wide. This is mounted horizontally, with two tapered square tangs at the ends turned up at right angles, and driven into the corresponding tapered holes in the stock. The wooden sole in front of the cutting edge is usually rounded for irons up to 2 in wide, and flat or square for larger sizes. Flat-fronted Spokeshaves are sometimes plated with brass or bone to resist wear. Used in many trades for cleaning up circular work.

Historical note. For such a common and versatile tool, the Spokeshave has a curiously obscure history. No known medieval or Renaissance illustration shows the tool. Neither Moxon (London, 1677) nor Diderot (Paris, 1763) mentions it, and the earliest representation so far known to us is the wooden Spokeshave included with coopers' tools in Smith's *Key* (Sheffield, 1816). This is practically identical with the tool in use today. Metal Spokeshaves were introduced about 1860–70 – see next entry.

According to W. L. Goodman's work on the sixteenth-century apprentice indentures (1972), coopers in Norwich and Bristol had spokeshaves' – a tool normally regarded as a wheelwright's tool. He goes on to discuss how the 'spoke' came to be added to the word 'shave' and suggests that it may have

done so as the result of an error: 'possibly the solution has been hidden in the *New English Dictionary* all the time. The reference is "1510. Stanbridge, Vocabula (W.deW.) Cj. *Radula*, a spokeshaue or a playne". But *radula* is perfectly good Latin for a handled scraper. Very likely the scholar who was translating the book either did not know this or it had slipped his memory for the moment and, searching his notes for a similar word, happened on *radius*, which in one sense is the spoke of a wheel.'

Variants include:

(*a*) *Ordinary Pattern* with Iron $1\frac{1}{2}$–5 in wide.

(*b*) *Miniature Spokeshaves* with $1\frac{1}{2}$ in iron. Used for very fine work.

(*c*) *Screwed and Plated.* The tangs of the cutter are fitted with nuts for adjustment, and part of the sole is plated to resist wear.

(*d*) *Single Handed.* The stock has only one handle. Used for working in otherwise inaccessible places.

(*e*) *Wheelwright's Spokeshave.* Similar to (*a*) above and made in beech, but about 14 in long with iron $2\frac{1}{2}$–$4\frac{1}{2}$ in wide. [Not illustrated]

(*f*) *Saddler's Spokeshave.* Of ordinary design, with 2–$2\frac{1}{2}$ in iron and usually screwed and plated as (*c*) above. Used for trimming the rough edges of thick leather harness. [Not illustrated]

(*g*) *Radius Spokeshave.* Sharply bent with 'winged' handles.

(*h*) *Spokeshave Irons.* Typical examples. The double irons are often seen on Chairmaker's Shaves.

Shave – Spokeshave (Metal) *Fig. 678 overleaf*

The stock of the metal Spokeshave is a casting of iron or brass (sometimes somewhat clumsily decorated) and shaped on the lines of its wooden prototype. The earliest illustration known to us appears in Goldenberg's catalogue (Alsace, 1875) under the caption 'Wheeler's iron spoke shave'. In this and other early examples the handles were straight and flat, but later they were raised in a smooth curve, sometimes perforated for lightness. The ordinary length is 10 in overall, with a $2\frac{1}{8}$ in cutter. This is in effect a small plane iron, with a straight or concave cutting edge bedded with the bevel underneath in the middle of the stock. Later models are often provided with a cap iron secured by a small round-head screw. In some patterns the mouth is adjustable for depth of cut; in others the cutter itself is adjustable. The face may be flat, round, concave, or convex.

A number of variants are illustrated.

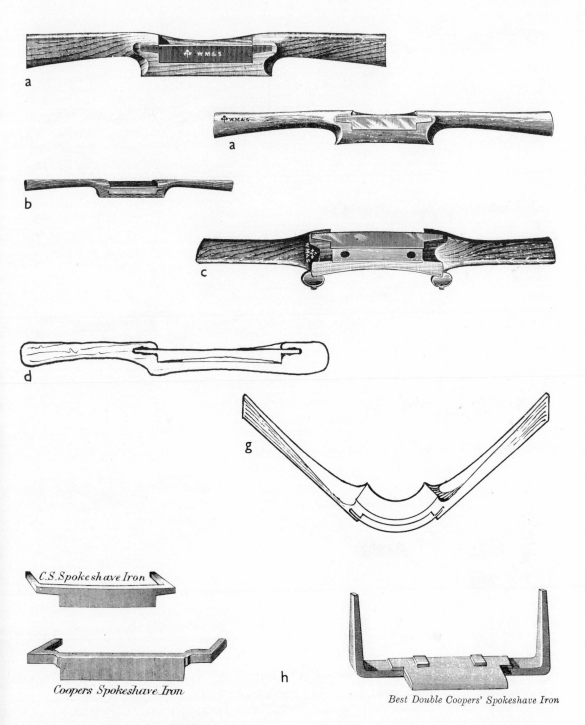

Fig. 677   Wooden Spokeshave: Variants

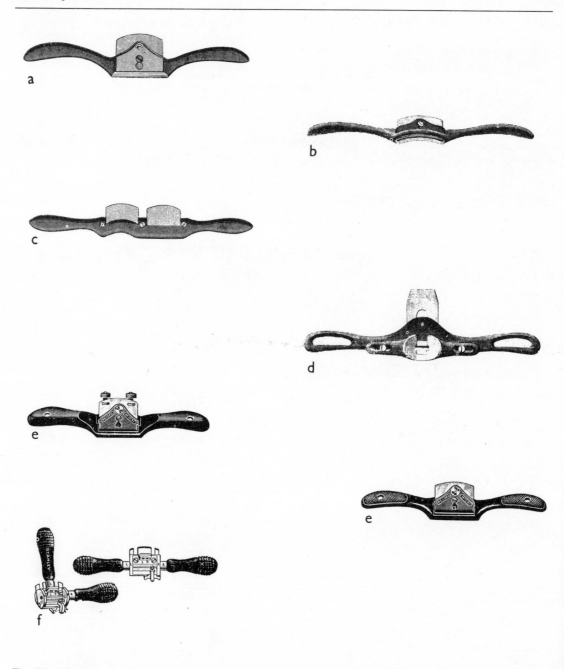

Fig. 678   Metal Spokeshave: Variants
  (*a*) With a straight iron
  (*b*) With a hollow iron
  (*c*) With combined hollow and round iron
  (*d*) With fences for chamfering ('Chamfering Shave')
  (*e*) Modern patterns
  (*f*) A modern Spokeshave with detachable handles, and a fence for rebating

462

Shave – Spokeshave (Circular) *Fig. 679*
An unconventional metal Spokeshave listed from the end of the nineteenth century until about 1910 by Millers Falls Company (U.S.A.) and others. It consists of a casting of circular section on to which a semi-tubular cutter is clamped with two screws. It has two wooden handles, either of which can be unscrewed to enable the tool to be used in cramped places. This Spokeshave was claimed to be able to work in a smaller circle than any other Shave, but tradesmen appear to have had difficulty in using it, and it gradually disappeared from the lists.

Fig. 679

Shave – Spokeshave (Four-Faced) *Fig. 680*
A flat Spokeshave with the iron bedded on a metal casting which includes a throat regulator presenting four alternative faces to the cutter: one flat, two oval or convex, and one concave. This may be further adjusted to give a wide or narrow throat. Either wooden handle can be unscrewed to give a single-handed tool for working in confined spaces. A general-purpose tool for circular work.

Fig. 680

Shave, Tub: see Inside Shave under *Shave, Cooper's.*

Shave, Wheelwright's: see under *Drawing Knife, Wheelwright's; Shave – Spokeshave* (Wood and Metal); *Shave, Jarvis; Shave, Nelson.*

Shaving Horse (Mare) *Fig. 681*
A low bench supported on splayed legs, on which the operator sits astride. There is a pedal-operated jaw which bears on a sloping platform to hold the work, thus leaving both hands free for shaping the workpiece, usually with a Drawknife.

Two of several kinds are illustrated below, and another under *Woodland Worker.*

(*a*) With a rocking frame, sometimes metal-toothed at the top.

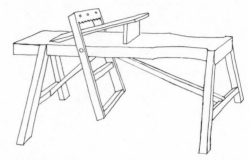

Fig. 681 (a)

(*b*) With a notched bar passing through the platform itself, the height and width of the vice opening being adjustable by pivoting the bar in one of a series of holes with a peg. This type is more often seen in Ireland and on the Continent of Europe; and it is illustrated by Mercer (U.S.A., 1929).

Shaving horses are used in many trades, perhaps most commonly by coopers, chairmakers, and woodland workers. The cooper normally uses it for smaller staves, while larger ones are shaped on the block. Chairmakers use it for the rough shaping of legs and stretchers before turning. Mr. L. J. Mayes (High Wycombe, 1948) informed us that it was also 'regularly employed by boys who used it to make wedges for the chairmakers. These were made from the 'slab', that is, the first cut from the tree made by the pit saw or the band mill, split out with beetle and splitting-out hatchet, and held in the shave-horse to be finished with the draw-shave. The old chairmakers would use no wedges other than those so made, reft with the grain and so free from the habit of snapping off before fully driven home – a vice associated with sawn wedges, for the saw took no account of the run of the grain. In *c.* 1910 a boy was paid one penny per gross for wedge making, and a skilled lad could make almost a gross an hour.'

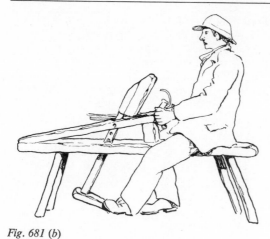

Fig. 681 (b)

Shaving Knife: see *Drawing Knife.*

Shears, Basket: see *Basket Maker* (*Osier*).

Shears, Brush Maker's *Fig. 682*
Large Shears used by brush makers for cutting and trimming brush bristles. See *Brush Maker.*

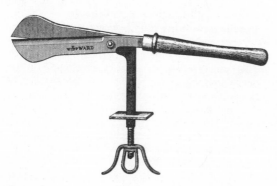

Fig. 682

Shears, Metal Working
Examples commonly found in woodworking shops are illustrated under *Metal-Working Tools.*

464

Shears, Tailor's *Fig. 683*
Often found in woodworking shops, perhaps because they were useful for cutting material used for coach trimming, van covers, etc.

Fig. 683

Sheave
The wheel in a pulley block. See *Ship's Block Maker.*

Shepherd's Bar: see *Hurdle Maker.*

Shift Stocks: see *Square, Bevel.*

Shingle Maker
A shingle is a thin piece of wood, about 4 in × 14 in, which is used as a roof tile. They are rectangular, but their lower ends are usually left thicker than the upper parts.

Shingles were commonly used in the Middle Ages for manor houses and churches. For ordinary homes thatch, and later slate or tiles, were cheaper. In America, however, shingling of roofs was still widely practised up to recent times. Church spires in England are occasionally re-shingled, particularly in the South-East.

Shingles are made of cleft oak, but Scotch pine is sometimes used and also red cedar, which is reputed to have an almost indefinite life. In making shingles, logs are cleft into halves with Beetle and Wedge. After the inner heart and outer sapwood are trimmed off, one man holds the Froe with its blade on the radius running to the centre of the tree, while another strikes it hard with a Mallet, so splitting off a shingle. The shingles are held on a Shaving Horse and trimmed to shape with a Drawing Knife. Holes are bored in each for pegs or copper nails to secure them to the roof battens.

Mercer (U.S.A., 1929) describes two home-made implements used in this trade. One is called a Shingle Punch and is used for punching holes through shingles instead of using a gimlet. The punch is mounted on a lever, which being hinged at one end to a strong wooden base, is forced down upon the shingle. The other is known as a Shingle Butter and consists of a wooden stand surmounted by a stout wooden arm, which, pivoted at one end and carrying a guillotine-knife, is brought down upon the butt-ends of shingles to trim or bevel them.

Other tools used include those described under the following entries: *Beetle; Hatchet, Shingling;* Slate Ripper under *Building Tools; Woodland Trades.*

Shingling Froe: see *Froe.*

Ship Maul: see *Hammer, Ship Maul.*

Ship's Blockmaker *Fig. 684*
The Block is a wooden housing containing one or more pulley wheels known as sheaves. The sheave revolves on a pin (pintle) fitted across the centre of the block, and it has a grooved rim for the reception of the rope. The Block is suspended by rope with an eye or hook at the top for attachment to the point of support. One purpose of a block and tackle – the name usually given to a system of pulley blocks and rope – is to apply pulling power in any required direction; another is to provide a mechanical advantage in order to lift a heavy load by hand. Thus, when two Blocks are used, each with one sheave, the mechanical advantage is doubled; with two pairs of sheaves, quadrupled, and so on.

The block and tackle is used for many lifting tasks, but the biggest demand came from ships. According to David Steel (London, 1794) a 74-gun ship required not less than 922 blocks.

Tough woods such as elm or ash are used for the shell. The sheaves are made of lignum vitae or other hardwood. The shells are sawn out from the solid and the corners taken off; the mortices are bored with an Auger, the intervening wood sawn out, and cleared with Chisel and Rasp. The hole for the pin is bored through the centre of the shell and the outside 'score' gouged out for the strap or rope by which it is suspended. The sheaves are turned on the lathe.

While being mortised, the Block is wedged on the face of a special stool (see *Clave*), and the outside of the Block is subsequently trimmed while held on a Clave Board. The broader end of this tapering board is wedged in the Clave and the narrower end is placed in one of the mortice holes in the Block, to support it.

Blockmaking machines were developed in the early years of the nineteenth century and some of them were still working at H.M. Dockyard, Portsmouth in 1969. They are described by K. R. Gilbert (London, 1965).

Tools used will be found under the following entries: *Auger, Blockmaker's; Chisel, Blockmaker's; Clave; Gouge, Blockmaker's.*

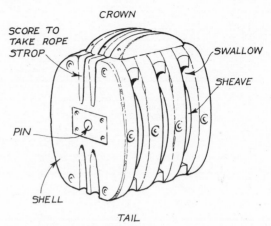

Fig. 684   Parts of a Ship's Block

Shipwright *Fig. 685*
The art of shipbuilding is very old. Boats and canoes were hollowed out from tree trunks in prehistoric times. Sea-going sailing ships, with sides formed of planks, were constructed by the Egyptians in about 2300 B.C., and three centuries after this men are known to have travelled by sea from the Mediterranean to Scotland. In the first century A.D., skill in shipbuilding was developed in the Scandinavian countries, in England, and also in China, where in the eighth century A.D. the stern-post rudder was first invented.

The magnificent full-rigged fighting and merchant ships of the fifteenth and sixteenth centuries were undoubtedly one of the greatest achievements of the shipwright's trade. They were perhaps surpassed only in grace and speed by famous sailing vessels like the *Cutty Sark* which raced across the world in the last days before steam. The first steamship to cross the Atlantic was built in Dover in 1826. From that time forward, iron began to replace wood in the construction of ships.

The introduction of power-saws and other machines into the shipyards took place only quite recently. 'Fifty years ago when I first went to work,' relates a retired shipwright, Mr. C. Bunday, 'the only machine in the Yard was a grindstone. Everything else was

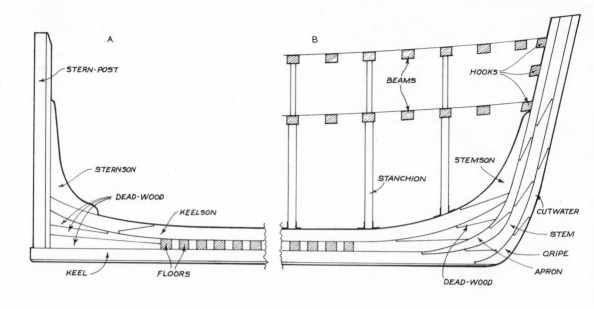

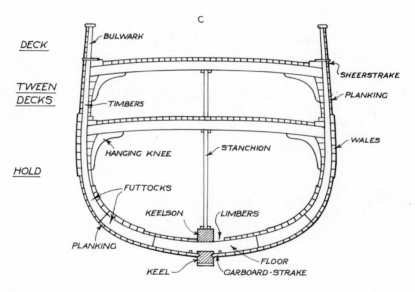

Fig. 685 Parts of a wooden vessel (diagrammatic)
    (*A*) After Part
    (*B*) Forward Part
    (*C*) Midship (sectional view)

done by hand.' (Bursledon, Hants, 1966.) Like other trades, traditional skills were often handed down within small family concerns. Mr. G. Worfolk, a Norfolk shipwright, put it this way: 'Me and my brother were apprenticed at fifteen. We've always worked on ships at sea and in dock. My father was a shipwright and my grandfather was a shipwright. We've always built boats, and we don't really know anything else.' (Kings Lynn, 1966.)

Allied Trades

The shipwright made the plan of the ship, prepared the timber, organised and carried out the construction of the vessel, and supervised its launching. At one time he also undertook the rigging and the making of anchors, masts, and spars. But during the last two centuries at least, these skills have gradually become separate trades. Some are described under the following entries: *Mast and Spar Maker; Sailmaker; Ship's Block Maker.* The trades of the anchorsmith and ropemaker are not included.

Timber

The most trusted and universal timber for building large vessels in England was oak. It provides the combined virtues of great strength and resistance to decay. It has been calculated by George Naish (1956) that a large warship required some 2000 oak trees, each of them needing a century to reach maturity. These trees could not have been grown on less than 50 acres of woodland, which would be left stripped.

Elm is the wood chiefly used for smaller vessels in England, with outer planks in wych elm or larch. Cedar and mahogany are used for many racing boats, and in recent years various laminated woods are also employed, culminating in the modern use of reinforced moulded plastics, including fibre-glass.

The Shipyard

Most large ships are built out-of-doors either on a slipway or in a dry dock. While being built, the ship rests on Keel and Bilge Blocks, and it is held steady in an upright position by Breast Shores and Bilge Shores. These are timber struts set up to span the distance between the sides of the ship and the ground, or between the ship and the dock walls.

Smaller vessels are often built under cover, and then pushed out of the workshop on a Greasy Slipper (a smaller version of the Launching Cradle) towards a sloping launchway, or to a low-loading vehicle for transport to the coast. A typical shipbuilding shed for smaller boats is about 50 yards in length. A bench may run the length of the shed on one or both sides. On one wall are hung the Moulds;

and against another wall stand sets of heavy Cramps which are shaped like giant carpenters' G-Cramps and are used for holding the planks against the timbers to which they are to be attached. But everything is overshadowed by the ship which is being constructed along the centre of the workshop and which towers upwards towards the high roof of the building. It is held steady by Breast Shores which greatly impede anyone walking along the sides of the workshop.

Building and Launching

Since early times there have been two common methods of planking the outside of a ship. In one, known as carvel-built, the planks meet edge-to-edge, and are secured to the frame of the ship by trenails or iron spikes. In the other process, known as clinker-built, the planks are overlapped, and their overlapping edges are fastened together with clenched metal rivets. The seams of carvel-built planks have to be caulked to render them watertight; but this does not apply to clinker-built boats which are said to be stronger weight-for-weight than carvel-built boats. In Britain small rowing and fishing boats are almost invariably clinker-built.

The ship's timbers are drawn out full size in the mould loft with chalk. Alternatively, the 'laying off', as the process is called, is done on a Scrive Board. This is a very large board on which the lines of the ship are incised with a Timber Scribe. The next step is the making of moulds from thin strips of wood, cut according to the drawing on the mould-loft floor. They are used as patterns when sawing out the ship's timbers, and some may be set up temporarily within a ship as a guide to the placing of the timbers.

Mr. G. P. B. Naish (Oxford, 1956) has described the building of sailing vessels in the sixteenth to seventeenth centuries; the following is an extract:

'The keel of a new ship was laid on blocks in a dry dock and the stem-post and stern-post were erected and scarfed on to the keel at each end. This was the heaviest work. Then the floor timbers were laid across the keel; the keelson was laid along the keel on top of the floor-timbers; and the keelson, floor-timbers, and keel were bolted together. The futtocks were next attached to the floor-timbers; these were the curved or compassing timbers that formed the curved sides of the ship. Clamps were heavy planks running horizontally on the inside of the timbers to support the ends of the deck-beams. Partners were strong pieces of timber bolted across the deck-beams to support the masts, the heels of which were to be stepped on top of the keelson. The frame was further held together with a multitude

467

of standing, lodging, and hanging knees, all made of oak. The shipwright always searched for crooks of timber, which he cut up most carefully to avoid waste.'

The process of launching is a highly skilled part of the shipwright's art. The vast weight of the ship has to be lifted, or rather eased off the Blocks on which it rested during construction, and transferred to the Launching Cradle on which it will ride on the short journey down the greased Slipway into the water. This lifting is done by pairs of folding wedges, known as Bilge Gluts, placed end to end at points around the hull. As the wedges are hammered inwards, the boat is lifted. Chocks (wedge-shaped blocks of wood) are used to prevent the cradle from moving before the ship is ready to be launched. The Cradle is also secured to the slipway by means of iron Dogs. When the time comes for launching, men carrying a Ship Maul are standing near each Dog. At the command 'Out Dogs' each man drives the tapered pin of the Maul under the bar of the Dog to knock it out.

Tools of the shipwright
From the evidence of the apprentice indentures of Bristol, Great Yarmouth & Southampton, 1547–1644 (W. L. Goodman, 1972), a typical shipwright's kit included the following:
Axe, Addes or Jennet (Adze); Handsaw; Chisel; Gouge; Shave (possibly a mastmaker's Drawing Knife); various Hammers including a Maul; Auger; Spike Gimlet; Wimble (Brace); Caulking Irons & a Caulking Mallet; and Reve Hook (a Rave Hook or Hoe used for raking out old pitch and rotted oakum before re-caulking).

The tools used in the trade today are described under the following entries:

Adze, American
Adze, Shipwright's
Auger:
  Coak Boring
  Deck Counter Boring
  Deck Dowelling
  Dodd's Pattern
  Expanding
  Long Pod
  Shipwright's
  Twist (Single Twist)
Awl, Boatbuilder's Piercing
Axe:
  Blocking
  Mast Maker's
  Shipwright's
  Newcastle Ship

Bevel, Boatbuilder's (see under *Square, Bevel*)
Bit, Deck Dowelling
Blockmaker (see *Ship's Blockmaker*)
Boat Grip (Tongs)
Brace, Shipwright's (Boat Sway)
Caulking Tools including:
  Caulking Box
  Caulking Irons (*Set, Bent, Making, Horsing, Reaming, Spike, Trenail, Jerry* and *Sharp Iron*)
  Caulking Mallet
  Hoe
  Pitch Mop and *Ladle*
  Oil Box
Calliper, Mast
Chisel:
  Blockmaker's
  Shipwright's
  Ship's Slice
Compass, Shipbuilder's
Cramp, Ship
Drawing Knife, Mastmaker's
Gimlet, Boat
Gimlet, Ship
Gouge, Blockmaker's
Hammer:
  Carpenter's
  Clench
  Coppering and *Sheathing*
  Maul (including Shackle-Pin Hammer)
  Spike Set
Mallet:
  Caulking Mallet (see *Caulking Tools*)
  Shipwright's Mallet (see *Mallet, Shipwright's*)
Marline Spike (see *Sailmaker*)
Mast and Spar Maker (Tools)
Poker, Mastmaker's
Plane:
  Mast and Spar
  Rounder, Trenail (or *Moot*)
  Shipwright's
Punch, Coppering
Punch, Shackle
Roving Iron
Sailmaker (Tools)
Saw, Shipwright's (Futtock Saw)
Scraper, Ship
Scrive Hook (see *Timber Scribe*)
Ship's Blockmaker (Tools)
Steam Chest
Trenail Tools
Trenail Rounder or Moot (see *Plane, Rounder, Trenail*)

Shive
A wooden disc used as a bung to stop the hole of a cask. See *Auger, Cooper's Bung Borer* for further details.

Shive Vice or Extractor: see Bung Removers under *Cellarman.*

Shoe
A term used in many trades for that part of a tool or implement which is designed to spread the load, e.g. the part of a Cramp or Lifting Jack that bears against the work or load. Or, in the case of a vehicle, the skid-pan brake, or the part of the brake which acts on the wheel.

Shoe Driver
A term used in U.S.A. catalogues for a Hoop Driver. See *Cooper (3) Hooping Tools.*

Shoe Peg Cutter: see *Knife, Bench.*

Shooting Board (Shuteing Board) *Fig. 686*
A wide board, 18 to 24 in (or more) long, with a rebate about 3 in wide for the Plane. A stop is fitted at one end. The lower edge of the rebate is usually under-cut to avoid inaccurate planing caused by an accumulation of dust in the corner. It is used for planing the edges of thin timbers perfectly straight when they are to be accurately fitted together, and particularly for trueing the square end-grain of work-pieces. In operation, the Plane is laid on its side and worked close up against the rebate which acts as a guide. (Other Shooting Boards are described under *Mitre Appliances.*)

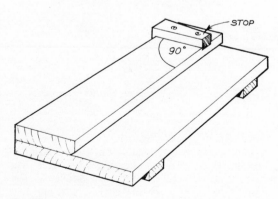

Fig. 686

Shot Rake: see Cork Removers under *Cellarman.*

Shoulder
Of wooden joints, spokes, etc. – the flat surface at the root of a tenon which, in the case of a spoke, takes the weight of the vehicle, and, in the case of a post, may support the weight of a roof.

Shovel Maker: see *Spade and Shovel Maker.*

Shrinkage Rule: see Pattern Maker's Contraction Rule under *Rule.*

Shutting
Welding, particularly of an iron tyre. See *Wheelwright (3) Tyring Tools.*

Side Hook: see *Bench Hook.*

Side Snipe: see *Plane, Side Snipe.*

Sill
An edge, e.g. the wooden edge of a saw pit or the wood or metal edge of a blacksmith's hearth.

Sizing Tool: see *Chisel, Turning.*

Skarsten Scraper: see *Scraper, Hook.*

Skate
Of a Plough or Grooving Plane – the metal plate on which the cutting iron is bedded, and which acts as a sole to the Plane. This plate is often turned up at the toe like a skate.

Skewed (Badgered)
Of a Plane – when the cutting iron is bedded at an angle to the length of the stock instead of lying square (e.g. see *Plane, Badger*). Of a Chisel – when the cutting edge is oblique.

Skewer: see *Upholsterer.*

Skid: see *Cellarman.*

Skillop: see *Auger, Cooper's Bung Borer.*

Slasher: see *Woodland Trades.*

Slater's Tools: see *Builder's Tools.*

Slip: see *Oilstone Slip.*

Slipped
Term applied to a Moulding Plane when a separate strip of wood is screwed to the side of the stock. This can be replaced after wear. See *Plane Maker.*

Sloper: see *Plane, Cooper's Chiv.*

Smith's Tools
A few of those commonly found in woodworking shops are described and illustrated under *Metalworking Tools; Wheelwright's (3) Tyring Tools; Screw Dies and Taps; Rasp, Two Handed; Wrench.*

Smoker-Back Knife: see *Shave, Chairmaker's.*

Smut Stick: see *Chalk Line and Reel.*

Snap Line: see *Chalk Line and Reel.*

Sneck
Of Plane cutting irons – the thickening or side extension at the top. This allows the iron to be adjusted by tapping the sneck with a Hammer instead of striking the stock of the Plane. See Cutting Iron under *Plane.*

Snipe Bill: see *Plane, Side Snipe.*

Snudgel: see *Grapnel.*

Soaking Trough: see *Basket Maker.*

Soldering Iron: see *Plumber's Tools.*

Sole
Of a Plane – the base. Of a wheel – the outside of the rim.

Sound Post Tool: see *Violin Maker.*

Spade and Shovel Maker *Fig. 687*
Wooden shovels are still made in small numbers for malting and for the fish markets; and wooden Spades, usually shod with iron like those of Roman times, are still made in small numbers for digging clay or mud. The blades are roughed out on the concave side with a hollowing Adze and finished with a Shave. Tools used include: *Adze, Chairmaker's Howel* (or similar); *Drawing Knife, Handle Maker's; Plane, Rounder; Shave, Shovel Maker's.* See also *Plane, Scoop Maker's.*

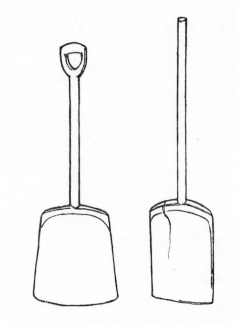

Fig. 687 Wooden Malt Shovel, and a Deck Scoop for Herrings

Span Dogs: see *Timber Handling Tools.*

Spanish Sneck: a type of Plane Iron, see *Plane, Cooper's Jointer.*

Spanner: see *Wrench.*

Spar Maker: see *Mast and Spar Maker.*

Speech *Fig. 688*
A term occasionally applied to a wheel hub fitted with spokes but without the felloes and tyre. The word is also sometimes given to a Spoke Set (see *Wheelwright's Equipment (2) Spoke Tools)* and to a Jack Staff (see *Millstone Dresser).*

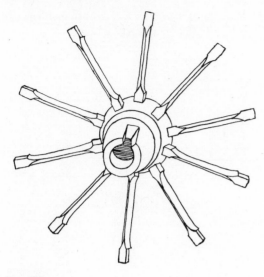

Fig. 688

Speech Bat: see Wheel Frame under *Wheelwright's Equipment (1) Workshop Furniture.*

Spider Mortice Bevel: see *Square, Coachbuilder's.*

Spiers (Tool Makers): see *Plane, Spiers and Norris Types.*

Spigot Cutter: see *Pipe and Pump Maker.*

Spike Set: see *Hammer, Ship's Spike Set.*

Spile
A tapered peg used to stopper the air hole made in the shive (bung) of a cask. (See *Auger, Cooper's Bung Borer.*)

Spile Borer: see *Gimlet, Brewer's.*

Spile Nippers or Grip: see *Cellarman.*

Spiral Tools
Tools sometimes referred to by this name include: *Auger, Twist; Drill, Archimedean; Screwdriver, Spiral Ratchet.*

Spitsticker: see *Wood Engraver.*

Splayed: see *Fantail.*

Spoke Shave: see *Shave – Spokeshave.*

Spoke Tools: see *Wheelwright's Equipment (2) Spoke Tools.*

Spoon
A term used for a number of spoon-like tools including: various *Augers* of pod or spoon shape; *Bit, Spoon;* Spoon Reamer under *Fishing Rod Maker.*

Spoon Maker *Fig. 689*
In the recent past wooden spoons and ladles were widely used in country homes and in the dairy. Most of them were made from sycamore. Some of the larger spoons or ladles have a hook on the end of the handle for hanging from the edge of a milk bucket, basin, or shelf. Spoons of similar design have been unearthed from prehistoric Swiss lake dwellings. The highly decorated 'Welsh' spoons were mostly carved as a pastime in the long winter evenings. Some were designed for use, but they were also made as love tokens.

Cleft from the log, the spoon billet is first rough-shaped with an Axe. The bowl of the spoon is then hollowed with a special hooked Knife and the back smoothed with a Spokeshave. The larger spoons or ladles were hollowed with a short-handled gouge-shaped Adze, and finished with the hooked Knife or Gouge.

Tools include some of those listed under *Woodland Trades* and also the curved knife described under *Knife, Hooked.* A one-handed Round Shave was sometimes used for hollowing; this is described under *Drawing Knife, Cooper's Round Shave.*

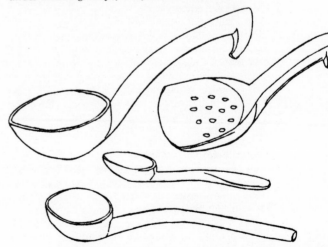

Fig. 689 Wooden spoons and ladle from Wales, and a Butter Scoop

Sprig Bit or Awl: see *Awl, Brad.*

Spring

A term applied to Moulding Planes. These are designed to be used when tilted over to one side; the amount of inclination is known as the 'spring'. See *Plane Maker; Plane, Moulding.*

Spur, Nicker, or Tooth

Of Planes, Gauges, etc.: a small knife-like point fitted vertically and designed to cut the fibres at the edge of a groove or rebate in advance of the plane iron. Of Augers and Bits: a similar point at the outer edge of the nose designed to cut the grain at the circumference. (See diagrams under appropriate tool entries.)

Spur-Wheel

A gear wheel, usually larger than the gear wheel (pinion) meshing with it on a parallel shaft.

Square

The following tools are included under this entry: Bevels, Mitre Squares, and Try Squares. Though differing in appearance, most of these tools (and their variants) consist of a blade (or tongue) set into a stock which is thicker than the blade itself. The stock is sometimes fitted with a spirit level (see *Level*). Though factory made from the eighteenth century onwards, those found in the older workshops were often made in wood by the tradesman himself and differ little in appearance from the Squares and Mitres that have survived from ancient Egypt and Rome. They are used for marking out, for testing the accuracy of workpieces, and for testing the angle of structures in building work.

Note: The Squares and Bevels described in subsequent entries are mostly factory-made examples. Like those illustrated in Smith's *Key* (Sheffield, 1816), many have rosewood or ebony stocks, often fitted with curiously elaborate, but not ungraceful, brass mounts which, together with the rivets, have the important function of securing blade and stock rigidly at the required angle.

Square, Bevel *Figs. 690; 691*

A straight stock with an adjustable blade which can be set as required for testing and setting out workpieces to any angle. The length of blade (or 'tongue') varies from about 7 to 15 in or longer.

Home-made examples were made entirely of wood, or had a wooden stock and a metal tongue. The end of the tongue was sometimes given a graceful ogee profile.

Factory-made examples were often made in rose-

472

wood or ebony with brass mounts like their counterparts among the Try-Squares.

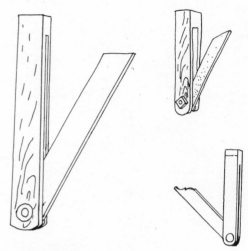

Fig. 690   The Bevel: home-made examples

(*a*) *Angle Bevel*. Factory-made version of the commonest type.

(*b*) *Boatbuilder's Bevel*. A stock of hardwood, 12 in long, sometimes graduated, and slotted from either end for thin wood or metal tongues, one long, one short. Used by shipwrights for measuring compound bevels of planks and frames where they lean off round the curve of the ship.

(*c*) *Canada Pattern Sliding Bevel*. With thumb or lever type adjusting screw.

(*d*) *Coachmaker's Bevel*. See *Square, Coachbuilder's.*

(*e*) *Horizontal Bevel*. See *Square, Coachbuilder's.*

(*f*) *Mason's Bevel* (*Shift Stocks*). Usually made in brass.

(*g*) *T-Bevel*. With the blade pivoted to the middle of the stock.

(*h*) *Sliding Bevel*. The blade is slotted about half way down the middle so that its effective length can be altered as well as its angle with the stock. The blade is designed to be set by means of a Screwdriver, or by a short lever or wing nut, or, as in the case of the modern version illustrated, by means of a thumb screw at the lower end of the handle.

(*i*) *Bricklayer's Bevel*. Like the Angle Bevel (see under *Square, Bevel*) but usually plated on the face of the stock and adjustable with a thumbscrew. Used when shaping bricks.

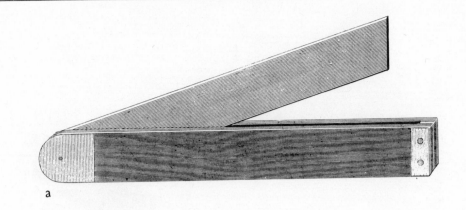

a

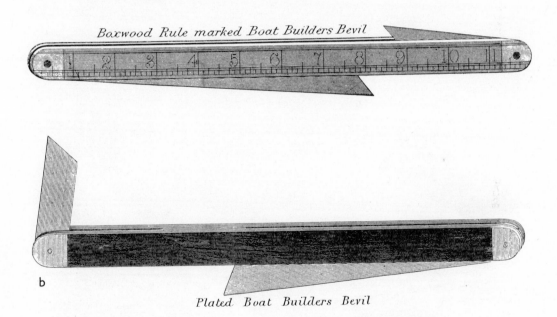

Boxwood Rule marked Boat Builders Bevil

b

Plated Boat Builders Bevil

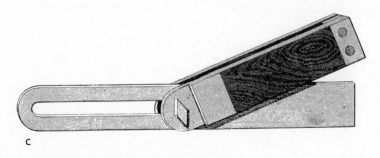

c

Fig. 691   The Bevel: factory-made examples—*continued overleaf*

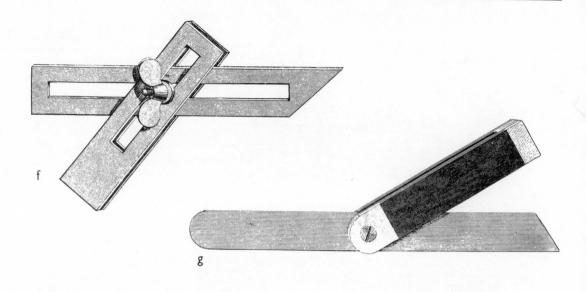

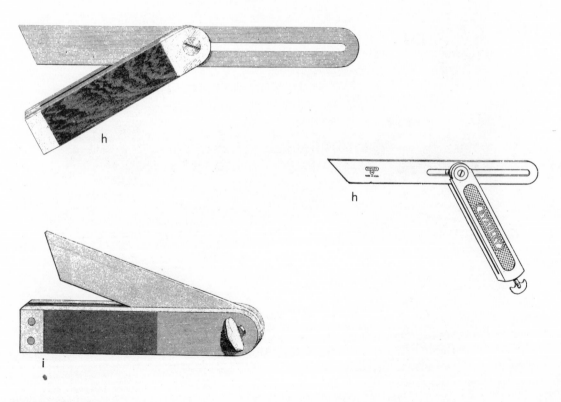

f

g

h

h

i

Fig. 691—continued

Square, Carpenter's, Roofing (Rafter or Framing Square) *Fig. 692*
This is the traditional Square of the carpenter, and is used for setting out roofing, staircase, and other carpentry framing and certain millwright's work. Illustrated by P. Nicholson (London, 1822), it is made of iron with one leg 18 in long and the other 12 in. The figures number from the right-angled corner outwards and are graduated in $\frac{1}{8}$ in, but unlike other iron squares, the figures are engraved to be read from inside the angle of the square.

The modern version of the Carpenter's or Roofing Square is a more complicated instrument altogether. It consists of a blade, about 24×2 in long, and a tongue $18 \times 1\frac{1}{2}$ in. The middle of the blade has groups of lines divided into 12 spaces, known as the 'board measure', while the middle of the tongue carries sets of figures called the 'brace measure'. The board measure is a form of ready reckoner for obtaining areas; the brace measure on the tongue gives the hypotenuse or 'brace' of a triangle, given the other two sides.

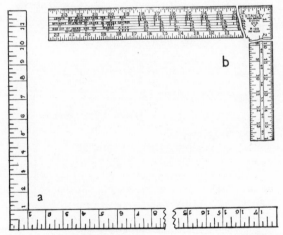

Fig. 692   Roof Square
       (*a*) As illustrated by P. Nicholson (London 1822)
       (*b*) A modern version

Square, Coachbuilder's *Fig. 693*
Squares and Bevels of the ordinary kind are used in this trade, and also the Plumb and other tools and Gauges described under *Level* and *Gauge, Wheelwright's*. In addition the following special Squares are employed:

(*a*) *Horizontal Square*
A gunmetal stock, about 3 in long and 1 in square, with a steel blade about 8 in long and $\frac{3}{4}$–1 in wide, pivoted at one end, and fixed at any angle relative to the face of the stock in the horizontal plane by means of a knurled screw. Home-made tools of a similar pattern are made of hardwood with the blade secured to the top of the stock with a wood screw.

One of the uses of this tool is to mark off a vertical tenon in a piece of wood (e.g. a coach pillar) set at an angle to the vertical. We have had some difficulty in understanding its function but the following is our interpretation of the information we have obtained from veteran coachbuilders, including the late Mr. Philip Clarke (Salford, 1967).

The Horizontal Square was used when a corner pillar of the carriage stood in an oblique plane, e.g. slanting outwards from the centre of the body. When a square piece of timber is cut at an angle to one of the diagonals, the cross-section will be a rhombus. But in coachwork it is often required that the two outer sides of this oblique-standing pillar should be square in the horizontal plane, so that the sides of the carriage body, though curved in a vertical direction, should be at right angles to one another. Accordingly the stock of the Horizontal Square is held horizontally on the face of the slanting pillar and the blade turned until it, also, is parallel with the horizontal plane. This indicates the amount of wood to be cut away to produce a square cross-section on a pillar standing in an oblique position.

Fig. 693 (*a*)

(*b*) *Spider Mortice Bevel*
A stock consisting of two triangular brass plates, about 3 in wide and $\frac{5}{32}$ in thick. These are screwed together at the corners over spacing pieces to take three steel blades, $\frac{1}{4}$ in wide, projecting about 1, $1\frac{1}{2}$, and 2 in respectively, and fixed at any angle by screws in the middle of the sides. In some patterns one of the blades is about 6 in long and slotted like

475

that of the Coachmaker's Mortice Bevel, thus forming a combination bevel and depth gauge. It is used for testing the angle of mortices when they have to be sunk at an angle to the face of a pillar.

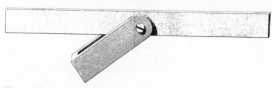

Fig. 693 (*d*)

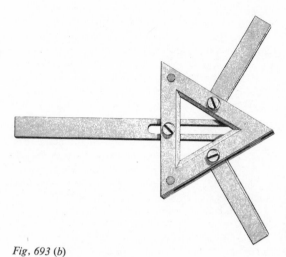

Fig. 693 (*b*)

(*c*) *Coachmaker's Mortice Bevel*
A simplified form of the Spider Mortice Bevel, used for similar work. It consists of a metal stock about 3 in long, slotted to receive a narrow steel blade. This is itself slotted for about half its length and can be adjusted to any angle and amount of projection.

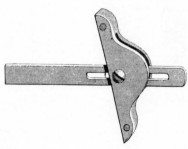

Fig. 693 (*c*)

(*d*) *Coachmaker's T-Bevel* (Iron Stock Bevel)
Like the ordinary woodworker's T-Bevel but usually made entirely in metal.

Square, Glazier's *Fig. 694*
A wooden T-Square, often with the face of the stock plated to resist wear. The graduated stem is made from 24 in to as long as 60 in. Used for measuring glass sheets before cutting.

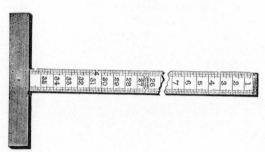

Fig. 694

Square, Mitre *Fig. 695*
A form of Square used for setting out and testing mitre joints. There are two common types:

(*a*) With the blade fixed at an angle of 45° with the stock. Home-made examples are mostly in wood throughout, but the factory-made version often has a rosewood or ebony stock with elaborate mounts, like its counterpart among the Try Squares.

Fig. 695 (*a*)

(*b*) A type more commonly found on the Continent of Europe and derived from those used in Roman times and before. It consists of a flat piece of wood or metal cut to form two arms at right angles, one of which forms an angle of 45° with the base. It can be used for testing right angles as well as mitres.

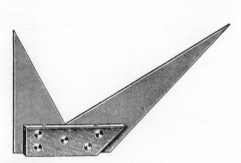

Fig. 695 (*b*)

Square, Radial (Centre Square; Circular Square; Round Square) *Fig. 696*

A flat piece of wood or metal with a rounded head and tapering blade. A number of stout pins or studs (or a right-angled fence) are set on the head in order to centre the blade when the Square is applied to the edge of a cylindrical workpiece. Radial lines can thus be drawn across circular work, for finding the centre point.

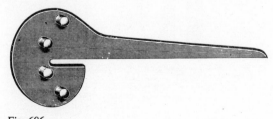

Fig. 696

Square Rimer or Rinder: see *Rimer, Hand.*

Square, Roofing and Framing: see *Square, Carpenter's.*

Square, Set *Fig. 697*

A triangular plate of wood or plastic used as a drawing instrument, usually in conjunction with a draughtsman's T-Square. The angles opposite the right angle are 45 and 45 or 60 and 30 degrees.

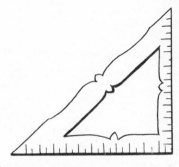

Fig. 697

Square, Tee

A wooden Square with a blade 3–4 ft long, used for marking straight lines on drawings. In use, the stock of the Square is held against the edge of the drawing board, while the blade is used as a guide for the pencil when making lines at right angles to the edge.

Square, Try *Fig. 698*

A Square in which the blade is set at right angles to the stock. Carpenter's Squares of the nineteenth century were large, often home made in wood or metal. P. Nicholson (London, 1822) writes that the Carpenter's Square is a rule of iron with two legs, one 18 and the other 12 in long. But in more recent years, tradesmen seem to have favoured the standard factory-made article with a thin steel blade, 3–24 in long, securely riveted into a brass-mounted rosewood or ebony handle. With his singular interest in tool design, Moxon (London, 1677) explains the construction of the Joiner's Square – in this case one made entirely of wood.

'The Reason why the Handle is so much thicker than the Tongue, is, because the Handle should on either side become a Fence to the Tongue. And the reason why the Tongue hath not its whole breadth let into the end of the Handle is, because they may with less care strike a line by the side of a thin than a thick piece: For if instead of holding the Hand upright when they strike a Line, they should hold it never so little inwards, the shank of a Pricker falling against the top edge of the Handle, would

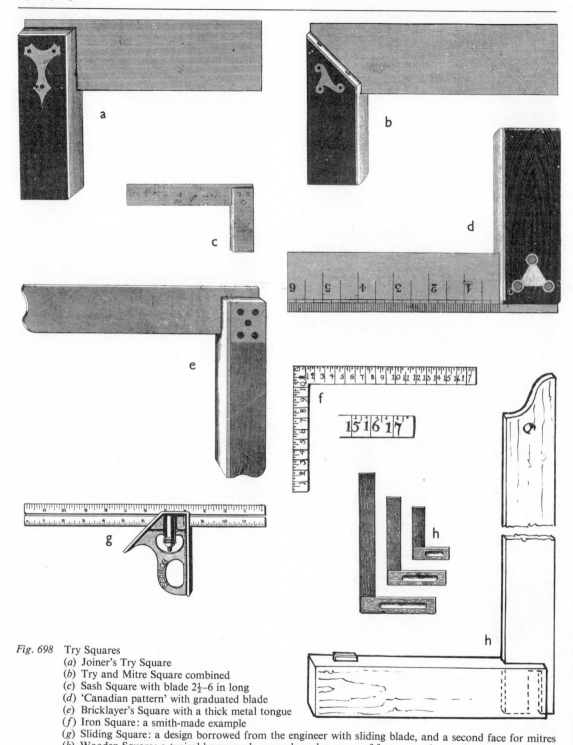

Fig. 698 Try Squares

    (*a*) Joiner's Try Square
    (*b*) Try and Mitre Square combined
    (*c*) Sash Square with blade 2½–6 in long
    (*d*) 'Canadian pattern' with graduated blade
    (*e*) Bricklayer's Square with a thick metal tongue
    (*f*) Iron Square: a smith-made example
    (*g*) Sliding Square: a design borrowed from the engineer with sliding blade, and a second face for mitres
    (*h*) Wooden Square: a typical home-made example and a group of factory-made wooden squares

throw the Point of a Pricker farther out than a thin Piece would: To avoid which Inconvenience, the Tongue is left about half an Inch out of the end of the Handle . . . .'

Variants are illustrated.

Stadda: see *Saw, Stadda.*

Stail Engine: see *Plane, Rounder.*

Stair Maker: see *Plane, Handrail, – Nosing, – Router; Saw, Stairbuilder's; Shave, Handrail.*

Stake
A name given to certain Anvils. See, for example, Anvil under *Cooper (3) Hooping Tools;* Saw-Setting Plate under *Saw-Setting Tools.*

Stake Driver: see *Beetle.*

Stamp Cutter: see *Print and Block Cutter.*

Stamp, Letter or Figure: see *Punch, Letter, or Figure.*

Stand Proud
A term used to describe a part which projects above the level of surrounding parts (as opposed to being flush or sunk).

Stanley Tools: see *Plane, Stanley-Bailey.*

Staple Clincher *Fig. 699*
An iron punch hollowed at one end and grooved to fit over the head of a staple. Staples were used to secure the side boards of certain vans and wagons to the vertical iron rods or wooden spindles which supported them. The tool was held over the staple head and rod while staple ends were clinched on the back of the side boards.

Fig. 699

Staple Puller *Fig. 700*
Formed like a small Hammer with a pointed pane. Used by fencers for removing and driving staples.

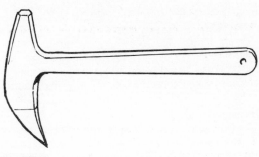

Fig. 700

Staple Vice: see *Vice, Woodworker's* (d).

Start or Starter: see Bung Removers (Flogger) under *Cellarman.*

Stave Block: see *Plane, Rounder.*

Stave Cramp: see Hooping Windlass under *Cooper (3) Hooping Tools.*

Stave Wrench: see *Flagging Iron.*

Steady: see *Turning Steady.*

Steam Bell
An old cask large enough to lower over casks during manufacture while they are steamed to soften the assembled staves before hooping.

Steam Chest (Steam Kiln; Steam Oven)
A box or cupboard, generally of wood, in which timber is heated by steam to render it more supple for bending. Used by boat builders to soften ship's timbers, and by coopers to soften the assembled staves of a cask before hooping. (The use of steam, rather than the Cresset, is necessary for thicker staves. See *Cooper (3) Hooping Tools.*)

Steel (used in tool making): see *Iron and Steel.*

Steel Toys
A term applied to small tools and other articles. See *Lancashire Tools.*

Steeling: see *Iron and Steel.*

Stencils
Metal plates in which holes are cut in the shape of letters, words, or numbers, through which paint or

ink can be applied (usually by brush) to the surface of the wood.

Stick

To 'stick' (or Scots: 'run') a moulding, is a joiner's term for making a profile with a Moulding Plane.

Sticking Board

A board used flat on the bench with an arrangement of pins and grooves for holding the wood when working the rebates and mouldings on sash bars. The wood is held in place with either a screw head (as illustrated under *Window Making*) or by points of nails which are driven in under the board so that their points project above. The rebates are worked in succession on each side; the resulting fillet is then inserted in the horizontal groove in the side of the board, and the moulding stuck on the face edges. Another type, for sticking beads, consists of a plain board with a groove in which the material is held vertically while the bead is stuck on the edge.

Stillson: see *Wrench* (2) *For Gripping Rounds*.

Stitch Mallet: see *Sailmaker*.

Stob: see *Awl, Brad*.

Stock

Main part of a tool, e.g. that which holds the blade, such as the body of a Plane or Shave. Also an alternative name for a Brace.

Stock Howel: see historical note under *Cooper* (1) *The Trade*.

Stock Knife: see *Knife, Bench*.

Stock Screw

Part of a Tyring Platform. See *Wheelwright's Equipment* (3) *Tyring Tools*.

Stockholm

A French name for a small Chiv. See historical note under *Cooper* (1) *The Trade*.

Stone Staff: see *Millstone Dresser*.

Stool: see under *Wheelwright's Equipment* (1) *Workshop Furniture;* also *Woodland Trades* (2) *Brakes*.

Stop Chamfer: see *Chamfer*.

Stoup: see *Plane, Cooper's Stoup*.

Stower: see *Plane, Stower*.

Straight Edge

A strip of well-seasoned hardwood, from 2 to 4 ft long and 4 in wide, with one accurately planed edge. Light steel straight edges are from 24 to 48 in long and $1\frac{1}{4}$ to 2 in wide. Used for testing a flat surface and for ruling lines.

Strake Tools: see *Wheelwright's Equipment* (3) *Tyring Tools*.

Stretcher: see *Upholsterer*.

Strike Block: see *Plane, Strike Block*.

Striking Iron: see *Millwright*.

Striking Stick: see Spoke Set under *Wheelwright's Equipment* (2) *Spoke Tools*.

Studdie: see Anvil under *Cooper* (3) *Hooping Tools*.

Sugar Cleaver: see *Hatchet, Sugar*.

Surform Tools (Shapers) *Fig. 701*

A trade name for a series of tools recently introduced from the U.S.A. consisting of a perforated plate with a rasp-like abrading surface, which can be flat, half-round, or circular. The plate is replaceable and is fitted to a flat channel-shaped stock. This stock may be handled for use like a File or made in the form of a Plane. These tools have largely replaced the Rasp and File for rough and rapid smoothing of wood, and have to some extent replaced the Roughing Plane. A special advantages is their freedom from clogging: there are openings between the teeth through which the chips fall.

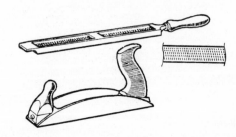

Fig. 701

Surveyor's Laths or Rods: see Folding Laths under *Rule.*

Swage
To form metal with a form or die, and also the name for the tool for doing this.

Sway: see *Brace.*

Sway Knife: see *Knife, Bench.*

Sweep
This term appears under the following entries:
 Brace: as an alternative name for a Brace, also a term for measuring the extent of the crank's movement (see diagram under *Brace*).
 Compass, Shipwright's: a name sometimes given to large wooden Compasses.
 Gauge, Wheelwright's and Coachbuilder's: a name for certain templets used in coachbuilding.

Sweep, Ramshorn *Fig. 702*
A large templet in thin wood or other material sometimes up to 3 ft overall and scroll-shaped. Used for setting out curves on the drawing board and also on the work itself, for instance, in coachbuilding.

Fig. 702

Sweep Stock
An alternative name for a Brace, e.g. the *Chairmaker's Brace.*

Swift: see Heading Swift under *Shave, Cooper's.*

Swing Dingle: see Cant Hook under *Timber-Handling Tools.*

Sword: see Bow under *Drill, Bow; Timber Girthing Tape and Sword.*

T

Used for cutting grooves along curved edges, for which purpose it is far less laborious to use than a Scratch Stock or Router. (cf. *Coachbuilder's Jigger*, page 393.)

T-Anvil: see Anvil under *Cooper* (*Hooping Tools*).

Tack Lifter or Claw (Tack Wrench) *Fig. 703*
A small curved claw fitted into a wooden handle. Used in many trades for lifting and removing tacks.

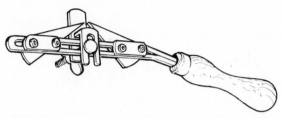

Fig. 704

Fig. 703

Tempering: see *Iron and Steel*.

Templet: see *Patterns, Templets, and Jigs; Plane Maker* (*Patterns and Jigs*); *Templet, Sash*.

Tail Stock
Movable back 'centre' of a lathe which can be shifted along to suit different lengths of work.

Tang
In Bits, the end of the shank which is held in the chuck of the Brace. In Chisels, Augers, etc. that part of the shank which is drawn down to fit into a handle. (A Cooper's Tang is a name sometimes given to a Hooping Dog. See *Cooper* (3) *Hooping Tools*.)

Tap
A screw-making tool (see *Screw Dies and Taps*) but the term is also applied to a peg or dowel rounder (see *Plane, Rounder*).

Tap Hole Borer or Burner: see *Auger, Cooper's Bung Borer*.

Tape: see *Rule, Flexible*.

Tapping Bit: see *Bit, Cock Plug*.

Tasting Tube: see Sampling Tube under *Cellarman*.

Tectool *Fig. 704*
The trade name given to a modern grooving tool. Two narrow cutters are mounted on a stock and slope in opposite directions to suit the grain.

Templet, Sash *Figs. 705; 727* under *Window Making*. These hardwood templets are used as a guide for cutting the mitre and 'scribing' the moulding on glazing bars etc., i.e. the fitting of one moulded profile over another.

(a) Saddle or Double Sash Templet
This is a piece of hardwood (usually beech), $7\frac{1}{2}$ in long and about $1\frac{1}{2}$ in square, with a groove shaped to a reverse of the sash moulding and designed to be laid over a sash bar.

There are two types, but they could be combined in the one tool. In the first type the end of the templet is cut square and shaped to the counter-profile of the moulding. This was probably used, as shown under *Window Making*, with the Sash Scribing Plane.

The second type, or the other end of the combined tool, is mitred on each side. It was laid over the sash bar and the projecting part of the moulding pared down flush with the mitre on the templet on each side. This gave an exact marking of the amount of material to be removed with the Sash Gouge or with the Coping Saw. This method eventually superseded the use of the Scribing or Coping Plane.

(b) Side or Single Sash Templet
This is the corresponding tool to (*a*) but designed for marking the scribed moulding on door or sash rails. It consists of one half of the grooved templet, with the appropriate counter-profile worked on one side. In the modern type both ends are cut at 45° to give the right-hand and left-hand mitres in the one

tool, the material being then removed with the appropriate Sash Gouge.

Saddle Templet

a

b

Side Templet

Fig. 705

Tenon: see *Mortice and Tenon Joint.*

Tenoning Machine: see *Mortising Machine.*

Terrier (Turrel)
A term used by Knight (U.S.A., 1877) for an Auger.

Thatch-Spar Maker (Other names for thatch spars include speaks, splints, sparrods, gads, brotches, ledgers, and roovers.)
Thatch spars are thin pieces of split wood, usually hazel, from 2 to 3 ft long and bent into the form of a staple. They are used for securing the longer spars ('liggers') which are laid across the exterior of the thatch to hold it down.
Thatch spars are not only made by the thatcher himself. They are often made as a spare-time job by other tradesmen. The tool used for trimming, splitting, and pointing the rods is most often a small curved knife made like a Bill Hook. (See *Woodland Trades* (1) for this and other splitting Tools.)

Thief
A name given to certain tools which in use are introduced, as it were, stealthily, e.g. see Thief Auger under *Auger, Cooper's Bung Borer;* Sampling Tube under *Cellarman.*

Thimble
For protecting the finger – see *Sailmaker.* For ropes – a metal eye, round or heart-shaped, inserted into a rope-ring (grommet) to make a durable loop.

Thixel (Thicksell; Thizle; Thixtell; Thyxtill)
An early term for Adze, and used occasionally in dialect until recent times.

Thole Reamer: see *Plane, Rounder.*

Thrift: see Mill Bill under *Millstone Dresser.*

Thrower: see *Froe.*

Throwing
Term for turning on a Lathe or moulding a pot on a wheel. For 'Throw' see diagram under *Brace.*

Thyxtill: see *Thixel.*

Ticketer
A chairmaker's term for a steel rod used for turning over the edge of a scraper to produce the cutting burr. See *Scraper.*

Tickler: see Bung Remover under *Cellarman.*

Tilter (Sprig Hammer)
A name given by Tomlinson (London, 1860) to a Hammer with a long head and square faces. It is illustrated as a glazier's tool used for sprigs. (See *Hammer, Sprig.*)

Tilting Haunch: see *Cellarman.*

Timber Girthing Tape and Sword. (Girthing Strap; Quarter-Girthing Tape) *Fig. 706*

(*a*) *Timber Girthing Tape*
A linen tape measure, $\frac{5}{8}$ in wide and up to 12 ft or more long, with a wire ring at the end. It is used for measuring the circumference of a round log in order to calculate the cubic contents. The most usual markings are a girth scale on one side, and feet and inches on the reverse. The girth scale is the actual measurement divided by four; this figure, together with the length of the log, is then applied to the 'Hoppus' table – the Timber Merchant's 'Ready Reckoner'—and the volume of the log thereby ascertained. (See References, Hoppus, E.)

(*b*) *Timber Girthing Sword*
A curved flexible steel rod, about 3 ft long fitted in a wooden handle and with a hook on one end. It is used for passing the Tape under a log lying on the ground. The sword is pushed under the log, hooked on to the end of the tape, and drawn back.

a

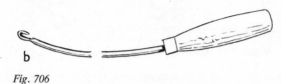

b

Fig. 706

Timber-Handling Tools (See also *Tree Feller*.) *Fig. 707*

The following tools are used for handling timber after felling:

(*a*) *Marking Hammer* See under *Hammer, Marking*.

(*b*) *Cant Hook* (Cant Dog; U.S.A.: Swing Dingle)
A strong wooden bar, about 5 ft long with a curved iron arm, hooked at one end and hinged at the other from a point about 7–10 in from the foot. Used by tree fellers and others for turning heavy logs when trimming the undersides, and for such tasks as dislodging a tree after felling, if it is caught in the crown of its neighbours.

The Peavey is similar to the Cant Hook but has a spike at the foot. According to the Chronicle of the Early American Industries Association (U.S.A., June 1967), it was invented by a blacksmith, Joseph Peavey, about 1870. A Gee Throw is a stout wooden Crow Bar with a pointed iron shoe, also used for handling Timber.

(*c*) *Ring Dog* (Cant Hook Dog; Log Hook)
A curved iron arm with a sharpened hook at one end and a ring at the other, about 20 in long overall. Often found in country workshops and saw pits, it is used for rolling over a heavy log. The hook-end is driven into the log, the tip of a crow bar is passed through the ring, and the log is levered over.

(*d*) *Woodhook* (Pulphook)
A sharp hook set on a handle, similar to a docker's Loading Hook. Used for handling pulpwood and pit-props.

(*e*) *Tongs*
Often made on the scissor principle with sharp inward pointing spurs to grip a log. Used for handling timber, as an alternative to (*f*).

(*f*) *Span Dogs*. (Drag Shackle; Span Shackle; Crotch Grabs)
A pair of sharp claws connected by chain or iron bar to a single ring. Used to grapple timber.

(*g*) *Sappie* (U.S.A.: Pickaroon)
A 10 in steel spike mounted like an Adze on the end of a 3 ft axe-haft.

Used for moving light timber or, after driving the point into a log, for pulling it along the ground.

Timber-Marking Tools
A list of tools used for marking or branding timber is given under *Marking Tools*.

Timber Scribe (Scriber; Race Knife; Scorer; Scrieve or Scrive Hook; Scriving Knife; Skiven Iron; Raze Knife) *Fig. 708 overleaf*
The simplest type consists of a steel blade with the end bent round to form a sharp gouge-like cutter; this excavates a groove (or 'race') when pulled toward the user. Others incorporate a central spike and a second cutter for making a circular groove, and an additional drag-knife for scribing numbers and letters. They are used by foresters, timber merchants, shipwrights, coopers, and other tradesmen to inscribe serial numbers, letters, or code marks on timber and wooden objects. Medieval carpenters used a Scribe (or a V-shaped Chisel) to mark the members of a wood-framed building. This was done to ensure that when the parts of the frame were transported to the site for erection, each tenon would be fitted into the mortice for which it was intended.

Special Uses. Shipwrights, who commonly call this tool a Scrive Hook, use it for marking out the plan of the ship's frame on the Scrive Board. The drawings are 'layed off' in full size on this board, which occupies the greater part of the floor. Since they are frequently walked over, any markings less permanent than an actual incision on the face of the board might become obliterated. Coopers sometimes use a Timber Scribe for cutting a croze groove in very small casks, especially during repair. A large size of Timber Scribe is used for rubber tapping. (See *Knife, Rubber Tapping*).

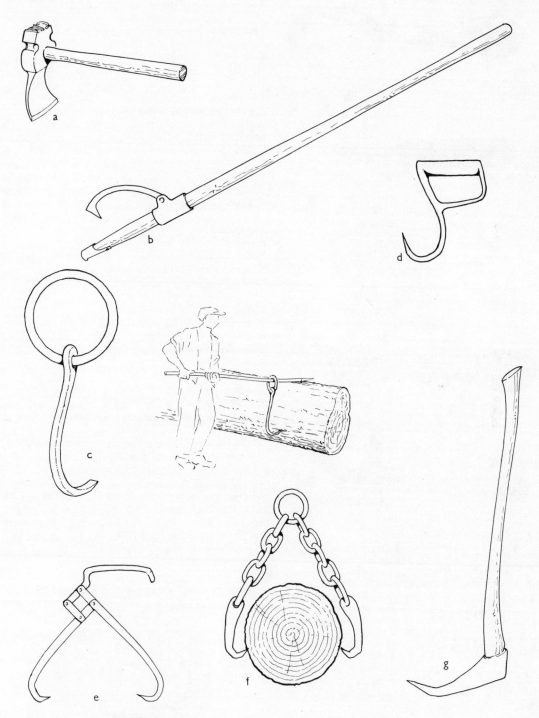

Fig. 707 Timber Handling Tools

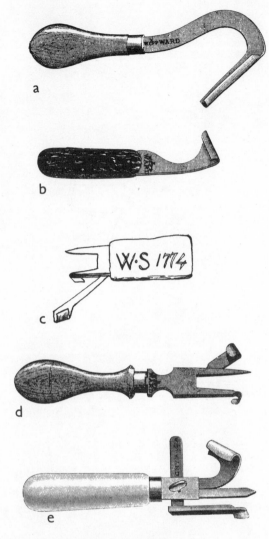

Fig. 708 Timber Scribe: variants
(*a*) With swan-necked drag-knife
(*b*) With folding drag-knife
(*c*) With point and cutter for inscribing circles, and folding drag-knife. (This example, dated 1774, came from a Suffolk workshop)
(*d*) As (*c*) but with a fixed drag-knife
(*e*) As (*d*) but with an adjustable cutter for inscribing circles of different sizes

Tine
A prong or tooth of a harrow, fork, or rake.

Tine Former and Horse: see *Rake Maker.*

Tint Tool (Tinter): see *Wood Engraver.*

Toat
Handle of a Bench Plane.

Tomahawk
A name sometimes given to the following tools:
   (*a*) Certain small Hand Axes or Hatchets when provided with a pick-poll.
   (*b*) A Mortice Knife described under *Hurdlemaker.*
   (*c*) A rough Scraper made from a worn-out plane iron (see *Scraper, Glue*).
   (*d*) Small Pick Axe type tools used by railway workers and others.

Tommy Bar
A rod of metal or wood inserted loosely through the handle or the head of a tool, roller, or windlass, in order to turn it. Another example is the loose bar used for turning the screw of a Bench Vice.

Tongs
These are usually large pincer-like tools, but the term is also applied to certain other tools described under: *Basket maker; Boat Grip; Cellarman; Timber-Handling Tools.*

Tongue
A projecting piece fitting into a corresponding groove or hole, e.g. a spoke tongue which fits into a hole in the felloe; or the tongue-and-groove of matched boarding. See also Tongue under *Woodwind Instrument Maker* and *Reed Maker.*

Tool Basket (Frail; Tool Bass; Tool Bag) *Fig. 709*
A large bag of canvas, woven rush, or bass, with rope handles, sometimes lined with sail cloth or leather. Used by tradesmen for carrying tools from job to job. Another type, with a lid, was known as a Workman's Hand or Dinner Basket. (See *Carrying Stick*.)

Painters' Nest Hand Basses.

Joiners' Tool Bass,
Web Bound and Canvas lined.

" Hardware "
Brown Canvas Tool Bass.

Fig. 709

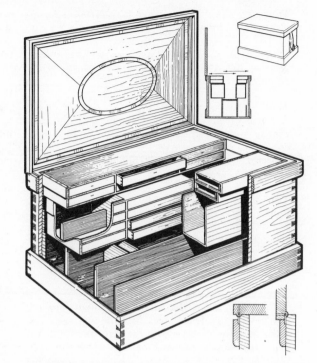

Fig. 710

Tool Chest *Fig. 710*
Between about 1780 and 1930 the joiner or Cabinet maker would often make himself a chest to contain his tools. Many of these were elaborately fitted and decorated. Edward Pinto (London, 1969) describes them as follows: 'Tool chests reached their zenith of size, elaboration and craftsmanship when woodwork attained its apex of quality in the eighteenth century. They are plain externally, strong and heavily built of pine, dovetailed at all angles, weathered, and well protected with paint, made to withstand rough travel by wagon and ship, and workshop usage. Interiors, however, are sometimes quite different – lined with mahogany, fitted with tool racks and nests of drawers, and occasionally as elaborately veneered and inlaid as the finest furniture. One of the most splendid examples is in the Victoria and Albert Museum.'
A typical nineteenth century example is illustrated.

Tool Grinding Rest: see *Grinding Appliances.*

Tool Holders: see *Awl, Pad; Grinding Appliances.*

Tool Pad: see *Awl, Pad.*

Tool, Parting: see *Chisel, Turning.*

Top Iron: see Cap Iron under *Plane.*

Torus
A form of moulding (see *Plane, Moulding*).

Toy
A term applied to small tools and other articles. See *Lancashire Tools.*

Tracer Bar: see Jack Staff under *Millstone Dresser.*

Tracing Wheel: see Traveller under *Wheelwright's Equipment (3) Tyring Tools.*

Track Tools: see list under *Railway Tools.*

Tram: see Wheel Frame under *Wheelwright's Equipment (1) Workshop Furniture.*

Trammel: see *Compass, Beam;* Radial Knife under *Drawing Knife, Cooper's Chamfering; Wheelwright's Equipment (2) Spoke Tools.*

Trap
A name given to a special Rounder Plane. See *Fishing Rod Maker.*

Traveller: see *Wheelwright's Equipment (3) Tyring Tools.*

Travice
Gilpin (Cannock) uses the term 'Cooper's Travice' in his catalogue of 1868 to describe a Bung or Shive Extractor (See Bung Removers under *Cellarman.*)

Travisher: see *Shave, Chairmaker's.*

Tree Feller *Fig. 711*
Until recent times, the village tradesman selected, purchased, and felled trees for his own use. The oldest tool for felling trees is the Axe, and though it has been largely replaced by the Cross-Cut Saw (and more recently by power-driven chain saws) it is still used to give the Saw a start. The process may be followed on *Fig. 711*:

(*a*) When the time comes for felling, the base of the trunk is 'rounded-up' with the Felling Axe as close to the ground as practicable. A series of horizontal cuts are made round the periphery, followed by vertical cuts from above, so removing the buttresses which swell out into the roots. A 'sink' or 'birdsmouth' is then chopped on the side on which the tree is to fall.

(*b*) The effect of (*a*) is to cause the tree to lean slightly in the direction of the sink, so bringing the fibres on the opposite side into tension. Thus when the Felling Saw is used at this side, it is not so liable to be trapped by the weight of the tree.

(*c*) The Saw may bind as the cut deepens, so Wedges are then driven into the kerf to free it. As the Saw and Wedges drive further and the tree begins to topple, the fellers step smartly back lifting their Saw clear as they do so.

The attitude adopted by the fellers varies. The most customary position is for both men to kneel, one on either side of the tree, facing away from the cut, with the Saw cutting towards them. But it is possible for them to face the other way or to work on their feet, bending over.

Tools used for felling and for the subsequent handling of the timber are described in the following entries:

Axe, Felling
Axe, Topping
Barking Iron
Calliper, Timber
Hammer, Marking
Lifting Jack
Log Stick Measurer (under *Rule*)
Saw, Cross-Cut
Timber Girthing Tape and Sword
Timber-Handling Tools, including *Cant Hook; Peavey; Ring Dog; Sappie or Pickaroon; Span Dogs; Tongs; Woodhook* (Pulp Hook)
Timber Scribe
Wedge, Tree Felling

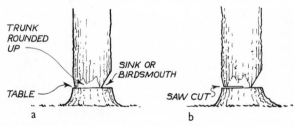
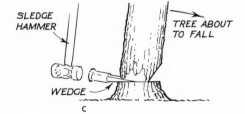

Fig. 711 Felling a Tree
(*a*) Making the 'sink' and 'Rounding up'
(*b*) Sawing through the trunk
(*c*) Wedging

Trenail Tools (Literally tree-nail pronounced trunnel)
For Trenail Maul see *Hammer, Ship Maul.*
For Trenail Iron see *Caulking Iron.*
For Moot or Trenail Rounder see *Plane, Rounder, Trenail.*
For Boring tools for Trenails see *Auger, Shipwrights.* (See also *Chisel, Trenail.*)

Trenails are round pegs of oak used for fastening together the parts of a ship and other structures. The best Trenails are split, not sawn, to ensure a continuous grain and so avoid fracture when being driven in. They were roughly trimmed to size, rounded with a Moot (see above), and then driven without grease (to avoid swelling). A common size was $12 \times 1\frac{1}{4}$ in, but much longer and thicker Trenails were also used. A heavier type of Trenail is called a Coak and is used for joining very heavy timbers, e.g. in the construction of Masts, Lock Gates, etc.

The 'pellets' used in shipbuilding should not be confused with Coaks or Trenails, which they may resemble in outward appearance. They are oak pegs of cross grain which are used for covering deck bolts; the cross grain allows the pellet to wear evenly with the deck. (See *Auger, Deck Dowelling.*)

Trepan and Trephine
Terms relating to the removal of a circular disc from the skull in surgery; but applied also to the cutting of discs in wood and metal work. See *Saw, Crown.*

Trestle (Trussel) *Fig. 652* under *Sawing Horse.*
This consists of a strong bed, about 4 in square and 30 in or more in length, supported on four splayed legs about 2 ft high. Trestles are used for supporting wood while being sawn, and in some trades for supporting the work in hand. For example, wheelwrights and coachbuilders use trestles on which to assemble vehicle bodies and undercarriages. A special trestle on which to stand furniture is illustrated under *Upholsterer.*

Trugger: see *Basket Maker; Trug.*

Truss Hoop: see *Cooper* (3) *Hooping Tools.*

Trying Tools
A term used in connection with certain tools, e.g. *Plane, Trying; Square, Try; Butter and Cheese Sampler.*

Tuner: see *Pianoforte Maker and Tuner.*

Tungsten Carbide: see under *Iron and Steel.*

Tuning Tools: see *Organ Builder; Pianoforte Maker and Tuner.*

Turnbuckle (Screw Shackle)
A long nut or buckle, screwed opposite hands at each end, and serving to connect and draw together two ends of a wire rope or bar; e.g. the tensioning of a saw blade in a Bow Saw.

Turner (Wood)
Wood turning was known to the Egyptians and also practised by the Greeks and Romans. At that time the work was held between two pointed centres and rotated by means of a strap or bow. (Devices of this type are still used in Cairo at the present time for turning beads etc.) An early improvement was the Pole Lathe, which according to Gordon Childe (Oxford, 1954) was known in Classical times. It is illustrated by Moxon (London 1677) and until recently was still in use by the chair bodgers of the Chilterns and by bowl and spoon turners. (See *Chairmaker* and *Lathe, Pole.*) By the middle of the eighteenth century it had largely been superseded in urban workshops by the treadle lathe with a crank and flywheel. For turning heavy work in hardwood such as elm wheel hubs, wheelwrights used a massive Lathe with the pulley driven by a belt from a large wheel turned by hand. (See *Lathe, Wheelwright's.*)

During the eighteenth and nineteenth centuries turning became a popular hobby among gentleman amateurs, using a treadle Lathe with ingenious accessories for various types of ornamental work. It was largely for this purpose that Holtzapffel made his celebrated Lathes and wrote his monumental work on the subject (London, 1847).

Wood turning may be divided into two main branches: centred work and face-plate work. In both cases the preliminary rounding and shaping is done with large Turning Gouges. In face-plate work the finishing is carried out with scraper tools; in centred work or spindle turning, with sets of Gouges, Chisels, and V-Parting tools; and for very hard, close-grained woods like box and ebony, with profiled scraper tools.

Tools used in ordinary wood turning are described under the following entries: *Calliper, Wheelwright's* etc.; *Chisel, Turning* (including *Turning Gouges*); Hook Tool, under *Chisel Turning; Lathes;* Sizing Tool under *Chisel, Turning.*

Turning Box: see *Moulding and Turning Box.*

Turning Steady
A name given to a wooden device which is mounted

on the lathe bed and used by chairmakers and others to steady slender or springy work while being turned. A wooden, hinged bar is held in contact with the back of the revolving work, and as this is reduced in size by turning, the vibration causes a wedge to slide down behind the bar and thus holds it in contact with the work.

Turning Tools: see *Chisel, Turning*.

Turning Web
A name sometimes given to a Saw of the Frame and Bow type, or to a very narrow Pit Saw.

Turnscrew: see *Screwdriver*.

Turrel
An obsolete term for an Auger surviving in the term Terrier used by Knight (U.S.A., 1877). Cotgrave (1611) describes the Turrel as 'th' Oager wherewith Coopers make holes for the barre-pins [bungs] of a peece of caske'.

Twist Bit
A general term in workshop practice for Bits with a spiral body. See Auger Bits under *Auger, Twist*.

Twybill or Two-Bill
Names sometimes given to certain two-bladed tools including *Axe, Double-Bitted; Axe, Grubbing; Axe, Mortice; Axe, Twybill;* Mortising Knife under *Hurdlemaker*.

Type: see Spoke Bridle under *Wheelwright's Equipment* (2) *Spoke Tools*.

Tyring Tools: see *Wheelwright's Equipment* (3) *Tyring Tools*.

U

Undertaker: see *Coffin Maker*.

'Universal' Tools: see *Combination Tools*.

Upholsterer (including the Carpet Layer) *Fig. 712*
The upholsterer's trade has developed from the attempt to increase the comfort of chairs or beds by providing fixed (rather than loose) cushioning material as part of the furniture itself. It is held in place by stretched fabrics or leather.

There are several types of upholstery, including:

1. Padding spread on a hard foundation.

2. Padding suspended on interwoven straps of webbing supported on the frame of the chair or other furniture.

3. Padding spread on stout canvas which is supported by helical springs.

The modern trend is to dispense with webbing and springs and to replace them with cushions of synthetic material such as foam rubber. The tools described below include those used by the carpet layer:

AWLS

(a) Garnish Awl (Entering Awl: Straining Awl)
Thicker at the base than the usual Awl, with a round, highly polished, pointed blade, about $2\frac{1}{2}$ in long and often fitted into a peg-top shaped handle. Used by carriage trimmers, upholsterers, and saddlers for enlarging holes in cloth and leather and for general purposes.

(b) Pritch Awl
A slender spike $1\frac{1}{2}$ in long with a shouldered portion immediately below it. Used by upholsterers and others for piercing and marking out leather and other materials, for example when fixing buttons. It was also used for marking out the glazed union cloth used for making blinds which, until recently, were commonly made by upholsterers.

(c) Square Awl (Birdcage Awl)
A tapered, pointed blade, square in cross-section. Used for boring hard wood. For further particulars see *Awl, Square*.

(d) Stabbing Awl (Stiletto) [Not illustrated]
Similar to the Garnish Awl but the blade is thinner.

Used by upholsterers and others for piercing tough material before using a needle.

(e) Trimmer's Awl [Not illustrated]
Similar to the Stabbing Awl, but with a longer blade ($2\frac{3}{4}$–$3\frac{1}{2}$ in). Used by carriage trimmers for piercing leather and other material.

CARPET-LAYER'S TOOLS

(f) Carpet Hammer
A long strapped head with a round face and a plain cross pane. In one pattern a claw is provided at the side or on the top of the head; in the other it is fitted by means of a ferrule to the end of the bulbous turned handle.

(g) Carpet Ringer
A blade shaped like a Spoon Bit set in a wooden handle. Rings are sometimes sewn on the back of a carpet, about 3 in from the edge, designed to drop over the heads of wood screws in the floor. The tool is used like a button hook. The edge of the carpet is turned up and the tool passed through one of the rings. The spoon end is caught behind the screw head and the ring is levered over the screw head; this stretches the carpet and secures it.

(h) Carpet Stretchers
Two types are illustrated. The older kind has a semi-circular head with a toothed edge on a handle about 2 ft 3 in long, with a mushroomed top. It is used for pushing and stretching a carpet towards the wall and round projections. The carpet layer uses the tool when kneeling, and he pushes the mushroom top with the side of the knee.

(*h1*) A modern Stretcher is used flat on the carpet. One end is a wooden block with teeth projecting into the carpet below it. On the back is a padded handle which, when pushed with the knee, forces the carpet away from the operator.

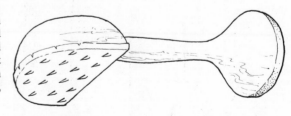

Fig. 712 (h1)

UPHOLSTERER'S HAMMERS
These light-weight hammers all have long strapped heads with a small round face, and have a claw at

491

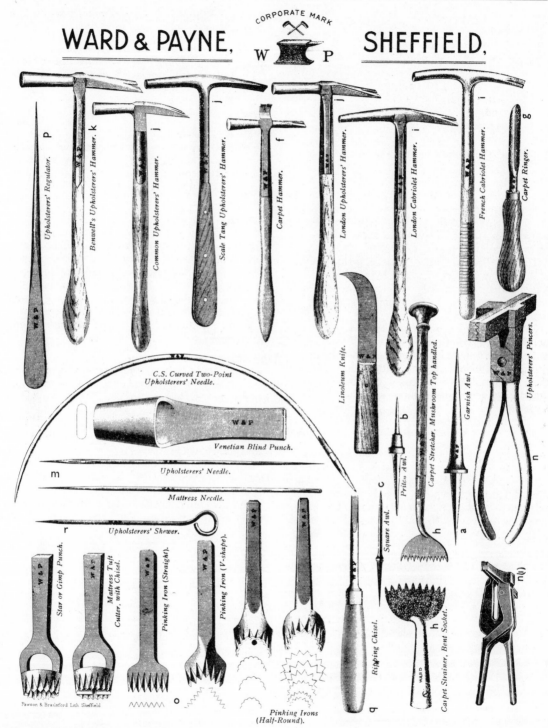

WARD & PAYNE, SHEFFIELD,

CORPORATE MARK
W P

Upholsterers' Regulator. **p**

Benwell's Upholsterers' Hammer. **k**

Common Upholsterers' Hammer. **j**

Scale Tang Upholsterers' Hammer.

Carpet Hammer. **f**

London Upholsterers' Hammer. **l**

London Cabriolet Hammer. **i**

French Cabriolet Hammer.

Carpet Ringer. **g**

Linoleum Knife.

C.S. Curved Two-Point Upholsterers' Needle.

Venetian Blind Punch.

Upholsterers' Needle. **m**

Mattress Needle.

Upholsterers' Skewer. **r**

Carpet Stretcher, Mushroom Top handled. **h**

Prikin Awl. **b**

Garnish Awl. **a**

Square Awl. **c**

Upholsterers' Pincers. **n**

Star or Gimp Punch.

Mattress Tuft Cutter, with Chisel.

Pinking Iron (Straight).

Pinking Iron (V-shape).

Pinking Irons (Half-Round). **o**

Ripping Chisel. **q**

Carpet Strainer, Bent Socket. **h**

n(l)

Pawson & Brailsford Lith Sheffield

Fig. 712 (a)–(r) Upholsterer's Tools

the end of the cross-pane. They are designed for nailing leather and cloth to furniture and coachwork and for saddlery. (See *Hammer, Lightweight* for their use in other trades.) Variants include:

(*i*) *Cabriolet Hammer*

The same tool as a Coach Trimmer's Cabriolet Hammer but used by upholsterers for 'show-wood' work; i.e. where the tacks are plainly visible along a narrow edge, when a Hammer with a very small face is needed to drive them in neatly.

According to Mr. C. Howes (1968) the upholsterer's apprentice took pride in having a Cabriolet Hammer of good quality and preferably of French make. The 'common' Upholsterer's Hammer was good enough for 'stuff-over' work, i.e. when the nails are covered by material or braid.

(*j*) *Common Upholsterer's Hammer*

This has a clawed pane which, being straight rather than curved, can be used for starting tacks. It is sometimes made with a double face, one of normal size and the other very small. The London Pattern has two ornamental 'nicks' where the straps meet the head.

(*k*) *Benwell's Hammer*

Similar to the common Upholsterer's Hammer but more elaborately finished. The head is ornamented with grooves and chamfers and the straps are long and pointed.

(*l*) *Scrim Hammer* [Not illustrated]

Similar to the common Upholsterer's Hammer, but with no claw on the end of the pane. Scrim is the thin canvas used for lining in upholstery work.

(*m*) *Needles*

These are made in various sizes, up to 18 in long, and often double pointed. They are used for sewing tufts on a mattress, buttoning chair seats, and similar work. The needle is pressed in plain-point first; the eyed end is not pulled right through but is returned at a different position, thus forming a loop in the stuffing. When sewing springs to webs and to hessian a Spring Needle is used, shaped to a fairly flat curve. The half-circular Mattress Needle is used for sewing covers etc. in awkward places.

(*n*) *Upholsterer's Pincers*

About 9 in long with the jaws extended on both sides, and serrated for holding material securely. At the centre of one jaw there is a projection (or 'hob') which acts as a fulcrum. Used for stretching the covers over the padding before nailing. Also used for stretching webbing if the web is too short to be held by a Web Strainer.

Another type, known as a Canvas Stretcher, is made on similar lines and for a similar purpose; but one jaw has a deep throat in which the edge or fold of the canvas rests. This enables the pincers to grip the canvas at some distance from the edge.

(*o*) *Pinking Irons and Punches*

All these Punches have serrated or scalloped edges. The Mattress Punch is round and used for cutting out leather tufts. The Pinking Irons are straight, V-shaped, or half-round, and are used for cutting ornamental designs in leather and other materials.

(*p*) *Regulator*

A steel spike, 6 to 12 in long, tapering to a sharp point at one end and flattened at the other. Used for adjusting the stuffing in a chair seat or arm, etc. The Regulator is poked through the hessian covering (which holds the stuffing in position) and is then moved with a circular or sideways motion in order to bring forward, re-arrange, or smooth out the stuffing.

(*q*) *Ripping Chisel*

Usually a round steel blade, $\frac{1}{2}$ in wide and 6 in long, set in a wooden handle and sharpened with a blunt edge like a screwdriver. Used by upholsterers for ripping off covers when remaking. It is also used with a Hammer for driving under the head of the tacks to remove them.

(*r*) *Skewer*

A spiked Skewer, from 3 to 5 in long, with a circular loop at the back. It is stuck into the seating etc. to hold the covers in place while setting out, and before nailing or sewing.

(*s*) *Trestle*

A pair of wooden Trestles, about 2 ft 6 in high and 3 ft long, made like carpenters' Sawing Trestles but provided with a wooden fillet round the top. Chairs and other furniture are stood on these trestles while being upholstered. The fillets prevent the legs from slipping off, and also serve to hold the small tools and nails.

Fig. 712 (s)

(*t*) *Tack Lifter or Claw.*
See separate entry under that heading.

WEB STRAINERS (Scots: Dwang; Web Stretcher)
This small wooden implement is one of the most important tools in the equipment of the upholsterer: it is in frequent use to give tension to the webbing which forms the foundation in most types of upholstery. There are two common types:

(*u*) A flat, bat-shaped piece of wood, about 10 in long with a rectangular aperture in the lower part. One end of the webbing is nailed in position; the other is looped through the slot in the Strainer, with the peg put through the loop to secure it. The Strainer is then levered over to stretch the webbing which, when taut, is nailed down to its point of attachment.

(*v*) Often made by the upholsterer himself, this is also bat-shaped, and is rebated at the square end. The 'grip' is a metal strip with the ends pivoted to the side of the Strainer. After one end of the webbing is tacked down, the free end is passed under the grip and around the bottom of the Strainer.

Note: In recent years a metal Strainer has been developed for stretching metal webs.

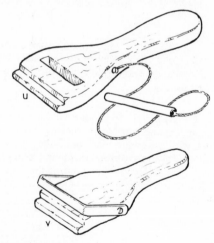

Fig. 712 (u)–(v)

Upright
One of the *Shaves* used in basket work (see *Basket Maker*.)

V-Block: see *Bevel Block*.

Veiner (Veining Tool): see *Chisel, Carving*.

Velincher: see Sampling Tools under *Cellarman*.

Veneering and Marquetry *Fig. 713 overleaf*

1. TERMS USED

Veneering. The practice of covering a good but comparatively uninteresting wood (the 'core') with a thin layer of more decorative wood. It has a long history, going back at least to Roman times.

Marquetry. This word is a general term for the various methods of making pictures and decorations formed by different coloured veneers, usually cut into a background of veneer.

Inlay. The flush insertion of bandings and decorative designs into the surface of veneers or solid wood. The insertions may be of wood, metal, or other materials.

Intarsia. The making of flush perspective pictures by using veneers of different colours or grains.

Buhl (or Boulle) Work. Named after a famous French cabinet maker (A. C. Boulle, 1642–1732), this term has come to be applied to cabinet work inlaid with ivory, mother-of-pearl, metal, and other materials other than wood.

2. VENEERING METHODS

Until quite recently veneer sheets were sometimes cut entirely by hand with a Saw, often at the saw pit, by means of a specially thin blade held rigid in a frame. But the more general method in the nineteenth century was the power-driven Circular Saw; later the Machine Knife was used for slicing out the veneer sheets; decorative veneers were sliced flat and others were rotary cut.

There are two methods of laying veneer, Hammer and Caul. In the former the veneer is glued, laid in position, and lightly damped. A warm Flat Iron is passed over it, and the Veneering Hammer worked over it zig-zag fashion from the centre outwards, pressing out surplus glue and forcing the veneer down. Caul veneering is used for veneer with a

difficult grain, for marquetry, for shaped surfaces, and usually for thicker saw-cut veneers. The Caul is a flat or shaped sheet of wood, sometimes faced with zinc. It is heated and cramped down over veneer, with paper interposed to avoid adhesion from squeezed-out glue. The surface is finally cleaned up with a Scraper and glasspaper.

3. TOOLS USED

(a) Veneer Press and Caul
A bench with a top adapted to take the hand screws which cramp down the Caul over the veneered surface.

The Caul is a rectangular plate, either flat or shaped to the contour of the work. It is applied warm to the veener and kept in position with the cramps until the glue has set.

Veneering Hammers (Flattener)

(b) In wood
A flat piece of wood with a short handle. The working edge, up to $4\frac{1}{2}$ in wide, is usually formed by an inserted strip of brass or aluminium about $\frac{1}{16}$ in thick. The word 'Hammer' is something of a misnomer, since it is not used for striking, but is used for pressing down veneers and squeezing out excess glue.

(c) In metal
Factory-made Veneering Hammers have an ordinary hammer head of the Warrington or Manchester type with round face and chamfered neck, with a thin cross pane flared out to a width of about 3 in or more.

(d) Flat Iron
The example illustrated is a Box Iron made *c.* 1900. The body is hollow and contains a loose piece of cast iron which is heated in the fire and placed inside. When applying Veneer with the Hammer, the Iron is worked over the damped surface of the veneer, in order to reheat and soften the glue beneath.

(e) Veneer Saw
A small blade with a convex edge and a wooden grip fixed to the back. Used with a straight edge for cutting veneers. (A Framed Saw for cutting veneer sheets is described under *Saw, Veneer*.)

(f) Knife
For Veneer. A short knife held in one hand and used to cut out and replace blemishes, to cut curved joints in veneer, and for cutting round a metal templet when insetting plain geometrical veneer inlays.

For Marquetry. A short knife fitted with a long curved handle, used widely on the Continent from

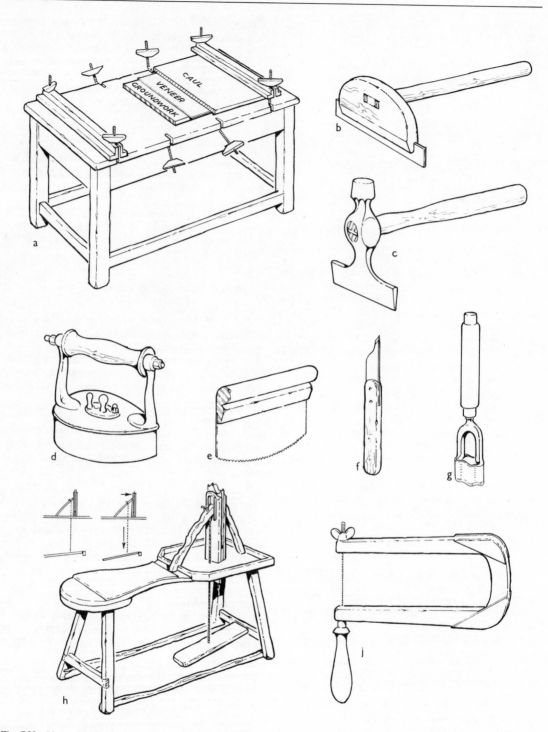

CAUL
VENEER
GROUNDWORK

Fig. 713 Veneering and Marquetry: tools and appliances

the sixteenth to the nineteenth century for cutting veneers for inlay and marquetry work. The long handle was held under the arm-pit or against the shoulder to apply pressure and control the tool. It was gradually superseded by the Fret and Jig Saw.

(g) Veneer Punch

With a cutting edge of irregular outline, this Punch is used to stamp out blemishes in the veneer such as the holes and cracks sometimes found in burr walnut. A corresponding piece of sound veneer is cut out with the same tool and inserted in the hole.

(h) Marquetry Donkey (Buhl Horse)

A special stool or 'Donkey', on which the operator sits astride. Used for cutting marquetry veneers with a Fret Saw. When the veneer (or packet of veneers) is placed between the jaws, it is held firm by pressure on the pedal, leaving one hand free to turn the packet as required, and the other to manipulate the Fret Saw. In the type of Donkey used by seventeenth and eighteenth century French cabinet makers, the curved jaws were fixed to a bench with the pedal below, the operator standing up to his work.

(j) Fret Saw

Used with the Donkey (see also separate entry under *Saw, Fret*).

Toothing Plane

Used for roughening surfaces to provide a key for the glue. See under *Plane, Toothing*.

Sand Tray

Used for shading light-coloured veneer in marquetry work. The sand is heated, and the piece of veneer, held in tweezers, is dipped into it. By only partially submerging the veneer, a brown shaded effect is produced.

Veneering Sandbag

A bag of a size to suit the work and filled with sand. Used for pressing down veneer on shaped surfaces, particularly on compound curves.

Vice

An instrument having jaws between which a work-piece can be gripped by tightening a screw, so leaving the hands free to work upon it. Besides the following entries, holding devices are described under *Cramp; Holdfast; Plane Maker; Woodland Trades*.

Vice, Cooper's (Head Vice; U.S.A.: Raising Iron) *Fig. 714*

Made like a coarse Gimlet with a key-like handle, this is not a Vice in the ordinary sense but is used by Coopers for holding up the head of a cask. The tool is twisted into the head in order to lift the head up into place. It is then unscrewed and removed. (There are two other methods of raising the head which are described under *Jumper*.)

Fig. 714

Vice, Hand *Fig. 715*

The hinged jaws, 4–6 in long, are kept apart by a spring and drawn together by a hand screw. Primarily a metal worker's tool but often used in woodworking shops for holding small objects or for grasping recalcitrant nuts or bolts.

Fig. 715

Vice, Leg: see *Vice, Woodworker's*.

497

Vice, Post: see *Vice, Woodworker's.*

Vice, Saw-Sharpening: see *Saw Sharpening: Holding Devices.*

Vice, Shive *Fig. 716*
A device for extracting shives. See Bung Removers under *Cellarman.*

a

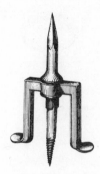

Fig. 716

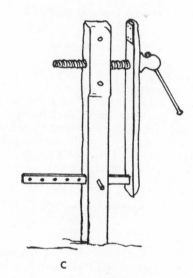

c

Vice, Staple: see *Vice, Woodworker's.*

Vice, Tail: see *Vice, Woodworker's.*

Vice, Woodworker's (Scots: Glaun) *Fig. 717*
The development of the woodworker's Vice is described under *Bench, Woodworker's.* Vices commonly found in woodworker's shops fall into the following classes:

(*a*) *Vices for holding the work against the edge of the bench top* (e.g. the common Bench Vice, as illustrated, but see also *Fig. 105* under *Bench*).
These usually consist of cast-iron jaws (chops or cheeks) faced with wood, maintained parallel to each other by steel slides, and closed with a steel screw. In the earlier types, still widely used, the cheeks, slides, and sometimes the screw itself are made of wood.

A different type has a long vertical wooden chop, about 4–5 in square, which carries a perforated strut at the lower end. The strut passes into a box or hole in the leg of the bench, where it is fixed with a peg which provides a movable fulcrum. This enables the jaws of the vice to hold either tapered or parallel work securely.

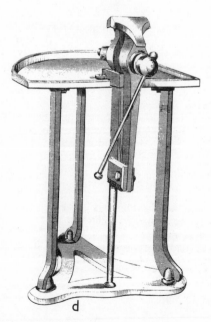

d

Fig. 717

(b) Vices for holding a workpiece flat on the top of the bench while its surface is worked upon (e.g. the Tail or German Vice, see *Fig. 105* under *Bench, Woodworker's*).

These are L-shaped blocks of wood, forming the right-hand corner of the bench top, which can be drawn in and out on a screw so that the workpiece is gripped between two stops – one set in a hole on the top of the Vice, the other in the bench top itself.

(c) Vices with chops rising above the level of the bench which will hold up a workpiece and leave a clear space all round (e.g. the Post Vice).

In this third class, the Vice is either free-standing as illustrated, or is mounted on a subsidiary bench. Such Vices are favoured by coachbuilders and wheelwrights for holding felloes while trimming them.

(d) An all-iron version of class (c) is the Leg (or Staple) Vice.

This is used primarily by smiths and other metal workers, but is commonly found in woodworking shops.

(e) Portable Vices which can be fixed to a bench top with a screw clamp. These are not firm enough to find favour in professional woodworking shops.

Violin Maker (Fiddle Maker) *Fig. 718 overleaf*

The immediate ancestors of the Violin were the Viols – the principal bowed instruments in use from the end of the fifteenth to the end of the seventeenth century. The parentage of the fiddle family has been ascribed to a bowed instrument of the early Middle Ages called a Rebec.

Materials. Sycamore is widely used in making a violin. For the back a ripple figure is preferred; it may be in one piece or jointed at the centre. The ribs, head, and neck are of sycamore also. Swiss pine is often used for the front, and the same wood

is used for the sound post and the bass-bar beneath the front. Pegs are of ebony, boxwood, or rosewood; the bridge spotted maple; the finger board is usually of ebony. The best violin bows are made from a resilient wood from Brazil known as Pernambuco.

Process. The front (belly) and back plates are sawn to outline with a Bow or Coping Saw, finished with the Violin Knife *(a)*, and inlaid round the edges with the use of the Purfling Tool *(c)–(e)*, followed by the Knife and Purfling Chisel *(b)*. (This purfling inlay is put in for decoration and to prevent the edges of the plates from splintering.) The belly and back are then fashioned to the characteristic contour with Gouge and Violin Plane *(f)* and finished with the Scraper *(n)* followed by glasspapering. Hollowing the inner surfaces is the next step, a Calliper Gauge *(m)* being used to test the thickness. The sound holes are cut with a Fret Saw or the Violin Knife, though some makers cut the small end holes with the Hole Piercing Tool *(k)*. The ribs or sides are bent round a heated Bending Iron and glued to the back. Internal blocks and linings follow, and the other parts are glued on in their proper order, the parts being held together during this process by various kinds of Clamps and Hand Screws, *(h)–(i)*. The head and neck are sawn out of one piece and the scrolled end formed with Carving Gouges. The holes for the pegs are tapered with the Reamer *(j)*, the sound post inserted with the Sound-Post Setting Tool *(l)*.

Among comprehensive and practical accounts of this trade see Heron-Allen (1884) and Varney (1958–1959) in the Bibliography and References.

Tools used include the following. (The term 'plate' is a name for the back and front of the violin.)

(a) Violin Maker's Knife

A rectangular blade, about $\frac{1}{2}$ in wide, held in a wooden handle, with the end skewed and bevelled on both sides. At least two are needed, one with a straight and the other with a round cutting edge.

Used for trimming the outline of the plates of a violin after sawing; also for cutting the sides of the purfling groove, after which the waste wood is picked out with the Purfling Chisel.

(b) Purfling Chisel or Picker

A very narrow chisel with a cranked blade like a Dog Leg Carving Chisel. Used to remove waste wood from the purfling groove.

(c), (d), (e) Purfling Tool

A form of cutting gauge used for marking out the groove for purfling. This is a narrow inlay round the edges of the plates which serves both to preserve the

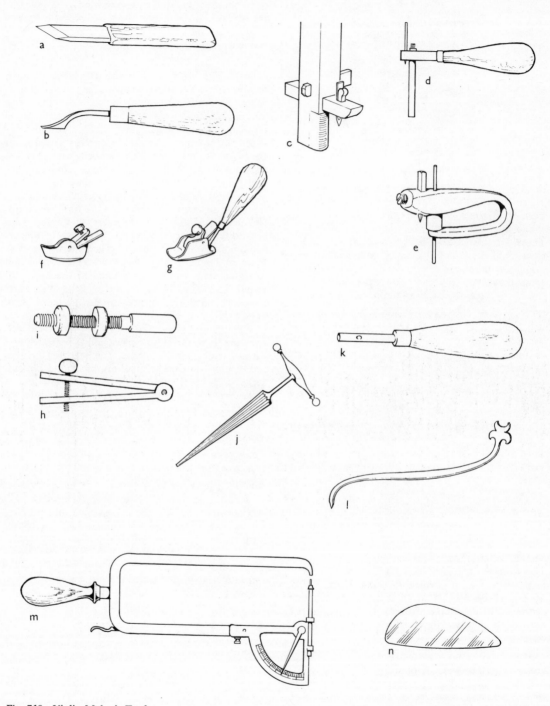

Fig. 718 Violin Maker's Tools

edges from splintering and as decoration. This tool marks the side of the groove; the cutting is done afterwards with the violin maker's knife.

Three types are illustrated: (*c*) made like a carpenter's marking or cutting gauge, (*d*) in metal with a metal rod for a fence, (*e*) a home-made example in iron with a wooden roller acting as a fence.

(*f*), (*g*) Planes
These thumb-type planes are described under *Plane, Violin*. They are used for shaping the back and front plates.

(*h*) Lining Clips
Two metal bars pivoted together at one end like a pair of compasses, with a thumbscrew bridging across them. Used for holding the lining while glueing them to the ribs. (The ribs are the sides of the violin; the linings are wood fillets glued round the inside joints of rib to plate.)

(*i*) Handscrew Clamps
Hardwood screws, about ¾ in. in diameter and 8 in long, with a turned handle and two round hardwood nuts. Used to cramp the top and bottom plates of the violin to the ribs.

(*j*) Rimer (Reamer)
One type has a tapered shank, about 6 in long, with straight cutting flutes; another is like a Taper Bit, with a half-funnel shaped blade. Both have short T-handles. Used for enlarging and tapering the holes in the peg-box of a violin in which the pegs are fitted.

(*k*) Sound-Hole Punch
A hollow steel cylinder with one end ground to a sharp circular edge, sometimes mounted in a wood handle. The small hole at the side is an escape for the wood chips. Used for piercing the small round holes at each end of the sound holes.

(*l*) Sound-Post Setting Tool
An S-shaped steel tool, about 9 in long, tapering to a point at one end and with two short prongs at the other. Used for inserting and removing the sound post when fitting it between the front and the back plates of the violin. The pointed end is used to spear the sound post and pass it through the sound hole; the hooked end is used to push or pull the sound post into the correct position.

(*m*) Calliper Gauge
This takes the form of ordinary wooden or metal Callipers, but the modern version illustrated has a U-shaped metal frame with a dial to register the thickness of the material being measured.

(*n*) Scraper
A thin steel plate of curved outline used for finishing the undulating surface of the violin plates.

Other Tools (not illustrated)

Arching Gauge
Wooden templates used for testing the contours when shaping the plates of a violin.

Bending Iron
A copper or iron bar about 8 in long, of rounded section at the head and square at the foot. Used for bending the ribs (sides) of a violin. The foot is held in the Vice and the head heated; the rib is lightly damped, held over the heated bar, and gradually bent to the required shape.

Bass Bar Clamps
Made like a clothes peg and varying in length from 5 to 8 in. Used for holding the Bass Bar in position when glueing it to the underside of the front plate. Another pattern is in metal with a screw adjustment.

Back or Crack Cramp
A length of mild steel rod bent to a narrow U-shape (like a miniature wheelwright's Sampson), about 12 in long and threaded at both ends to take wing nuts which are tightened over a metal plate. Used in jointing a two-piece back of a violin, or for mending a crack in the plates.

Voicer's Tools: see *Organ Builder*.

W

Wagon Builder's Tools
A term applied to certain tools used in making farm and railway wagons. See *Wheelwright; Railway Tools; Bit, Wagon Builder's; Chisel, Wagon Builder's.*

Waney Edge
The unsquared edge of a board as cut from the log, e.g. one on which the bark may still be attached.

Washer Cutter *Fig. 719*
These tools were used mainly by wheelwrights and coachbuilders for cutting out the leather washers which are fitted on the axles of carriages and vans between the inside of the axle box and the shoulder of the axle. They were fitted for quietness and to keep the oil in. Variants include:

(a) Washer-Cutter Bit
Made for use in a Brace, they consist essentially of a central post with a tapered square shank to fit in the Brace at one end and a point at the other. A rectangular bar, about 5 in long, slides through the lower end at right angles, and is fixed in the required position with a thumbscrew. The bar carries a spur cutter at the end which is also fixed with a thumb-screw. In another pattern two L-shaped spur cutters run through the central shank at right angles.

(b) Handled Washer Cutter
Often home made, this tool is similar to *(a)* above, but is provided with a short cross-handle.

(c) Washer-Cutter Brace
An iron Brace with a wide sweep, the lower arm of which is rectangular in cross section. A pointed Bit, and two knives for cutting the washers, can be fixed at any required distance along the lower arm.

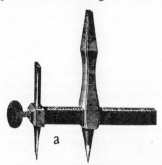

Fig. 719

Waving Engine
An implement described and illustrated by Félibien (Paris, 1676) and by Moxon (London, 1677). It consists of a board, about 5 ft long and 7 in wide, on which is mounted a sliding jaw used for pulling the strip of wood to be treated under a vertical cutter or scraper. By means of a guide pattern of the wave-moulding placed under it, the strip rose or fell in relation to the cutter so that its 'upper side receive the Form of the several Waves on the under side of the Rack and also the Form, or Molding, that is on the edge of the bottom of the Iron, and so at once the Ringlet will be both molded and waved.' (Moxon.) The so-called ripple-mouldings on early picture frames may have been made in a machine of this kind.

Wax Tongs: see *Cellarman.*

Web (Span Web; Tab Web): see *Saw, Web.*

Web, Calico Printer's (Doctor's Web; Calico Web)
A thin blade of steel, about $33 \times 2\frac{1}{2}$ in, with holes punched through the ends. Though made by saw makers it has a smooth edge without teeth. Used for scraping away superfluous colour and loose threads from the blocks or cylinders of a Calico Printing Machine.

Web Strainer: see *Upholsterer.*

Wedge Bittle: see *Beetle.*

Wedge Making
Wedges of hardwood, usually of oak, are made in the workshops for such purposes as fixing the box-bearing in a wheel hub; driving into the tongues of spokes, after assembly; driving into the top tenons of chair legs etc. Some account of their making will be found under *Axe, Coachbuilder* and *Shaving Horse.*

Wedge Spanner: see *Wrench (1) For Gripping Nuts.*

Wedge, Tree-Felling *Fig. 720; Fig. 711* under *Tree Feller*
Large iron Wedges, similar to the Wood Splitting Wedge but sometimes provided with a socket to take a short wooden butt. They were used by tree fellers to drive into the saw cut behind the Saw, in order to lift the trunk clear of the Saw. This prevented the Saw from seizing and gave the trunk a slight lean in the direction of the fall.

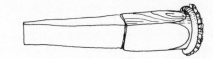

Fig. 720

Wedge, Wood-Splitting (Cleaving Wedge) *Fig. 721*
A solid steel wedge, 4–9 in long, often fluted about three-quarters of the way down on both faces. Used extensively for riving or splitting timber, particularly home-grown hardwoods such as oak. The purpose of the flutes is not clear; but they are said to relieve the pressure of air which builds up when driving into sappy wood.

Fig. 721

Whalebone Gauge: see Spoke Set under *Wheelwright's Equipment (2) Spoke Tools.*

Wheel Brace: see *Drill, Hand.*

Wheeler: see Mallet under *Sailmaker.*

Wheelwright (The tools used by wainwrights are included; but see also *Coachbuilder.*) *Figs. 722; 723*
According to the late Professor V. Gordon Childe (Oxford, 1954), wheeled vehicles were used in Lower Mesopotamia *c.* 3500 B.C. The wheels were wooden discs made up from three wooden planks cut to form a circle and clamped together by a pair of transverse wooden struts; a swelling, left in the middle of the central plank, formed a raised hub round the axle hole.

Spoked wheels – an outstanding achievement of the wood-worker's skill – are first represented *c.* 2000 B.C. in painted clay models from Near Eastern countries. Finely made wheels of light construction were being made in Egypt and Crete *c.* 1500 B.C., and two hundred years later in China. About 500 B.C. Celtic wainwrights in the Rhineland were making wheels of surprisingly modern appearance, and their art spread to northern Europe and to Britain. In some instances they made the rim from a single length of timber, bent into a circular form with heat; the ends were scarfed and overlapped, and the junction held by a metal swathe. It is interesting to note that similar bent rims were being made in the United States in 1909 and were regarded as a modern development.

Wheels and vehicles of many different kinds were made throughout the Roman Empire, but little is known of their development from that time until the twelfth century, when ready-made pairs of wheels could be bought in English market towns.

Fig. 722 Parts of a typical Wagon. (Nomenclature from Sussex and Surrey. Alternative names come from other southern counties of England.)

Body framework
1. Crook (Sole)
2. Side (Hindsole)
3. Front dware (Fore shutlock)
4. Middle dware (Middle Cross Bed)
5. Hind dware (Hind shutlock)
6. Summer

Body supports
7. Shore staff (False stuck)
8. Hindstaff
9. Strouter (Standard)
10. Clip (Side Iron)
11. Body spindle (Stretcher)
12. Stay (Rave stay)

Rails and Panelling
13. Top body lade (Top rave; Rail)
14. Out rave (If bowed – Wheel bow)
15. Middle rave
16. Panel board (Side board)
17. Bridle (Fore buck; Head piece)
18. Front panel board (Headboard)
19. Tail board or hawk

Undercarriage
20. Top pillow (Top bolster)
21. Bottom pillow (Carriage bolster)
22. Axle bed, or if wood (Ex bed)
    Axle Tree
23. Master pin
24. Hind bolster
25. Wagon pole (Tail pole; Tongue pole; Swimmer pole)
26. Brace (Spreader)
27. Hound (Guide)
27A Hound shutter
28. Slider (Sweep)

Wheels
33. Arm
34. Nave (Stock; Hub)
35. Nave bond (Nave hoop)
36. Spoke
37. Felloe
38. Tyre
39. Strake
40. Linch pin
40A Axle collar or collet
41. Stopper
42. Clip

Shafts
29. Rod (Shaft)
30. Rod key (Shicklebar; Shuth; Shutter)
31. Rod brace
32. Draught bar (Limmer bolt; Shaft bar)
(For two horses abreast, two pairs of shafts are fitted on a Splinter-bar on the fore-carriage in this position)

504

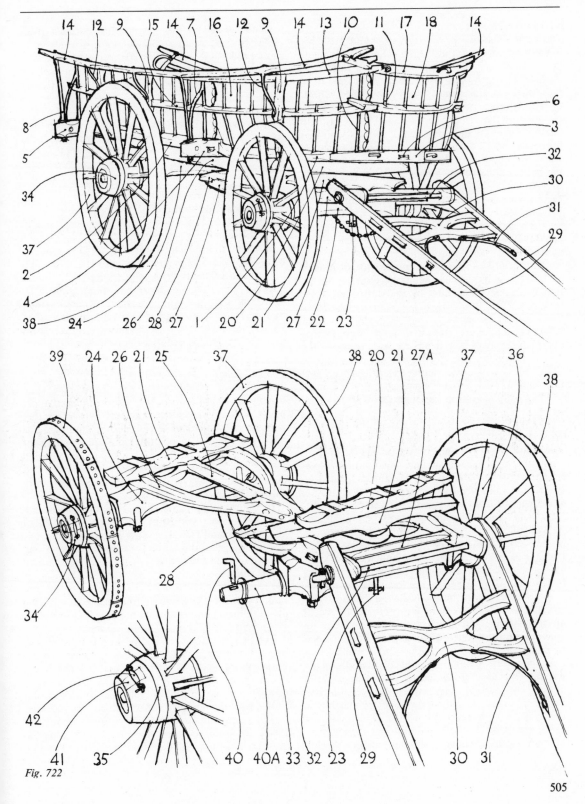

Fig. 722

Dished Wheels

During the last two hundred years – and probably longer – many cart and wagon wheels were made in the form of a flat cone – a process known as 'dishing' in which the spokes are set to emerge from the hub at a slight angle. As these dished wheels turn, the lowest spoke, which momentarily takes the full load of the vehicle, is vertical. This is brought about by bending the axle-arms downwards, so that the spoke nearest the ground is always perpendicular, while the upper spokes lean outwards, away from the body of the vehicle. This arrangement enables the vehicle to be built wider and consequently to carry a greater load. Another advantage of 'dish' is the added resistance to lateral thrusts; for, confined by the tyre, the cone-shaped circle of the spokes behaves like an arch and resists the sideways shocks which occur when transporting heavy loads over a rough or rutted surface. The process of dishing is achieved first by cutting the spoke mortices in the hub at an angle, and later by the shrinkage of the iron tyre. (See *Wheelwright's Equipment* (2) *Spoke Tools* and (3) *Tyring Tools.*)

When the spokes are 'staggered' on the hub, so that alternative spokes are made to lean towards the centre, thus forming two cones facing one another, the wheel is said to be 'double dished'. This was a later development and was done when it was desired to mount the wheels on horizontal axle-arms.

Wheelwright's Work

The wheelwright was not only a maker of wheels. He made wagons, carts, ploughs, agricultural implements, hay rakes, and many other things needed by the village and surrounding farms. In addition, the wheelwright often supervised the felling of local trees, and where a millwright was not available, he was called in to repair windmills and watermills. Some workshops had their own smith who made iron tyres, bonds for the hubs, and the iron cart-furniture and fittings for the vehicle. But in most instances, this was done by the local blacksmith.

A basic feature of both wheelwright's and wain-wright's work was strength. Joints had to be mechanically strong in themselves, with no reliance on glue to hold them together, for glues were weakened by damp.

In spite of the paramount need for strength, many wagons and carts are beautiful, and their graceful lines are enhanced by the practice of stop-chamfering. Done originally to lighten the load for the horse, this consists of a bevel cut with a Drawing Knife along the edge of the exposed frame and under-carriage, stopping short of the joints where the full

thickness is needed. (A note on this process is given under *Drawing Knife, Wheelwright's*. A similar finishing process can be seen on the connecting rod of a steam locomotive, on the shoulders of 'Lancashire' made Callipers, and on the forged frames of early bicycles.)

By the beginning of the nineteenth century wheelwrights became established in many villages and towns in England. And their finest products, the Box and Bow Wagons, were still being made in the early years of the present century.

Materials

The chief woods used are:

For the frame and undercarriage of wagons and carts – oak or ash.

For the wheels – oak for the spokes; elm (or occasionally oak) for the hubs; ash (or wych-elm) for the felloes, i.e. the rim of the wheel.

Oak is preferred for the spokes on account of its great strength; the hubs are made of elm because its tough and twisted grain prevents splitting when the spokes are driven into the mortice holes; ash is used for the felloes owing to its resilience.

The Workshop

The buildings belonging to a wheelwright's shop usually surround a yard in which are placed those pieces of equipment which could survive the weather and are convenient to use outdoors. These include the Tyre Benders – possibly three generations of them, Post, Lever, and the relatively modern geared Roller; and a Grindstone may be found mounted between two heavy posts under a lean-to or partly sheltered by a tree.

In the main shop, the Wheel Pit may be near a window or door for good light. The Wheel Stool and Chopping Block are set in a central position; and on one side, the Lathe for turning hubs. Along one wall is a narrow bench often made from a thick plank with its waney (uneven) edge left showing at the front. Immediately behind the bench is the tool rack filled with Chisels and Gouges. Hanging above are Templets and other gear. (See *Wheelwright's Equipment* (1) *Workshop Furniture.*)

The Saw Pit may be found in the workshop itself, but more often in a separate shed, adjacent to the timber store where boards and logs (for the hubs) lie for their long seasoning before use. (See *Sawyer.*) The Smithy is separated from the woodworking shops, with the Tyring Platform placed just outside.

The paintshop is sometimes housed separately. Inside are found the chest of drawers containing the powdered colours, and the Mulling Slab (see *Painting*

Equipment). The paint brushes are cleared of paint by slapping them on any convenient wooden surface, such as the door, leaving several inches of accumulated paint in many colours adhering to it.

The Process of wheel making may be briefly described as follows:

Hub. Elm logs are cut into lengths, and after several years' seasoning are turned on the Lathe to the diameter required. The mortice holes for the foot of the spokes are marked out and excavated with heavy Chisels including the Bruzz.

Spokes. Cleft from well-seasoned oak, the spokes are roughly dressed with an Axe and trimmed with a Drawing Knife and Spokeshave. The foot is tenoned to fit the mortice hole in the hub. The spokes are driven into the hub while it rests on the Wheel Pit or Wheel Horse. The Spoke Set Gauge is applied to check the angle of dish given to each spoke, and their alignment is adjusted with the help of the Spoke Bridle. Lastly, the upper ends of the spokes (the tongues) are cut to fit the hole in the felloes. (See *Wheelwright's Equipment* (2) *Spoke Tools*.)

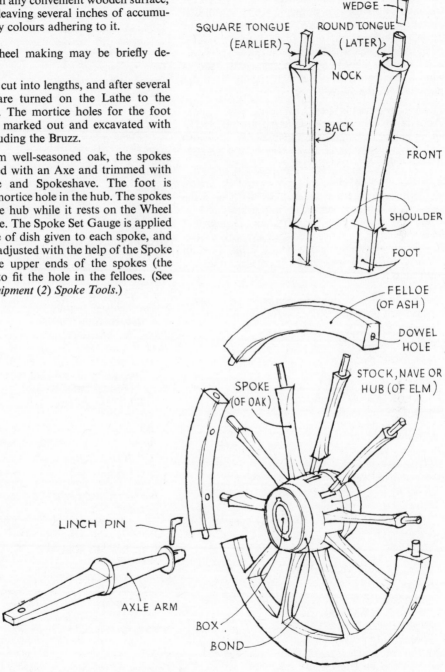

Fig. 723   Parts of a Wheel

Felloes. After sawing to shape with a Bettye Saw the felloes (which make up the rim of the wheel) are trimmed with Adze and Drawing Knife, and finally planed. Holes for the spoke tongues are bored and dowels inserted in the ends of each felloe to bind them in line. The wheel is laid on the Wheel Stool and the felloes tapped on to the tongues of the spokes while the spoke ends are drawn together for this purpose with the Spoke Dog.

Tyring. The wheel is shod with either a continuous iron tyre or with separate shoes called strakes. The tyre is put on hot, and on being quenched it shrinks and draws all the parts of the wheel together. (See *Wheelwright's Equipment (3) Tyring Tools.)*

Axle Box. The last process is boring out the centre of the hub to take a wooden axle, or to take the cast-iron box which serves as a bearing for iron axles. This is done with special Boxing Gouges and Bent Chisels or, in more recent times, with a Hub Boring machine. Finally the wheel is hung on an axle and the axle-box centred with wedges until the wheel runs true.

Special Tools used by wheelwrights and coach-builders will be found under the following entries:

Adze, Wheelwright's
Auger, Wheelwright's
Axe, Coachmaker's
Axe, Wheelwright's
Axle Keys (under *Wrench, Coachbuilder's*)
Bar Bender (under *Wrench, Bar Bending*)
Bettye (under *Saw, Bettye*)
Bit, Carriage Makers
Bit, Hollow
Bit, Spoke Trimming
Bit, Wagon Builder's
Brace, Wagon Builder's
Burning Rods
Carriage Trimming Tools (under *Cramp, Carriage Trimmer's; Hammer, Coach Trimmer's; Upholsterer*)
Chisels:
  Bent
  Boxing (under *Gouge, Wheelwright's*)
  Bruzz
  Coachmaker's
  Mortising (*Wheelwright's*)
  Wagon Builder's
  Wedging
Clip Wrench (under *Wrench, Coachbuilder's*)
Cramp, Carriage Trimmer's
Drawing Knife (*Wheelwright's*)
Furniture, including *Mortising Cradle; Bench; Chopping Block; Mandrel; Shaft Stand; Wheel Stool,*

etc.; *Wheel Cradle; Wheel Frame,* etc. (under *Wheelwright's Equipment (1) Furniture*)
Gauges, Wheelwright's (under *Gauge, Wheelwright's and Coachbuilder's; also Wheelwright's Equipment (2) Spoke Tools; Squares and Bevels* (*Coachbuilder's*)
Gouge, Coachbuilder's
Gouge, Wheelwright's
Hammers:
  Coach Trimmer's
  Dolly
  Framing
  Sledge
Hub Boring Engine
Jarvis (under *Shaves*)
Lifting Jack
Lathe, Wheelwright's
Mandrel (under *Wheelwright's Equipment (1) Furniture*)
Nelson (under *Shaves*)
Painting Equipment, including *Brushes; Muller and Slab; Maulstick; Painting Horse; Paint Mill*
Planes:
  Centre-Board
  Coachbuilder's (including *Door Check; Door Rabbet; Side Chamfer; Tee Rabbet Planes*)
Rasp, Two-Handed
Router, Coachbuilder's (including *Beading, Boxing, Corner, Grooving* (*Fence*), *Jigger, Moulding, Listing, Pistol, Reeding, Side Cutting*)
Samson (under *Wheelwright's Equipment (3) Tyring Tools*)
Saw, Bettye
Shaves:
  Jarvis
  Nelson
  Spoke Shave
Smith's Tools:
  Bar Bender (under *Wrench, Bar Bending*)
  Mandrel
  Metal-Working Tools
  Rasp, Two-Handed
  Wheelwright's Equipment (3) Tyring Tools
  Wrenches
Spoke Tools (including *Spoke Set; Spoke Guard; Spoke Dog; Shaves; Spoke Bridle; Spoke Extractor; Spoke Holders* (under *Wheelwright's Equipment (2) Spoke Tools*))
Square, Coachbuilder's (including *Horizontal Square; Mortice Bevel; Spider Mortice Bevel*)
Staple Clincher
Traveller (under *Wheelwright's Equipment (3) Tyring Tools*)
Tyring Tools (including *Oven and Fire Forks; Samson* (*Strake Cramp*); *Strake-Nail Claw; Strake Dogs;*

Traveller; Tyre Benders; Tyring Dogs; Tyre Lifters; Tyring Platform (all under *Wheelwright's Equipment* (3) *Tyring Tools*))
Vice, Post (under *Vice, Woodworker's*)
Washer Cutter
Wrench, Coachbuilder's (including *Axle Key; Bar Bender* (*Step Wrench*); *Clip Wrench*)

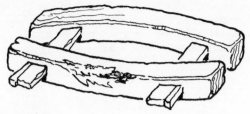

Fig. 724 (a)

Wheelwright's Equipment: (1) WORKSHOP FUR-NITURE *Fig. 724*
(A description of the workshop itself is given under *Wheelwright* and *Wainwright*.)

Workshop furniture and appliances include the following:

(*a*) *Mortising Cradle* (Nave or Hub Cradle)
A heavily made stool or frame designed to hold a wheel hub while the mortice holes are cut. In one example illustrated four wedges hold the hub steady; in the other, the hub is held in an adjustable frame (made from old felloes) which can be laid on the bench.

(*b*) *Wheel Stool* (Wheel Block)
A strong low stool with an open top on which a wheel could be 'thrown' when under repair or when fitting the felloes. A peg (known as a Stool Pin) was sometimes fitted on the front on which a wheel could be hung slanting towards the floor where its rim rested on a Wheel Cradle. The three-cornered example is made from a forked branch; the other is made from two wheel felloes. Both types are common in the south of England.

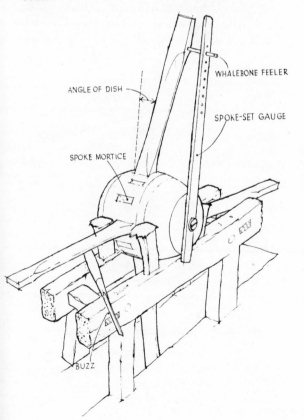

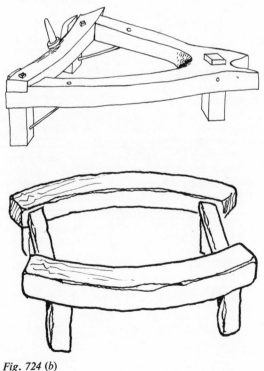

Fig. 724 (b)

(c) Wheel Cradle

A simple frame, often made from two felloes, joined across at both ends. This was laid on the floor to steady the lower side of a wheel when lying at an angle from the Stool Pin in the Wheel Stool.

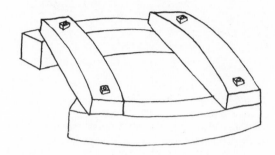

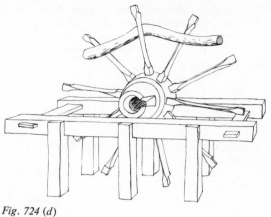

Fig. 724 (d)

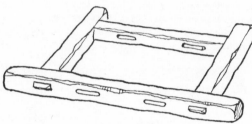

Fig. 724 (c)

(d) Wheel Frame (Wheel Horse; Wheel Stand; Speech Bat; Tram)

A substantial frame in which the parallel beams which form the bed could be drawn apart to accommodate wheels of larger size. Used for holding a wheel upright during repair, but if the shop possessed a Wheel Pit it was used only for the smaller wheels. A Spoke Bridle shown on the spokes is used for adjusting the alignment of the spokes (see *Wheelwright's Equipment* (2) *Spoke Tools*).

(e) Mandrel (Cone Mandrel; Sugar-Loaf Mandrel)

A cone made in iron (but occasionally in stone) up to 4 ft high, used for trueing up the circular iron bonds which bind the wheel hubs. After the bond has been made on the anvil, it is placed on the Mandrel and hammered down until perfectly round, and also splayed, to make it fit the slightly tapered face of the hub.

Fig. 724 (e)

Other equipment and furniture include:

Bench. This is usually about 18 in wide, and made from a 4 in plank, often with the rough natural edge left showing at the front.

Chopping Block. Used for chopping out spokes and wedges, and for similar tasks.

Jack. Both wooden and metal Jacks, usually of the lever type, are used when removing wheels (see *Lifting Jacks*).

Lathe. Earlier examples are made almost entirely of wood (see *Lathe, Wheelwright's*).

Painting Horse. A wooden post and bracket on which is hung a wheel while being painted (see *Painting Equipment*).

Saw Pit. Every wheelwright's shop needed its own Saw Pit (see *Sawyer*).

Shaft Stand. A wooden post, about 3 ft high, set on a heavy wooden base. On top of the post is a short iron spike. Used for holding up the shafts during repairs, painting, etc.

Trestle. Two or three pairs are used on which to assemble vehicle bodies and undercarriages. (See illustration under *Sawing Horse.*)

Vice. Wheelwrights use the same vices as other woodworkers, but often a Post Vice as well. (See *Vice, Woodworker's.*)

Wheel Pit. A narrow trench sunk in the floor of the workshop, about 5½ ft long by 9 in wide and 3 ft deep, lined with brick or timber. Used for holding a large wheel when under repair or when 'speeching the stock', i.e. driving the spokes into the hub (see *Hammer, Sledge*). In the centre of the sills are semi-circular depressions to hold the hub.

Wheelwright's Equipment: (2) SPOKE TOOLS
Fig. 725 overleaf
Before the use of machine-turned spokes the timber for spokes (usually oak) was riven from the log, not sawn, so that any weakness would be immediately noticed. The splitting was done with a Froe, the wood was then trimmed with a Draw Knife and finally smoothed with a Spokeshave or Jarvis. These and other relevant tools are described under *Bit, Spoke-Trimmer; Draw Knife, Wheelwright's; Froe; Shave, Jarvis;* and *Shave, Spokeshave.* Other tools used exclusively for making and fitting spokes include the following:

(*a*) *Spoke Bridle* (Spoke Cramp; Allan Jobson (London, 1953), when describing examples of this tool being used in Suffolk, calls it a 'Type' or 'Bucker'.)
A yoke-shaped stick, about 2½ in thick, usually made from an oak or ash branch and used when driving spokes into the hub. It was 'woven' behind the spoke being driven, and in front of the two adjoining spokes, and hammered down if it was necessary to bring that spoke forward; or when

inserted in the reverse way, to force the spoke back. It was used in conjunction with the Spoke Set. (See also note on driving spokes under *Hammer, Sledge*.)

(*b*) *Spoke Set* (Set Stick; Speech; Whalebone Gauge)
This tool is used to gauge the amount of 'dish', i.e. the amount by which the spokes are made to lean outwards from the hub, and to check their alignment when being driven into the hub. It is a wooden bar, about 2 ft 6 in long, with a number of small holes, about 1 in apart, drilled through the upper part. The bottom of the bar is fastened to the centre of the front end of the hub with a coach-screw, upon which it can be revolved. The Gauge is used in conjunction with the Spoke Bridle, by which the angle of the spoke can be adjusted as it is driven into the hub.

A piece of whalebone or flexible cane, known as a 'Feeler', is inserted into one of the holes in the stick at a height which corresponds to the shoulder at the top of the spoke. The feeler is adjusted to gauge the distance from the front of the hub to the face of the spoke, less the predetermined amount allowed for dish, and then wedged firmly in place. When being driven in, each spoke must just touch the end of the whalebone; if out of line, it can be forced back with the help of the Spoke Bridle. The object of the feeler being flexible is to permit the Spoke Set to be moved round from spoke to spoke. Being pliant, the whalebone springs past the face of any spoke which may be too forward. If the Gauge were rigid and there were two forward spokes, it would get locked between them. A Sussex wheelwright (1950) told the author that whalebone could be 'salvaged from old umbrella ribs or from a pair of mother's discarded corsets'.

Note: The process of driving spokes into a hub with a Sledge Hammer is described under *Hammer, Sledge.*

A narrow stick about 3 ft long, known as a *Striking Stick*, tapered off on one side near the foot, is used in conjunction with the Spoke Set for gauging the correct angle of the mortice holes in the hub. The foot of the stick is placed in the mortice hole against its front end, its tapered foot allowing room for the fingers to hold it there. The top end of the stick will contact the feeler of the Spoke Set.

(*c*) *Spoke Guard*
A rounded piece of wood, hollowed out inside and made to fit over the tongue of a finished spoke. Used to protect the tongue while the spoke is being driven into the hub.

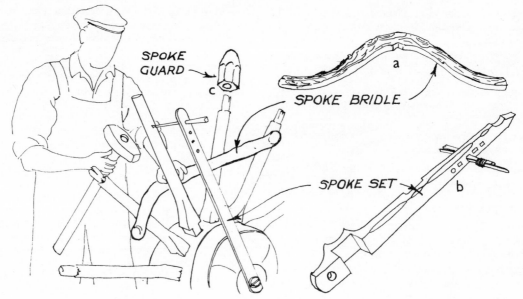

Fig. 725 (*a*)–(*c*) Spoke-driving appliances (diagram)

(d) *Spoke Dog* (Spoke Lever)

A wooden lever, about 3 ft 6 in long, fitted with a hooked iron bar which passes through a slot at its lower end. Used for straining two spokes together so that their tongues will enter the felloe. This is necessary because the divergence of the spokes results in the tongues being wider apart at their extreme ends than at the shoulders. The hook of the Spoke Dog is put round the spoke already in position in the felloe, and the lever end is placed on the far side of the spoke which needs cramping in. The lever rests on the wheelwright's shoulder as he crouches over the wheel; when he raises his shoulder under the lever, the spokes are drawn together. As soon as the spoke tongue is in line with the spoke-hole, the felloe is quickly driven on. (When the spokes are very short, the spoke tongues are tapered at their outer ends so that they draw together when the felloes are hammered over them.)

(e) *Spoke Extractor* (Spoke Drawer; Spoke Remover)

An iron ring and three wooden wedges are used for extracting old spokes from the hub. The ring is placed over the spoke it is desired to remove, and the wedges driven through in an upward direction until the spoke is held fast. Then the wedges are hammered upward until the whole bunch begins to move, drawing out the old spoke with them.

(f) *Spoke Fitter* (Spoke Tongue Gauge)

A piece of wood, about 8–12 in long, roughly shaved down at the ends for handles and with a hole in the middle. Several were kept with different sized holes for testing the correct diameter of round spoke tongues.

(g) *Spoke Trammel* (Spoke Length Gauge)

A stick, about 4 ft long, with a sliding marking-point which can be set at different heights from the base by means of a wedge or screw. Used for checking the height of spoke shoulders above the hub.

(h) *Spoke Holders*

Several different holding devices are made by wheelwrights to hold a spoke while it is being trimmed with Shave or Draw Knife. All are designed to allow the spoke to be turned as the work proceeds. The following are examples found in the home counties.

1. *The Fiddle*. The spoke is pivoted between lathe-like centres, one of which is set upon a handscrew. It can thus be held tight enough for trimming and yet turned conveniently by the user's hand. (The same device is used by handle makers.)

2. *Shave Peg*. A peg with a metal spike at one end. The tapered end of the peg is held in a hole in the wall. The spoke to be trimmed with a Draw Knife is held between the spike-end and the operator's chest.

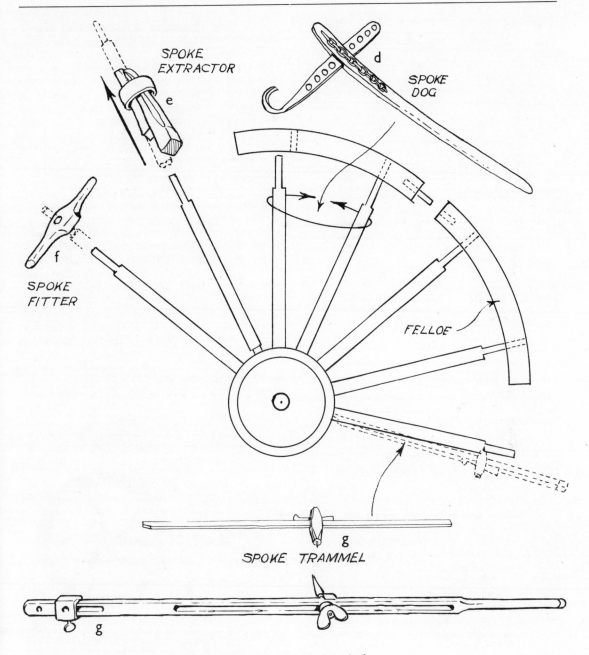

Fig. 725 (*d*)–(*g*) Spoke-fitting appliances (diagram)—*continued overleaf*

513

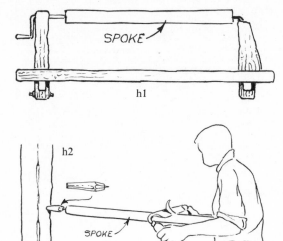

Fig. 725 (h1)–(h2)

3. W. H. Pyne (London, 1845) illustrates a wheelwright trimming a spoke (presumably with a Drawing Knife) which is held on a leaning board like the Heading Board described under *Cooper* (2) *Furniture*.

(i) Spoke Bits
Used when shaping the tongue and shoulder of a Spoke. See *Bit, Hollow; Bit, Spoke Trimming*.

Wheelwright's Equipment: (3) TYRING TOOLS
Fig. 726
Though hooped iron tyres were used in Europe about 500 B.C. and probably earlier, medieval wheels were made both with and without iron tyres. Strakes, which were often used instead of tyres, persisted until the end of the nineteenth century. Strakes are short, curved 'shoes', which were nailed over the junction between two felloes, and were long enough to stretch from the centre of one felloe to the centre of the next. They were preferred when making very large wheels, because the bending and fitting of a wide hooped tyre is a very heavy and difficult operation.

Another tyring process was the cutting off, shortening, and re-welding of a tyre if the wheel shrank or if the tyre became enlarged owing to being constantly rolled on hard roads. This was known as 'cutting

and shutting'. (It used to be said of a man who was considered slovenly in his dress, 'he needs cutting and shutting'.)

Tyres were heated before being forced on to the wheel, and were then quenched by pouring on water. The resulting contraction of the tyre pulled the joints of the wheel together. This shrinking process was also used to increase the amount of dish – when this was required.

The following are some of the tools used when fitting tyres or strakes:

(a) Traveller (Tyre Runner; Wheel or Tyre Measurer; U.S.A.: Follower or Tracing Wheel)
A wheel or disc pivoted in the fork of a short handle. It was most often smith made and is found in a surprising variety of sizes and designs, in wood or metal, or in a combination of both. The factory-made Traveller illustrated is all-iron, is graduated on the rim, and is provided with an adjustable arm to mark the number of turns. Travellers are used for measuring the length of iron bar required for making the tyre. The metal bar is laid on the ground, and a chalk mark is made on the rim of the Traveller, and another on the rim of the untyred wheel. With the chalk mark as starting point, the Traveller is run round the wheel rim and the number of turns noted. It is then run along the bar on the ground for an equal number of turns to obtain the correct length, allowance being made for 'shutting' the ends of the tyre, and for the expansion of the tyre when heated.

Note. Travellers are occasionally used by other tradesmen, e.g. by millwrights when fitting a millstone with an iron hoop.

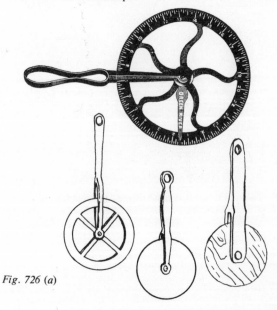

Fig. 726 (a)

514

(b) Tyre Oven and Fire Fork

Tyres and strakes were heated before being fitted. This could be done by building a fire on the ground, but many shops had a Tyre Oven (also known as a Hooping Grate or Tyre Furnace). This was a brick-built oven with an iron door and chimney above. Fire Forks with long handles were used for making up the fire or for lifting strakes and small tyres out of the fire.

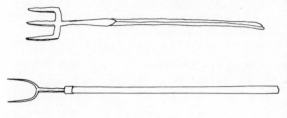

Fig. 726 (b)

(c) Tyre Benders

The iron bars used for making tyres or strakes were bent cold on Benders of various types including:

1. *Post Bender* A primitive device found in many of the older shops. A block of wood, rounded above, is bolted to a post, with a wood or metal 'stop' above it. The iron is inserted between the block and stop and pulled downwards sufficiently to give the desired curvature. A piece of curved iron bar, e.g. an old strake, was sometimes fixed to the post instead of the wooden block.

2. *Lever type* These are made in various designs based on the following system. The bar is passed under the cross-bar at the front, and wedges of the size required are fitted to the hinged plate beneath it. When the lever is pressed down, the hinged plate forces the tyre in an upward curve. The circumference of the tyre can be altered by altering the setting of the connecting link between the lever and the hinged plate.

3. *Roller type* A factory-made machine mounted on a cast-iron stand and operated by a cranked handle. There are many varieties but most work on a simple three-roller system. The bar is passed over two rollers, with a third 'idle' roller fitted centrally above, which can be forced downwards between them. The two lower rollers are geared to turn in unison by means of the cranked handle; they are serrated in order to drive the bar along under the centre roller which, according to its setting, gives the required curvature to the bar.

4. *Bending Block and Dolly* [Not illustrated] When a tyre comes off the Tyre Bender, or when an old tyre becomes loose and comes in for refitting ('cutting and shutting'), the ends often spring apart when cut and have to be brought together for welding. This is done on an oak stump, about 2 ft 6 in high, with the top hollowed out. The tyre is held across the hollow and struck with a hammer until the ends close. H. R. Bradley Smith (U.S.A., 1966) illustrates an iron anvil resembling an oriental pillow with a concave top, which served a similar purpose.

If, after bending, the ends of the tyre overlap too far, an operation in reverse to that described above is carried out. The unjoined tyre is laid on the ground and an 18 lb hammer, known as a Dolly, is held against the inside of the tyre opposite the opening. The outside of the tyre is then struck with a Sledge Hammer, at a point just opposite the Dolly. This causes the join in the tyre to open.

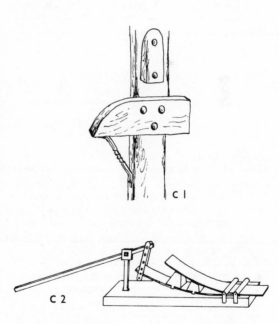

Fig. 726 (c1)–(c2)—continued overleaf

515

Fig. 726 (c3)

(d) Tyring Dogs (U.S.A.: Pulling Tongs)
A pair of iron bars, about 3 ft long, with forked ends. Used for forcing the previously heated tyre over a wheel when lying on the Tyring Platform. One prong of the fork is hooked over the edge of the tyre, and the other prong is used as a lever against the rim of the wheel to draw the tyre over it, if necessary helped by blows from a Hammer.

Fig. 726 (d)

(e) Tyre Lifter
Small tyres can be taken from the fire with ordinary smith's tongs, or with special tongs with their jaws bent inwards. But larger tyres are too heavy to be carried by this means, and Tyre Lifters of various designs were made for this purpose. That illustrated is 3 ft long and has open forked ends which fitted over the top edge of the tyre and held by friction. Made in sets of three, so that a heavy tyre could be carried and lowered into position by three men.

Fig. 726 (e)

516

(f) Tyring Platform (Ringing Plate: Wheel Platform)
A circular plate of iron (occasionally of stone) up to 6 ft 6 in diameter and $1\frac{1}{2}$ in thick, with an 18 in diameter hole at the centre and bedded at ground level. Wheels are laid face downwards on it for tyring. A Stock-Screw, which protruded from the centre hole, holds the wheel down on the platform, but loose enough to allow the hub to rise up slightly as the heated tyre contracts. By this means the wheel becomes slightly cone-shaped, i.e. dished. The amount of dish required can be controlled by adjusting the nut on the Stock-Screw.

(g) Strake Tools
Strakes are curved iron shoes that were sometimes fitted instead of a continuous tyre. (See *Wheelwright* above.)
   1. *Samson* (Strake cramp) A rectangular iron frame, up to 4 ft 6 in long by about 9 in across. (Though usually U-shaped, one with L-shaped arms from a Sussex smith's shop is also illustrated here.) Both cross-bars, including the movable one which is forced down the arms of the Cramp when the nuts are turned, have their inner edges bevelled to catch behind the head of the strake-nail left protruding for this purpose.
   There is some difference of opinion among Wheelwrights about the way in which strakes were fitted. Wheelwrights whom we have consulted were not themselves in charge of the work. Their account is based on having watched the operation performed by their father or his men – a dramatic memory of heated strakes being hammered on to the burning wheel and later quenched under a cloud of steam. But the purpose of the Samson is clear: it was to draw adjacent felloes together in the absence of a continuous tyre. The tyre, as Sturt remarks, 'did the work far better', though he added, 'There was not much left for mortal men to tighten after Samson had pulled the joints together' (Cambridge 1942).
   The late George Weller told us (1949) that the joint between adjacent felloes was tightened by forcing a wedge into the joints at the end of adjoining felloes and also by the contraction of the hot strake after cooling. But the wedging process could not be applied to the last joint since all the other joints were by now covered by strakes. Instead, the Samson was laid over the wheel and its cross-bars hooked over strake nails left protruding for this purpose in adjacent strakes. The nuts of the Samson were then tightened, and the last strake was fitted before taking the Samson off the wheel.
   2. *Strake Dog* (Strake Cramping Iron) Iron Cramps with rings for levering over the strake with

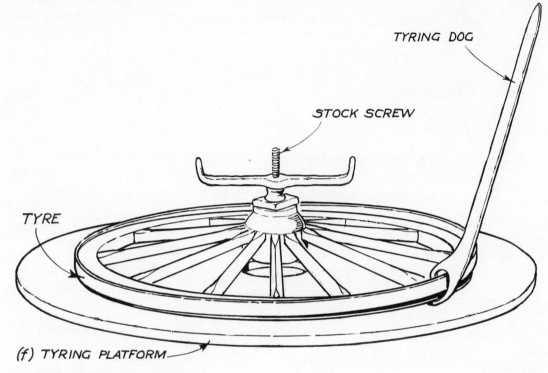

TYRING DOG

STOCK SCREW

TYRE

(f) TYRING PLATFORM

Fig. 726 (f) Tyring Platform (diagram)

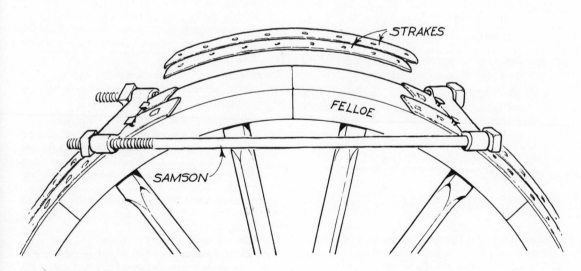

STRAKES

FELLOE

SAMSON

Fig. 726 (g) Strake 'Samson' in use

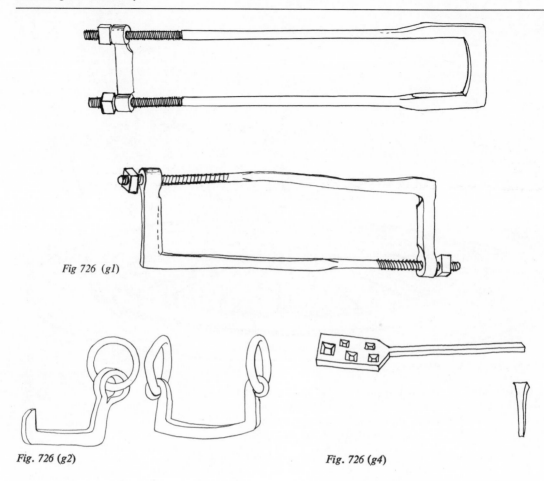

Fig 726 (g1)

Fig. 726 (g2)

Fig. 726 (g4)

a Crow Bar. Used for holding the strakes flat on the sole of the wheel while nailing them on.

3. *Strake-Nail Claw* A heavy iron claw about 10 in long. Used for extracting strake-nails when taking off old strakes.

Fig. 726 (g3)

4. *Strake-Nail Heading Tool* An iron plate containing countersunk holes to correspond with the square, tapered heads of different sizes of strake nails. Used by the smith when forming the heads.

Wheelwright's Machinery

Towards the end of the nineteenth century hand and power driven machines were developed for wheel-making. These include:

(*a*) *Hub Boring and Boxing Machine* Made like the machine tool known as a Horizontal Boring Machine with a face plate large enough to take the whole wheel. See also *Hub Boring Engine*.

(*b*) *Wheel turning machine* A lathe capable of taking a wheel on the face plate.

(*c*) *Hub Mortising Machine* Like the lever-operated Mortising Machine, but with a special bed adapted for holding a hub.

(*d*) *Spoke Lathes*, usually automatic.

(*e*) *Spoke Tonging and Felloe Boring Machine* A form of lathe on which the wheel is mounted on the bed instead of the tool turret. The wheel can be moved towards a Hollow Auger or Drill which is held in the lathe chuck.

(*f*) *Tyring Platform* With special gear for drawing on the tyre, and for lowering the wheel into a water tank for cooling.

Whetting Block: see *Saw Sharpening: Holding Devices.*

Whittle: see *Knife, Cooper's Whittle.*

Width-Maker: see Chip Plane under *Plane, Scaleboard.*

Wimble (Whimble; Wymble)
A name sometimes given to a *Brace* (see note under that entry).

Winding Strips: see *Boning Strips.*

Windlass: see *Cooper* (3) *Hooping Tools.*

Window Making *Fig. 727 overleaf*
This work, now regarded as coming exclusively within the province of the joiner, includes the making of glazed frames which may be side-hinged (known as casement windows), sliding (known as sash windows), pivoted, or fixed. A similar process is applied to the making of other frames, e.g. that of a panelled door.

Nowadays the work is mostly done by machinery, but when making sashes by hand, after setting out and cutting the mortices and the cheeks of the tenons, the rails and stiles are rebated and moulded on the Bench or in the Vice. The bars are usually worked on the Sticking Board. The glazing rabbet is worked with a Fillister Plane and the moulding stuck (i.e. moulded) on one side with the corresponding Sash Moulding Plane. The bar is then turned over and the other side worked in the same manner.

The moulded and finished sash bar is mitred and scribed with the help of Sash Templets. Scribing in this sense is the process of fitting the irregular moulded profile of one bar over the corresponding profile of another. This irregular profile was at one time cut with a Sash Scribing Plane but was later cut with a Sash Gouge.

Some of the tools and equipment used in window making are illustrated. Other tools connected with window making are described under the following entries:
Bit, Sash
Chisel, Astragal
Chisel, Sash Pocket
Gouge, Sash

Mouse
Plane, Sash Moulding
Plane, Sash Scribing or Coping
Plane, Sash Templet Making
Plane, Throating
Sash Router (under *Router, Metal; Router Sash*)
Sash Fillister (under *Plane, Fillister*)
Saw, Coping
Sticking Board
Templet, Sash

Wine Fret: see *Gimlet, Wine Fret.*

Wing Nut (Fly nut; Butterfly nut; Thumb screw)
A nut with protruding ears or wings for turning by hand.

Witchet (or Widget)
A name used mainly in the U.S.A. for a *Rounding Plane.* (See *Plane, Rounder.*)

Womell (Womble; Womill)
Obsolete form of Wimble (See under *Brace.*)

Wood Boring Machine (Angular Boring Machine; Boring Machine Auger) *Fig. 728*
A hand-operated Machine Auger for boring holes in wood at any required angle. The Auger is fitted in a vertical spindle mounted on a frame which can be set at an angle with the workpiece by means of thumbscrews acting on quadrants mounted on the base on each side of the frame. The Auger-spindle is driven by two cranks through bevel gears, one of which can be made to engage with a rack which lifts the Auger out of the hole, when required. Augers or

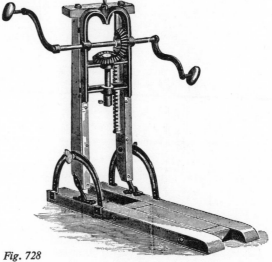

Fig. 728

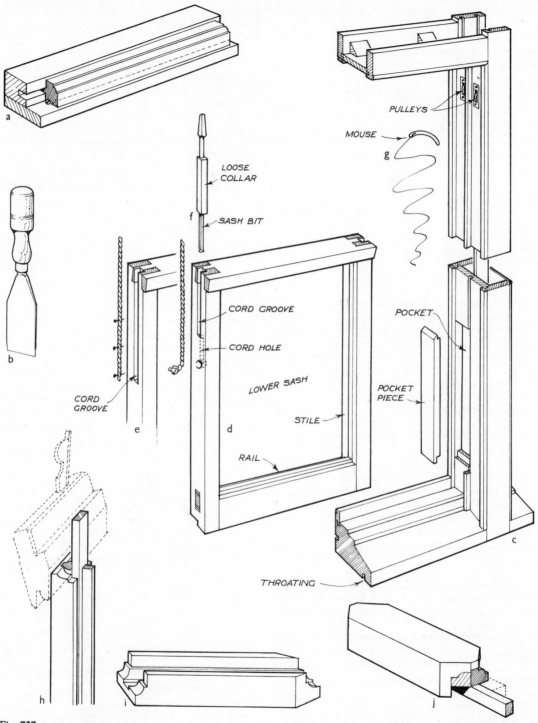

LOOSE
COLLAR

SASH BIT

f

PULLEYS

MOUSE

g

CORD GROOVE

CORD HOLE

LOWER SASH

STILE

RAIL

POCKET

POCKET
PIECE

CORD
GROOVE

e

d

c

THROATING

a

b

h

i

j

Fig. 727

Fig. 727 Window Making: tools and equipment
(*a*) Sticking Board for holding the wood when making sash bars
(*b*) Sash Pocket Chisel for cutting out pockets in frame
(*c*) Frame for sliding sashes
(*d*) Lower sliding sash with cord secured by a knot
(*e*) Alternative to (*d*), with cord secured by nails.
(*f*) Sash Bit with collar to guide it
(*g*) Mouse for reeving the sash cord over the pulley
(*h*) Use of Sash Scribing Plane – probable method
(*i*) Saddle Templet (combined mitre and scribing)
(*j*) Saddle Templet (mitre)

Bits of the twist type, of ½–2 in diameter, with round shanks, were made for this machine. Used by carpenters for boring large holes for mortices, e.g. for tenon joints in trimmer joists and similar work.

Wood Carver
During the medieval period, when furniture and interior fittings were mostly of oak and other home-grown hardwoods, the principal form of decoration was carving. The art reached a high level in England during the seventeenth century, when it was much used by architects such as Wren and when the famous woodcarver Grinling Gibbons (1648–1721) was at work. By this time wood carving had become a separate trade, and though carving declined during the nineteenth century, most cabinet-making shops of any size still employed a carver at that time.

From trade catalogues of the nineteenth and early twentieth century it is clear that many amateurs took up carving as a hobby. Special sets of tools were made, basically the same as those used by the tradesman, but generally lighter, with polished handles, and incidentally more expensive.

Tools used by woodcarvers are described under:
Carver's Bench Screw
Carver's Clip
Chisel, Carving
Cramp, Fretwork, Buhl, or Carver's
Knife, Chip Carving
Mallet, Carver's
Plane, Router
Punch, Carver's
Riffler

The Wood Engraver *Figs. 730; 731*
Engraving on wood is the oldest method used for printing pictures or books. According to Joseph Needham (Cambridge, 1954) printing with wood or metal blocks was carried on in China in A.D. 740.

The method was used in Europe in the fifteenth century, and continued as the chief means of decorating and illustrating books until recent times.

The end grain of a hardwood (e.g. box wood) is used for the detailed lines of a wood engraving; softer woods such as pear, holly, or pine, cut along the grain, can be used for woodcuts with simpler and broader lines. (See *Print and Block Cutter*.)

The Tools
In general, the finer lines of wood engravings are made with chisels known as Gravers (see (*a*)–(*d*) below). The broader incisions of a woodcut are cut with knives. A typical Graver is illustrated in Fig. 730. It has a blade about 4–5 in long which removes a sliver of wood when pushed with the hand. It is designed to cut across the grain and is held at a very low angle to the block being cut. The blade may be straight, or slightly bowed (bellied). The handles are made in many patterns – balloon-shaped, peg-top, but more commonly mushroom-shaped, often with the lower side removed to enable the Graver to be held at a low angle. The face is ground at an angle of about 45°.

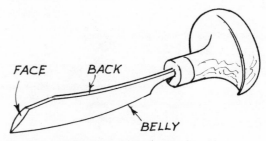

FACE   BACK

BELLY

Fig. 730 A typical Wood Engraver's Graver

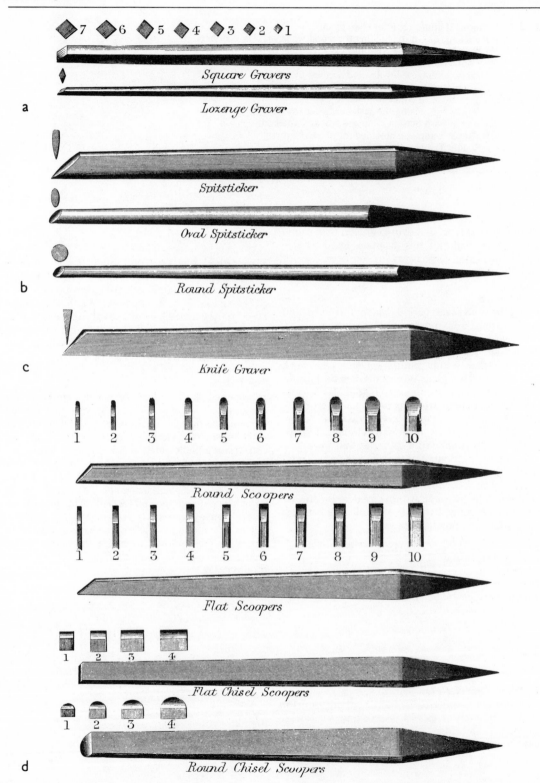

7 6 5 4 3 2 1

Square Gravers

Lozenge Graver

a

Spitsticker

Oval Spitsticker

Round Spitsticker

b

Knife Graver

c

1 2 3 4 5 6 7 8 9 10

Round Scoopers

1 2 3 4 5 6 7 8 9 10

Flat Scoopers

1 2 3 4

Flat Chisel Scoopers

1 2 3 4

Round Chisel Scoopers

d

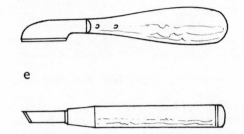

e

Fig. 731 Wood Engraver's Tools (Sheffield Illustrated List, 1888)

Types of Engraving Tools include (*Fig. 731*):

(*a*) *Burin*
This is the oldest and simplest engraving tool and is used for cutting straight lines of varying depth in wood or metal.

(*b*) *Spitsticker or Spitstick* (Bullsticker)
Shield-shaped, oval, or round faced, used for cutting curved incisions of varying depth. The Bullsticker is the same as the Spitsticker but it has bulging sides for a rapid spread in the width of the cut.

(*c*) *Tint Tools* (Tinters; Knife Gravers)
A knife-shaped face, used mainly for making parallel lines of even width, i.e. to produce 'tints' or shades of even tone. The same tool with a serrated face will engrave 6–10 lines at one stroke.

(*d*) *Scorper* (Scauper; Scooper)
Round or square faced and sometimes made with a bent blade. Used for clearing out spaces.

(*e*) *Knives*
Two kinds are used for making woodcuts:
The European: the blade has a rounded back like a pen knife and is held as one holds a pen.
The Japanese: the blade has a chisel-like cutting edge, either skewed or diamond pointed, and is held like a dagger.
Note: Chisels and Gouges of the carving type are also used.

Wood Hook: see *Timber Handling Tools*.

Wood Horse: see *Sawing Horse*.

Wood Turner: see *Turner*.

Woodcut Tools: see *Wood Engraver*.

Woodland Trades (1) (Also known as Coppice Trades) *Figs. 732; 733; 734*
These trades include those woodland and village industries which use as their chief raw material the stems and branches of young trees grown in a wood which has been 'coppiced'; that is, the stems cut to ground level in winter or early spring and new shoots allowed to grow from the stump. The trees usually dealt with in this way include ash, sweet chestnut, osier willow, field maple, and hazel; but some country tradesmen, e.g. the chair-bodgers, use beech. The stems are cut with a rotation of from one to sixteen years, according to the type of material required, and vary from $\frac{1}{2}$ to about 8 in diameter at the butt.

Though not, perhaps, requiring the developed skills of the carpenter or wheelwright, the woodland worker produces objects of beautiful design which, like the gate-hurdle, stand up to years of hard usage in the open. The nature of their work is well described by H. L. Edlin (London, 1949): 'Always the woodmen's work was carried on in the shadow of the trees that grew to form the stuff of their trade, for timber is heavier to haul than its products, and for many purposes it must be worked whilst still freshly felled or green. So they grew up, as they remain, true woodlanders, living in intimate touch with the wild life of the open air.'

Writers on these trades include {see List of References}: H. L. Edlin (1949); J. Arnold (1968); K. S. Woods (1949); J. G. Jenkins (1965); and F. Lambert's booklet *Tools and Devices for Coppice Crafts* (Evans Bros., 1957), on whose illustrations some of the sketches shown below are based.

Brief particulars of the following trades that use coppice material are given under their appropriate entries:
Basket Maker – Osier, Spale, and Trug
Broom Maker
Brush Maker
Chairmaker (*Bodger*)
Clog-Sole Maker
Handle Maker
Hoop Maker
Hurdle Maker, Gate and Wattle
Lath Maker
Rake Maker
Shingle Maker
Spoon Maker
Thatch-Spar Maker

Typical Tools used in the woodland trades are illustrated below. The more specialised tools will be found under their appropriate trade, e.g. Basket Maker; Hurdle Maker, and some are described in greater detail under their own entries.

Fig. 732 Harvesting Tools
Some of the tools commonly used for harvesting the material are illustrated below.

(*a*) *Axe*

(*b*) *Bill Hook* An example of the Hertfordshire type with a hook on the back which is useful for picking up the next piece to be chopped. Bill hooks are also used for splitting.

(*c*) *Slasher or Hedge Bill* Used mainly for clearing undergrowth etc.

Crook stick A naturally grown pulling hook sometimes used with the left hand for pulling the boughs over, while wielding the Bill Hook with the right hand.

Saws See *Saw, Cross Cut* etc.

Fig. 733 Splitting Tools

(*a*) *Froe* The basic tool of the woodland and coppice worker. Used for cleaving or splitting poles for hurdles, stakes, etc. (For alternative names see under *Froe*.

(*b*) *Mallet* For driving a Froe. A maul or cudgel is shown in use.

(*c*) *Cleaving Knives* (*River*) Some are made like Pruning Knives, others like miniature Bill Hooks. (See also *Lath Maker*.)

(*d*) *Cleaving Adze* Sometimes used for splitting. The sides of the blade are sharpened as well as the front.

(*e*) *Cleaver* (Bond Splitter) An egg-shaped piece of hardwood, similar to the basketmaker's Cleave but larger. The end is sharpened with three cutting edges and used by, among others, broom and hoop makers for splitting withies and rods.

See also entries under *Beetle; Bill Hook; Thatch-Spar Maker; Wedge*.

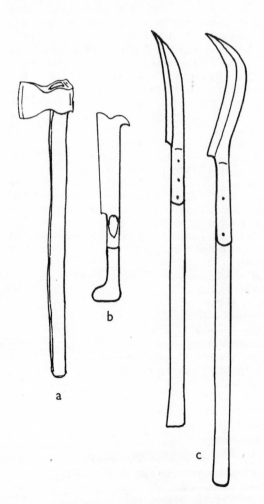

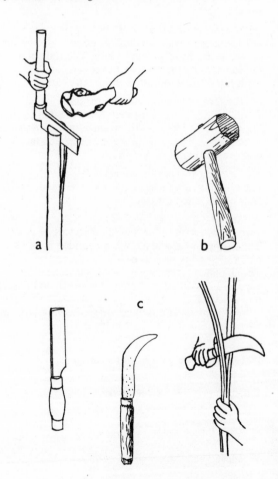

Fig. 732 Woodland trades: harvesting tools

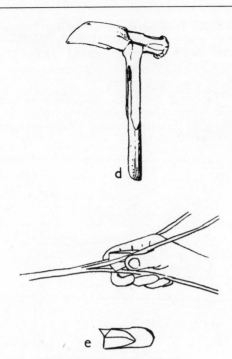

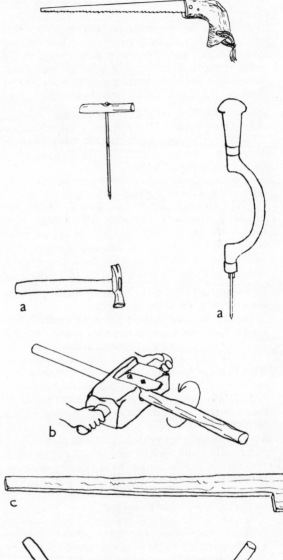

e

Fig. 733 Woodland trades: splitting tools

Fig. 734 Miscellaneous Tools

(*a*) *Drawing Knife* etc. A basic tool of the woodland trades, used for trimming, shaping, or pointing; also shown are the Gimlet, Saw, Hammer, and Brace.

(*b*) *Rounder* For rounding handles, pegs, etc. See also *Plane, Rounder; Handle Maker*.

(*c*) *Measuring Stick* Many kinds are made by the tradesman himself for measuring the length of a workpiece, e.g. when making chair legs or tent pegs in large numbers.

(*d*) *Woodman's Grip* Two strong poles, about 3 ft 6 in long, connected by a length of rope at a point about a foot from the lower end. Its purpose is to compress rods and stakes into bundles for tying up before carting away.

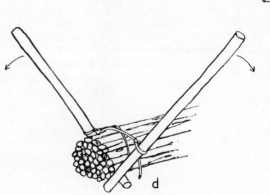

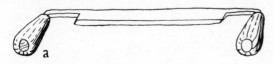
a

Fig. 734 Woodland trades: miscellaneous tools

Woodland Trades (2): **Brakes** (sometimes spelt 'Break') *Fig. 735*

A Brake is a wooden frame designed to hold a pole or billet while being split or trimmed. These holding devices, which are probably of great antiquity, are made in many different patterns. They are almost invariably home made, and are usually erected in the open air. The following are typical examples:

(a) Cleaving Brake (U.S.A.: Frow Horse)

An essential feature of these Brakes are two horizontal bars, one slightly lower than the other, between which the pole to be cleft can be wedged. Two examples are shown below.

1. is made from a natural fork supported on two posts driven into the ground in the open.

2. is a foot-operated Brake, about 8 ft high, and comes from the workshop of an Oxfordshire hurdle maker. The top bar is pivoted at one end weighted with a stone at the other, thus holding down the work which is laid on the lower bars. The work is released by lifting the stone with the foot pedal.

(b) Shaving Brakes (called Peeling or Rinding Brakes if used mainly for removing bark)

These are designed to hold the work firmly, while leaving both hands free for trimming with a Drawing Knife. The following are common types:

1. Two short posts set firmly in the ground about 2 ft 6 in apart. One has a natural fork at the top; the other is notched to provide a bed for the work.

2. This is designed for shaving slender material, such as hoops, or spelks for baskets. A stout post is set in the ground, leaning away from the operator, and supported from behind by two sloping legs. A bar, on which the hoop or spelk is to lie while being shaved, lies in a trench on the top end of the post. When this bar is forced downwards, it acts as a Vice by closing on a short cross-bar, thus holding the top end of the material to be shaved.

3. This is the familiar Shaving Horse, used by coopers, chairmakers, and many other tradesmen besides woodland workers. (See separate entry under *Shaving Horse*.)

4. A more sophisticated foot-operated Shaving Brake is mounted on a post about 3 ft high (not illustrated). The lower jaw of the vice is the flat top of the post itself; the upper jaw is a bar of iron pivoted to an upright screwed to one side of the post. One end of the bar carries a counterweight; the other end is attached by a metal rod to a pedal at the foot of the post. The work is held by pressure on the pedal.

See also Bending Brake (or Setting Pin) under *Handle Maker; Shaving Horse;* Peeling Brake under *Basket Maker;* Mortising Stool under *Hurdle Maker* (2) *Gate Hurdles*.

Sources of drawings Fig. 735

a(1) and *b*(1) after F. Lambert's *Tools and Devices for the Coppice Crafts* (Evans Bros., 1957)

b(2) after J. Geraint Jenkins' *Traditional Country Craftsmen* (Routledge and Kegan Paul, 1965)

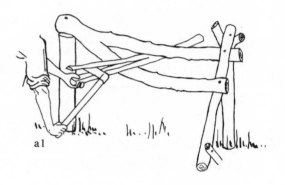

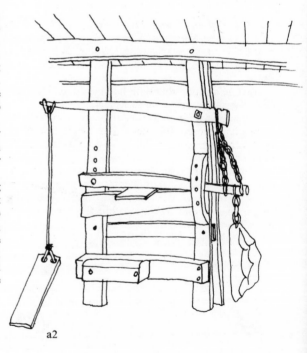

a1

a2

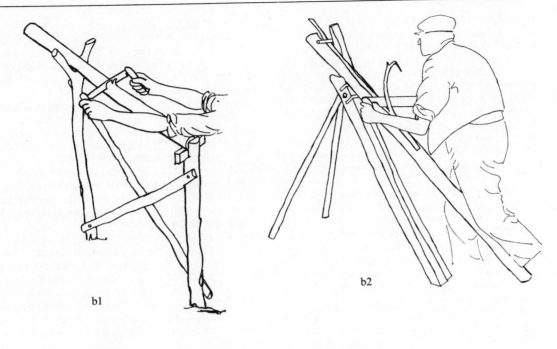

b1

b2

b3

Fig. 735 Woodland trades: Brakes

Woodwind Instrument Maker and Reed Maker *Fig. 736 overleaf*

Woodwind instruments – the recorders, flutes, clarinets, oboes, and bassoons – are made by turning in a lathe. Subsequently a hole is bored longitudinally down the centre, and lateral holes are bored at measured intervals down the length of the pipe to produce different notes.

Woods include cocos, South African blackwood, rosewood, and maple. Plastic materials are being increasingly used for recorders and for the simpler instruments. The keys of the best instruments are cast in nickel silver and finished by hand.

Apart from the Lathe, most of the work is done by hand and requires a high degree of skill. The boring is done with long Augers and Reamers. The shaping of bassoons, after initial turning, is done with hollow-soled Planes and Wood Rasps.

Tools Used for Boring

Woodwind instruments are bored down the centre, first with long Augers and afterwards, if a tapered bore is required, with a long Reamer. The following tools for boring bassoons were seen in the works of Messrs Boosey & Hawkes in London (1970).

A pilot hole is first bored with an Auger about $\frac{1}{4}$ in. in diameter, made like a long Bell-Hanger's Shell Gimlet but without the usual in-bent nose. Instead, the shell body is cut square at the end (or at a slight angle) and this is said to be best for boring the end grain. Owing, presumably, to the shape of its cross-section, this tool was referred to as a 'Dee Bit'.

The next stage is to open out the pilot hole with a series of shell-type Augers. These have spoon-shaped noses, are 18–20 in long, and are made in

527

various diameters up to 1 in. If the bore is to be tapered, this is done with Reamers of various designs. One of 3 ft and over in length, used for boring bassoons, has a single cutter made by turning the edge of a V-shaped groove which runs the length of the Reamer. Only one cutting groove is normally provided, since the effort of turning the Reamer would be too great if there were more. Some of the shorter Reamers consist of a tapered brass cone which is made in two halves. A steel knife, or rather scraper, is clamped between the half cones, protruding just sufficiently to scrape out the hole to size.

For cylindrical bores a cutter-bar is used. This consists of a steel rod with a short platform cut in its length, on which is screwed a steel cutter. This cutter extends beyond the diameter of the bar to the extent needed to finish the hole to size.

The lateral finger-stop holes of woodwind instruments are bored with a bit on which is mounted a cutting wing. This 'counter-bores' a level surface to provide a seating for the pad of the key.

Tools Used for Making the Reeds Fig. 736

The sound waves that come from woodwind instruments, such as the oboe and bassoon, are generated in the tube by the vibration of a double reed in the mouthpiece. The reeds are made from a special cane, much of which comes from the Mediterranean coast near Nice. Single reeds for clarinets and saxophones are cut from cane of much larger diameter.

Many musicians make their own reeds with tools designed for scraping and paring the reeds to shape. The illustrations below are based on the tools used by Mr. Anthony Aspden of St. Albans, Hertfordshire.

(*a*) *Cane Splitter* An 8 in rod fitted with vanes for cutting the cane into three parts.

(*b*) *Gouging Machine* A purpose-made tool consisting of a platform about 7 in long, on which the cane is bedded, and a small metal Hollowing Plane mounted on a sliding bar above it. Its purpose is to hollow out the cane and leave it of uniform thickness. Formerly this work was done with a Gouge.

(*c*) *Guillotine* Usually mounted on the same base as the gouging machine and used for cutting the cane to length.

(*d*) *Bridge or Easel* A cylindrical rod of hardwood made in varying lengths up to 3 in, scored and shouldered to act as a measure. Used first as a gauge in cutting the cane to length; then to indicate where the cane is to be scored at the centre for folding over on itself.

(*e*) *Hollow Scrapers* One is fitted with two circular blades, the other with a single scraper. Both are used for scraping the inside of the gouged cane to give it a fine finish.

(*f*) *Reed Knife* A straight knife, about 9½ in long overall, usually hollow ground and with a thick back. Two are used: one with the edge turned over for scraping the cane down to the correct thickness; the other sharpened for paring the cane on the shaper to its exact contour.

(*g*) *Shaper* After hollowing and cutting to length, the folded cane is held over the end of this tool by a spring clamp. Surplus cane is then cut away down to the steel sides of the shaper, leaving the reed correctly shaped and ready to be parted at the fold and finally wired and mounted on the Mandrel.

(*h*) *Mandrel* A tapered metal rod, about 6 in long, used for holding the reed to check the bore of the staple and to support it while the reed is bound on. (The staple is the conical brass tube, covered with cork at one end, on which the reed is bound and which connects the reed with the instrument.)

(*i*) *Billot or Cutting Block* A hardwood round block, about 1¼ in across, used when trimming and cutting off the tip of the reed.

(*j*) *Tongue or Plaque* A thin, oval-shaped steel or plastic plate, about 1½ in across, which is slipped between the blades of the reed to support it while scraping and filing.

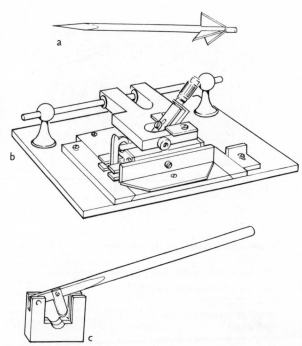

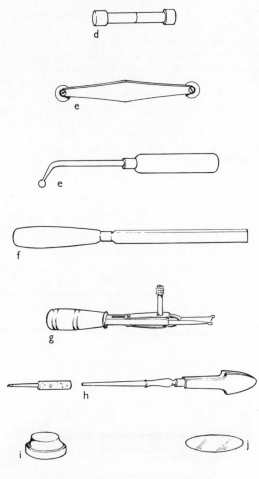

Fig. 736 Tools for making Reeds for Woodwind Instruments

Bench; Bench Hook etc.
Bending Tools
Bevel Block
Block
Boning or Winding Strips
Bracket
Calliper
Carrying Stick
Carver's Bench Screw etc.
Compass
Cooper: Furniture etc.
Cramp
Dog
Donkey's Ear (see Mitre Appliances)
Dovetail Marker
Glue Plate
Glue Pot etc.
Grease Pot
Grinding Appliances
Holdfast
Horse
Level
Lifting Jacks
Lighting
Lubricating Tools
Mitre Appliances
Moulding and Turning Box
Oil Stone etc.
Painting Equipment
Patterns, Templets, and Jigs
Pot hanger
Rounding Cradle
Rule
Samson, see Wheelwright's Equipment (Tyring Tools)
Saw Sharpening and Holding Devices
Sawing Horse, Box etc.
Shaving Horse
Shooting Board
Squares
Sticking Board
Templets
Tool Basket
Tool Chest
Trestle
Tyring Equipment, see under Wheelwright's Equipment
V-Block
Vices
Wheelwright's Equipment

Work Bench: see *Bench, Woodworker's.*

Workshop Equipment
Appliances and Furniture used in the Woodworking Trade.

(*a*) Particulars of workshop equipment are given under many trade entries, e.g. *Cellarman; Chairmaker; Plane Maker; Saw Sharpener; Shipwright; Wheelwright; Woodland Trades.*

(*b*) Other equipment will be found under individual entries including the following:

Wrecking Bar
A name given to Crow Bars of various kinds used for stripping or pulling down wooden structures.

Wrench
The term Wrench is applied both to the tools that turn nuts or pipes and to the quite different tools used for bending rods or bars. The term Spanner is applied only to tools used for turning nuts and bolts. Wrenches commonly found in woodworking shops are described below.

Wrench: (1) FOR GRIPPING NUTS *Fig. 737*

(*a*) *Fixed Spanners*
These were usually smith made and often S-shaped. The factory-made examples are commonly drop-forgings.

(*b*) *Wedge Spanner* (Slip Wrench; Shifting Spanner)
The lower jaw, which slides up or down the shaft of the spanner, is locked by an iron wedge. This is probably the first attempt at producing an adjustable grasp, and is still preferred by some tradesmen because it can be hammered tight on a worn or ill-shaped nut. The larger sizes, up to 30 in long, that were used by millwrights, were sometimes provided with a ring at the lower end of the handle. A rope was attached to the ring for pulling on a recalcitrant nut. Both home- and factory-made examples are illustrated.

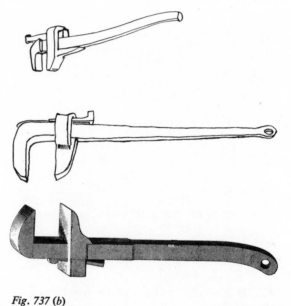

Fig. 737 (b)

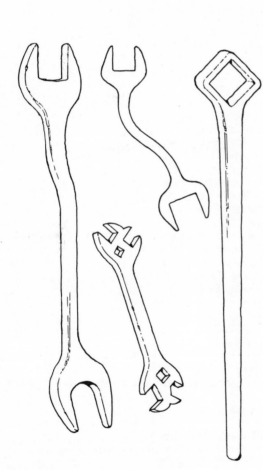

Fig. 737 (a)

(*c*) *Screw Wrench* (Bolt Clam; Coach Wrench; Monkey Wrench; Screw Hammer)
The lower jaw of this Wrench is moved up and down in various ways, e.g. by turning a nut on the screwed stem of the Wrench or on the short stem fitted parallel to it; or by twisting the handle of the Wrench itself, in which the screwed stem operates.

1. This smith-made example was sometimes called a 'Screw Hammer' because the upper jaw of the Wrench was made in the form of a Hammer and could be used as such.

2. These factory-made examples appear in toolmaker's catalogues of the early nineteenth century in almost modern form. They were made in sizes from 6 to 18 in long overall.

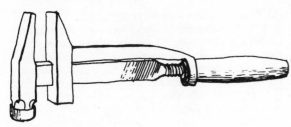

Fig. 737 (c1)

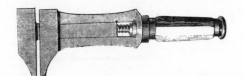

Fig. 737 (c2)

(*d*) *Pincer Spanner*

This was an attempt to make an adjustable Spanner without screw or wedge. The two halves of the tool interlock at the serration on jaw and stem. It is a factory-made tool, probably of mid-nineteenth century origin.

Fig. 737 (d)

Wrench: (2) FOR GRIPPING ROUNDS (Pipe Wrenches) *Fig. 738*

Much ingenuity is shown by the many designs of adjustable Wrenches intended for gripping a round rod or pipe. Though designed primarily as Pipe Wrenches, these tools are commonly used in woodworking shops for gripping or turning worn nuts etc. The adjustable patterns are designed to increase the grip the harder they are pressed. Today, many of these Wrenches have been superseded by one in which the Wrench grips by means of a chain which is tightened round the pipe.

The following factory-made examples are illustrated:

(*a*) *Footprint Wrench* (The name Footprint derives from the trade mark of the makers Thos. R. Ellin of Sheffield.)

(*b*) An earlier Pipe Wrench of the 'Footprint' type

(*c*) *Stillson Wrench* (The term is taken from the name of an early maker of Pipe Wrenches in the U.S.A.). A type which is still popular today.

(*d*) *Crocodile* (or *Bulldog*) *Wrench* – an attempt to dispense with moving parts

Fig. 738 (a)

531

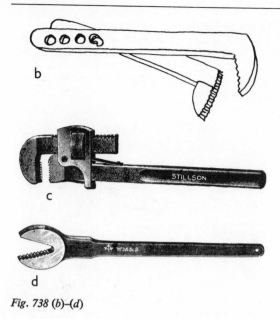

b

c

STILLSON

WM&S

d

Fig. 738 (b)–(d)

Wrench, Bar Bending (Step Wrench; Twisting Wrench) *Fig. 739*
A strong iron bar, about 2–4 ft long, with two prongs set at right angles at one end. Used for bending iron bars and rods, e.g. by coachbuilders for bending steps and brake fittings.

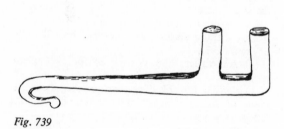

Fig. 739

Wrench, Bed: see *Bed Key*.

Wrench, Clip: see *Wrench, Coachbuilder's*.

Wrench, Coachbuilder's *Fig. 740*
The ordinary Screw Wrench is commonly called a Coach Wrench, but is used in many other trades. The following are the special Wrenches used by coachbuilders, wheelwrights, and carriage owners.

(*a*) *Axle Key* (Axle Cap Wrench)
A hexagon Spanner – an ancestor of the elegant drop-forged bi-hexagon Ring Spanner of the modern garage mechanic. They were used for turning the axle caps that were screwed on the outside ends of factory-made axle-boxes.

1. Drabble Axle Key
2. Mail Axle Key
3. Nut Axle Key
4. Collinge Axle Key
5. Mandrel on which Axle Keys of different sizes were forged

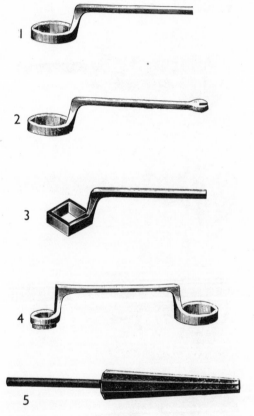

Fig. 740 (a)

532

(*b*) *Adjustable Axle Cap and Nut Wrench*
Variants include:

    1. Home-made examples, one with wedge adjustment, the other with a tapered slot to take different sized nuts.

    2. 'Petch's' pattern   } typical factory-made
    3. 'Railway' pattern } examples
    4. Hinged Vice pattern

(*c*) *Clip Wrench* (Clip Dog; Clip Tongs)
An iron bending tool about 14 in long. Two forms are illustrated, (*c*1) with a hinge and (*c*2) with a fixed jaw. The end of one prong is recessed, and the other pointed and sometimes hooked. The tool is used when fitting the clips which secure the leaf springs to coach or cart axles. The legs of the clip tend to spread when driven over the springs. The Wrench is then used for drawing the legs of the clip together and easing them through the holes in the plate to which they are bolted. The hooked jaw of the Wrench is held in one bolt hole in the plate, while the concave-ended jaw pushes a leg of the clip inwards until it slips into its allotted hole. The factory-made example (*c*3) shows the method of operation.

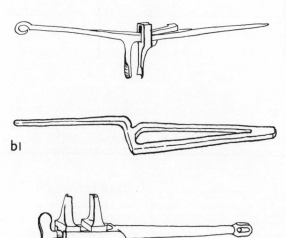

bl

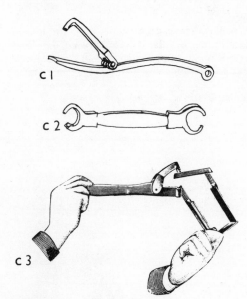

c l

c 2

c 3

Fig. 740 (c)

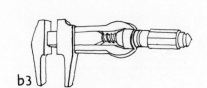

b2

b3

(*d*) *Step Wrench* (*Fig. 740* under *Wrench, Bar Bending*)
A Bar Bending Wrench, usually smith-made, used for straightening or adjusting the iron work for the steps and brakes on carriages and carts. These fittings were usually bought in, and had to be bent to fit.

Wrench, Monkey
A term often applied to a Screw Wrench or Spanner. Mercer (U.S.A., 1929) indicates that the term has been used since about 1860; but there is evidence of earlier usage in the catalogue of Richard Timmins

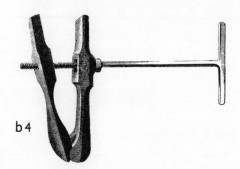

b 4

Fig. 740 (b)

533

(Birmingham, *c.* 1855), in which the engravings appear to have been made in the late eighteenth century.

Wrench, Nut: see *Bit, Nut Wrench.*

Wrench, Pipe: see *Wrench (2) For Gripping Rounds.*

Wrench, Stave: see *Flagging Iron.*

Wrench, Step: see *Wrench, Bar Bending.*

Wrench, Tap: see *Metal-Working Tools; Screw Dies and Taps.*

Wrench, Twisting: see *Wrench, Bar Bending.*

Wrest

A name often given to tools used for twisting or bending, e.g. to a Bar Bender (*Wrench, Bar-Bending*); to a Saw Set (*Saw-Setting Tools*); or to the Wrest used for straightening plane irons (*Plane Maker*).

Wrong Iron: see *Barking Iron.*

Y

'Yankee' Tools

A name sometimes given to tools described under the following entries: *Axe, Felling, Wedge type; Drill, Archimedean; Screwdriver, Spiral Ratchet.*

Mr. F. Seward has informed us (1969) that aside from the colloquial use of 'Yankee' in this country for anything American, it has, to his knowledge, been used as a recognised trademark for tools made by the following tool makers:

(*a*) Collins & Company, Hartford, Conn. – 'Collins' Yankee Felling Axes'.

(*b*) North Bros. Mfg. Co., sometime of Philadelphia, Pa. – 'Yankee' Spiral Ratchet Screwdrivers.

(*c*) The Stanley Works, New Britain, U.S.A., and now also in Sheffield. This firm took over North Bros. Mfg. Co. and they still use 'Yankee' as a brand name for their Spiral Ratchet Screwdrivers.

Yarning Tool or Chisel

An all-iron, cranked bar made like a Cold Chisel. Used by plumbers for caulking pipe joints.

BIBLIOGRAPHY AND REFERENCES

Abreviations

J   Article from a journal.

P   Printed book or pamphlet.

T   Trade catalogue, price list, directory, or works tool list.

V   Visit to, or correspondence with, the person or firm named, during the period 1940–73. These entries also signify the author's thanks for help and information given.

*   Now deceased.

———————●———————

P   Abell, Sir Westcott. *The Shipwright's Trade.* Caravan Book Service, New York. 1962.

T   Addis, J. B. & Son, Ltd., Sheffield. (Woodcarver's chisels) *c.* 1920.

P   Admiralty. *Manual of Seamanship.* H.M. Stationery Office, London. 1951. (See also Dockyard, H.M.)

Amman, Jost. See *Schopper, Hartman.*

Apprentice Indenture Records. See Goodman, W. L., 1972.

PV   Arnold, J. *The Shell Book of Country Crafts.* Baker, London. 1968.

P   ——*The Farm Waggons of England and Wales.* Baker, London. 1969.

T   Arnold & Sons, London. 1885 and 1886. Surgical and veterinary instruments.

P   Ashton, T. S. *An Eighteenth-century Industrialist, Peter Stubs of Warrington, 1756–1806.* University Press, Manchester. 1939.

V   Aspden, Anthony (St. Albans). Musician. Reed making for woodwind instruments.

T   Atkin & Sons, Birmingham. 1894. Tools for many trades.

JV   Bagshawe, T. W. 'Rake and Scythe-Handle Making in Bedfordshire and Suffolk', *Gwerin* (Blackwell, Oxford) 1956. Vol. I. No. 1 p. 34.

P   Bailey, William. *The Advancement of Arts, Manufactures, and Commerce; or, Descriptions of the useful machines and models.* London. 1772.

T   Baird & Tatlock Ltd., London. 1910, 1927, 1954. Chemical and scientific apparatus.

T   Ball, George, Rugby. 1898. Agricultural implements and vehicles.

V   Barnsley, Edward (Foxfield, Hants.). Furniture maker.

T   Barnsley, George Ltd., Sheffield. 1890. Shoemaker's, tanner's, and currier's tools.

V   Basset, K. H. (Washington, U.S.A.). Cooper's tools in the U.S.A.

P   Beedham, R. John. *Wood Engraving.* Faber, London. 1921.

T   Belknap Hardware & Manufacturing Co., Kentucky, U.S.A. 1940. Tools for many trades.

T   Bennett, J. *Artificier's Lexicon.* London. 1837.

T   Berg, E. A. Mfg. Co. Ltd., Eskilstuna, Sweden. 1939. Tools for many trades.

P   Bergeron, H. *L'Art du Tourneur.* Paris. 1816.

P   Biggs, John R. *The Craft of Woodcuts.* Blandford Press, London. 1963.

P   Blanckley, Thomas Riley. *A Naval Expositor . . . explaining the works and terms of art belonging to building . . . and fitting a ship for sea.* London. 1750.

P   Blythe, Ronald. *Akenfield, Portrait of an English Village.* Allen Lane, The Penguin Press, London. 1969.

V   Boosey & Hawkes Ltd. (London). Musical instrument makers.

T   Boswell, Peter & Sons, London. *c.* 1920. Motor, coach, and van ironmongery.

T   Bowles, John J. Hartford, Conn., U.S.A. *c.* 1840 (Price List). Planes and joiner's tools.

TV   Brades, Skelton & Tyzack Ltd., Oldbury. *c.* 1963. Scythes and hooks.

T   Brades, The, Ltd. See Hunt (William) & Sons.

P   Bradley Smith, H. R. *Blacksmith's and Farrier's Tools at the Shelburne Museum.* Vermont, U.S.A. 1966.

V   Bradwell, Wm. (Besses O'th Barn, Lancs.). Clogmaker.

J   Brazier, G. W. 'The Craft of the Cooper', *Practical Education and School Crafts* (Southampton). September 1961, page 6.

V   Brightwell, J. B. (Boston, Lincs.). Sailmaker.

T   Bristol Wagon & Carriage Works Co. Ltd., Bristol. 1906. Agricultural implements and vehicles.

T   British Railways. A leaflet entitled 'Tools and appliances in general use by the various regular Platelaying, Relaying, Draining and Ballast gangs etc'. Drawing No. 40, L. & N.W.R. London. *c.* 1900.

P   British Standards Institution. Leaflets 1953–65 on various hand tools, including Hand Saws; Metal-bodied Planes; Augers and Auger Bits; Axes and Hatchets; Chisels and Gouges; Bit Braces. London. 1953–65.

T   Brombacher, A. F. & Co. Inc., New York, U.S.A. 1922. Cooper's and gauger's tools.

T   Brown Bros. Ltd., London. 1939. Garage tools and equipment. Motor-car accessories.

P   Brunet, R. *Manuel de Tonnellerie.* J.-B. Baillière et Fils, Paris. 1948.

T   Buck, G. London. *c.* 1860 (Price List). Saws, Files, Planes, etc.

T   Buck & Hickman Ltd., London. 1907, 1935. Tools for many trades.

TV   Buck & Ryan Ltd., London. *c.* 1925, 1935. Tools for many trades.

J   *Building Trades Journal* (formerly Illustrated Carpenter and Builder). London. 1945–50. See also Weller, George.

V*   Bunday, C. E. (Bursledon, Hants.) Shipwright. Employed up to *c.* 1960 by A. H. Moody and Son Ltd., Boatbuilders, Southampton.

V   Bysouth, A. G. (Baldock, Herts.) Wheelwright.

T   Callow, W. H. & Lake, J. L. Ltd., London. 1909, 1915, 1925. Wheelwright's and coach-builder's equipment and accessories.

T   Cam, James, Sheffield 1787–*c.* 1835. A Price list with illustrations attached to Smith's *Key to the various manufactories of Sheffield* (1816) in the possession of the Sheffield City Library. (See Smith, Joseph.)

J   Carlson, Robert H. 'Nomenclature and Variations of Parts of the Spiral Auger Bit', *The Chronicle of Early American Industries* Vol. XXI No. 2. June 1968.

PV   Carrington, Noel. *Popular Art in Britain.* Penguin, London. 1945.

V   Carter, Harry (Oxford University Press). Printer's tools.

V*   Casbon, George (Barley, Herts.). Wheelwright.

V   Chalkley, J. F. (London). Authority on various trades.

P   Chapelle, H. I. *Boat Building.* W. W. Norton, New York. 1941.

T   Chesterman, James & Co. Ltd., Sheffield. 1925. Rules, measuring tapes, and gauges.

PV   Childe, Gordon. *The Story of Tools.* Cobbett, London. 1944.

     ——'Wheeled Vehicles', *A History of Technology* (Vol. I) Oxford. 1954.

J   *Chronicle, The.* Journal of the Early American Industries Association, Williamsburg, Virginia, U.S.A. 1950–71.

T   Churchill, Charles & Co. Ltd., London. 1937. Tools for many trades.

V*   Clarke, Philip. Coachbuilder at Messrs. Joseph Cockshoot & Co. Ltd. (Manchester).

J   Coleman, J. C. 'The Craft of Coopering', *Journal of the Cork Historical and Archaeological Society*, Cork, Ireland, 1944, Part 2 – Vol. XLIX, No. 170. p. 1.

V*   Collier, Arthur (London). Tool merchant and authority on early tools.

T   Cooper, William, Ltd., London. 1890, 1910. Gardener's tools, horticultural providers, and portable buildings.

T   Cooperage Industry, District Industrial Council. Price List of various operations, including Dry Work. London *c.* 1913.

TV   Copley, A. London. *c.* 1905, 1925. Stone mason's, sculptor's, and building worker's tools.

T   Corcoran, Bryan Ltd., London. 1949. Flour miller's implements and accessories.

P   Corkhill, Thomas. *A Glossary of Wood.* Nema Press, London. 1948.

P   Cotgrave, Randle. *A dictionarie of the French and English tongues.* London. 1611.

T   Coubro & Scrutton Ltd., London. *c.* 1912. Ship's fittings, tools, and equipment.

J   *Country Life.* Articles on trades. London. 1948–1970.

J   *Countryman.* 'Rescuing the Past' and articles on trades. Burford. 1950–70.

P   Coventon, W. *Woodwork Tools and their use.* Hutchinson, London. 1953.

T   Crosskill, W. & Sons, Beverley. 1894. Agricultural implements, vehicles, and wheels.

T   Crowden & Garrod Ltd., London. 1898. Domestic hardware and ironmonger's sundries.

V*   Curtis, W. H. late Director of the Curtis Museum, Alton, Hampshire.

TV   Davey & Co. Ltd., London. 1936. Ship's fittings and appliances. Naval hardware.

P   Davey, Norman. *History of Building Materials.* Phoenix House, London. 1961.

V   Davis & Son (Harpenden, Herts.). Piano repairers.

JV   Dickinson, H. W. 'Origin and Manufacture of Wood Screws', *Transactions of the Newcomen Society*, Vol. XXII, 1941, p. 79.

J   ——'Historical account of Besoms and Brushmaking', *Transactions of the Newcomen Society*, Vol. XXIV, 1943, p. 99.

P   Diderot, D. and D'Alembert, J. le R. (Eds) *Encyclopédie ou dictionnaire raisonné des sciences, des arts, et des métiers.* Paris. 1751–1772.

T   Directories:
Federation of British Hand Tool Manufacturers. *Buyer's Guide*. Sheffield. 1957.

Pigot and Co. *Commercial Directory*. London. 1826.

Gales and Martin. *A Directory of Sheffield*. Sheffield 1787 (Reprinted in facsimile by Pawson and Brailsford, Sheffield 1889).

TV Disston, Henry & Sons, Inc., Philadelphia, U.S.A. *c*. 1915. Saws and Files.

P   Disston, Henry & Sons. *The Saw in History* and *Saw and Tool Manual*. H. K. Porter, Pittsburgh, Pen. U.S.A. 1925.

V   Dixon Clarke Ltd. (London). Coopers.

V   Dixon, F. A. (National Coal Board, Gateshead). Mining Carpentry.

T   Dixon, Joseph, Tool Co. Ltd., Walsall. *c*. 1950. Saddler's tools.

V   Dockyard, H.M. (Portsmouth). Victory workshop; mast maker's and block-making shops.

V   Dornum & Sons (Salcombe, Devon). Boatbuilders.

T   Dottridge Bros. Ltd., London. *c*. 1916. Undertaker's equipment, coffins, etc.

P   *Douglas's Encyclopaedia* (Butcher's and other food trades). William Douglas, London. 1924.

T   Drew, C. & Co., Kingston, U.S.A. *c*. 1920. Ship's Caulking Chisels and other tools.

PV Edlin, H. L. *Woodland Crafts in Britain*. B. T. Batsford, London. 1949. Reissued 1973 by David and Charles, Newton Abbot.

T   Ellin, T. R. (Footprint Works) Ltd., Sheffield. Pipe wrenches and other tools.

V   Elliot, James (Harpenden, Herts.). Railway track layer's tools.

P   Ellis, George. *Modern Practical Joinery*. London 1902.

    ——*Modern Practical Carpentry*. Batsford, London. 1915.

TV Elwell, Edward Ltd., Wednesbury, Staffs. 1947, 1965. Slashers, hooks, spades, axes, etc.

P   Evans, E. E. *Irish Folk Ways*. Routledge, London. 1957.

PV Evans, George Ewart. *Ask the Fellows who Cut the Hay*, 1956.

    ——*The Pattern under the Plough*, 1966.

    ——*The Farm and the Village*, 1969.

    ——*Where Beards Wag All*, 1970. Faber and Faber, London.

P   Evelyn, John. *Sylva, or a Discourse of Forest Trees*. 1664.

P   Falconer, William. *A New Universal Dictionary of the Marine*. London. 1815.

T   Farmiloe, T. & W. Ltd., London. *c*. 1910. Painter's and plumber's requisites, sanitary appliances, pumps, etc.

T   Fearn, John Ltd., Sheffield. 1964. Cooper's tools.

P   Feldhaus, F. M. *Die Sage*. Berlin. 1921.

P   Félibien, A. *Principe de l'Architecture* . . . Paris. 1676.

JV Fenton, A. 'Ropes and Rope-making in Scotland', *Gwerin*, 1961. Vol. III No. 3 p. 142 and No. 4 p. 200.

T   Féron & Cie., Paris. *c*. 1950. Woodworking planes and other tools.

T   Flather, David & Sons Ltd., Sheffield. 1900. Tools for many trades.

T   Fletcher, H. J. & Newman, Ltd., London. 1959 and 1966. Piano tools and accessories.

T   Fleuss, Peter, Küllenhahn, Germany. 1855. Tools for many trades.

V   Flower's Brewery (Luton) now Whitbread (London) Ltd. Former cooper: K. J. Kilby.

J   Folk Life Society. *Folk Life* (formerly *Gwerin*) Journal of the Society for folk life studies, The Folk Life Society, St. Fagan's Castle, Cardiff 1955–71.

P   *Forestry*. Young Farmers' Booklet No. 20. Evans, London. 1955.

T   Forestry Commission. Descriptive list of tools for forestry. London. 1946.

V   Fowle, J. A. (London). Saw doctors and sharpeners.

PV Freeman, Charles. *Luton and the Hat Industry*. Luton Museum. 1953.

P   Freese, Stanley. *Windmills and Millwrighting*. University Press, Cambridge. 1957.

V   Frost, Wm. C. (Harpenden, Herts.). Basket maker.

P   Fussell, G. E. *The Farmer's Tools 1500–1900*. Andrew Melrose, London. 1952.

T   Gallenkamp, A. & Co. Ltd., London. *c*. 1925. Laboratory equipment.

T   Gas Light and Coke Company. List of tools used in Gas Production. *c*. 1930–40.

PV Gilbert, K. R. *The Portsmouth Blockmaking Machinery*. Science Museum, London. 1965.

PV Gilding, Bob. *The Journeyman Coopers of East London*. Ruskin College, Oxford. 1971.

V   Gilling, Percy & Son (Copenhagen Wharf, London). Coopers and Vat Builders.

TV Gilpin (Wm.) Senr. & Co. (Tools) Ltd., Cannock. 1868, 1946 & 1970. Axes, adzes, and tools for many trades.

TV Goddard, J. & J., London. *c.* 1950. Piano and organ tools and fittings.

T   Goldenberg & Co., Alsace, France. 1875, 1901, 1904, and 1950. Tools for many trades, and tools made for export to many different countries.

T   Goodell-Pratt – see Millers Falls Co.

PV Goodman, W. L. *A History of Woodworking Tools.* Bell, London. 1964.

P   ——*Woodwork. From Stone Age to Do-it-yourself.* Blackwell, Oxford. 1962.

P   ——*British Plane Makers from 1700.* Bell, London. 1968.

J   ——'Some Elizabethan Woodworkers and their tools', *The Journal of the Furniture History Society.* Vol. VII. 1971.

J   ——'Woodworking Apprentices and their Tools in Bristol, Norwich, Great Yarmouth and Southampton 1535–1650', *Industrial Archaeology.* Vol. 9 No. 4, Nov. 1972.
    Note Mention of Mr. Goodman's name in the text also relates to private correspondence.

T   Greaves, William & Sons. See Turton, Thomas.

PV Greenhill, Basil. *The Merchant Schooners.* Percival Marshall, London. 1951.

T   Griffiths, Hannah. Norwich, 1842–75. Plane Makers (price list).

TV Griffiths, Horace. Norwich, 1900–48. Plane Makers (price list).

J   *Gwerin.* See Folk Life Society.

P   Halliwell, J. O. *A Dictionary of Archaic and Provincial Words.* London. 1847.

T   Hampson & Scott, 'Equine Album'. Walsall. *c.* 1900. Saddler's tools, saddlery, harness, and stable equipment.

P   Hampton, C. & J. Ltd. (Publishers). *Planecraft.* Sheffield, 1950.

T   Hampton, C. & J. Ltd., Sheffield. See Record Tool Works.

P   Hankerson, Fred Putnam. *The Cooperage Handbook.* Chemical Publishing Co., New York. 1947.

TV Hardypick Ltd., Sheffield. Mining Tools.

V   Hawley, Kenneth (Sheffield). Tool merchant and authority on Sheffield tool makers.

PV Hayward, Charles, H. *Practical Veneering.* 1949.

P   ——*Tools for Woodwork.* 1960.

P   ——*Practical Wood Carving and Gilding.* 1963.

P   ——*Woodworking Joints.* 1950. Evans Brothers, London.

P   Hellyer, S. S. *Principles and Practice of Plumbing.* Bell, London. 1905.

P   Hennell, Thomas. *The Countryman at Work.* Architectural Press, London. 1947.

P   ——*Change in the Farm.* The University Press, Cambridge. 1936.

T   Herbert & Sons, Ltd., London. 1930, 1934. Butcher's tools and equipment.

P   Heron-Allen, Edward. *Violin-Making.* Ward Lock, London. 1885.

P   Hibben, Thomas. *The Carpenter's Tool Chest.* Lippincott, Philadelphia, U.S.A. 1933.

P   Himsworth, J. B. *The Story of Cutlery.* Benn, London. 1953.

P   Hine, H. J. *Farm Implements.* Young Farmer's Club Booklet No. 7. Evans, London. 1958.

V   Holland and Holland Ltd. (London). Gun-makers and Gunstockers.

P   Holmes, Charles M. *The Principles and Practice of Horseshoeing.* The Farriers' Journal, Leeds. 1928.

P   Holtzapffel, Charles. *Turning and Mechanical Manipulation.* Holtzapffel & Co., London. 5 vols. 1846–47.

P   Hommel, R. P. *China at Work.* John Day, New York. 1937.

P   Hoppus, E. *Practical Measuring made easy to the Meanest Capacity, by a New Set of Tables.* (The solid or superficial content of timber, stone, board, glass, etc.). London. 1761–1846.

P   Howard, A. L. *Timbers of the World.* London. 1920.

T   Howarth, James & Sons, Sheffield. 1884. Tools for many trades.

PV Howes, C. *Practical Upholstery.* Evans, London. 1950.

V   Hughes, E. (St. Albans). Fishing-rod maker's tools.

P   Hummel, Charles, F. *English Tools in America: The Evidence of the Dominys.* Winterthur Portfolio, 1965. Winterthur Museum, Delaware, U.S.A.

TV Hunt (William) & Sons, The Brades Ltd., Birmingham. 1905, 1924, 1942. Tools for many trades.

T   Isler, C. & Co. Ltd., London. 1912–46. Well-sinker's tools and equipment.

T   Jagger, Albert, Walsall. *c.* 1910–15. Coach builder's tools, equipment, and accessories. Harness, lamps, etc.

T   James, John & Sons, Redditch. 1948. Sail-maker's, upholsterer's, and leather worker's awls and needles.

V   Javeleau, H. R. (St. Albans). Hurdle maker.

PV Jenkins, J. Geraint. *The English Farm Wagon.* Oakwood Press, Reading. 1961.

——*Traditional Country Craftsmen.* Routledge, London, 1965.

V Jenkinson, R. (Harpenden, Herts.). Plumber.

PV Jobson, Allan. *Household and Country Crafts.* Elek, London. 1953.

V Joice, A. H. (Basingstoke). Coachbuilders.

P Jones, P. d'A. and Simons, E. N. *Story of the Saw. 1760–1960.* Spear & Jackson Ltd., Sheffield. 1960.

V Kelly, J. M. General Manager of Wm. Gilpin, Senr. & Co. (Tools) Ltd., Cannock.

PV Kilby, Kenneth. *The Cooper and his Trade.* Baker, London. 1971.

V* King, F. T. (Barley, Herts.). Blacksmith and farrier.

P Kipping, Robert. *Elementary Treatise on Sails and Sailmaking.* Wilson, London. 1862.

P Knight, Edward H. *Practical Dictionary of Mechanics.* Published in the U.S.A. in 1877 and afterwards by Cassel, Petter and Galpin in London. (A supplementary volume is dated 1884.)

P Lambert, F. *Tools and devices for the Coppice Crafts.* Young Farmers' Club Booklet No. 31. Evans, London. 1957.

PV Lambeth, Reginald, C. *Some Former Cambridgeshire Agricultural and other Implements.* The Cambridge Folk Museum. 1941.

TV Langley, W. & Co., London. *c.* 1930 and 1950. Cooper's tools and appliances.

V Latham, F. H. Director, Samuel Latham Ltd. (Birmingham). Brushmakers.

P Leadbetter, Charles. *The Royal Gauger; or Gauging Made Perfectly Easy.* E. Wicksteed. London. 1755.

V Lee, W. Robert, Director, James Lee & Son (Midhurst, Sussex). Millwrights.

P Legros, E. See Liège Museum.

V Leon, J. & Co. Ltd. (London). Wholesale Tobacconists.

P Leslie, Robert C. *The Sea-Boat. How to Build, Rig and Sail Her.* Chapman and Hall, London. 1892.

V Lewington, Neil. A collector of early trade tools.

T Lewis, Samuel & Co. Ltd., Dudley. 1908. Nails and general ironmongery.

P Liège Museum, Belgium. Musée de la Vie Wallonne. *Enquêtes du Musée de la Vie Wallonne,* including articles on the cooper, clogger, handle-maker, and sawyer. By E.

Legros, E. Remouchamps, and others. 1926–1949.

P Lillico, J. W. *Blacksmith's Manual Illustrated.* The Technical Press, London. 1944.

V* Lines, H. C. (Harpenden, Herts.). Blacksmith.

TV Loftus, W. R. Ltd., London. 1890–1900. Cooper's, cellarman's, and bottler's tools; barfitter's requisites.

P London Museum. Medieval Catalogue. 1940.

P Loudon, J. C. *An Encyclopedia of Agriculture.* Longman, Rees, Orme, Brown and Green, London. 1831.

V Lowrie, W. P. & Co. Ltd. (Glasgow). Coopers.

J Lu, Gwei-Djen, Salaman, R. A. and Needham, J. 'The Wheelwright's Art in Ancient China.' *Physis.* Florence, 1959. Vol. I. p. 103 and 196.

TV Lusher & Marsh, Norwich. 1949. Wooden shovels and spades.

V Luth, Walter F. (Chicago, U.S.A.). Mast maker.

V Mandel, Edward (London). Piano making and repairs.

TV Marples, Wm. & Sons Ltd., Sheffield. 1903, 1921, 1938, and 1965 (List of American Tools 1909). Tools for many trades.

T Martin, A. & Son, Birmingham. Brush makers.

V Martin, Malachi (Barley, Herts.). Hurdle maker.

PV Massingham, H. J. *Country Relics.* The University Press, Cambridge. 1939.

T Mathieson, Alex & Sons, Ltd., Glasgow. 1900, 1933 and 1957. Tools for many trades.

PV Mayes, L. J. *The History of Chairmaking in High Wycombe.* Routledge, London. 1960.

T Melhuish, Richard & Sons Ltd., London. 1885, 1912, 1913. Tools for many trades.

P Mercer, Henry, C. *Ancient Carpenters' Tools.* The Bucks County Historical Society, Doylestown, Pennsylvania, U.S.A. 1929.

T Millers Falls Company (incorporating Goodell-Pratt Co.) Greenfield, U.S.A. 1935. Tools for many trades.

T Mills Bros., London. 1901, 1911. Tools for masons, sculptors, and stone yards.

V* Mitchell, E. (Oakington, Cambs.). Ladder maker.

J Morgan, F. C. 'The Craft of the Wooden Pump Maker.' *Transactions of the Newcomen Society,* London. 1942–43. Vol. XXIII. p. 59.

P Moxon, Joseph. *Mechanick Exercises: or the Doctrine of Handy-Works. Applied to the Arts of Smithing Joinery Carpentry Turning Bricklayery.* London. 1677.

V Murison, David (Edinburgh). Editor, *The Scottish National Dictionary.*

P McCullock, Dean Walter F. *Wood Words. A comprehensive Dictionary of Loggers' Terms.* Oregon Historical Society, U.S.A. 1958.

P McDermaid, N. J. *Shipyard Practice as applied to Warship Construction.* Longmans, London. 1918.

V McKears, R. H., Director, William Ridgeway & Son Ltd. (Sheffield). Auger and bit makers.

PV Naish, G. P. B. 'Ships and Shipbuilding', *A History of Technology*, Oxford. 1956. Vol. III. p. 493.

T Nash, Isaac & Sons Ltd. (Brades and Nash Tyzak Industries Ltd. Stourbridge. 1899, 1960. Farm and garden tools, scythes, and hooks.

V Naval Yards, see Dockyard, H.M.

PV Needham, Joseph, and Wang Ling. *Science and Civilisation in China.* The University Press, Cambridge. 7 Vols. 1954–65.

J The Newcomen Society for the study of the history of Engineering and Technology. *Transactions*, London. 1920–60.

P Nicholson, M. A. (from the original of P. Nicholson). *The Carpenter and Joiner's Companion.* H. Fisher, R. Fisher and P. Jackson, London. 1826.

P Nicholson, P. *Mechanical Exercises*, London. 1812.

P ——*The New Practical Builder*, London. 1822.

P Norman, H. and J. *The Organ Today.* Barrie & Rockliff. 1966.

V Norris, J. O. H., Director, Messrs. Joseph Cockshoot & Co. Ltd., Coachbuilders (Manchester).

T Norris, T. & Son, London. 1900–35. Metal-bodied planes.

T Norton & Gregory Ltd., London. 1935. Drawing instruments.

T Nurse, Chas. & Co. Ltd., London. 1891, 1934. Tools for many trades.

P Oakley, Kenneth, P. *Man the Tool-Maker.* British Museum, London. 1958.

V O'Connor Fenton, Alan, Director of Wm. Marples and Sons Ltd., Toolmakers (Sheffield).

V Offord & Son Ltd. (London). Coachbuilders.

TV Osborne, Garrett, Nagele, Ltd. London, 1970. Wigmaker's and hairdresser's sundries.

J O'Sullivan, John C. 'Wooden Pumps', *Folk Life*, Vol. 7, 1969. p. 101.

P *Oxford English Dictionary* (12 Vols.). Clarendon Press, Oxford. 1933.

PV Pain, F. *The Practical Wood Turner.* Evans, London. 1957.

P Parish, Rev. W. D. *A Dictionary of the Sussex Dialect.* Farncombe & Co., Lewes. 1875.

V* Parker, J. F. (Bewdley). Authority on early trades.

T Parnall & Sons, Ltd., London. 1899. Tools, sundries, and equipment for the grocery and allied trades.

T Pascal Atkey & Son, Ltd., Cowes, Isle of Wight. *c.* 1930. Yacht fittings and chandlery.

V Paterson, Robin (San Francisco, U.S.A.). Shipwright.

T Pawson and Brailsford Ltd., see Sheffield Illustrated List.

T Peck, Stow & Wilcox Co., U.S.A. *c.* 1915. Carpenter's braces.

V Peirce, S. (Southampton). Bar fitter and pewterer.

V Peters, L. C., Print and Block cutter at Messrs. Arthur Sanderson & Sons Ltd. Wallpaper manufacturers, Perivale, Middlesex.

P Petrie, Sir W. M. Flinders. *Tools and Weapons.* Constable, London. 1917.

T Peugeot Frères, Paris. 1880. Saws.

P Philipson, John. *The Art and Craft of Coachbuilding.* Bell, London. 1897.

V Phillip's Brewery (Royston, Herts.). Former cooper: H. J. Bonner.

V Philpot, Harold (Michigan, U.S.A.). Wagon construction.

T Pigot & Co's Commercial Directory. London. 1826.

T Pike, George, Ltd., Birmingham. *c.* 1925. Contractor's tools and tools for railways, mines, etc.

PV Pinto, Edward H. *Treen and other wooden bygones.* Bell, London. 1969.

T Plane Maker's List of Prices. Price list of parts and fittings, cost of repairs, etc. Printed by E. Bower, 256 Bradford Street, Birmingham. 1871.

T Plumpton, George & Co. Ltd., Warrington. 1937 and 1966. Linesman's, telegraph, and electrician's tools.

V Plumrose Ltd. (Ebberup, Denmark). Dairy products.

V Port of London Authority (London). Wine vaults and cooperage.

T Preston, Edward & Sons Ltd., Birmingham. 1914. Rules, levels, and other tools.

P Pugh, Reginald. *Factory Training Manual.* Management Publications Trust Ltd., Bath. 1941.

V Pyne, J. F. Ltd. (London). Piano-Key Makers.

P Pyne, W. H. *Picturesque Groups for the embellishment of Landscape.* London. 1845.

T Rabone, John & Sons Ltd., Birmingham. 1910, 1954. Rules, levels, and other measuring tools.

Railways. See British Railways.

P Rålamb, Åke Classon. *Skeps Byggerij eller Adelig Ofrings* (An illustrated book on ship building). Stockholm. 1691. Facsimile issued by Sjohistoriska Museet, Stockholm. 1943.

V Reading Museum of English Rural Life (Reading). Card index of tools.

T Record Tool Works (now C. & J. Hampton Ltd.). Sheffield. 1938. Planes and spoke shaves.

V Renshaw, E. S. (Ford Motor Co.). Authority on metals.

P Richardson, M. T. *Practical Carriage Building.* M. T. Richardson, New York. 1903.

T Richter, V., Benesov, Czechoslovakia. 1937. Tools for many trades.

TV Ridgway, Wm. & Sons Ltd., Sheffield. 1954. Booklet 'Wood Boring'. 1966. Augers and bits.

V Ripley, D. (St. Albans). Shipbuilding.

V Roberts, Sir Harold (Late Chief Inspector of Mines). Mining carpenter's tools.

JV Roberts, Kenneth D., and Jane W. Various articles on wooden planes. *Chronicle of Early American Industries.* 1966–71.

P Rose, Walter. *The Village Carpenter.* The University Press, Cambridge. 1937.

Rural Industries Bureau. Library List. Rural Industries Bureau, London. 1950.

T Rushbrooke, G., Ampthill, Beds. 1887. Butcher's tools and fittings.

J Russell, J. 'Millstones in Wind and Water Mills,' *Transactions of the Newcomen Society,* 1944, Vol. XXIV (1943–45) page 55. (See also Wailes, Rex.)

T Russell Jennings Mfg. Co., U.S.A. *c.* 1950. Auger bits.

P Salaman, R. A. 'Tradesmen's Tools *c.* 1500–1850,' *A History of Technology.* Oxford, 1954. Vol. III.

J ——'Tools of the Shipwright 1650–1925', *Folk Life,* 1967, Vol. 5, p. 19.

J ——'Pincher Jack: A Travelling Blacksmith', *Gwerin,* 1960, Vol. III. No. 1, p. 18.

TV Salmen, A. B. Ltd. London. 1949. Wooden planes.

P Salzman, L. F. *English industries in the middle ages; being an introduction to the industrial history of medieval England.* Constable, London. 1913.

P ——*Building in England down to 1540.* Clarendon Press, London. 1952.

V Sanderson, Arthur, and Sons, (Perivale). Wallpaper Manufacturers. See Peters, L.C.

T Schmidt, Peter Ludwig, Elberfeld, Germany. *c.* 1890 and 1900. Tools for many trades.

P Schopper, Hartman. *De omnibus illiberalibus sive mechanicis artibus, humani ingenii . . .* (Illustrations by Jost Amman). Frankfurt am Maine. 1568.

V Seward, F. (London). Authority on English and American tools of the early twentieth century.

T *Sheffield, A Directory of, 1889.* See Directories.

T Sheffield Illustrated List, The. (F. Edition) Pawson and Brailsford, Sheffield. 1888.

T Sheffield Illustrated List, The. (G. Edition) Pawson and Brailsford, Sheffield. 1910.

T Sheffield Standard List, The. R. T. Barras, Sheffield. 1862.

P Shelburne Museum. *Woodworking Tools.* By F. H. Wildung. Shelburne Museum, Shelburne, Vermont, U.S.A. 1957. (See also Bradley Smith, H.R.)

T Shepherd & Sons Ltd., Burscough, Lancs. 1956. Tool handles.

TV Shetack Toolworks Ltd., London. *c.* 1953. Plumber's and builder's tools.

V Siegal, Bob, Jnr. (Mequon, U.S.A.). Sabot making.

PV Singer, Charles. *Technology and History.* Oxford University Press, London. 1952.

P Singer, Holmyard and Hall. *A History of Technology.* 5 Vols. The Clarendon Press, Oxford. 1954.

TV Skelton, C. T. & Co. Ltd., Sheffield. *c.* 1950. Spades, forks, garden, and edge tools.

V* Skinner, J. Sawyer employed *c.* 1930 by Messrs. William Weller and Son, wheelwrights and undertakers, Sompting, Sussex.

P Sloane, Eric. *A Museum of Early American Tools.* Wilfred Funk, New York. 1964.

V Sly, R. J. (Harpenden, Herts.). Upholsterer.

T Smith, Joseph (Referred to in the text as *Smith's Key*). 'Explanation or KEY, to the various Manufactories of Sheffield with Engravings of each article, designed for the utility of Merchants, Wholesale Ironmongers, and Travellers.' Sheffield. 1816.
Note: The original is in the Sheffield City Library together with a Price List and addi-

tional plates marked 'James Cam or Marshes and Shepherd.' James Cam's dates are 1787 – *c.* 1835.

T *'Smith's Key.'* See Smith, Joseph.

TV Sorby, Robert & Sons Ltd., Sheffield. *c.* 1875 and 1907. Tools for many trades.

V Sparks, W. N. (London). Shipbuilders and mast makers.

V* Spary, C. (Markyate, Herts.). Wheelwright.

TV Spear and Jackson Ltd., Sheffield. 1880, 1937, and 1970. Saws and tools for many trades.

T Spiers, Stewart. Ayr, Scotland. 1909 and *c.* 1935. Metal-bodied planes.

P *Spons' Mechanics' Own Book, a Manual for Handicraftsmen and Amateurs.* E. & F. N. Spon, London. 1893.

T Spratt, Jack, Music Co., Stamford, U.S.A. 1960–64. Reed-making tools and supplies.

T Stahlschmidt Tool Co. Ltd., Cronenfeld, Germany. 1911. Tools for many trades.

TV Staines Co. Ltd., London. 1937. Kitchen tools and equipment.

TV Staniforth, Thos. & Co. Ltd., Sheffield. 1960. Scythes, sickles, garden, and edge tools.

T Stanley Rule and Level Co., New Britain, U.S.A. 1902 (Now Stanley Works). Planes, rules, and other tools.

P Stanley Works Ltd. *A brief History of the Woodworker's Plane.* The Stanley Works, Sheffield. *c.* 1965.

TV Stanley Works, U.S.A. and London Ltd., 1910–1965. Tools for many trades.

T Starrett, L. S. Company, London. 1930. Precision tools for measuring.

P Steel, David. *The Elements and Practice of Rigging and Seamanship.* London. 1794. 2 vols.

P ——*The Art of Sailmaking as practised in the Royal Navy.* London. 1796.

P ——*The Art of making Masts, Yards, Gaffs, Booms, Blocks and Oars.* London. 1797.

T Steel Nut, The, and Hampton, Joseph, Ltd., Wednesbury. 1903. Cramps, screw jacks, etc.

P Stephens, Henry. *The Book of the Farm.* 2 vols. Blackwood, Edinburgh. 1851.

T Stormont Archer, Ltd., Sheffield. 1965. Chisels etc.

TV Stubs, Peter, Ltd., Warrington. *c.* 1845, 1945, 1947, 1949. Files, screwing, and other tools.

P Sturt, George. *The Wheelwright's Shop.* The University Press, Cambridge. 1942.

P Sutherland, William. *The Ship-Builder's Assistant, or some Essays towards compleating the Art of Marine Architecture.* London. 1711.

V Talton, V. (Melton Mowbray). Ironmonger.

TV Taylor, Henry Ltd., Sheffield. *c.* 1950–1970. Woodcarver's chisels and tools.

P Templeton, William. *The Millwright and Engineer's Pocket Companion.* Simpkin, Marshall & Co., London. 1839.

TV Temporal (William) Ltd., Sheffield. 1969. Awls, knives, etc.

V* Thomas, F. W. (Sheerness). Shipwright.

T Tilling, G. J. & Sons Ltd., Southampton. *c.* 1905. Sailmaker's and shipwright's tools, chandlery, etc.

T Timmins, Richard & Sons, Birmingham. *c.* 1850 (with eighteenth-century plates). Tools for many trades and steel toys.

TV Tingley, Thomas, Ltd., London. *c.* 1913–1940. Cart and coach ironmongery, wheels, etc.

P Tomlinson, Charles. *Cyclopaedia of Useful Arts and Manufactures.* 9 vols. George Virtue, London. 1853.

P ——Illustrations of Trades. London. 1860.

P Trades. *The Book of English Trades and Library of the useful arts.* G. & W. B. Whittaker, London. 1824.

T Turner, Naylor & Marples, Sheffield. *c.* 1870. Tools for many trades.

T Turton, Thomas & Sons Ltd. (Successors to Wm. Greaves & Sons). Sheffield *c.* 1900. Tools for many trades.

P Tusser, Thomas. *Five hundreth pointes of good husbandrie.* London. 1573. There is a modern edition prepared for *Country Life* by Dorothy Hartley (1931).

T Tyzack, Joseph & Son Ltd., Sheffield. 1879, 1945. Tools for many trades.

T Tyzack, W. Sons & Turner Ltd., Sheffield. *c.* 1910. Saws etc.

P Varney, E. H. 'Making a Violin', *The Woodworker*, Feb. 1958–Feb. 1959.

T Varvill & Sons, York. *c.* 1870. Planes.

J Vialls, Christine M. 'The Casting of Surfaces for Textile Hand Block Printing,' *Transactions of the Newcomen Society*, Vol. XLI, 1968–69, p. 69.

PV Wailes, Rex. *The English Windmill.* Routledge & Kegan Paul Ltd., London. 1954.

J Wailes, Rex, and Russell, John. 'Windmills in Kent,' *Transactions of the Newcomen Society* 1955, Vol. XXIX (1953–55) p. 221.

PV Waiting. H. R. *Farm Waggons.* Unpublished manuscript, *c.* 1948.

V Walker, J. W. & Sons Ltd. (Ruislip, Middlesex). Organ builders.

V   Walker, Philip (London). Authority on wood-working tools and their history.

V*   Ward, W. (Harpenden, Herts.). Blacksmith.

TV   Ward & Payne Ltd., Sheffield. *c.* 1900 and 1911. Tools for many trades.

V*   Ward, Stanley (Letchworth). Authority on early trades.

PV   Waterer, John W. *Leather in Life, Art and Industry.* Faber, London. 1946.

V   Weber, C. Earl. (H. K. Porter Co. Pittsburgh, U.S.A.). Saw making.

V*   Weller, George W. (Sompting, Sussex). William Weller & Son, Wheelwrights and Under-takers.

     Weller, George W. *The Illustrated Carpenter and Builder.* London:

J   *The Village Undertaker.* Sept.–Oct. 1951, p. 1402.

J   *Craft of the Wheelwright.* June–July 1950. p. 970.

J   *The Village Workshop – an account of the workaday life of a Country Builder and Wheelwright.* Oct.–Dec. 1953, p. 2352.

V*   Whitcher, Jack (Aldeburgh). Shoemaking.

V   White, Ian, Musician (London). Violin making.

P   White, K. D. *Agricultural Implements of the Roman World.* Cambridge University Press. 1967.

P   White, Lynn Jr. *Medieval Technology and Social Change.* Oxford University Press. 1962.

T   Whitehouse, Cornelius & Sons, Ltd., Cannock. 1930. Tools for many trades.

V   Wightman, Jesse E. (Chelmsford). Millwright.

J   Wildung, Frank H. 'Making Wood Planes in America'. *Chronicle of the Early American Industries Association,* Vol. VIII, Number 2, April 1955. See also Shelburne Museum.

PV   Wilkerson, Jack. *Two Ears of Barley. Chronicle of an English Village.* Priory Press, Royston. 1969.

T   Williams, Thomas, London. 1900 and 1904. Butcher's and slaughterhouse equipment.

V*   Winser, J. K. (Petersfield, Hants.). Building methods.

P   Woods, K. S. *Rural Crafts of England.* Harrap, London. 1949.

V   Worfolk Bros. (Kings Lynn, Norfolk). Ship-wrights.

V   Wright, J. (Arnprior, Scotland). Carpenter and wheelwright.

P   Wright, Joseph. *The English Dialect Dictionary.* 6 vols. Henry Frowde, London. 1898–1905.

V*   Wright, R. W. (Leverstock Green, Herts.). Wheelwright.

V   Wyllie, H. M. (Harpenden). Rubber tapping tools.

P   Wymer, Norman. *English Country Crafts.* 1946.

P   ——*English Town Crafts.* 1949. B. T. Batsford Ltd., London.

TV   Wynn, Timmins & Co. Ltd., Birmingham. 1890, 1892. (See also Timmins, Richard, above). Tools for many trades.

T   Wynn, W. & C. 'List of Prices to Pattern Book of Articles manufactured by W. & C. Wynn'. Birmingham. 1810. Tools for many trades and steel toys.

P   Young, Francis. *Every Man His Own Mechanic.* Ward, Lock and Bowden, Ltd., London. 1893.